Advances in FOOD AND NUTRITION RESEARCH

VOLUME 65

Marine Medicinal Foods: Implications and Applications - Animals and Microbes

Edited by

SE-KWON KIM

Marine Bioprocess Research Center,
Department of Chemistry,
Pukyong National University,
Busan, Republic of Korea

ELSEVIER

AMSTERDAM • BOSTON • HEIDELBERG • LONDON
NEW YORK • OXFORD • PARIS • SAN DIEGO
SAN FRANCISCO • SINGAPORE • SYDNEY • TOKYO
Academic Press is an imprint of Elsevier

Academic Press is an imprint of Elsevier
225 Wyman Street, Waltham, MA 02451, USA
525 B Street, Suite 1900, San Diego, CA 92101-4495, USA
32 Jamestown Road, London NW1 7BY, UK
Radarweg 29, PO Box 211, 1000 AE Amsterdam, The Netherlands
Linacre House, Jordan Hill, Oxford OX2 8DP, UK

First edition 2012

ISBN: 978-0-12-416003-3
ISSN: 1043-4526

For information on all Academic Press publications
visit our website at elsevierdirect.com

12 13 14 15 10 9 8 7 6 5 4 3 2 1

CONTENTS

CONTRIBUTORS

Numbers in parentheses indicate the pages on which the authors' contributions begin.

- Ira Bhatnagar
  Department of Chemistry, Pukyong National University, Busan, Republic of Korea; Marine Bioprocess Research Center, Pukyong National University, Busan, Republic of Korea (*353*)
- Hee-Guk Byun
  Department of Marine Biotechnology, Gangneung-Wonju National University, Gangneung, Republic of Korea (*47*)
- T. Caradec
  Laboratoire ProBioGEM, IUT A, EA 1026, Université Sciences et Technologies, Lille, Villeneuve d'Ascq, France (*479*)
- L. Catiau
  Laboratoire ProBioGEM, IUT A, EA 1026, Université Sciences et Technologies, Lille, Villeneuve d'Ascq, France (*479*)
- Byeong-Dae Choi
  Department of Seafood Science and Technology, Gyeongsang National University, Tongyeong, Korea (*11*)
- Yeung Joon Choi
  Department of Seafood Science and Technology, Gyeongsang National University, Tongyeong, Korea (*11*)
- B. Cudennec
  Laboratoire ProBioGEM, IUT A, EA 1026, Université Sciences et Technologies, Lille, Villeneuve d'Ascq, France (*479*)
- Mahanama De Zoysa
  College of Veterinary Medicine, Chungnam National University, Yuseong-gu, Daejeon, Republic of Korea (*153*)
- Pradeep Dewapriya
  Marine Biochemistry Laboratory, Department of Chemistry, Pukyong National University, Busan, Republic of Korea (*137*)
- Anupam Giri
  Department of Food Science and Technology, Tokyo University of Marine Science and Technology, Tokyo, Japan (*73*)

- **Yoshihiko Hayashi**
  Department of Cariology, Graduate School of Biomedical Sciences, Nagasaki University, Nagasaki, Japan (*107*)
- **S. W. A. Himaya**
  Department of Chemistry, Pukyong National University, Busan, Republic of Korea (*297*)
- **Mamoru Igarashi**
  Department of Host Defense and Biochemical Research, Juntendo University, Graduate School of Medicine, Tokyo, Japan (*337*)
- **Takeshi Ikeda**
  Department of Cariology, Nagasaki University Hospital, Nagasaki, Japan (*107*)
- **Yoshikazu Inoue**
  Lipid Lab., Yasuda Onoue-cho, Kakogawa city, Okayama, Japan (*31*)
- **R. Jaiganesh**
  Department of Biotechnology, School of Bioengineering, SRM University, Kattankulathur, Tamil Nadu, India (*269, 389*)
- **Jae-Young Je**
  School of Food Technology and Nutrition, Chonnam National University, Yeosu, Republic of Korea (*121, 321*)
- **Joong-Kyun Jeon**
  Department of Marine Biotechnology, Gangneung-Wonju National University, Gangneung, Republic of Korea (*47*)
- **Won-Kyo Jung**
  Department of Marine Life Science, Chosun University, Gwangju, Republic of Korea; Marine Bio Research Center, Chosun University, Gwangju, Republic of Korea (*287*)
- **Nishan S. Kalupahana**
  Department of Animal Science, University of Tennessee, Knoxville, Tennessee, USA; UT Obesity Research Center, University of Tennessee, Knoxville, Tennessee, USA; Department of Physiology, Faculty of Medicine, University of Peradeniya, Peradeniya, Sri Lanka (*211*)
- **Kyong-Hwa Kang**
  Marine Bioprocess Research Center, Pukyong National University, Busan, Republic of Korea (*353*)
- **Fatih Karadeniz**
  Department of Chemistry, Pukyong National University, Busan, Republic of Korea (*223*)

- Se-Kwon Kim
  Department of Chemistry, Pukyong National University, Busan, Republic of Korea; Marine Bioprocess Research Center, Pukyong National University, Busan, Republic of Korea; Department of Marine Biotechnology, Gangneung-Wonju National University, Gangneung, Republic of Korea; Department of Chemistry, Marine Biochemistry Laboratory, Pukyong National University, Busan, Republic of Korea; Department of Chemistry, Marine Biochemistry Laboratory, Pukyong National University, Daeyoung 3-dong, Nam-Gu, Busan, Republic of Korea (*1, 47, 121, 137, 223, 235, 249, 261, 287, 297, 321, 353, 409, 415, 423, 495*)
- Hideki Kishimura
  Research Faculty of Fisheries Sciences, Hokkaido University, Minato, Hakodate, Hokkaido, Japan (*437*)
- Zenya Koyama
  Department of Cariology, Nagasaki University Hospital, Nagasaki, Japan (*107*)
- Maheshika S. Kurukulasuriya
  Department of Animal Science, Faculty of Agriculture, University of Peradeniya, Peradeniya, Sri Lanka (*457*)
- Jung Kwon Lee
  Department of Marine Biotechnology, Gangneung-Wonju National University, Gangneung, Republic of Korea (*47*)
- Yong-Xin Li
  Department of Chemistry, Marine Biochemistry Laboratory, Pukyong National University, Busan, Republic of Korea (*409*)
- A. Malshani Samaraweera
  Department of Animal Science, Faculty of Agriculture, University of Peradeniya, Peradeniya, Sri Lanka (*457*)
- Norihito Maru
  Department of Chemistry, Faculty of Science, Kanagawa University, Hiratsuka, Japan (*171*)
- Naima Moustaid-Moussa
  Department of Animal Science, University of Tennessee, Knoxville, Tennessee, USA; UT Obesity Research Center, University of Tennessee, Knoxville, Tennessee, USA (*211*)
- Isao Nagaoka
  Department of Host Defense and Biochemical Research, Juntendo University, Graduate School of Medicine, Tokyo, Japan (*337*)

- **Dai-Hung Ngo**
  Department of Chemistry, Pukyong National University, Busan, Republic of Korea (*249*)
- **Chamilani Nikapitiya**
  Department of Fisheries, Animal and Veterinary Science, University of Rhode Island, Kingston, Rhode Island, USA (*363*)
- **Osamu Ohno**
  Department of Chemistry, Faculty of Science and Technology, Keio University, Yokohama, Japan (*185*)
- **Toshiaki Ohshima**
  Department of Food Science and Technology, Tokyo University of Marine Science and Technology, Tokyo, Japan (*73*)
- **Ramjee Pallela**
  Marine Bioprocess Research Center, Pukyong National University, Busan, Republic of Korea (*1*)
- **Vitor H. Pomin**
  Complex Carbohydrate Research Center, University of Georgia, Athens, Georgia, USA; Program of Glycobiology, Institute of Medical Biochemistry, Federal University of Rio de Janeiro, Rio de Janeiro, Brazil (*195*)
- **R. Ravallec**
  Laboratoire ProBioGEM, IUT A, EA 1026, Université Sciences et Technologies, Lille, Villeneuve d'Ascq, France (*479*)
- **Koji Sakamoto**
  Koyo Chemical Co., Ltd., Tokyo, Japan (*337*)
- **N. S. Sampath Kumar**
  Department of Biotechnology, School of Bioengineering, SRM University, Kattankulathur, Tamil Nadu, India (*269, 389*)
- **N. Satya Vijaya Kumar**
  Nanotechnology Research Center, SRM University, Kattankulathur, Tamil Nadu, India (*269*)
- **Mahinda Senevirathne**
  Marine Bioprocess Research Center, Pukyong National University, Busan, Republic of Korea (*235, 495*)
- **K. F. S. T. Silva**
  Department of Animal Science, Faculty of Agriculture, University of Peradeniya, Peradeniya, Sri Lanka (*457*)

- **Nalin Siriwardhana**
  Department of Animal Science, University of Tennessee, Knoxville, Tennessee, USA; UT Obesity Research Center, University of Tennessee, Knoxville, Tennessee, USA (*211*)
- **Kiyotake Suenaga**
  Department of Chemistry, Faculty of Science and Technology, Keio University, Yokohama, Japan (*185*)
- **Koretaro Takahashi**
  Faculty of Fisheries Sciences, Hokkaido University, Hakodate, Japan (*31*)
- **Daisuke Uemura**
  Department of Chemistry, Faculty of Science, Kanagawa University, Hiratsuka, Japan (*171, 185*)
- **Quang Van Ta**
  Department of Chemistry, Marine Biochemistry Laboratory, Pukyong National University, Busan, Republic of Korea (*261*)
- **Janak K. Vidanarachchi**
  Department of Animal Science, Faculty of Agriculture, University of Peradeniya, Peradeniya, Sri Lanka (*457*)
- **Thanh-Sang Vo**
  Department of Chemistry, Pukyong National University, Busan, Republic of Korea (*249*)
- **Shizuka Yamada**
  Department of Cariology, Graduate School of Biomedical Sciences, Nagasaki University, Nagasaki, Japan (*107*)
- **Kajiro Yanagi Guchi**
  Department of Cariology, Nagasaki University Hospital, Nagasaki, Japan (*107*)
- **Chen Zhang**
  School of Medicine, Tongji University, Shanghai, China; Department of Chemistry, Pukyong National University, Busan, Republic of Korea; Key Laboratory of Molecular Enzymology and Engineering of the Ministry of Education, College of Life Science, Jilin University, Changchun, China; Marine Bioprocess Research Center, Pukyong National University, Busan, Republic of Korea (*415, 423*)

PREFACE

It is a real fact that the importance of marine organisms as sources of new substances is growing. With marine species comprising approximately a half of the total global biodiversity, the sea offers an enormous resource for novel compounds, and it has been classified as the largest remaining reservoir of natural molecules to be evaluated for medicine. Various kinds of substances have been obtained from marine organisms because they are living in a very exigent, competitive, and aggressive surrounding entirely different from the terrestrial environment in many aspects, a situation that demands the production of quite specific and potent active substances.

Owing to the immense diversity of animals and microbes in the marine environments, and their almost untouched capacity to produce natural products, the importance for marine pharmaceutical and nutraceutical industries was realized on a broad basis by the scientific communities recently. This has strengthened worldwide research activities on the exploration of bioactive compounds from marine organisms for pharmaceutical use, as well as the development of other valuable products, such as enzymes, nutraceuticals, and cosmeceuticals.

The growing need for novel substances for the treatment of severe human diseases such as cancer, microbial infections, and inflammatory processes, combined with the recognition that marine organisms provide a rich potential source of such substances, demands intensive search for new substances from marine organisms. Modern marine biotechnology has moved its focus to microbes and animals, and encompasses the discovery of new pharmaceuticals. In particular, bacteria and fungi associated with marine organisms such as sponges, corals, and algae are potent producers of biological active substances with prominent activities not only against pathogenic bacteria, fungi, and viruses but also against tumor cells.

Sea animals are rich in soluble dietary fibers, proteins, minerals, vitamins, antioxidants, phytochemicals, and polyunsaturated fatty acids, with low caloric value. Polysaccharides from marine animals have been reported to possess biological activities with potential medicinal values in addition to their current status as a source of dietary fibers and prebiotics. Moreover, they have a lot of dietary fiber, which lowers blood cholesterol, and iodine, which improves metabolism, vascular and cardiac action, body temperature, and perspiration regulation, and are effective in preventing hypertension, heart disease, and arterioscleroses. Further, current research has proved their potential medicinal value against various diseases, including allergy, cancer, diabetes, hypertensive, oxidative stress, inflammation, thrombosis, obesity, lipidemia, and other degenerative disorders. Hence, a large number of bioactive components from marine organism have been identified with potential uses in various areas including functional food, pharmaceutical, and cosmeceutical.

This book is completed with the help of many invaluable contributors around the world. Therefore, I would like to thank them all for their valuable time and effort invested in this book. This book is an attempt to describe the present and future prospects of marine animals and microbes as medicinal foods because the remedies containing natural ingredients from marine animals go back to the ancient era. The nutritional elements and biological activities centering around the reported information on marine animals were discussed with the view of helping the personnel in the academia and also for general public. Since the synthetically derived materials show some side effects, huge attention from general public and academia on the products from natural-based materials including marine animals and microbes have been gained due to their prominent biological activities and abundance. Therefore, this book discusses the importance of biologically active compounds from marine animals under five sections.

Chapter 1 familiarizes the readers with the prospects of marine organism-derived substances in relation to medicinal value. It covers the present and future prospects of marine organisms. Chapters 2–11 describe the sources of marine bioactives. Individual chapters cover the importance and values of some frequently used marine animal species. Chapters 12–23, cover the medicinal effects of bioactives from marine animals. Chapters 24–29 discuss the value of microbes as medicinal foods. Finally, Chapters 30–32, discusses the applications of marine animals-derived materials in medicinal and nutraceutical industries in relation to human health.

SE-KWON KIM

CHAPTER 1

Medicinal Foods from Marine Animals: Current Status and Prospects

Se-Kwon Kim[*,†,1] and **Ramjee Pallela**[†]

Contents

Abstract

The lifestyle of human being is changing day by day toward the simplified and more convenient way of living. Human wellbeing is majorly dependent on the daily food habits that are in accordance with the habits of individual community and the surrounding environments. Although the food habits are simplified and fashioned according to the current lifestyle, many of the Asians are still showing much importance to the naturally derived and traditional foods. One such medicinally important natural source is the foods from marine organisms, which are an important growing notion for the development of marine nutraceuticals and functional foods. In this context, we have already brought the recent trends and applications of marine algal (macro and micro) foods in my previous volume. The current preliminary chapter of this book volume on marine animals and microbes describes about the prospects of

* Department of Chemistry, Pukyong National University, Busan, Republic of Korea
† Marine Bioprocess Research Center, Pukyong National University, Busan, Republic of Korea
1 Corresponding author: Se-Kwon Kim, *E-mail address:* sknkim@pknu.ac.kr

Advances in Food and Nutrition Research, Volume 65
ISSN 1043-4526, DOI: 10.1016/B978-0-12-416003-3.00001-9

various marine animals and their derived substances/materials as medicinal foods. In addition, this chapter encourages the new researchers as well as various health communities to implement the marine animal-based medicinal foods and their applications.

I. INTRODUCTION

Oceans are the Earth's most valuable natural resources, which provide majority of the food in the form of fish and shellfish (about 200billion pounds are caught each year). Oceans are the richest resource for numerous biomedically important chemical as well as the protein molecules, which are ranging from the general medication purposes to the specific therapeutic applications (Jimeno, 2002; Vo and Kim, 2010). While the use of floristic resources in traditional medicine has been widely explored, there is a paucity of information regarding the utilization of faunistic resources (Alves and Rosa, 2006), and as such, the medical value of animal species has not been included in the calculations of economic value of biodiversity by the World Conservation Union (IUCN) (Alves *et al.*, 2009; Pearce and Moran, 1994).

Science is being useful to humans day by day to cure the latest ailments, and the development of medicine from sea becoming the best nowadays because of the conventional belief that ''life originated from sea.'' Asians supplement their regular diet with sea food, which provides them more proteinaceous and medicinally valuable ingredients. Healthy and nutritious life of human beings depend on the food and agricultural products they use, and *in lieu* to this, search of medicinal foods from marine sources is effective and continuous. The present day importance of focusing on marine medicinal substances has been procured from past experiences of use and applications of marine resources, for example, old Chinese medicinal records mentioned more than 200 species of marine organisms that are extensively been used in previous decades, whereas very few species have been under the medicinal usage. In general, until recently, very few researchers were cognizant of the value of marine animals in studying fundamental processes of relevance to human beings. However, as the thirst of exploring marine animal foods is increasing, there have been a number of articles describing the special attributes of these marine animals and forecasting noteworthy benefits from their increased use in medically oriented research.

The impact of marine food is directly proportional to the rate of consuming each level of marine food chain. For example, the phytoplanktons, seaweeds/grasses are consumed by the next level of ecological pyramid, the herbivores, which consists of zooplanktons, parrot fish, oyster, shrimp, clams, crabs scallops, tilapia, lobster, etc. Oceans carnivores (squids, sardines, snappers, catfish, herming, lobster, pollock, sablefish, soles, striped bass, etc.) are the next level occupiers in the food chain, which dependent on other carnivores and herbivores like zooplankton; however, they ultimately serve for the nourishment of bigger and faster animals that are presented on the next level of the food chain, the top predators. The top level predators along with the subsequent food levels are well explored by humans from past decades, but the medicinal significance has not been put forward toward the sustainable and renewable way of ocean's exploration and exploitation. Table 1.1 represents few of the medicinally valuable marine species along with the ones reported earlier by Alves *et al.* (2009).

Hence, the value of the marine animals in terms of medicinal and nutraceutical applications has to be brought in such a manner that the consumption of frequently available marine animals can be directed for best human diet practices. In the current chapter, the role of various marine animals as medicinal foods has been covered and in addition, the possible recommendation of utilizing these marine animals in regular diet has been advised toward the health benefits of human kind.

II. FISH

The oceans have been explored for thousands of years and became the integral part of human society. Especially, fish have been very important to the world economy for many years occupying the bigger position in marine animal consumption, starting with the Viking trade of cod till the diverse fisheries found in many countries like Lofoten, Europe, Italy, Portugal, Spain, and India. Fisheries of today provide more than 16% of the total world's protein in many developing nations Marine Bio. Fisheries are still enormously important to the economy and wellbeing of communities. The most common species making up the global fisheries are herring, cod, tuna, salmon, anchovy, mullet, flounder along with other organisms like shrimp, squid, crab, lobster, scallops, and oyster. Among them, especially cod is used for consumption in fresh as well as frozen forms. Atlantic herring is used for canning, fishmeal, and fish oil. The Atlantic menhaden is used for fishmeal and fish oil, and Alaska pollock is not only consumed, but also used for fish paste to simulate crab. The Pacific cod has recently been used as a substitute for Atlantic cod which has been overfished. Another important medicinally valuable fish is tuna (*Thunnus obesus*), one of the largest commercially canned fishery products in the world, and annually, more than 4million metric tons of tuna are harvested worldwide and approximately 3 million tons of canned tuna are produced (Choi *et al.*, 2011). Even shark fin soup is traditionally regarded as beneficial for health in East Asia, and its status as an "elite"

TABLE 1.1 Few marine species that are consumed and proposed as medicinal food

Name of the species	Consuming part	Remedy for diseases
Electrophorus electricus (Linnaeus, 1766)—electric eel	Fat	96 Headache, acne, skin spots, wounds, rheumatism, and pain
Hippocampus reidi (Ginsburg, 1933)—longsnout seahorse	Whole animal	Asthma, epilepsy, and alcoholism
Thunnus obesus (Lowe, 1839)—bigeye tuna	Whole animal	Blood pressure, stroke and cardiovascular disease, Alzheimer's disease, arthritis, cancer, macular degeneration
Gadus macrocephalus (Tilesius, 1810)—cod fish	Liver fats	Cardiovascular disease, heart muscle function, skin diseases
Theragra chalcogramma—Alaska pollock (close cousin of cod fish)	Liver fats and meat	Blood pressure, stroke, and other heart disease
Rhincodon Typus (Smith, 1828)—shark	Liver, cartilage, fins, meat	Immunity disorders, cancer, arthritis, psoriasis, and other allergic reactions
Thaleichthys pacificus (Richardson, 1836)—ooligan fish	Fat	Skin complications
Several species of Coregonidae and also squid and crab	Fish paste/jelly	Nutritional disorders
Arraia—unidentified species of ray	Flesh	Weakness
Echinaster brasiliensis (Müller and Troschel, 1842)—starfish	Whole animal	Asthma and alcoholism
Echinaster echinophorus (Lamarck, 1816)—starfish	Whole animal	Asthma and alcoholism
Luidia senegalensis (Lamarck, 1916)—starfish	Whole animal	Asthma and alcoholism
Mellita quinquiesperforata (Leske, 1778)—starfish	Whole animal	Asthma
Oreaster reticulatus (Linnaeus, 1758)—starfish	Whole animal	Asthma and alcoholism

TABLE 1.1 *(continued)*

Name of the species	Consuming part	Remedy for diseases
Echinometra lucunter (Linnaeus, 1758)—malvaric	Whole animal	Asthma
Caiman crocodilus (Linnaeus, 1758) LR/II—common cayman	Skin	Asthma, allergies, and epilepsy
Pteriomorphia sp. (Beurlen, 1944)—mussel	Whole animal	Hypertension and stroke
Haliotis asinina (Linnaeus, 1758)—abalone	Whole animal	Cough and immune disorders

dish has led to huge demand with the increase of affluence in China. Naturally fermented fish and their derived peptides after enzyme treatment gain much importance in healing various diseases like Alzheimer's disease, hypertension, cancer, viral infections, bone disorders, etc. (Kim and Mendis, 2006; Vo and Kim, 2010; Wijesekara and Kim, 2010).

Additionally, fish have so many medicinal values in their composed forms viz., oils, fats, proteins/peptides, carbohydrates, etc. Marine collagen, especially fish collagen, is evaluated for its potential application as an alternative to the immunogenic mammalian collagen. In addition, the employment and the further development of sophisticated fish and marine mammal hunting techniques allowed intensive gathering, preservation, and subsequent use during times of food scarcity (Mos *et al.*, 2004). Similarly, the oils from fish gained tremendous value in the medicine for centuries. Middle age records show that cod liver oil was prescribed for rickets. Stored surplus resources also served in trade, which flourished among several coastal regions, for example, ooligan grease, a fat rendered from ooligan fish (*Thaleichthys pacificus*), was used by some coastal First Nations as an ingredient in the preparation of food items, in the preservation of fruits, wood, and leather, and as a medicinal ointment for skin rashes.

Other types of fish products are essentially the minced fish flesh. A steamed and molded fish paste called "Kamaboko" (fish jelly) is famous in Japan since sixteenth century. Further, the mechanization for flesh preparation and mincing is advanced, and several fish meat analogues have been developed to use as traditional medicine (Whittle, 1998).

III. CRUSTACEANS

Shell fish has been contributing a lot among the marine resources to provide significant medicinal prospects to humans. Crustaceans, by possessing a vast diversity in their species, play a major role in the development of food and nutritional industries worldwide. The medicinal impact of these sources from crustaceans is the basis for many of the novel nutraceuticals and functional foods. For example, shells of shrimp and crab are majorly used for the derivation of chitin, chitosan, chitooligosaccharides (COS), and further derivatives, which have been used for various biomedical significances (Kim, 2010). Chitin and chitosan are natural biopolymers found in many natural organisms. In fact, chitin is the most important natural polysaccharide (second after cellulose), whereas chitosan is a derivative, which can be extracted from external skeleton of an octopus or some sea fishes with the chemical name 2-acetamido-2-deoxy-D-glucose-*N*-acetyl-D-glucosamine. It has many biological applications in pharmaceutical, nutraceutical, and cosmeceutical industries and even in food and agriculture sectors. Hypocholesterolemic effects of the chitosan and COS have attracted much attention as this property can be applied widely in the biomedical field (Vidanarachchi *et al.*, 2010). However, according to an early study, chitosan oligomers have not shown a cholesterol-lowering activity (Sugano *et al.*, 1988). However, Kim *et al.* (1998) have carried out an experiment for the investigation of the effects of hydrophilic COS on the liver functions of mice. They have found that the COS can partially reduce the elevated serum cholesterol levels after a diet containing cholesterol. After the 6-week experimental period, the serum cholesterol level has been reduced by 23% in the group of mice fed with a diet containing 1% COS and 3% cholesterol than the group fed with 3% cholesterol alone (Kim *et al.*, 1998). This reflects the involvement of the orally administered COS in the partial reduction of elevated serum cholesterol levels.

The rich proteins from shrimp and crab are a positive notion nowadays for the supplementing of high ranked nutrition for many of the patients of nutrition depletion. The collagen and other protein disorders can be overcome by the intake of these crustaceans regularly. However, the financial limitations and the inadequacy of fishing profound disturbances due to the human population and pollution; many of the developing nations are underutilize these animals. In addition, chitin and its derivatives are vastly known for their biomedical importance; however, the utilization of the crustacean foods is restricted mostly to Asian countries, and many of the developed, developing, and underdeveloped countries are henceforth advised well for the proper implementation of these medicinally valuable resources.

IV. MOLLUSKS AND ECHINODERMS

Mollusks along with the crustaceans are also widely sought to participate in the world's consumption of marine food. Fermented marine food sauces such as blue mussel sauce and oyster sauce possess bioactive peptides which play a major role as Angiotensin I converting enzyme inhibitors that indirectly suppress hypertension (Wijesekara and Kim, 2010). Hence, the mollusk-derived proteins and other macromolecules are highly valuable to indicate these organisms as medicinally valuable food sources.

Abalones are other important marine animals which have significant medical importance. Besides the use of soft body of the animal in food markets, dried abalone is sold among herbs and other items in medicine shops in Hong Kong, Singapore, and elsewhere in Southeast Asia. Ground and processed abalone shells are used as a calcium supplements both for humans and animals. In China, where food and medicine have long been considered more or less the same thing, abalone is highly valued for its healing powers. It is considered especially well for treating eye disease, particularly the powder made by grinding up the shells. Paolin, a drug made from abalone juice, is an effective inhibitor of penicillin-resistant strains of bacteria. The flat shell of the abalone, with its iridescent inside, is powdered and taken orally to improve vision, to remove keratoses (cataracts), and to improve such conditions as hemeralopia (Man and mollusk).

More importantly, oyster juice has been found to have antiviral properties, which might be directing a way to develop drugs. In Vietnam, traditional medicine has a wide variety of uses for shells: powdered oyster shell is taken to treat acid indigestion, fatigue, and to stop hemorrhage. It is also sprinkled over open wounds and boils. Powdered pearls from oysters are used as a topical eye medicine, and it has been scientifically proven to have some anti-inflammatory effects on a painful condition called conjunctivitis, where the surface of the eye becomes red and sore. Cuttlefish bones are used as a remedy for rickets, as a healing agent in the treatment of gastrointestinal troubles, as a local antihemorrhagic, and as an antiseptic in the cases of inflammation of the middle ear.

Octopuses, squid, cuttlefish, and nautiloids are also other common food source for many of the Asians because of their importance of potentiality in medical applications. More specifically, octopus and squids are well explored in Asian countries due to their traditional medical values.

V. OTHER MARINE ANIMALS OF MEDICINAL VALUE

As mentioned earlier, marine animals are the richest source of various biologically and biomedically important compounds/materials/substances. One such traditional and ancient medicinally important chordate is seahorse, which plays a fundamental role as an ingredient in various therapies for a variety of disorders including asthma, arteriosclerosis, incontinence, impotence, thyroid disorders, skin ailments, broken bones, heart disease, as well as to facilitate childbirth and even as an aphrodisiac. The seahorse *Syngnthus acus* L. is a very important traditional Chinese medicine from sea, which possesses amino acids, proteins, trace elements, poly-carbon non-saturation fat acids, etc. (Li *et al.*, 2001). In addition, *S. acus* has sexual hormones, anticancer, and fatigue-resisting effects. It can also improve the immunity and enhances the systolic strength of heart muscle. It can be widely used in many fields such as food science, medicine, and aquatic industries.

There are several other species of marine origin, which have significant medical uses. This featured chapter, however, depicts very limited information to recommend the readers for the sustainable exploration and exploitation of the marine animals toward medicinal benefits.

VI. CONCLUSION

The ocean plays a critical role in removing carbon from the atmosphere and providing oxygen. It regulates Earth's climate and is an increasingly important source of biomedical organisms with enormous potential for fighting disease. Seafood consumption makes tremendous impact on the ocean and its interdependents. However, today's health, safety, and sustainability considerations direct to determine the best seafood choices for human survival at the best. This interactive guide compiles all the information you need to continue to eat healthfully while lowering your seafood footprint. It is very important to explore not only the sea plants but also the marine animals to find out whether the consumed favorite food represents the high rank of sustainability, biocompatibility, with medicinal values, as well its place in the food chain. Hence, the exploration and exploitation of the marine animals toward their medical use is a greater deal to understand why we must keep the ocean healthy for future generations.

REFERENCES

Alves, R. R. N. and Rosa, I. L. (2006). From cnidarians to mammals: The use of animals as remedies in fishing communities in NE Brazil. *J. Ethnopharmacol.* **107**, 259–276.

Alves, R. R. N., Neto, N. A. L., Brooks, S. E., and Albuquerque, U. P. (2009). Commercialization of animal-derived remedies as complementary medicine in the semi-arid region of Northeastern Brazil. *J. Ethnopharmacol.* **124**, 600–608.

Choi, J. I., Kim, J. H., and Lee, J. W. (2011). Physiological properties of tuna cooking drip hydrolysate prepared with gamma irradiation. *Process Biochem.* **46**, 1875–1878.

Jimeno, J. M. (2002). A clinical armamentarium of marine-derived anti-cancer compounds. *Anticancer Drugs* **13**(Suppl. 1), S15–S19.
Kim, S. K. (2010). Chitin, Chitosan, Oligosaccharides and Their Derivatives: Biological Activities and Applications. CRC Taylor & Francis, USA (pp. 666).
Kim, S. K. and Mendis, E. (2006). Bioactive compounds from marine processing byproducts—A review. *Food Res. Int.* **39**, 383–393.
Kim, S. J., Kang, S. Y., Park, S. L., Shin, T., and Ko, Y. H. (1998). Effects of chitooligosaccharides on liver function in the mouse. *Korean J. Food Sci. Technol.* **30**, 693–696.
Li, C., Zou, G., Bian, H., and Ju, X. (2001). Summarization of studies on Chinese marine medicinal animal *Syngnthus acus*. *Zhong Yao Cai* **24**, 686–688.
Man and mollusk. http://www.manandmollusc.net/beginners_uses/3.html.
Marine Bio. http://marinebio.org/oceans/ocean-resources.asp.
Mos, L., Jack, J., Cullon, D., Montour, L., Alleyne, C., and Ross, P. S. (2004). The importance of marine foods to a near-urban first nation community in coastal British Columbia, Canada: Toward a risk-benefit assessment. *J. Toxicol. Environ. Health A* **67**, 791–808.
Pearce, D. and Moran, D. (1994). The Economic Value of Biodiversity. IUCN—The World Conservation Union, Earthscan Publications, London (p. 172).
Sugano, M., Watanabe, S., Krishi, A., Izume, M., and Ohtakara, A. (1988). Hypocholesterolaemic action of chitosans with different viscosities in rats. *Lipids* **23**, 187–193.
Vidanarachchi, J. K., Kurukulasuriya, M. S., and Kim, S. K. (2010). Chitin, chitosan, and their oligosaccharides in food industry. *In* "Chitin, Chitosan, Oligosaccharides and Their Derivatives: Biological Activities and Applications", (S.-K. Kim, Ed.), pp. 543–560. CRC Taylor & Francis, USA.
Vo, T. S. and Kim, S. K. (2010). Potential anti-HIV agents from marine resources: An overview. *Mar. Drugs* **8**, 2871–2892.
Whittle, K. J. (1998). Marine organisms as food, forage, industrial, and medical products. *In* "Fisheries and Aquaculture", (P. Safran, Ed.). Food and Agricultural Sciences, Engineering and Technology Resources, Vol. I. EOLSS Publishers Co., IOM, UK.
Wijesekara, I. and Kim, S. K. (2010). Angiotensin-i-converting enzyme (ace) inhibitors from marine resources: Prospects in the pharmaceutical industry. *Mar. Drugs* **8**, 1080–1093.

CHAPTER **2**

Nutraceutical Functionalities of Polysaccharides from Marine Invertebrates

Byeong-Dae Choi[1] and **Yeung Joon Choi**

Contents

Abstract

Many researchers are seeking functional materials from marine resources. These marine resources can be used as traditional food additives, and specifically, these are based on polysaccharides. To date, there is a big opportunity to develop new high-value added products with indispensable functional characteristics, which can be used in nutraceuticals either as additives or supplements. Also, a crossover in the pharmaceutical market may be established. Some glycosaminoglycans (GAGs) mimetic-type molecules are already being utilized in the field of nutrition as well as in the cosmetics industry. This chemical is used as a dietary supplement to maintain the structure and function of cartilages, for the relief of pain caused by osteoarthritic joints, and can also be used as an anti-inflammatory agent. Recently, in relation to the prevalence of mad cow disease and avian influenza, the production of GAGs from marine invertebrates offers new market opportunities as compared with that obtained from bovine or avian livestock.

Department of Seafood Science and Technology, Gyeongsang National University, Tongyeong, Korea
[1] Corresponding author: Byeong-Dae Choi, *E-mail address:* bdchoi@gnu.ac.kr

Advances in Food and Nutrition Research, Volume 65
ISSN 1043-4526, DOI: 10.1016/B978-0-12-416003-3.00002-0

I. INTRODUCTION

The discovery of *O*-linked β-*N*-acetylglucosamine (*O*-GlcNAc) more than 28years ago disproved the long-held dogma that protein glycosylation is restricted to the luminal compartments of the secretory machinery and to the cell surface and extracellular matrix (Hart *et al.*, 1989). Early studies of *O*-GlcNAc's subcellular localization in rat hepatocytes established that it is highly concentrated at the nuclear envelope, particularly at the nuclear pore complex, but is also abundant and widespread within chromatin (Holt and Hart, 1986). Aside from the biosynthetic intermediates, *O*-GlcNAc modification, where β-linked GlcNAc is attached to Ser or Thr residues of cytosolic/nuclear proteins, is the best-known example of cytosolic/nuclear glycosylation. This modification is found to occur exclusively in the cytosol or nucleus, and unlike other types of glycosylation, it remains as a single GlcNAc residue through a β-glycosidic linkage to Ser/Thr residues and is not elongated further by the addition of other sugars (Altmann *et al.*, 2001). The biological functions of *O*-GlcNAcylation in cytosolic events can be summarized as follows: (1) it inhibits protein phosphorylation on Ser/Thr residues, by reciprocal site occupation (Wang *et al.*, 2007); (2) it affects protein degradation, by influencing PEST sequences known as the rapid degradation signal motif (Cheng *et al.*, 2000) or by directly modifying the proteasome complex to decrease the proteasomal function (Sumegi *et al.*, 2003); (3) it regulates intracellular localization of the carrier proteins (Guinez *et al.*, 2005); (4) it is involved in protein interaction, some *O*-GlcNAcylated transcription factors are known to interact for transactivation (Hiromura *et al.*, 2003); and (5) it affects the activity of transcription factors or repressors (Yang *et al.*, 2002).

Glycosaminoglycans (GAGs) are large, complex carbohydrate molecules that are linear, negatively charged, and composed of disaccharide repeating units (Gandhi and Mancera, 2008). These molecules are sometimes known as sulfated mucopolysaccharides because of their viscous, lubricant properties, as found in mucous secretions. GAGs are present on all animal cell surfaces and in the extracellular matrix (Medeiros *et al.*, 2000). Owing to the variability in sulfate substitution, all GAGs display considerable sequence heterogeneity, and it is believed that structural differences are responsible for highly specific interactions of GAGs with other macromolecules (Coombe and Kett, 2005). Their strategic location and highly charged nature make them important biological players in cell–cell and cell–matrix interactions that take place during normal and pathological events, including organogenesis in embryonic development, cell recognition, adhesion, migration, regulation of growth factor action, wound repair, lipid metabolism, neural development and regeneration, and initiation and modulation of inflammation (Laabs *et al.*, 2005; Sugahara *et al.*, 2003;

Taylor and Gallo, 2006; Whitelock and Iozzo, 2005). GAGs include chondroitin sulfate (CS), dermatan sulfate (DS), keratan sulfate (KS), heparin, and heparan sulfate (HS; Nadanaka and Kitagawa, 2008; Nader *et al.*, 2004; Volpi, 2007; Yamada and Sugahara, 2008). Their disaccharide repeating units are composed of hexosamine (D-glucosamine or D-galactosamine) and either hexuronic acid (D-glucuronic or L-iduronic acid) or galactose (as in KS). DS, HS, and heparin contain both glucuronic acid and iduronic acid units, whereas CS has glucuronic acid as the only hexuronic acid. In the tissue, GAGs are covalently bound to a protein core forming a structure known as proteoglycan (Didraga *et al.*, 2006).

CS, in particular, is composed of alternate sequences of D-glucuronic acid and differently sulfated residues of *N*-acetyl-D-galactosamine linked by β(1→3) bonds. Depending on the disaccharide nature, CS with different carbohydrate backbones is known. Chondroitin-4-sulfate, CSA, is constituted by 4-sulfated disaccharides [(1→4)-*O*-(D-glucopyranosyluronic acid)-(1→3)-*O*-(2-*N*-acetamido-2-deoxy-D-galactopyranosyl-4-sulfate)]. Chondroitin-6-sulfate, CSC, is mainly composed of a disaccharide unit sulfated in the position 6 of the *N*-acetyl-D-galactosamine [(1→4)-*O*-(D-glucopyranosyluronic acid)-(1→3)-*O*-(2-*N*-acetamido-2-deoxy-D-galactopyranosyl-6-sulfate)] (Fig. 2.1). However, even if the known CS samples are mainly composed of various percentages of these two kinds of disaccharide units, monosulfated in position 4 and monosulfated in position 6 of the *N*-acetyl-D-galactosamine, disaccharides with a different number and position of sulfate groups can be located, in various percentages, within the polysaccharide chains. For example, the nonsulfated disaccharide is

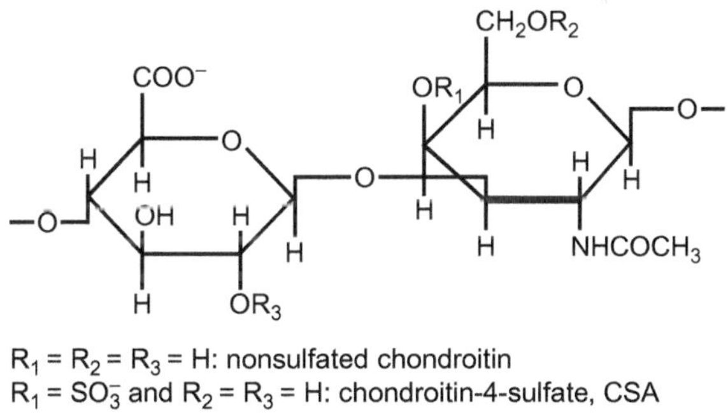

$R_1 = R_2 = R_3 = H$: nonsulfated chondroitin
$R_1 = SO_3^-$ and $R_2 = R_3 = H$: chondroitin-4-sulfate, CSA
$R_2 = SO_3^-$ and $R_1 = R_3 = H$: chondroitin-6-sulfate, CSC
$R_2 = R_3 = SO_3^-$ and $R_1 = H$: chondroitin-2,6-disulfate, CSD
$R_1 = R_2 = SO_3^-$ and $R_3 = H$: chondroitin-4,6-disulfate, CSE
$R_1 = R_3 = SO_3^-$ and $R_2 = H$: chondroitin-2,4-disulfate, CSB
$R_1 = R_2 = R_3 = SO_3^-$: trisulfated chondroitin

FIGURE 2.1 Structures of disaccharides forming chondroitin sulfate. Adapted from Volpi (2007). With permission from Wiley Publisher.

generally present in low amounts in the CS backbone, while the monosulfated disaccharide in position 2 of the glucuronic acid is very uncommon in this natural polymer. On the contrary, disulfated disaccharides having two sulfate groups *O*-linked in various positions, such as 2 of D-glucuronic acid and 6 of *N*-acetyl-D-galactosamine (disaccharide D), or in position 4 and 6 of *N*-acetyl-D-galactosamine (disaccharide E) or 2 of the uronic acid and 4 of *N*-acetyl-D-galactosamine (disaccharide B; Fig. 2.1) may be present in the CS backbone in various percentages in relation to specific animal sources. Further, these heterogeneous structures are responsible for the various and more specialized functions of these polysaccharides (Volpi, 2007).

The CS, such as the other GAGs, is linked to the core protein through a tetrasaccharide junction, composed of xylose, galactose, and glucuronic acid. The whole assemblage, is generally referred to as proteoglycan and is widely distributed in connective tissues both in the cellular membrane (Yanagishita and Hascall, 1992) and in the extracellular matrix (Iozzo, 1998). In this last case, aggrecan is the principal CS proteoglycan in the cartilage, where it is found associated with hyaluronic acid, collagen, and other proteins; this multimeric aggregate confers to the tissue its hydrate gel-like and elastic characteristics. The role of aggrecan is to draw water into the extracellular matrix which swells and expands acquiring its compressive resilience. Aggrecan molecular modifications are regulated by many cellular and extracellular events; however, individual age appears to play an important role on cartilage composition, influencing the sulfation pattern of chondroitin and proteolytic cleavage of aggrecan yielding an extracellular matrix with a reduced amount of poorly sulfated chondroitin. These events play a central role in joint degenerative diseases such as osteoarthritis and rheumatoid arthritis (Hardingham and Bayliss, 1990). In addition, cartilage breakdown products with antigenic properties are observed and their release into synovial fluid induces synovial inflammation (Volpi, 2006). Pharmacological management of the osteoarthritis is based on the use of analgesics, steroidal and nonsteroidal anti-inflammatory drugs, in combination with CS polysaccharides, resulting in the increasing demand of this material and in the search of new sources which are able to provide this polysaccharide with a sulfation pattern compatible with human physiology (Gargiulo *et al.*, 2009).

II. SOURCES OF POLYSACCHARIDES

A. Sea cucumbers

The body wall of sea cucumber is easily and greatly hardened by heat, acid treatment, or even by handling stimulation under physiological conditions. Chemical and rheological analyses of the body wall have not provided any clues about the mechanism involved (Motokawa, 1981, 1982). It has been argued that the GAG of the sea cucumber's (*Stichopus japonicus*) body wall is involved in the cation-dependent change of the toughness of its connective tissue (Kariya *et al.*, 1990). Also, it has been shown that the specific viscosity of the GAG solution was affected differently depending on the valence and species of cation. Interestingly, the mode of change in the viscosity of this GAG solution was quite different from those reported for known GAGs, such as CS and DS. The sea cucumber GAG is not digested by both chondroitinase ABC (EC 4.2.2.4) and AC (EC 4.2.2.5), although in addition to neutral sugar (fucose), the GAG has a sugar backbone identical to CS which contains glucuronic acid (GlcA) and *N*-acetylgalactosamine (GalNAc).

Such unique physicochemical properties of the sea cucumber GAG seem to be related to its unusual structure. Vieira and Mourao (1988) demonstrated the presence of two fucopyranosyl residues linked glycosidically through position (1→4) and/or (1→2) for the fucose branch in the GAG from the sea cucumber *Ludwigothurea grisea*. The two fucopyranosyl residues of the fucose branch were linked (1→3) only. In addition, about 20% of total fucose branches were linked glycosidically through the *O*-3 position of a GlcA moiety (the remaining 80–90% were assumed to be linked through GalNAc). Fucose-branched chondroitin sulfate E (CSE) was prepared from the body wall of sea cucumber, *S. japonicus* (Kariya *et al.*, 1997). The purified GAG was chemically desulfated, followed by carboxyl reduction. Intact, desulfated, and desulfated/carboxyl-reduced GAGs fractions were subjected to per-*O*-methylation. GC-MS analyses of the resultant partially methylated alditol acetates demonstrated that fucose branch is formed by two fucopyranosyl residues linked glycosidically through position (1→3), and that the fucose glucuronic acid branches are almost equimolar.

The sea cucumber, *Apostichopus japonicus* is a traditional food and of high demand in China. In recent years, farming and sea ranching of *A. japonicus* have grown into a prosperous economic sector in northern China, where 2–3 billion juveniles and 60,000 ton of sea cucumbers were produced each year (Yang *et al.*, 2005). The rapid expansion of sea cucumber farming and high farming intensity resulted in serious diseases; mainly skin ulceration syndrome, which is highly infectious and lethal to this animal (Zhang *et al.*, 2006). Application of antibiotics and chemicals in aquaculture leads to the spread of drug-resistant pathogens and deleterious environmental consequences (Reilly and Kaferstein, 1997; Sun *et al.*, 2007). Therefore, finding alternatives to antibiotics and chemicals is very urgent and important for *A. japonicus* farming. Using immunostimulants is a promising area in aquaculture because they are biocompatible, biodegradable, environment-friendly, and safe for humans (Luo, 2007). Among potential immunostimulants, β-glucan and mannan oligosaccharide (MOS) were used with many kinds of fish diets (Chansue *et al.*, 2000), and both have commercial products available for aquaculture, such as MacroGard (Biotec Pharmacon, Tromso, Norway), DVAQUA (Diamond V Mills, Inc., IA, USA), and Bio-Mos (Alltech, USA). Many feeding trials and *in vitro* tests have shown that β-glucan and MOS are able to enhance the resistance of aquatic animals against infections as well as immune capacities such as phagocytosis, superoxide anion production, and lysozyme activity in shrimps and fishes (Dugenci *et al.*, 2003; Zhao *et al.*, 1990).

Collagen fibrils from the dermis of the sea cucumber *Cucumaria frondosa* are aggregated *in vitro* by the dermal glycoprotein stiparin (Trotter *et al.*, 1996). Under physiological ionic conditions, stiparin appears to be both necessary and sufficient to cause fibrils to aggregate (Trotter *et al.*, 1997). Trotter *et al.* (1999) reported the initial biochemical and biophysical characterization of a sulfated glycoprotein from *C. frondosa* dermis that binds stiparin and inhibits its fibril-aggregating activity. This inhibitory glycoprotein, which has been named "stiparin-inhibitor," has the highest negative charge density of all the macromolecules extracted from the dermis. SDS-PAGE reveals three ~31 kDa bands which stain with Alcian blue but not with Coomassie blue. Analytical ultracentrifugation indicates a native molecular weight of 62 kDa. Transmission electron microscopy of rotary-shadowed molecules shows curved rods about 22 nm long. The glycoprotein does not bind collagen fibrils but does bind with stiparin in a 1:1 stoichiometry. The binding of stiparin-inhibitor to stiparin prevents the binding of stiparin to collagen fibrils. The carbohydrate moiety produced by papain digestion of the glycoprotein retains all of its inhibitory activity. The carbohydrate moiety of the inhibitor is dominated by galactose and sulfate. The metabolism of both orally and parenterally administered exogenous GAGs is not well understood. An understanding of this metabolism is essential in exploiting new therapeutic applications (Dawes *et al.*, 1991). The GAG isolated from a holothurian (*S. japonicus*) is a unique fucose-branched CS that has an anticoagulant activity (Table 2.1; Vieira and Mourao, 1999). An undesirable characteristic of this GAG is that it induces platelet aggregation (Li and Lian, 1988). In an effort to reduce this property, a partially depolymerized holothurian glycosaminoglycan (DHG) has been prepared by oxidative depolymerization with hydrogen peroxide (Suzuki *et al.*, 1991). DHG, a fucosyl CS chain, was isolated from sea cucumber and was intravenously and orally administered to experimental animals. After intravenous injection, clearance of DHG, as measured by postcolumn HPLC, displayed complex kinetics that were not dose dependent. DHG was excreted unchanged in the urine. No degradation products of DHG were detected by either gel filtration or anion exchange HPLC at any time in the plasma, indicating

TABLE 2.1 Methylation analysis of intact, desulfated, and desulfated/carboxyl-reduced fractions prepared from the sea cucumber *S. japonicas* GAG

			Molar ratio[c]		
Peak	PMAA[a]	t_R[b]	Intact	Desulfated	Desulfated/ carboxyl-reduced
1	2,3,4-Fuc	1.00	ND[d]	0.21 (21)	0.21 (21)
2	2,4-Fuc	1.19	0.16 (16)[e]	0.45 (45)	0.25 (25)
3	2,3-Fuc	1.22	0.08 (8)	0.03 (3)	0.04 (4)
4	3,4-Fuc	1.33	0.23 (23)	0.07 (7)	0.03 (3)
5	3-Fuc	1.41	0.17 (17)	0.05 (5)	0.03 (3)
6	Fuc	1.43	0.18 (17)	0.04 (4)	0.04 (4)
7	2,3,6-Glc	1.50	ND	ND	0.15 (15)
8	2,6-Glc	1.64	ND	ND	0.04 (4)

Adapted from Kariya *et al.* (1997). With permission from Elsevier Publisher.
[a] Partially methylated alditol acetate with methoxy groups at the positions shown.
[b] Retention time on a SP2330 capillary column relative to 2,3,4-tri-*O*-methyl-fucitol.
[c] The molar ratios were based on the peak area.
[d] Not detected.
[e] Numbers in parentheses represent percentage distribution of PMAAs.

that administered DHG cannot undergo catabolic degradation in mammals. Anion exchange chromatographic behavior of DHG excreted into the urine after oral administration showed that partial desulfation might occur through intestinal absorption. After oral administration of DHG (50 mg/kg), 0.1% of the dose was found in the urine collected for 24h. More than 5% of intravenously administered DHG (1mg/kg) was excreted into the urine in 24h. These results suggest that orally administered macromolecules such as DHG are absorbed in the gastrointestinal tract (Imanari *et al.*, 1999).

CSE obtained from sea cucumber markedly enhanced plasminogen activation by tissue plasminogen activators (t-PAs) and urinary plasminogen activator (u-PA) *in vitro*; in the presence of 10mg/ml of CSE, the potentiation factors of single chain of t-PA, two chains of t-PA, and u-PA were 400, 140, and 130, respectively. Though the potentiation activity of CSE gradually decreased when it was depolymerized by chondroitinase ABC, the specific disaccharide from CSE still showed significant activity. GAG from sea cucumber, which possesses a very similar core structure to CSE, but has additional sulfated fucose branches, exhibit very weak activity. The minimal structural requirement in CSE to enhance plasminogen activation by plasminogen activators is GlcUAβ1-3GalNAc(4S,6S) and that additional branching sugars deter the activity (Sakai *et al.*, 2000).

B. Ascidians

Ascidians are marine invertebrates and the closest living relatives of vertebrates. Tadpole-type larvae of ascidians represent the simplest chordate body plan (Satoh and Jeffery, 1995). However, ascidians do not produce cartilage or mineralized bone at any stage of their lives. These types of endoskeleton have developed in vertebrates after their divergence from ascidian groups. Therefore, ascidians are expected to have only a minimal number of genes involved in the enzymatic modification of GAGs. Proteoglycans are glycoproteins consisting of many unbranched polysaccharide chains associated with a core protein. CS-containing proteoglycans (CSPGs) are ubiquitously found in the extracellular matrix and on the cell surface (Tetsukawa *et al.*, 2010). In vertebrates, CSPGs are major components of cartilage where they mainly serve as a cushion. The cell adhesion molecule integrin requires CSPG on the cell surface to bind fibronectin (Moyano *et al.*, 1999). In some cases, CSPGs promote neurite outgrowth (Clement *et al.*, 1998), while in others, they function as a repulsive signal to inhibit axonal growth (Masuda *et al.*, 2004). A type of CS binds to a heparin-binding neurotrophic factor, midkine, and inhibits its function (Ueoka *et al.*, 2000).

The antithrombotic effect of *Styela plicata* heparin was evaluated in two models of thrombosis. One of the models, widely used for venous thrombosis, is based on the evaluation of thrombus formation induced by two major factors, activation of coagulation and stasis (Vogel *et al.*, 1989). The resultant venous thrombus is composed of fibrin and red blood cells. In the arteriovenous shunt model, thrombosis is initiated by platelet adherence to a silk thread anchored in the shunt, and both activation of platelets and coagulation contribute to thrombus formation. *S. plicata* heparin was much less potent to inhibit thrombosis than mammalian heparin, and complete thrombus inhibition was not achieved, even at a concentration of fourfold higher than that required for mammalian heparin to inhibit 100% thrombosis. Therefore, heparin preparations obtained from the body of *S. plicata* would have a safer therapeutic action in the treatment of arterial thrombosis than mammalian heparin (Santos *et al.*, 2007). Scully *et al.* (1988) reported that the ability of DS to accelerate AT-factor Xa interaction or HCII-T interaction is considerably increased in oversulfated naturally occurring forms and concluded that the possession of high charge density is not the only requirement for the interactions, and the spatial position of the sulfates are also of importance. With this, Pavao *et al.* (1998) have demonstrated that 4-*O*-sulfation of the galactosamine unit is essential for the anticoagulant activity of natural DSs isolated from marine invertebrates. Andoniades *et al.* (2002) have demonstrated that the specific activity of the DS of high purity increased three times after a resulfation under mild conditions. All these results suggest that alterations in the sulfation pattern of the GAGs, and particularly DS, could be an important factor for the efficient control of the thrombogenic process and the multifactorial thrombotic events (Calabrese *et al.*, 2004).

Inflammation is considered to be an important component of tumorigenesis, although the underlying mechanisms remain largely unknown (Coussens and Werb, 2002). Interestingly, benign tumor cells induce inflammatory response in the host and collaborate in establishing the tumor through a process called desmoplasia (Mareel and Leroy, 2003). Essentially, all of the elements that constitute the inflammatory response participate in the host reaction, which could, therefore, have an atrophic purpose for tumor cells (Arias *et al.*, 2005). Nuclear factor κB (NF-κB) is a ubiquitous and well-characterized protein responsible for the regulation of complex phenomena, with a pivotal role in controlling cell signaling in the body under certain conditions. Among other functions, NF-κB controls the expression of genes encoding the proinflammatory cytokines, for example, interleukin-1 (IL-1), tumor necrosis factor-R (TNF-R), chemokines such as IL-8, macrophage inflammatory protein-1R (MIP-1R), and monocyte chemoattractant protein-1 (MCP-1), adhesion molecules such as vascular cell adhesion molecule (VCAM), and inducible enzymes such as cyclooxygenase-2 (COX-2), all of which play critical roles in controlling inflammatory processes (Aggarwal, 2004). CS is a GAG, which is naturally present in the extracellular matrix of articular cartilage (Fioravanti and Collodel, 2006). Recently, several studies indicated that CS played an important role in anti-inflammation and anticancer activities, and a few studies were performed to elucidate the underlying molecular mechanism. Legendre *et al.* (2008) reported that CS inhibited proinflammatory gene expression such as COX-2 and iNOS in chondrocytes; however, they did not provide detailed information of how CS suppresses the proinflammatory factor expression. Recently, Campo *et al.* (2008) found that CS inhibited collagen-induced NF-κB activation. However, the precise mechanism of CS was not investigated thoroughly either. Xu *et al.* (2008a) investigated the anti-inflammatory effects of CS on TPA-induced inflammation in mouse skin *in vivo* and explored the underlying molecular mechanism (Fig. 2.2). The results showed that CS suppressed TPA-induced edema, expression of cyclooxygenase-2, vascular cell adhesion molecule-1, and Akt signaling in the mouse skin. Therefore, it may be a promising strategy to inhibit the NF-κB activation in order to reduce the tumor formation. In addition, CS inhibited TNF-α-induced NF-κB activation and subsequent vascular cell adhesion molecule and inducible nitric oxide synthase expressions by blocking Akt signals in JB6 cells (Xu *et al.*, 2008b).

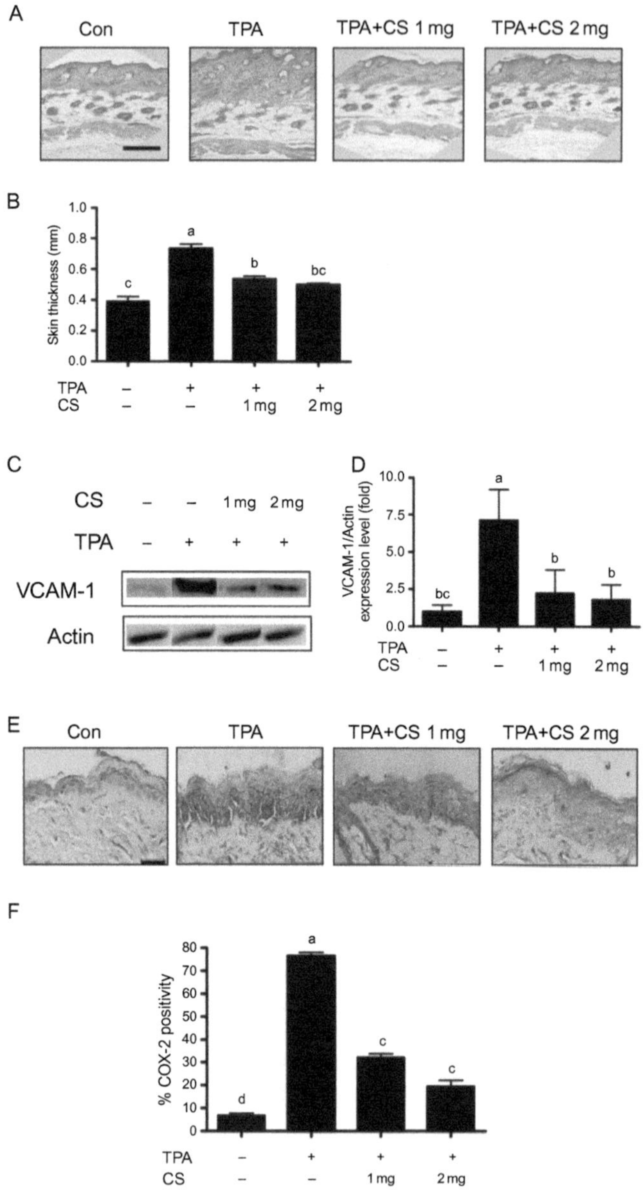

FIGURE 2.2 Inhibitory effects of CS on TPA-induced edema and expression of VCAM-1 and COX-2 in mouse skin. Mice were treated topically with TPA (200 μL of 10 nmol of TPA dissolved in acetone) in the presence or absence of CS (1 or 2 mg). Control animals were treated with acetone alone. (A) Skin section was stained with hematoxylin and eosin (original magnification, ×100; bar, 100 μm); (B) thickness of the skin (the thickness of the ear of each mouse was determined by averaging the values measured at five independent

C. Sea urchins

Astragalus membranaceus (Fisch) Bge. var. *mongolicus* (Bge.) Hsiao, a specialty crop in China, has been used in preparing functional foods such as soups and teas. It has been recognized for potential detoxification, diuresis, antiperspirant, myogenic, and anti-aging effects (Qi *et al.*, 2006; Wang *et al.*, 2006). Astragalus polysaccharides (APS) are considered as a group of possible bioactive components contributing to the beneficial effects of Astragalus. Several extractions of APS have been prepared and investigated for their monosaccharide compositions and molecular weights since the first one was purified from the roots of *A. membranaceus* in the 1980s (Fang and Wagner, 1988; Fang *et al.*, 1982; Li *et al.*, 2009; Masashi *et al.*, 1992). The APS was extracted and purified from the roots of *A. membranaceus* and characterized for its chemical structure and potential health properties. The APS was composed of α-D-glucose residues with the estimated equivalent dextran molecular weight of 2.07×10^4 Da. Periodate oxidation analysis, 1D and 2D NMR spectroscopy demonstrated that APS has repeating (1→4)-linked backbone with a (1→6)-linked branch every 10 residues. The APS possess scavenging activities against hydroxyl radicals and hydrogen peroxide and showed chelating effect on ferrous ions. The APS was also able to bind cholic and chenodeoxycholic acids *in vitro*. In addition, APS was able to stimulate activity of purified mouse B cells without promoting T cell proliferation. These data provided information for future development of APS as a nutraceutical (Niu *et al.*, 2011).

Malignancy of tumor and cancer cells is one of the most serious problems faced by cancer patients. This accounts for the eventual 12% of all deaths worldwide according to estimates by the World Health Organization (WHO; Ferlay *et al.*, 2001). Hepatocarcinoma is a kind of tumor with both a high incidence and lethality rate. The inhibitory effect and evaluation of immunological mechanisms of a polysaccharide that was isolated from *Strongylocentrotus nudus* eggs (SEP) against hepatocellular carcinoma in H22-bearing mice were performed by determining its

---

regions of the cross section); (C) Western blot analysis of VCAM-1 expression in mouse skin; (D) bands of interests were further analyzed by densitometer; (E) immunohistochemical measurement of COX-2 in the mouse skin (dark brown color indicates COX-2 expression; original magnification, ×200; bar, 100 μm); (F) comparison of COX-2 labeling index in the mouse skin (COX-2-positive staining was determined by counting five randomly chosen fields per section, determining the percentage of DAB--positive cells per 100 cells at ×400 magnification). Different letters (a–c) denote statistical difference ($P<0.05$). Each bar represents the mean±SE ($n=3$). CON, control; TPA, TPA-treated group; TPA+CS, TPA-treated mice in the presence of CS (1 or 2 mg). Adapted from Xu *et al.* (2008a). With permission from ACS Publisher. (For color version of this figure, the reader is referred to the Web version of this chapter.)

effects on the growth of transplanted tumors and immune response in H22-bearing mice. ICR mice inoculated with mouse hepatoma carcinoma cell lines H22 were treated with SEP at doses of 4, 8, 16 mg/kg/day for 12 days. The effects of SEP were measured via the growth of the transplanted tumors, splenocyte proliferation, T lymphocyte counts, CTL activity, and the production of cytokines from splenocytes and the levels of serum Ig in tumor-bearing mice. In addition, the effects of SEP on Erk phosphorylation in mouse splenocytes and on the transcriptional activity of NFAT in Jurkat T cells were also investigated. Wang *et al.* (2011) showed that SEP significantly inhibited the growth of transplanted tumors in mice. SEP could not only remarkably enhance splenocyte proliferation, $CD4^+$ and $CD8^+$ T cell numbers as well as CTL activity, but it also elevated IL-2 and TNF-α secretion as well as IgA, IgM, and IgG levels in the serum. Further, the activation of Erk phosphorylation and the NFAT promoter by SEP promoted the transcription and expression of downstream gene IL-2. In conclusion, our study demonstrates that SEP effectively inhibits hepatocellular carcinoma *in vivo* via enhancement of host immune system function, and it could be a potential therapeutic drug for hepatocarcinoma.

D. Nudibranchs

Nudibranchs are some of the most beautiful animals found in the oceans. Their delicate bodies are usually brightly colored, often in striking multi-hued geometric patterns (Behrens, 1980). The hard shell used by most of their molluscan relatives for protection, which leaves their sensory rhinophores and oxygen-gathering branchial plumes decoratively exposed on their dorsums are absent in nudibranchs. They have large feet that provide locomotion, but only at speeds that are slow relative to potential predators. Nudibranchs tend to be found in shallow water habitats, where they frequently sit out in the open, blatantly exposing their vulnerability. Despite their apparent lack of physical attributes and behavioral patterns suited for defensive purposes, nudibranchs have few documented predators (Edmunds, 1968; Thompson 1960a,b). Astute field observations and some simple antifeedant experiments led marine biologists to propose that chemicals provide an invisible protective armor for these soft-bodied molluscs. In book "*Between Pacific Tides*" (Ricketts *et al.*, 1968), the naturalist Ricketts wrote that "Many nudibranchs, but especially the dorids, have a penetrating fruity odor that is pleasant when mild but nauseating when concentrated. Undoubtedly, this odor is one of the reasons why nudibranchs seem to be left strictly alone by their predatory animals."

Mucus functions in many invertebrate physiological processes and also influences the structuring of the community and the ecosystem. Molluscan mucus is mostly water. The remaining components are proteins, carbohydrates, and lipids. The predominant carbohydrates in the GAG fraction of *Arion ater* mucus were identified as glucosamine, iduronic acid, and galactose. The pattern of digestion of *A. ater* mucus with a series of polysaccharide lyases was consistent with its GAG fraction having the structure (βGlcA1-4αGlcN1-4αIdA1-4αGlcN1-4)$_n$. Some GAG chains appeared to be linked to protein through GalN and/or GlcN, which would constitute novel protein–saccharide linkages, but others were linked through another, as yet unidentified, sugar (not xylitol or mannitol; Cottrell *et al.*, 1994). Mucus is probably released in dehydrated form in distinct, membrane-bound packages, which then absorbs water. Functional mucus is probably formed by mixing of mucins from different types of glands. Under small deformities, hydrated mucus is a viscoelastic solid, able to function as a rope. As stress increases, it yields to become a liquid which can return to the solid state once the stress is released. It is these properties that allow the locomotion in molluscs on what is seemingly called as an adhesion. On dehydration, the strength and stiffness of mucus production have been studied quantitatively by various methods, some gravimetric and some calorimetric using pedal, fecal, pseudofecal, and hypobranchial mucus: there are much spatial and temporal variations, though (Davies and Hawkins, 1998).

Mucus or slime is a gel exuded from the internal and external surfaces of many animals, both vertebrates and invertebrates. It is primarily composed of protein–sugar conjugates (glycoproteins) and water, although it may contain a wide range of dissolved materials such as salts, enzymes, antibodies, antibiotics, and various components of the cells which produce it. Its primary roles are surface lubrication and protection, although it is also used for capturing food particles, as defense against predators, and resistance to desiccation, among of its other functions (Eylers, 2008). Mucus plays a vital role in feeding. In filter-feeding bivalves, mucus aids the transport of food from gill to mouth and is employed to cleanse the mantle cavity of particles rejected by the labial palps. In gastropods, mucus nets and bags are used to trap food prior to ingestion and some groups roll their prey in mucus to prevent its escape. Pedal mucus may be ingested after it has become studded with organic material and perhaps act as a fertilizer for microbial growth (Davies *et al.*, 1990). A copious secretion of epithelial mucus is used to isolate molluscs from their environment, and mucus may also serve as an ionoregulator. Mucus may also contain specific products to render the animal poisonous, distasteful, or irritating. Agglutinin and lysozyme have been found in mucus from marine molluscs. Mucus secretion can present a considerable drain of energy (up to 70% of consumed energy). The fate of molluscan mucus is largely unknown and probably makes a considerable contribution to

POM in inshore waters, although it is readily degradable by marine microbes. Given the persistence of mucus, densities of benthic gastropods and their motility patterns, much of the gastropod-inhabited benthos is likely to be covered for most of the time with a layer of pedal mucus (Davies *et al.*, 1990).

Invertebrates that lack GAGs synthesize sulfated polysaccharides such as spirulan from bivalve molluscs (Burson *et al.*, 1956), horatin from annelids (Rahemtulla and Lovtrup, 1974), sulfated L-galactans from Styela (Pavao *et al.*, 1994), and Kakelokelose, a sulfated mannan from another ascidian (Riccio *et al.*, 1996). These polysaccharides, showing some structural features similar to GAGs, may well carry out the same or similar critical biological functions that GAGs do. Hovingh and Linker (1998) shown that CS is the major GAG present in the heart, mantle, and kidney of *Helix aspersa*. HS is present as a minor component mainly in heart and mantle. This HS differs somewhat from the major mammalian GAG, that is, the total sulfate content is lower and there is considerably less *N*-sulfated and more *N*-acetylated hexosamine. In *Aplysia*, "heparin" appeared to be the major GAG in the foregut and a minor component in the Ctenidium. The resistance of this heparin-like polymer to heparinase, however, indicates that it may be a highly sulfated HS or it may belong to the category of heparin–HS-like GAGs.

Mammalian mucus can exhibit abrupt stress softening (Celli *et al.*, 2007), a rheological behavior attributed to the extended conformation, polyelectrolye nature, and liquid crystal propensity of its mucin molecules (Celli *et al.*, 2007; Hyun *et al.*, 2002). Slug mucus also exhibits this behavior. The slug mucus of *Lehmannia valentiana* does not contain mucins, but its Sm proteins contain A-domains, and A-domains are often found in systems that respond abruptly to hydrodynamic shear. Similarly, slug mucus exhibits a discontinuous response to shear stress. Below a critical strain (~0.15 s), the mucus behaves as an elastic solid, but above this threshold, it transforms into a viscous liquid; uniquely, a short period of rest (<1 s) is all that is required to return the hydrogel to its initial state (Cottrell *et al.*, 1993; Denny, 1989; Ewoldt *et al.*, 2007). The ability of their pedal mucus to undergo rapid and successive "yield-heal" cycles, and thereby act as a fluid ratchet, is what permits slugs and other gastropods to engage in adhesive locomotion (Denny, 1989; Ewoldt *et al.*, 2007). Given the precedents above, it is tempting to suggest that a strain-dependent (and rapidly reversible) change in the conformation and binding properties of the A-domains in Sm-family proteins contributes to the distinctive rheology of slug mucus (Li and Graham, 2007). Moreover, GAGs are reported to be available in some kind of sea slugs (Fig. 2.3).

FIGURE 2.3 Glycosaminoglycans resources of some kinds of sea slugs. (A) *Melibe viridis*, (B) *Aplysia kurodai*, (C) *Glossodoris rufomarginata*. Adapted from http://www.umiushi.info/. (For color version of this figure, the reader is referred to the Web version of this chapter.)

REFERENCES

Aggarwal, B. B. (2004). Nuclear factor-κB: The enemy within. *Cancer Cell* **6**, 203–208.

Altmann, F., Fabini, G., Ahorn, H., and Wilson, I. B. (2001). Genetic model organisms in the study of N-glycans. *Biochimie* **83**, 703–712.

Andoniades, S. E., Schenone, M., Recondo, E. F., and Fernandez De Recondo, M. E. (2002). Influence of the degree of sulfation in the biological activity of dermatan sulate and chondroitin sulfate C. *Anal. Asco. Quimica Argentina* **90**, 21–26.

Arias, J. I., Aller, M. A., and Arias, J. (2005). The use of inflammation by tumor cells. *Cancer* **104**, 223–228.

Behrens, D. W. (1980). Pacific Coast Nudibranchs. Sea Challengers, Los Osos.

Burson, S. L., Fahrenbach, M. J., Frommhagen, L. H., Riccardi, B. A., Brown, R. A., Brockman, J. A., Lewry, H. V., and Stokstad, E. L. R. (1956). Isolation and purification of mactins, heparin-like anticoagulants from *Mollusca*. *J. Am. Chem. Soc.* **78**, 5874–5878.

Calabrese, G. C., Alberto, M. F., Tubio, R., Marani, M. M., Fernandez de Recondo, M. E., Lazzari, M., and Recondo, E. F. (2004). A small fraction of dermatan sulfate with significantly increased anticoagulant activity was selected by interaction with the first complement protein. *Thromb. Res.* **113**, 243–250.

Campo, G. M., Avenoso, A., Campo, S., D'Ascola, A., Traina, P., and Calatroni, A. (2008). Chondroitin-4-sulphate inhibits NF-κB translocation and caspase activation in collagen-induced arthritis in mice. *Osteoarthr. Cartil.* **16**, 1474–1483.

Celli, J. P., Turner, B. S., Afdhal, N. H., Ewoldt, R. H., McKinley, G. H., Bansil, R., and Erramilli, S. (2007). Rheology of gastric mucin exhibits a pH-dependent solgel transition. *Biomacromolecules* **8**, 1580–1586.

Chansue, N., Ponpornpisit, A., Endo, M., Sakai, M., and Satoshi, Y. (2000). Improved immunity of tilapia *Oreochromis niloticus* by CUP III, a herb medicine. *Fish Pathol.* **35**, 89–90.

Cheng, X., Cole, R. N., Zaia, J., and Hart, G. W. (2000). Alternative O-glycosylation/O-phosphorylation of the murine estrogen receptor β. *Biochemistry* **39**, 11609–11620.

Clement, A. M., Nadanaka, S., Masayama, K., Mandl, C., Sugahara, K., and Faissner, A. (1998). The DSD-1 carbohydrate epitope depends on sulfation, correlates with chondroitin sulfate D motifs, and is sufficient to promote neurite outgrowth. *J. Biol. Chem.* **273**, 28444–28453.

Coombe, D. R. and Kett, W. C. (2005). Heparan sulfate-protein interactions: Therapeutic potential through structure–function insights. *Cell. Mol. Life Sci.* **62**, 410–424.

Cottrell, J. M., Henderson, I. F., Pickett, J. A., and Wright, D. J. (1993). Evidence for GAGs as a major component of trail mucus from the terrestrial slug, *Arion ater* L. *Comp. Biochem. Physiol.* **104B**, 455–468.

Cottrell, J. M., Henderson, I. F., and Wright, D. J. (1994). Studies on the glycosaminoglycan component of trail mucus from the terrestrial slug, *Arion ater* L. *Comp. Biochem. Physiol. B* **107**, 285–296.
Coussens, L. M. and Werb, Z. (2002). Inflammation and cancer. *Nature* **420**, 860–867.
Davies, M. S. and Hawkins, J. (1998). Mucus from marine molluscs. *Adv. Mar. Biol.* **34**, 1–71.
Davies, M. S., Jones, H. D., and Hawkins, S. J. (1990). Seasonal variation in the composition of pedal mucus from *Patella vulgata* L. *J. Exp. Mar. Biol. Ecol.* **144**, 101–112.
Dawes, J., McLaren, M., Forbes, C., Belch, J. J. F., Lane, D. A., Bray, B., McEwen, J., Houin, G., and Gianese, F. (1991). The pharmacokinetics of dermatan sulphate MF701 in healthy human volunteers. *Br. J. Clin. Pharmacol.* **32**, 361–366.
Denny, M. W. (1989). Invertebrate mucous secretions: Functional alternatives to vertebrate paradigms. *Symp. Soc. Exp. Biol.* **43**, 337–366.
Didraga, M., Barroso, B., and Bischoff, R. (2006). Recent developments in proteoglycan purification and analysis. *Curr. Pharm. Anal.* **2**, 323–327.
Dugenci, S. K., Arda, N., and Candan, A. (2003). Some medicinal plants as immunostimulant for fish. *J. Ethnopharmacol.* **88**, 99–106.
Edmunds, M. J. (1968). Acid secretion in some species of Doridacea (Mollusca, Nudibranchia). *Proc. Malac. Soc. Lond.* **38**, 121–133.
Ewoldt, R. H., Clasen, C., Hosoi, A. E., and McKinley, G. H. (2007). Rheological fingerprinting of gastropod pedal mucus and synthetic complex fluids for biomimicking adhesive locomotion. *Soft Matter* **3**, 634–643.
Eylers, J. P. (2008). Mucus and slime: Structure and rheology of natural polysaccharides. *Encyclo. Materials: Sci. Technol.* 5848–5850.
Fang, J. N. and Wagner, H. (1988). The chemical structure of polyglucosan of *Astragalus membranaceus*. *Acta Chinic. Cinica.* **46**, 1101–1104.
Fang, S. D., Chen, Y., Xu, X. Y., Ye, C. Q., Zhai, S. K., and Shen, M. L. (1982). Studies of the active principles of *Astragalus mongholicus* Bunge I. Isolation, characterization and biological effect of its polysaccharides. *Org. Chem.* **2**, 26–31.
Ferlay, J., Bray, F., and Pisani, P. (2001). GLOBOCAN 2000: cancer incidence, mortality and 524 prevalence worldwide, version 1.0. IARC CancerBase, No. 5 Lyon, USA: IARC 525 Press.
Fioravanti, A. and Collodel, G. (2006). In vitro effects of chondroitin sulfate. *Adv. Pharmacol.* **53**, 449–465.
Gandhi, N. S. and Mancera, R. L. (2008). The structure of glycosaminoglycans and their interactions with proteins. *Chem. Biol. Drug Des.* **72**, 455–489.
Gargiulo, V., Lanzetta, R., Parrilli, M., and De Castrol, C. (2009). Structural analysis of chondroitin sulfate from *Scyliorhinus canicula:* A useful source of this polysaccharide. *Glycobiology* **19**, 1485–1491.
Guinez, C., Morelle, W., Michalski, J. C., and Lefebvre, T. (2005). O-GlcNAc glycosylation: A signal for the nuclear transport of cytosolic proteins? *Int. J. Biochem. Cell Biol.* **37**, 765–774.
Hardingham, T. and Bayliss, M. (1990). Proteoglycans of articular cartilage: Changes in aging and in joint disease. *Semin. Arthritis Rheum.* **20**, 12–33.
Hart, G. W., Haltiwanger, R. S., Holt, G. D., and Kelly, W. G. (1989). Glycosylation in the nucleus and cytoplasm. *Annu. Rev. Biochem.* **58**, 841–874.
Hiromura, M., Choi, C. H., Sabourin, N. A., Jones, H., Bachvarov, D., and Usheva, A. (2003). YY1 is regulated by O-linked N-acetylglucosaminylation (O-glcNAcylation). *J. Biol. Chem.* **278**, 14046–14052.
Holt, G. D. and Hart, G. W. (1986). The subcellular distribution of terminal *N*-acetylglucosamine moieties. Localization of a novel protein–saccharide linkage, O-linked GlcNAc. *J. Biol. Chem.* **261**, 8049–8057.

Hovingh, P. and Linker, A. (1998). Glycosaminoglycans in two mollusks, *Aplysia californica* and *Helix aspersa*, and in the leech, *Nephelopsis obscura*. *Comp. Biochem. Physiol. B* **119**, 691–696.

Hyun, K., Kim, S. H., Ahn, K. H., and Lee, S. J. (2002). Large amplitude oscillatory shear as a way to classify the complex fluids. *J. Non-Newtonian. Fluid Mech.* **107**, 51–65.

Imanari, T., Washio, Y., Huang, Y., Toyoda, H., Suzuki, A., and Toida, T. (1999). Oral absorption and clearance of partially depolymerized fucosyl chondroitin sulfate from sea cucumber. *Thromb. Res.* **93**, 129–135.

Iozzo, R. V. (1998). Matrix proteoglycans: From molecular design to cellular function. *Annu. Rev. Biochem.* **67**, 609–652.

Kariya, Y., Watabe, S., Ochial, Y., Murata, K., and Hashimoto, K. (1990). Glycosaminoglycan involved in the cation-induced change of body wall structure of sea cucumber *Stichopus japonicus*. *Connect. Tissue Res.* **25**, 149–159.

Kariya, Y., Watabe, S., Kyogashima, M., Ishihara, M., and Ishii, T. (1997). Structure of fucose branches in the glycosaminoglycan from the body wall of the sea cucumber *Stichopus japonicas*. *Carbohydr. Res.* **297**, 273–279.

Laabs, T., Carulli, D., Geller, H. M., and Fawcett, J. W. (2005). Chondroitin sulfate proteoglycans in neural development and regeneration. *Curr. Opin. Neurobiol.* **15**, 116–120.

Legendre, F., Bauge´, M., Roche, R., Saurel, A. S., and Pujol, J. P. (2008). Chondroitin sulfate modulation of matrix and inflammatory gene expression in IL-1-stimulated chondrocytes-study in hypoxic alginate bead cultures. *Osteoarthr. Cartil.* **16**, 105–114.

Li, D. and Graham, L. D. (2007). Epidermal secretions of terrestrial flatworms and slugs: *Lehmannia valentiana* mucus contains matrilin-like proteins. *Comp. Biochem. Physiol. B* **148**, 231–244.

Li, J. Z. and Lian, E. C. (1988). Aggregation of human platelets by acidic mucopolysaccharide extracted from *Stichopus japonicus* Selenka. *Thromb. Haemost.* **59**, 435–439.

Li, R., Chen, W. C., Wang, W. P., Tian, W. Y., and Zhang, X. G. (2009). Extraction, characterization of Astragalus polysaccharides and its immune modulating activities in rats with gastric cancer. *Carbohydr. Polym.* **78**, 738–742.

Luo, R. (2007). Induction of immunity substance in *Penaeus chinensis* by Chinese herbal medicine. *Oceanol. Limnol. Sin.* **28**, 573–577.

Mareel, M. and Leroy, A. (2003). Clinical, cellular and molecular aspects of cancer invasion. *Physiol. Rev.* **83**, 337–376.

Masashi, T., Noriko, S., and Naoko, O. (1992). A reticuloendothelial system-activating glycan from the roots of *Astragalus membranaceus*. *Phytochemistry* **31**, 63–66.

Masuda, T., Fukamauchi, F., Takeda, Y., Fujisawa, H., Watanabe, K., Okado, N., and Shiga, T. (2004). Developmental regulation of notochord-derived repulsion for dorsal root ganglion axons. *Mol. Cell. Neurosci.* **25**, 217–227.

Medeiros, G. F., Mendes, A., Castro, R. A. B., Bau, E. C., Nader, H. B., and Dietrich, C. P. (2000). Distribution of sulfated glycosaminoglycans in the animal kingdom: Widespread occurrence of heparin-like compounds in invertebrates. *Biochim. Biophys. Acta* **1475**, 287–294.

Motokawa, T. (1981). The stiffness change of the holothurian dermis caused by chemical and electrical stimulation. *Comp. Biochem. Physiol.* **73C**, 41–48.

Motokawa, T. (1982). Rapid change in mechanical properties of echinoderm connective tissues caused by coelomic fluid. *Comp. Biochem. Physiol.* **73C**, 223–229.

Moyano, J. V., Carnemolla, B., Albar, J. P., Leprini, A., Gaggerol, B., Zardi, L., and Garcia-Pardo, A. (1999). Cooperative role for activated $\alpha4\beta1$ integrin and chondroitin sulfate proteoglycans in cell adhesion to the heparin III domain of fibronectin. *J. Biol. Chem.* **274**, 135–142.

Nadanaka, S. and Kitagawa, H. (2008). Heparan sulphate biosynthesis and disease. *J. Biochem.* **144**, 7–14.

Nader, H. B., Lopes, C. C., Rocha, H. A. O., Santos, E. A., and Dietrich, C. (2004). Heparins and heparinoids: Occurrence, structure and mechanism of antithrombotic and hemorrhagic activities. *Curr. Pharm. Des.* **10**, 951–966.

Niu, Y., Wang, H., Xie, Z., Whent, M., Gao, X., Zhang, Z., Zou, S., Yao, W., and Yu, L. (2011). Structural analysis and bioactivity of a polysaccharide from the roots of *Astragalus membranaceus* (Fisch) Bge. var. *mongolicus* (Bge.) Hsiao. *Food Chem.* **128**, 620–626.

Pavao, M. S. G., Rodrigues, A. M., and Mourao, P. A. S. (1994). Acidic polysaccharides of the *Ascidian styeloplicata.* Biosynthetic studies of the sulfated L-galactans of the tunic and preliminary characterization of a dermatan sulfate-like polymer in body tissues. *Biochem. Biophys. Acta* **1199**, 229–237.

Pavao, M., Aiello, K., Werneck, C., Silva, L., Valente, A. P., Mulloy, B., Colwell, N. S., Tollefsen, D. M., and Mourão, P. A. (1998). Highly sulfated dermatan sulfates from ascidians. Structure versus anticoagulant activity of these glycosaminoglycans. *J. Biol. Chem.* **273**, 27848–27857.

Qi, L. W., Yu, Q. T., Li, P., Li, S. L., Wang, Y. X., Sheng, L. H., and Yi, L. (2006). Quality evaluation of Radix Astragali through a simultaneous determination of six major active isoflavonoids and four main saponins by high-performance liquid chromatography coupled with diode array and evaporative light scattering detectors. *J. Chromatogr. A* **1134**, 162–169.

Rahemtulla, F. and Lovtrup, S. (1974). The comparative biochemistry of invertebrate mucopolysaccharides-II *Nematoda annelida. Comp. Biochem. Physiol.* **49B**, 639–646.

Reilly, A. and Kaferstein, F. (1997). Food safety hazards and the application of the principles of the hazard analysis and critical control point (HACCP) system for their control in aquaculture production. *Aquac. Res.* **28**, 735–752.

Riccio, R., Kinnel, R. B., Bifulco, G., and Scheuer, P. J. (1996). Kakelokelose a sulfated mannose polysaccharide with anti-HIV activity from the Pacific tunicate didemnum molle. *Tetrahedron Lett.* **37**, 1979–1982.

Ricketts, E. F., Calvin, J., and Hedgpeth, J. W. (1968). Between Pacific Tides. Stanford University, Standford.

Sakai, T., Kyogashima, M., Kariya, Y., Urano, T., Takada, Y., and Takada, A. (2000). Importance of GlcUAb1-3GalNAc(4S,6S) in chondroitin sulfate E for t-PA- and u-PA-mediated Glu-plasminogen activation. *Thromb. Res.* **100**, 557–565.

Santos, J. C., Mesquita, J. M. F., Belmiro, C. L. R., Carolina da Silveira, B. M., Viskov, C., Mourier, P. A., and Pavão, M. S. G. (2007). Isolation and characterization of a heparin with low antithrombin activity from the body of *Styela plicata* (Chordata-Tunicata). Distinct effects on venous and arterial models of thrombosis. *Thromb. Res.* **121**, 213–222.

Satoh, N. and Jeffery, W. R. (1995). Chasing tails in ascidians: Developmental insights into the origin and evolution of chordates. *Trends Genet.* **11**, 354–359.

Scully, M. F., Ellis, V., Seno, N., and Kakkar, V. V. (1988). Effect of oversulphated chondroitin and dermatan sulphate upon thrombin and factor Xa inactivation by antithrombin III or heparin cofactor II. *Biochem. J.* **254**, 547–551.

Sugahara, K., Mikami, T., Uyama, T., Mizuguchi, S., Nomura, K., and Kitagawa, H. (2003). Recent advances in the structural biology of chondroitin sulfate and dermatan sulfate. *Curr. Opin. Struct. Biol.* **13**, 612–620.

Sumegi, M., Hunyadi-Gulyas, E., Medzihradszky, K. F., and Udvardy, A. (2003). 26S proteasome subunits are O-linked N-acetylglucosamine-modified in *Drosophila melanogaster. Biochem. Biophys. Res. Commun.* **312**, 1284–1289.

Sun, Y. X., Wang, J. Q., Wang, T. T., Xue, J. P., Liu, G., and You, J. S. (2007). A review: Defense mechanism in sea cucumber. *Fish. Sci. China* **6**, 358–361.

Suzuki, N., Kitazato, K., Takamatsu, J., and Saito, H. (1991). Antithrombotic and anticoagulant activity of depolymerized fragment of the glycosaminoglycan extracted from *Stichopus japonicus* Selenka. *Thromb. Haemost.* **65**, 369–373.

Taylor, K. R. and Gallo, R. L. (2006). Glycosaminoglycans and their proteoglycans: Host-associated molecular patterns for initiation and modulation of inflammation. *FASEB J.* **20**, 9–22.

Tetsukawa, A., Nakamura, J., and Fujiwara, S. (2010). Identification of chondroitin/dermatan sulfotransferases in the protochordate, *Ciona intestinalis*. *Comp. Biochem. Physiol. B* **157**, 205–212.

Thompson, T. E. (1960a). Defensive acid-secretions in marine gastropods. *J. Mar. Biol. Assoc.* **39**, 115–122.

Thompson, T. E. (1960b). Defensive adaptations in opisthobranchs. *J. Mar. Biol. Assoc.* **39**, 123–134.

Trotter, J. A., Lyons-Levy, G., Luna, D., Koob, T. J., Keene, D. R., and Atkinson, M. A. L. (1996). Stiparin: A glycoprotein from sea cucumber dermis that aggregates collagen fibrils. *Matrix Biol.* **15**, 99–110.

Trotter, J. A., Salgado, J. P., and Koob, T. J. (1997). Mineral content and salt-dependent viscosity in the dermis of the sea cucumber *Cucumaria frondosa*. *Comp. Biochem. Physiol.* **116A**, 329–335.

Trotter, J. A., Lyons-Levy, G., Chino, K., Koob, T. J., Keene, D. R., and Atkinson, M. A. L. (1999). Collagen fibril aggregation-inhibitor from sea cucumber dermis. *Matrix Biol.* **18**, 569–578.

Ueoka, C., Kaneda, N., Okazaki, I., Nadanaka, S., Muramatsu, T., and Sugahara, K. (2000). Neuronal cell adhesion, mediated by the heparin-binding neuroregulatory factor midkine, is specifically inhibited by chondroitin sulfate E. *J. Biol. Chem.* **275**, 37407–37413.

Vieira, R. P. and Mourao, P. A. (1988). Occurrence of a unique fucose-branched chondroitin sulfate in the body wall of a sea cucumber. *J. Biol. Chem.* **263**, 18176–18183.

Vogel, G. M., Meuleman, D. G., Bourgondien, F. G., and Hobbelen, P. M. (1989). Comparison of two experimental thrombosis models in rats effects of four glycosaminoglycans. *Thromb. Res.* **54**, 399–410.

Volpi, N. (2006). Advances in chondroitin sulfate analysis: Application in physiological and pathological states of connective tissue, and during pharmacological treatment of osteoarthritis. *Curr. Pharm. Des.* **12**, 639–658.

Volpi, N. (2007). Analytical aspects of pharmaceutical grade chondroitin sulfates. *J. Pharm. Sci.* **96**, 3168–3180.

Wang, S. C., Shan, J. J., Wang, Z. T., and Hu, Z. B. (2006). Isolation and structural analysis of an acidic polysaccharide from *Astragalus membranaceus* (Fisch.) Bunge. *J. Integr. Plant Biol.* **48**, 1379–1384.

Wang, Z., Pandey, A., and Hart, G. W. (2007). Dynamic interplay between O-linked N-acetyl-glucosaminylation and glycogen synthase kinase-3-dependent phosphorylation. *Mol. Cell. Proteomics* **6**, 1365–1379.

Wang, M., Wang, H., Tang, Y., Kang, D., Gao, Y., Ke, M., Dou, J., Xi, T., and Zhou, C. (2011). Effective inhibition of a *Strongylocentrotus nudus* eggs polysaccharide against hepatocellular carcinoma is mediated via immunoregulation *in vivo*. *Immunol. Lett.* **141**, 74–82.

Whitelock, J. M. and Iozzo, R. V. (2005). Heparan sulfate: A complex polymer charged with biological activity. *Chem. Rev.* **105**, 2745–2764.

Xu, C. X., Jin, H., Chung, Y. S., Shin, J. Y., Lee, K. H., Beck, G. R., Palmos, G. N., Choi, B. D., and Cho, M. H. (2008a). Chondroitin sulfate extracted from ascidian tunic inhibits phorbol ester-induced expression of inflammatory factors VCAM-1 and COX-2 by blocking NF-κB activation in mouse skin. *J. Agric. Food Chem.* **56**, 9667–9675.

Xu, C. X., Jin, H., Chung, Y. S., Shin, J. Y., Woo, M. A., Lee, K. H., Palmos, G. N., Choi, B. D., and Cho, M. H. (2008b). Chondroitin sulfate extracted from the *Styela clava* tunic suppresses TNF-α-induced expression of inflammatory factors, VCAM-1 and iNOS by blocking Akt/NF-κB signal in JB6 cells. *Cancer Lett.* **264**, 93–100.

Yamada, S. and Sugahara, K. (2008). Potential therapeutic application of chondroitin sulfate/ dermatan sulfate. *Curr. Drug Discov. Technol.* **5**, 289–301.

Yanagishita, M. and Hascall, V. C. (1992). Cell surface heparan sulfate proteoglycans. *J. Biol. Chem.* **267**, 9451–9454.

Yang, X., Zhang, F., and Kudlow, J. E. (2002). Recruitment of O-GlcNAc transferase to promoters by corepressor mSin3A: Coupling protein O-GlcNAcylation to transcriptional repression. *Cell* **110**, 69–80.

Yang, H. S., Yuan, X. T., Zhou, Y., Mao, Y. Z., Zhang, T., and Liu, Y. (2005). Effects of body size and water temperature on food consumption and growth in the sea cucumber *Apostichopus japonicus* (Selenka) with special reference to aestivation. *Aquac. Res.* **36**, 1085–1092.

Zhang, C. Y., Wang, Y. G., and Rong, X. J. (2006). Isolation and identification of causative pathogen for skin ulcerative syndrome in *Apostichopus japonicus*. *J. Fish. Sci. China.* **1**, 118–123.

Zhao, K. S., Mancini, C., and Doria, G. (1990). Enhancement of the immune response in mice by *Astragalus membranaceus* extracts. *Immunopharmacology* **20**, 25–34.

CHAPTER 3

Marine By-product Phospholipids as Booster of Medicinal Compounds

Koretaro Takahashi*,[1] and **Yoshikazu Inoue**†

Contents

Abstract

Marine phospholipids are defined as phospholipids containing docosahexaenoic acid (DHA) or eicosapentaenoic acid that would be more effective than fish oil, which is mostly composed of triacylglycerol, in exerting health benefits. Marine phospholipids would boost the effect of both the health-beneficial hydrophilic and the hydrophobic compounds such as cell differentiators, anticancer compounds, and antiobesity compounds. When marine phospholipids are served as liposomal drinks, they would be more effective than adding into solid foods or feeds. As long as the liposome bilayer is basically composed of marine phospholipids, they would promote the encapsulated functional compounds. And this is the principal advantage of choosing marine phospholipids as liposomal membrane. Bioconversion of marine phospholipid would also be advantageous in delivering DHA into the desired tissue.

* Faculty of Fisheries Sciences, Hokkaido University, Hakodate, Japan
† Lipid Lab., Yasuda Onoue-cho, Kakogawa city, Okayama, Japan
[1] Corresponding author: Koretaro Takahashi, *E-mail address:* kore@fish.hokudai.ac.jp

Advances in Food and Nutrition Research, Volume 65
ISSN 1043-4526, DOI: 10.1016/B978-0-12-416003-3.00003-2

For example, lysophosphatidylserine obtained through phospholipase D-mediated transphosphatidylation and phospholipase A_1 or *sn*-1 positional specific lipase-mediated partial hydrolysis seemed to be the most effective chemical form in delivering DHA into brain.

I. INTRODUCTION

Marine (fisheries) by-products are rich in docosahexaenoic acid (DHA) or eicosapentaenoic acid (EPA). Although the most abundant DHA- or EPA-rich lipid class in most of the marine by-products is triacylglycerol (TG), some of the by-product sources are rich in DHA- or EPA-containing phospholipids. And these DHA- or EPA-containing phospholipids are often called marine phospholipids. Well-known sources for DHA-containing phospholipids are squid skin, squid muscle, squid connective tissues, and testis as well as ovary of fish and shellfish. On the other hand, recently known sources for EPA-bounded phospholipids are starfish (*Asterias amurensis*) internal organs and gonads. Typical sources for both DHA- and EPA-abundant phospholipids are those of giant scallop internal organs, salmon heads, and krills. We can say that marine phospholipids are rich in fisheries by-products or poorly used marine bioresources.

II. MARINE PHOSPHOLIPID MUST BE MORE EFFECTIVE THAN FISH OIL TG ON HEALTH BENEFITS

Health benefits of DHA or EPA have been widely recognized. However, DHA or EPA itself is just a moiety of lipid molecules such as TG or phospholipids. TG is a typical hydrophobic compound, while on the other hand, phospholipids are amphiphilic compounds. For this reason, phospholipid form DHA and EPA are considered to be much more bioavailable than those of TG forms. In fact, Galli *et al.* (1992) has borne out that phospholipid form polyunsaturated fatty acid dosing should result in a higher polyunsaturated fatty acid content in plasma than TG form polyunsaturated fatty acid dosing. It has been discussed in detail that marine phospholipids must be more efficient than marine TG (i.e., fish oil) in delivering $n-3$ polyunsaturated fatty acids to desired tissues (Lu *et al.*, 2011). Fish oils are well known to decrease the levels of total cholesterol, blood TG content, and low density lipoprotein (LDL), while on the other hand increase high density lipoprotein (HDL) in blood. However, in practical cases, decreasing the TG content and increasing the HDL content in blood cannot be achieved by moderate intake of fish oil. Large amount of fish oil administration is necessary for this purpose

TABLE 3.1 Lipid content in the serum of rats fed with DHA- or EPA-containing triacylglycerol (TG) and phosphatidylcholine (PC)

	EPA-PC	EPA-TG	DHA-TG	Control
HDL cholesterol	64.9 ± 5.5^{a}	54.1 ± 10.1^{b}	53.6 ± 9.2^{b}	57.4 ± 6.0^{b}
LDL cholesterol	4.5 ± 0.4^{b}	5.2 ± 0.6^{a}	5.1 ± 0.6^{a}	5.1 ± 0.5^{a}
Total cholesterol	55.1 ± 3.2^{b}	61.9 ± 5.3^{a}	64.4 ± 6.1^{a}	66.0 ± 7.8^{a}
Free cholesterol	11.1 ± 1.7	13.0 ± 1.0	11.9 ± 3.0	12.6 ± 1.3
Cholesterol ester	43.9 ± 2.3^{b}	48.9 ± 5.9^{b}	52.4 ± 5.3^{a}	53.6 ± 7.3^{a}
Phospholipid	139.9 ± 18.1^{a}	102.4 ± 13.7^{c}	108.0 ± 16.3^{c}	124.7 ± 13.3^{b}

mg/dL serum.
Shina *et al.* (2008).
Different superscript letters are significant different ($P<0.05$).

compared to marine phospholipids. In fact as shown in Table 3.1, we recently observed that among marine lipids, for example, the phosphatidylcholine (PC) form EPA (EPA-PC), TG form DHA (DHA-TG) and EPA (EPA-TG), EPA-PC was the only lipid molecular species effective in increasing the level of HDL with moderate amount of intake. Krill oil contains approximately 35% phospholipids. The effect of krill oil on serum lipids and markers of oxidative stress and inflammation have recently been studied (Ulven *et al.*, 2011). Molecular species of krill phospholipids are also well studied (Winther *et al.*, 2011). It was observed that the metabolic effects of krill oil rich in marine phospholipids was comparable to fish oil even if the EPA+DHA dose in the krill oil was 63% of the fish oil. A typical health benefit of fish oil is the prevention of overproduction of arachidonic acid-derived eicosanoids which often increase the risk of cancer, thrombosis, allergy, and other diseases. In this sense, marine phospholipids would be more effective when they are served in same amount. Moreover, Di Marzo *et al.* (2010) showed that DHA in the brain increased significantly after krill oil administration, when compared to control animals. It is interesting to note that $n-3$ fatty acids linked to phospholipids may be differently distributed in the body compared to $n-3$ fatty acids in other lipid molecular forms. Wijendran *et al.* (2002) found that efficacy of dietary arachidonic acid provided as phospholipid as substrate for brain arachidonic acid accretion in baboon neonates is much higher than TG. And this also tells that phospholipid form must be more effective than TG form in reaching the desired polyunsaturated fatty acid to the desired tissue when administrated. There are many kinds of phospholipid classes. Recently, it was observed that among the DHA-containing phospholipids, as substrates for DHA accretion in fetus rat brain, lysophosphatidylserine (LPS) form DHA seemed to be the most effective DHA carrier. As depicted in Table 3.2, the percentage of DHA in fatty acid composition of the rat fetus brain fed with LPS-added diet was

TABLE 3.2 Fatty acid composition of the fetal brain lipid in pregnancy rat administrated phospholipid samples by probe (%)

Fatty acid	DHA-PS	DHA-LPS	DHA-PC	DHA-LPC	Control
C14:0	2.3±0.2	2.1±0.2	2.7±0.7	2.2±0.1	2.2±0.1
C16:0	38.7±0.9	38.2±0.7	38.0±1.5	38.4±0.6	38.8±0.4
C18:0	16.9±1.6	18.4±1.5	16.9±1.9	18.5±1.6	16.9±1.0
C18:1	18.4±1.6	16.2±1.8	19.0±2.5	16.7±1.9	17.7±1.0
C18:2	1.2±0.1	1.4±0.9	1.2±0.4	1.1±0.2	1.0±0.1
C20:4	12.9±0.4	12.2±0.4	12.5±0.7	13.0±0.5	12.9±0.2
C22:6	9.5±1.3	11.5±1.2[a]	9.8±1.1	10.1±1.2	9.5±0.8

Onoyama *et al.* (2011).
Datum with superscript letter is significant different (*P*<*0.05*).

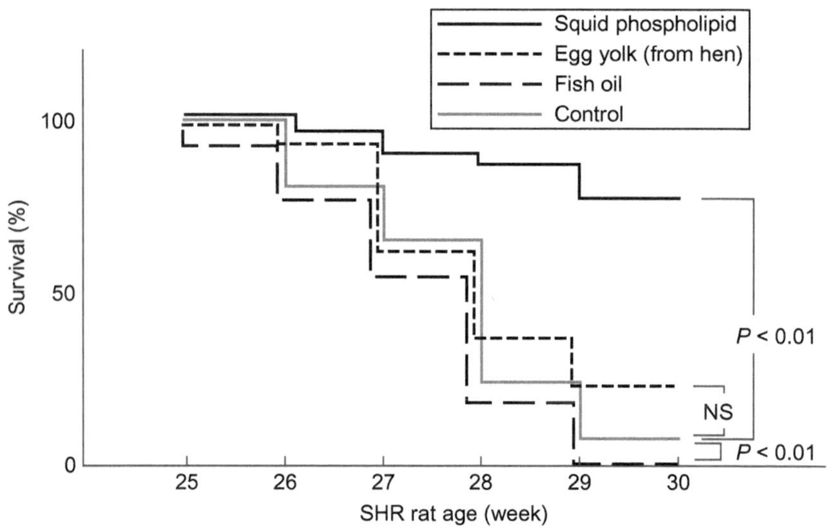

FIGURE 3.1 Survival of SHR rats (easily die by apoplexy) served with phospholipid form DHA (squid PC), fish oil, and egg yolk PC (Inoue, 2001).

the highest. This LPS form DHA can be obtained through bioconversions using phospholipase D to convert DHA-containing PC into phosphatidylserine (PS) and then using phospholipase A_1 or the position *sn*-1 specific lipase to convert PS into LPS.

A typical feature that tells the importance of the backbone of phospholipid molecule which DHA bounds is the prevention effect on apoplexy. A combination of phosphoryl base group with DHA must be crucial to exert antiapoplexy function as shown in Fig. 3.1. Among fish oil TG, egg yolk phospholipid, and squid phospholipid which is rich in *sn*-2

DHA-bounded phospholipids, the squid phospholipid is the only functionally available lipid chemical form in preventing apoplexy (Inoue, 2001).

III. BOOSTING EFFECT ON CANCER CELL DIFFERENTIATION

Applying marine phospholipids rich in DHA and EPA must also be beneficial for the treatment of cancer. Because DHA-containing phospholipids have been recognized as lipids that promote cancer cell differentiation induced by some cell differentiators. When cancer cell differentiates, it becomes mortal. And that should help the therapy of cancer. Tochisawa (1997a) showed that DHA-containing phospholipids promote cancer cell differentiation induced by retinoic acid (Fig. 3.2), and Hosokawa *et al.* (2001) showed that DHA-containing phospholipids promote cancer cell differentiation induced by dibutyryl cyclic adenosine monophosphate on nonfixative cancer cells such as leukemia HL-60 cell line. On the other hand, Hossain *et al.* (2006) showed that DHA-containing phospholipids promote cancer cell differentiation induced by butylate for fixative cancer cells such as colon cancer. However, on fixative cancer cells such as colon cancer, PS with the same fatty acid combination exerted more effectively than PC on the promotional effect. Hossain *et al.* (2006) also showed that the suppression of colorectal cancer (Caco-2) cells by the marine phospholipids may not only caused by the promotional effect of cancer cell differentiation but also by the induction of apoptosis with other bioactive compounds as borne out through enrichment factor comparison experiments which is an index of

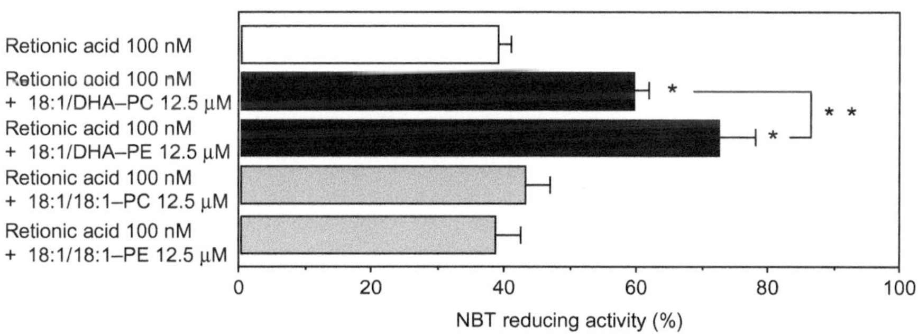

FIGURE 3.2 Effect of retinoic acid (RA) and various lipids on cell differentiation of HL-60 cells (Tochisawa *et al.*, 1997). HL-60 cells (5×10^4) were incubated with the individual lipids for 24 h, then RA was added followed by incubation for additional 72 h. Nitroblue tetrazolium (NBT) reducing activity in HL-60 cells was assayed. Data shown as mean value±SD. *$p<0.01$ versus control. **$p<0.01$ versus RA.

DNA fragmentation relative to control. The enrichment factor was significantly high in squid PC and much higher in squid PC-bioconverted PS when combined with sodium butylate. Unfortunately, PS is poorly found in natural sources except brain and nerve tissues. For this reason, transphosphatidylation of marine PC to PS should become critical when obtaining DHA- or EPA-bounded PS in a reasonable price. The promotional effects on cell differentiation compounds must be very useful in cancer therapy, because DHA-containing phospholipids, the marine phospholipids, should decrease side effects of the anticancer drugs that seriously impair quality of life. For example, large amount of retinoic acid is usually prescribed as an anticancer drug for the lives of patients, and with too much retinoic acid, side effect seriously impairs the quality of life of the patients. Burns and Spector (1987) showed that uptake of Adriamycin was greater in DHA-enriched L1210 cells compared to other unsaturated fatty acid-enriched cells. This was also true for mitoxantrone (Burns *et al.*, 1988). In those experiments, free fatty acid form DHA was used which is no doubt less bioavailable than the phospholipid forms. There have been several reports that explain at least in part the mechanisms of how DHA-containing phospholipids exert promotional effects, as seen above, on functional compounds. Stillwell *et al.* (1993) considered that increase in permeability of the organelle biomembranes by the fusion of DHA-bounded phospholipid should be the reason why the cells become susceptible to medicinal substances. Tochisawa (1997) carried out a TNBS-quenching method by using DHA-PE and considered that easy incorporation of that phospholipid into the cell membrane and then increase in fluidity of the cell membrane owing to the property of DHA might be one reason why the HL-60 cells become susceptible to retinoic acid. They also observed that simultaneous addition of marine phospholipids with retinoic acid to the cell lines is not effective in promoting the cell differentiation unlike the preincubation with marine phospholipids and considered that DHA-containing phospholipids should be incorporated into the cells in advance to the incorporation of retinoic acid. On the other hand, Larsson *et al.* (2004) considered that DHA may also influence on gene expression, transcription factor activity, and signal transduction pathways. For example, 1-oleoyl-2-DHA-containing PE was shown to facilitate upregulating expressions of c-*jun* mRNA and c-Jun protein, while it was effective in downregulating c-*myc* mRNA (Hosokawa, 2004; Fig. 3.3). c-Jun protein is in part responsible for cell differentiation process, and c-Myc protein is responsible for cell cycle progression. Therefore, it is considered that phospholipids containing DHA may at least in part suppress leukemia by modulating the oncogene expressions. Moreover, Tochisawa (1997) got the evidence that protein kinase C is involved in the promotional effect on cell differentiation. He showed that the promotional effect was obviously impaired by protein kinase C inhibitor. He also showed that when HL-60 cell line is preincubated

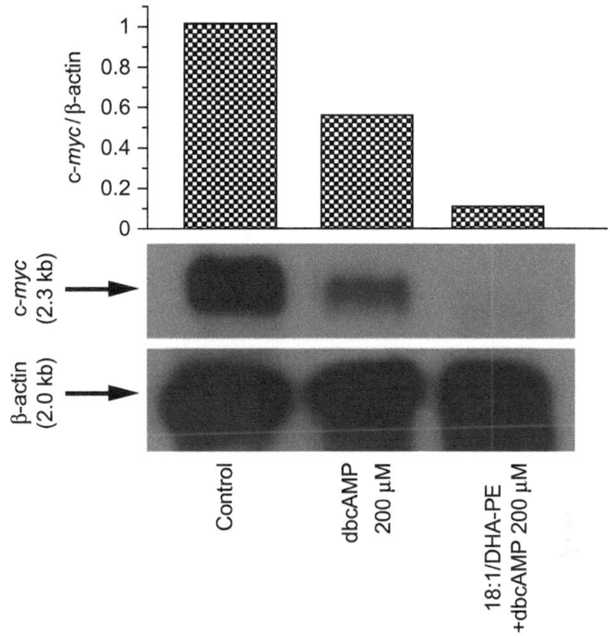

FIGURE 3.3 Downregulation of *c-myc* mRNA in HL-60 cells treated with dibutyryl cAMP (dbcAMP) and PE form DHA (Hosokawa, 2004). HL-60 cells (5×10^4) were incubated with (dbcAMP) after preincubation with 50 μM 1-oleoyl, 2-docosahexaenoyl-phosphatidylethanolamine for 24 h. Nitroblue tetrazolium (NBT) reducing activity in HL-60 cells was assayed. Data shown as mean value±SD. *$p<0.01$ versus control. **$p<0.01$ versus RA. (For color version of this figure, the reader is referred to the Web version of this chapter.)

with marine phospholipid prior to the addition of retinoic acid, the expression of mRNA of retinoic acid receptor α increases, which suggests that marine phospholipids should go into nucleus, and then enhances the susceptibility to retinoic acid. Konishi (2003) showed that DHA-containing PS may promote Caco-2 cell differentiation induced by sodium butylate via upregulation of tumor growth factor β-1 as shown in Table 3.3. On the other hand, polyunsaturated fatty acids are known to regulate peroxisome proliferators-activated receptor (PPAR) which is involved in cell differentiation, cell proliferation, and inflammatory responses (Grimaldi, 2001; Vamecq and Latruffe, 1999). DHA was found to induce apoptosis in vascular smooth muscle cells by activation of PPAR-α and p38 mitogen-activated protein kinases (Diep *et al.*, 2000).

TABLE 3.3 TGF β-1 expression of Caco-2 cells treated with DHA-PS and sodium butyrate

	TGF β-1 (ng/mg protein)
Control	213±9
DHA-PS	210±4
Sodium butylate (0.5mM)	269±8*
DHA-PS+sodium butylate	350±13*,**

Data are shown as means±SD (n=5). *p<0.01 versus control, **p<0.01 versus sodium butylate.
PS, phosphatidylserine; TGF, tumor growth factor.

IV. BOOSTING EFFECT ON CANCER SUPPRESSION

Marine phospholipids in the form of liposomes were borne out to be also useful in boosting the effect of water-soluble anticancer compounds such as β-glucan. Murakawa *et al.* (2007) carried out an *in vivo* study on mice to see the promotional effect on cancer tumor suppression when *Agaricus blazei* Murrill water extract (rich in β-glucan) encapsulated marine phospholipid is orally taken as liposomal drink. In their study, myeloma sp2 cells were first implanted (3.0×10^5 cells/BALB/c male mouse). After 20 days, mice with comparable tumor sizes were selected and divided into four groups: (i) control (only water served), (ii) 0.4mg/mL squid phospholipid liposome (PC/PS/cholesterol(Chol)=9/1/10 molar ratio) alone served as drink, (iii) 0.5mg/mL *A. blazei* Murrill water extract alone served as drink, and (iv) 0.4mg/mL squid phospholipid liposome (PC/PS/Chol=9/1/10 molar ratio) with 0.5mg/mL *A. blazei* Murrill water extract partially encapsulated served as drink. These drinks were orally administrated for 3weeks. The administrated drink amount per mouse was approximately 5mL per day. The results showed that by serving the mice with only low concentration of squid phospholipid liposome or *A. blazei* Murrill water extract individually, no appreciable suppression against cancer growth was observed. However, when both the squid phospholipid and *A. blazei* Murrill water extract were combined, suppression became noticeable as shown in Fig. 3.4. To see the reproducibility of this promotional effect on mushroom immune activator, they did another *in vivo* study with some modifications. Half number of myeloma sp2 cells (1.5×10^5 cells/BALB/c male mouse) compared to the prior experiment was implanted. Then after 20days, mice with comparable size were selected and divided into the four groups that are (i) control (only water served), (ii) 1.0mg/mL squid phospholipid liposome (PC/PS/Chol=9/1/10 molar ratio) alone served as drink, (iii) 1.0 mg/mL *A. blazei* Murrill water extract alone served as drink, (iv) 1.0mg/mL squid phospholipid liposome (PC/PS/Chol=9/1/10 molar ratio)

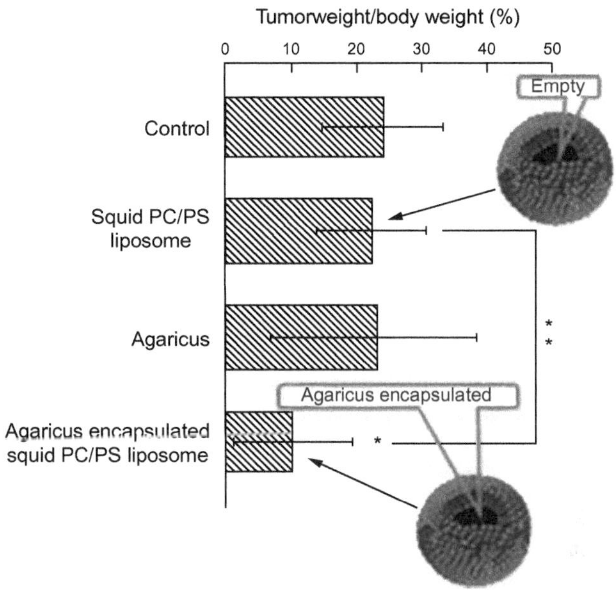

FIGURE 3.4 Tumor weight (%) against body weight of myeloma sp2-bearing BALB/c mice. Samples were administered 7days after myeloma sp2 tumor (3.0×10^5 cells) implantation. Marine phospholipid based liposomes (0.4mg/mL) were orally administered from day 0 to 13 and *Agaricus* (0.5mg/mL) from day 0 to 28. AeL: *Agaricus* (0.5 mg/mL) encapsulated liposomes (0.4mg/mL) were also administered from day 0 to 28 (n=7). (For color version of this figure, the reader is referred to the Web version of this chapter.)

with 1.0mg/mL *A. blazei* Murrill water extract in the form of simple mixture served as drink, and (v) 1.0mg/mL squid phospholipid liposome (PC/PS/Chol=9/1/10 molar ratio) with 1.0mg/mL *A. blazei* Murrill water extract partially encapsulated served as drink. These drinks were orally administrated for 3weeks. The orally administrated liposomal drink amount per mouse was approximately 5mL per day. As shown in Fig. 3.5, *A. blazei* Murrill water extract alone and squid phospholipid alone served groups show moderate tumor suppression with total administrations being approximately 105mg/mouse throughout the experimental term. When both *A. blazei* Murrill water extract and squid phospholipid were administrated simultaneously in a simple mixture form with the same amount of the individual components, boosting effect of squid phospholipid on sp2 myeloma cancer tumor suppression was observed. Similarly, when *A. blazei* Murrill water extract was partially encapsulated with squid phospholipid liposome with total administrations being the same amount with the above simple mixture form (105mg/mouse), the boosting effect on cancer tumor suppression by *A. blazei* Murrill water

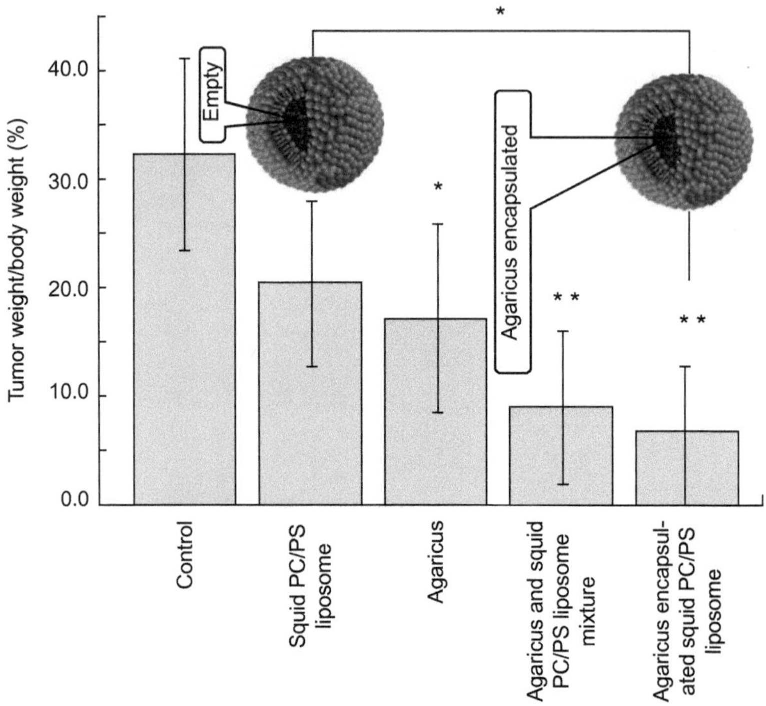

FIGURE 3.5 Effect of marine phospholipid based liposomes and *Agaricus* on tumor weight (%) against body weight of myeloma sp2-bearing BALB/c mice (Murakawa *et al.*, 2007). Myeloma sp2 tumor cells (1.5×10^5 cells/mice) were implanted into mice. Twenty days after implantation of the cells, liposomes (1.0 mg/mL), *Agaricus* (0.5 mg/mL), AaL: *Agaricus* (0.5 mg/mL)+liposomes (1.0 mg/mL) mixture, and AeL: *Agaricus* (0.5 mg/mL) encapsulated liposomes (1.0 mg/mL) were administered for 21 days. The results are the means±SD ($n=7$). Asterisks indicate significant difference as compared to control (*$p<0.05$, **$p<0.001$). (For color version of this figure, the reader is referred to the Web version of this chapter.)

extract was more enhanced. Though there was no significant difference in tumor sizes between the simple mixture form administrated group, that is, group (iv) and the β-glucan partially encapsulated form administrated group, that is, group (v), the promotional effect seemed more superior in the form of encapsulated. In fact, although it was only one, the tumor-disappeared mouse was seen in the partially encapsulated form liposomal drink served group. If we succeed in encapsulating more amount of *A. blazei* Murrill water extract in the marine phospholipid liposomes, there should be a significant difference in tumor sizes between groups (iv) and (v). Thus, it was concluded that *A. blazei* Murrill water extract encapsulated marine phospholipid may be useful in myeloma sp2 therapy.

In the early 1990s, the mucoadhesive function of chitosan has received much attention for prolonging the residence time of dosage forms at the absorption site (Gupta *et al.*, 1990). It has been reported that the stability of chitosan-coated liposomes in stimulated gastric fluid was significantly higher as compared to uncoated liposomes. It has been proposed that chitosan-coated liposomes might be utilized as a bioadhesive intestinal delivery system (Filipović-Grčić *et al.*, 2001). Thus, it can be considered that chitosan-coated marine phospholipid may be beneficial in promoting antitumor activity of marine phospholipid or chitosan itself. It is anticipated that chitosan-coated marine phospholipid liposomes may also promote the anticancer effects by prolonging the retention on intestinal wall. Therefore, the tumor suppression effect of marine phospholipids with chitosan on myeloma sp2 tumor-bearing mice was investigated. Weanling BALB/c mice were divided into five groups. Control (only water served), 1.0mg/mL squid phospholipids liposome (PC/PS/Chol=9/1/10 molar ratio) alone served, 5.0mg/mL chitosan alone served, chitosan (5mg/mL)+squid phospholipid (1mg/mL) liposome mixture served, and chitosan-coated (5mg/mL) squid phospholipid liposome (1mg/mL) served were treated as groups (i), (ii), (iii), (iv), and (v), respectively.

The orally administered liposomal drink amount per mouse was approximately 5mL per day. Results (Hossain, 2007) showed that chitosan alone and squid phospholipid alone served groups showed slight tumor-associated weight loss (Fig. 3.6) and tumor suppressions (Fig. 3.7) throughout the experiment term. There was a significant cancer tumor-associated weight loss suppression (Fig. 3.6) as well as tumor suppression effect (Fig. 3.7) when both chitosan and squid phospholipid were administrated simultaneously in a simple mixture form and chitosan-coated liposome form. We may, thus, say that marine phospholipids not only promotes the effect of anticancer drugs and functional compounds but might also moderately suppress cancer-associated body weight loss.

V. BOOSTING EFFECT ON ANTIOBESITY COMPOUNDS

By inserting hydrophobic antiobesity compounds into the liposomal bilayers, marine phospholipids would boost the effect of the inserted hydrophobic antiobesity compounds. When marine phospholipids are served as liposomal drinks, they would be more effective than adding into solid foods or feeds. These facts were borne out by Okada *et al.* (2011). They carried out the following experiment. Brown seaweed (*Undaria pinnatifida*) lipid containing fucoxanthin (UL) encapsulated into scallop midgut gland phospholipids (PL) liposomes were prepared to see the promotional effect of marine phospholipid liposome on antiobesity. Animal model used in their study was 3-week-old male diabetic-obese

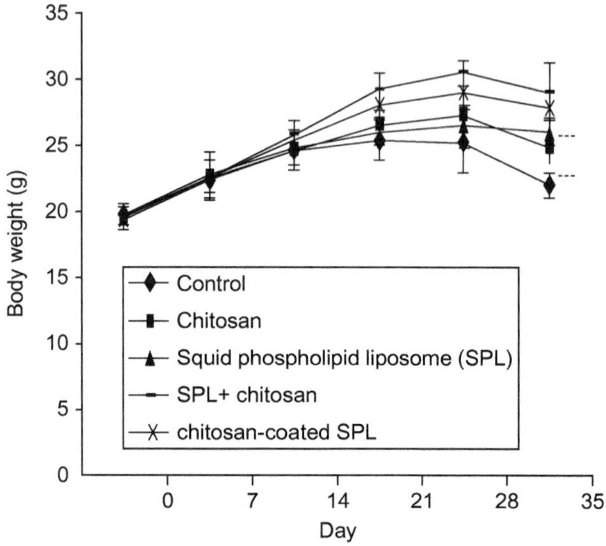

FIGURE 3.6 Effect of squid meal liposome and chitosan samples on the body weight of myeloma sp2-bearing BALB/c mice. Twenty days after implantation of the cells, liposomes (1.0 mg/mL), chitosan (5.0 mg/mL), chitosan (5.0 mg/mL)–liposomes (1.0 mg/mL) mixture, and chitosan-coated (5.0 mg/mL) liposomes (1.0 mg/mL) were administered for 35 days. The results are the means±SD (n=12). Asterisks indicate significant difference as compared to control and liposome (***p<0.001).

mice model KK-A^y mice. Each group received different combinations of lipid drinks, that is, 0.2% UL, 0.3% PL (scallop midgut gland phospholipid), 0.2% UL encapsulated 0.3% PL liposomes. The same amount of total intake of UL and PL in the form of feed were also set up for comparison. And as a result, the total intake of the PL and UL were almost the same between the drink form serving and feed form serving as illustrated in Fig. 3.8. Animals were sacrificed after a 4 weeks experimental feeding period, and adipose tissues and organs were dissected and then weighed. The mesenteric white adipose tissue (WAT) weight was reduced in all experimental groups as compared to the control group. This reduction was significantly different (p<0.05) from the control group for mice receiving the UL-encapsulated PL liposomal drink. Total WAT is composed of perirenal WAT, retroperitoneal WAT, mesenteric WAT, epididymal WAT, and gluteal WAT. Only the UL-encapsulated PL liposome (9.02±1.23 g/100 g BW) exhibited a significant weight reduction of the total WAT compared to the control group (10.1±0.41 g/100 g BW). Uncoupling protein 1 (UCP1) mRNA expression levels and UCP1 expression were determined by RT-PCR (real-time polymerase chain reaction) analysis and Western blotting analysis, respectively. UCB1 is known to

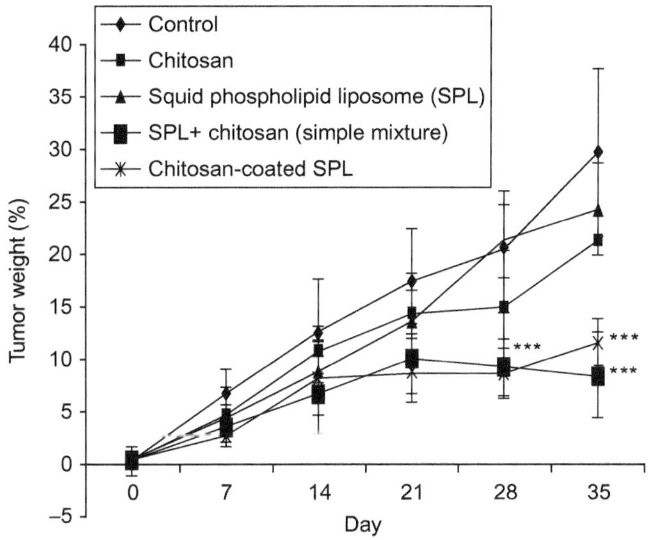

FIGURE 3.7 Effect of squid meal phospholipid liposome and chitosan samples on the tumor weight of myeloma sp2-bearing BALB/c mice. 20 days after implantation of the cells, liposomes (1.0 mg/mL), chitosan (5.0 mg/mL), chitosan (5.0 mg/mL)+liposomes (1.0 mg/mL) mixture, and chitosan-coated (5.0 mg/mL) liposomes (1.0 mg/mL) were administered for 35 days. The results are the means±SD (n=12). Asterisks indicate significant difference as compared to control and SPL (***$p<0.001$).

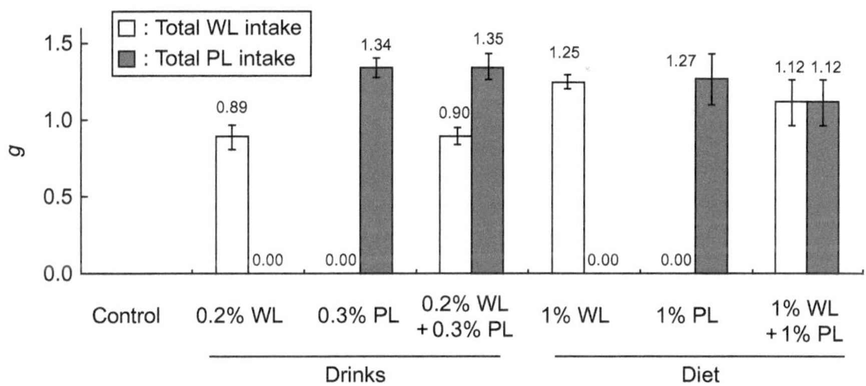

FIGURE 3.8 Total sample intake of KK-A^y mice during the experimental period. For the abbreviations, refer to the text.

decrease WAT by being upregulated to generate heat. Both UCP1 (Fig. 3.9) and UCP1 mRNA expression (Fig. 3.10) in epididymal fat on mice fed with the UL-encapsulated PL liposomes were significantly higher than in the control group. Although the both groups receiving

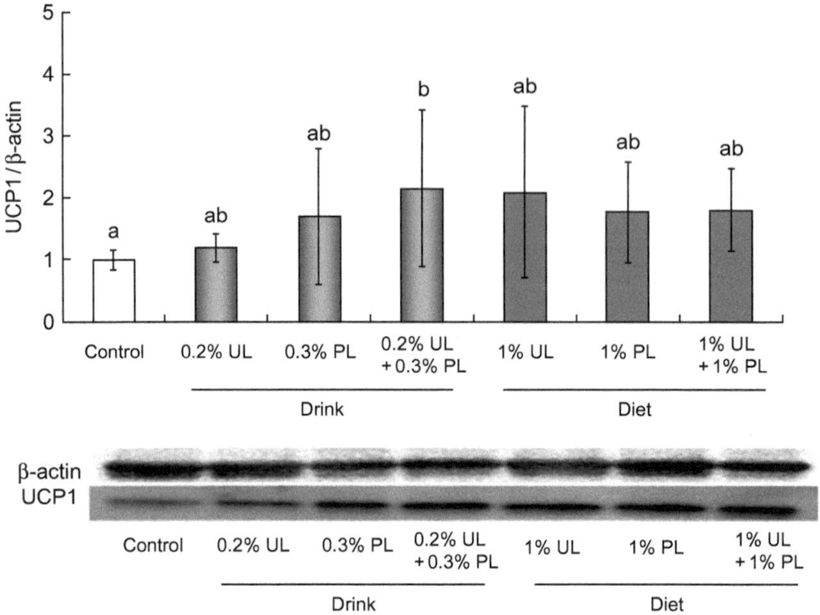

FIGURE 3.9 Western blotting analysis of uncoupling protein 1 (UCP1) in epididymal WAT and relative expression level of UCP1 protein compared to β-actin (Okada *et al.*, 2011). Columns with different superscript letters are significantly different ($p<0.05$).

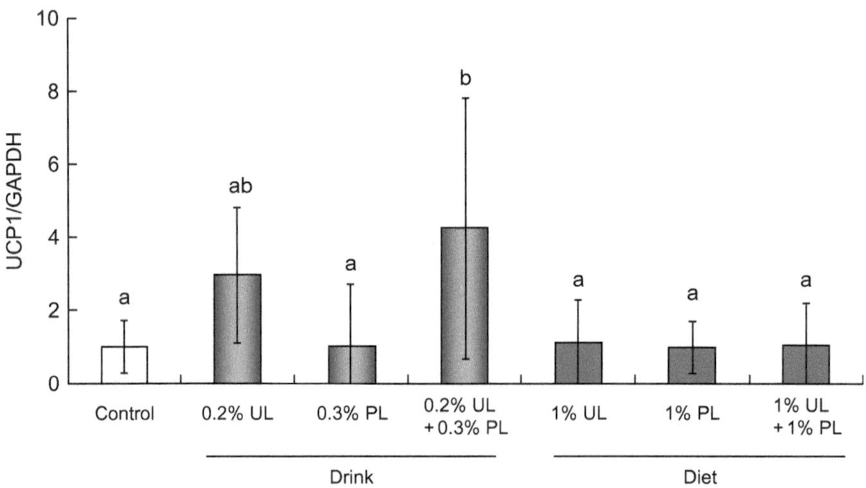

FIGURE3.10 UCP1 mRNA expression levels in epididymal WAT (Okada *et al.*, 2011). Expression of UCP1 mRNA was estimated by quantitative real-time polymerase chain reaction. Relative values were presented as the ratio of UCP1 mRNA to GAPDH mRNA.

0.2% UL and the 0.2% UL-encapsulated 0.3% PL liposome groups exhibited higher UCP1 mRNA expression levels than the control, these levels were significantly different only in the liposome served group being 4.23 times higher than that of control group. These results suggest that UL and PL work in an additive manner to reduce obesity, resulting in a greater antiobesity effect than that which would be expected at these reduced concentrations. And the PL (the marine phospholipid) liposomal form must be more effective than the feed form in serving to exert antiobesity function. Though fucoxanthin is no doubt an excellent functional compound to reduce WAT, limited supply has been a bottle neck. The supply of β-glucan is also limited. Marine phospholipids especially in the form of liposomes should help to boost the effect of precious functional compounds such as it has been seen.

REFERENCES

Burns, C. P. and Spector, A. A. (1987). Membrane fatty acid modification in tumor cells: A potential therapeutic adjunct. *Lipids* **22**, 178–184.

Burns, C. P., Haugstad, B. N., Mossman, C. J., North, J. A., and Ingraham, L. M. (1988). Membrane lipid alteration: Effect on cellular uptake of metoxantron. *Lipids* **23**, 393–397.

Di Marzo, V., Griinari, M., Carta, G., Murru, E., Ligresti, A., Cordeddu, B. K., Giordano, E., Bisogno, T., Collu, M., Batetta, B., Sanna, F., Uda, S., *et al*. (2010). Dietary krill oil increases docosahexaenoic acid and reduces 2-arachidonoylglycerol but not *N(italic)*-acylethanolamine levels in the brain of obese Zucker rats. *Int. Dairy J.* **20**, 231–235.

Diep, Q. N., Touyz, R. M., and Schiffrin, E. L. (2000). Docosahexaenoic acid, a peroxisome proliferator-activated receptors-α ligand induces apoptosis in vascular smooth muscle cells by stimulation of p38 mitogen-activated protein kinase. *Hypertension* **36**, 851–855.

Filipović-Grčić, J., Škalko-Basnet, N., and Jalšenjak, I. (2001). Mucoadhesive chitosan-coated liposomes: Characteristics and stability. *J. Microencapsul.* **18**, 3–12.

Galli, C., Sirtori, C. R., Mosconi, C., Medini, L., Gianfranceschi, G., Vaccarino, V., and Scolastico, C. (1992). Prolonged retention of doubly labeled phosphatidylcholine in human plasma and erythrocytes after oral administration. *Lipids* **27**, 1005–1012.

Grimaldi, P. A. (2001). Fatty acid regulation of gene expression. *Curr. Opin. Clin. Nutr. Metab. Care* **4**, 433–437.

Gupta, P. K., Leung, S. H. S., and Robinson, J. R. (1990). Bioadhesive/mucoadhesives in drug delivery to the gastrointestinal tract. *In* "Bioadhesive Drug Delivery Systems", (V. Lenaerts and R. Gurny, Eds), pp. 65–92. CRC Press, Boca Raton.

Hosokawa, M. (2004). Chapter 10. Improvement of biological response of cells by marine lipids. *In* "Suisangaku-shirizu: Marine Functional Lipids: Sources, Functionalities, and Applications", (K. Takahashi, Ed.), Vol. 142, pp. 145–155. Kouseisha Kouseikaku, Tokyo (in Japanese).

Hosokawa, M., Sato, A., Ishigamori, H., Kohno, H., Tanaka, T., and Takahashi, K. (2001). Synergistic effects of highly unsaturated fatty acid-containing phosphatidylethanolamine on differentiation of human leukemia HL-60 cells by dibutyryl cyclic adenosine monophosphate. *Jpn. J. Cancer Res.* **92**, 666–672.

Hossain, Z. (2007). Marine complex lipid as a bioavailable material for colon carcinoma and myeloma suppression. "A dissertation submitted to the graduate school of fisheries sciences, Hokkaido University", pp. 1–136.

Hossain, Z., Konishi, M., Hosokawa, M., and Takahashi, K. (2006). Effect of polyunsaturated fatty acid-enriched phosphatidylcholine and phosphatidylserine on butyrate-induced growth inhibition, differentiation and apoptosis in Caco-2 cells. *Cell Biochem. Funct.* **24**, 159–165.

Inoue, Y. (2001). Recent developments of polyunsaturated fatty acids and their derivatives related research. *New Food Industry* **43**, 22–26 (in Japanese).

Konishi, M. (2003). Studies on the molecular structure of highly polyunsaturated phospholipids that promotes the induction of colorectal cancer cells induced by cell differentiater. "A master's thesis submitted to the graduate school of fisheries sciences, Hokkaido University", pp. 1–115.

Larsson, S. C., Kumlin, M., Ingelman-Sundberg, M., and Wolk, A. (2004). Dietary long-chain n-3 fatty acids for the prevention of cancer: A review of potential mechanisms. *Am. J. Clin. Nutr.* **79**, 935–945.

Lu, F. S. H., Nielsen, N. S., Timm-Heinrich, M., and Jacobsen, C. (2011). Oxidation stability of marine phospholipids in the liposomal form and their applications. *Lipids* **46**, 3–23.

Murakawa, K., Fukunaga, K., Tanouchi, M., Hosokawa, M., Hossain, Z., and Takahashi, K. (2007). Therapy of myeloma *in vivo* using marine phospholipid in combination with *Agaricus blazei* Muril as an immune respond activator. *J. Oleo Sci.* **56**, 179–188.

Okada, T., Mizuno, Y., Sibayama, S., Hosokawa, M., and Miyashita, K. (2011). Antiobesity effects of *Undaria* lipid capsules prepared with scallop phospholipids. *J. Food Sci.* **76**, H2–H6 Nr. 1.

Onoyama, K., *et al.* (2011).The Japanese Society of Fisheries Sciences, Spring Meeting 2011, Program and Abstract, p. 120, Tokyo.

Shina, *et al.* (2008).47th Annual Meeting of the Japan Oil Chemist's Society, Abstract, p. 224, Tokyo.

Stillwell, W., Ehringer, W., and Jenski, L. J. (1993). Docosahexaenoic acid increases permeability of lipid vesicles and tumor cells. *Lipids* **28**, 103–108.

Tochisawa, K. (1997). Effect of docosahexaenoic acid containing phospholipids on HL-60 human promyelocytic leukemia cell differentiation and their mechanism involved. "A master's thesis submitted to the graduate school of fisheries sciences, Hokkaido University", pp. 1–136.

Tochisawa, K., Hosokawa, M., Kohno, H., Odashima, S., and Takahashi, K. (1997). Effect of phospholipids containing docosahexaenoic acid on differentiation and growth of HL-60 human promyelocytic leukemia cells. *J. Jpn. Oil Chem. Soc.* **46**, 383–390.

Ulven, S. M., Kirkhus, B., Lamglait, A., Basu, S., Elind, E., Haider, T., Berge, K., Vik, H., and Pedersen, J. I. (2011). Metabolic effects of krill oil are essentially similar to those of fish oil but at lower dose of EPA and DHA, in healthy volunteers. *Lipids* **46**, 37–46.

Vamecq, J. and Latruffe, N. (1999). Medical significance of peroxisome proliferator-activated receptors. *Lancet* **354**, 141–148.

Wijendran, W., Huang, M. C., Diau, G. Y., Boehm, G., Nathanielsz, P. W., and Brenna, J. T. (2002). Efficacy of dietary arachidonic acid provided as triglyceride or phospholipid as substrates for brain arachidonic acid accretion in baboon neonates. *Pediatr. Res.* **51**, 265.

Winther, B., Hoem, N., Berge, K., and Reubsaet, L. (2011). Elucidation of phosphatidylcholine composition in krill oil extracted from *Euphausia superba*. *Lipids* **46**, 25–36.

CHAPTER 4

Characterization of Bioactive Peptides Obtained from Marine Invertebrates

Jung Kwon Lee, Joong-Kyun Jeon, Se-Kwon Kim, and **Hee-Guk Byun**[1]

Contents

Department of Marine Biotechnology, Gangneung-Wonju National University, Gangneung, Republic of Korea

[1] Corresponding author: Hee-Guk Byun, *E-mail address:* hgbyun@gwnu.ac.kr

Advances in Food and Nutrition Research, Volume 65
ISSN 1043-4526, DOI: 10.1016/B978-0-12-416003-3.00004-4

Abstract

Bioactive peptides as products of hydrolysis of diverse marine invertebrate (shellfish, crustacean, rotifer, etc.) proteins are the focus of current research. After much research on these muscles and by-products, some biologically active peptides were identified and applied to useful compounds for human utilization. This chapter reviews bioactive peptides from marine invertebrates in regarding to their bioactivities. Additionally, specific characteristics of antihypertensive, anti-Alzheimer, antioxidant, antimicrobial peptide enzymatic production, methods to evaluate bioactivity capacity, bioavailability, and safety concerns of peptides are reviewed.

I. INTRODUCTION

Bioactive peptides with beneficial effects for human health have been found that are marine in origin. Several bioactive peptides were isolated from various invertebrate marine organisms. Marine invertebrates are animals that inhabit a marine environment and are invertebrates (they lack a vertebral column). In order to protect themselves, they may have evolved a shell or a hard exoskeleton, but this is not always the case. Invertebrate muscle cells, however, comprise two major cell classes: striated and smooth. Striated muscle can be subdivided according to the type of striation into transversely striated (like that of vertebrate striated muscle) and obliquely striated. As bioactive peptides have already been discovered in various vertebrate muscles, and as the structures of vertebrate and invertebrate muscles are very similar, invertebrates/insects may be new sources of bioactive peptides. The basic proteins of vertebrate and invertebrate muscles are actin, myosin, and collagen (Oota and Saitou, 1999). Components of proteins in marine foods contain sequences of bioactive peptides which could exert a physiological effect in the body. Moreover, some of these bioactive peptides have been identified to possess nutraceutical potentials that are beneficial for the promotion of human health. Recently, the possible roles for food-derived bioactive peptides in reducing the risk of cardiovascular diseases have been explored (Erdmann *et al.*, 2008). Bioactive peptides usually contain 3–20 amino acid residues, and their activities are based on their amino acid composition and sequence (Pihlanto-Leppala, 2000). These short chains of amino acids are inactive within the sequence of the parent protein, but can be released during gastrointestinal digestion, food processing, or fermentation. Marine-derived bioactive peptides have been obtained widely by enzymatic hydrolysis of marine proteins and have been shown to possess many physiological functions including antimicrobial (Liu *et al.*, 2008), antioxidant (Mendis *et al.*, 2005), and antihypertensive activities (Lee *et al.*, 2011). The goal of this chapter is not to present a complete panorama of marine bioactivity peptides but to show that new materials in marine substances provide new solutions for tackling some of the major public health problems of the twentieth century. These include hypertension, oxidation damage, Alzheimer's disease (AD), and antibacterial issues. These are correlated with aging populations, especially in advanced countries.

II. PREPARATION OF BIOACTIVE PEPTIDES FROM INVERTEBRATES

There is great potential in the marine bioprocessing industry for converting and utilizing products from most marine organisms and marine food products as valuable functional ingredients. There has been increasing interest in the utilization of marine invertebrates, and novel bioprocessing technologies are being developed to isolate bioactive substances from marine food products. The substances contain antihypertensive, antioxidatives, and anti-Alzheimer's properties, and can be used as functional foods and nutraceuticals. Development of these functional ingredients involves certain biotransformation processes through enzyme-mediated hydrolysis in batch reactors. Membrane bioreactor technologies equipped with ultrafiltration membranes are recently emerging for the bioprocessing and development of functional ingredients. It is considered a potential method for utilizing marine food products efficiently (Kim and Mendis, 2006; Kim and Rajapakse, 2005; Kim *et al.*, 2006). Biologically active peptides can be generated from precursor proteins in multiple ways, including enzymatic hydrolysis (either by digestive enzymes or enzymes derived from microorganisms and plants) and microbial fermentation (Kim *et al.*, 2001). Enzymatic hydrolysis is a particularly famous method for obtaining bioactive peptides from organism tissue by *in vivo* hydrolysis of protein sources using appropriate proteolytic enzymes. The physicochemical conditions of the reaction media, such as temperature and pH of the protein solution, must then be adjusted in order to optimize for the activity of the enzyme used. Proteolytic enzymes from microbes, plants, and animals can be used for the hydrolysis process of marine proteins to develop bioactive peptides (Simpson *et al.*, 1998). Further, alcalase, α-chymotrypsin, neutrase, papain, pepsin, and trypsin have been used for the hydrolysis of marine invertebrate muscle under optimal conditions of pH and temperature (Lee *et al.*, 2009). Moreover, one of the most important factors for producing bioactive peptides with the desired functional properties is the molecular weight (MW) of the bioactive peptide (Deeslie and Cheryan, 1981). Therefore, for efficient recovery and in order to obtain bioactive peptides with both the desired molecular size and functional properties, the use of an ultrafiltration membrane system is suitable. An ultrafiltration membrane system equipped with the appropriate MW cutoff is effective in separating peptides with desired MWs from fish protein hydrolysates (Jeon *et al.*, 1999). In order to obtain functionally active peptides, a common method used sorts enzymes, which allows for sequential enzymatic digestion. Moreover, the ultrafiltration membrane system has been able to obtain serial enzymatic digestions using a multistep recycling membrane reactor combined with the ultrafiltration membrane system to separate fish protein hydrolysates based on their MWs (Fig. 4.1) (Byun and Kim, 2001). Additionally, it is possible to obtain serial enzymatic digestions in a system using a multistep recycling membrane reactor combined with an ultrafiltration membrane system to separate marine-derived bioactive peptides (Byun and Kim, 2001; Kim and Mendis, 2006). This membrane bioreactor technology equipped with ultrafiltration membranes is a newly emerging technology for the development of bioactive compounds and is considered a potential method for utilizing marine proteins as value added nutraceuticals with beneficial health effects.

Bioactive peptides can be extracted and purified with these technologies, which vary from simple to complex. Following this, the isolation of bioactive peptides, oligosaccharides, fatty acids, enzymes, water-soluble minerals, and biopolymers for biotechnological and pharmaceutical applications is possible. Further, some of these bioactive peptides have been identified to possess nutraceutical potentials that are beneficial for human health.

III. ANTIHYPERTENSIVE ACTIVITY

A. The mechanism of hypertensive

The renin–angiotensin system (RAS) constitutes one of the most important hormonal systems in the physiological regulation of blood pressure. Indeed, deregulation of the RAS is considered a major factor in the development of cardiovascular disease, and blockade of this system offers an effective therapeutic regimen. Originally defined as a circulation or endocrine system, multiple tissues expressed as a complete local RAS are compelling evidence for angiotensin I and angiotensin II, the primary peptides that can affect the system. ACE is a dipeptidyl carboxypeptidase that catalyzes the conversion of angiotensin I (decapeptide) to angiotensin II (octapeptide), inactivates the antihypertensive vasodilator bradykinin, and increases blood pressure (Fig. 4.2). Inhibition of ACE activity leads to a decrease in the concentration of angiotensin II and consequently reduces blood pressure (Skeggs *et al.*, 1957).

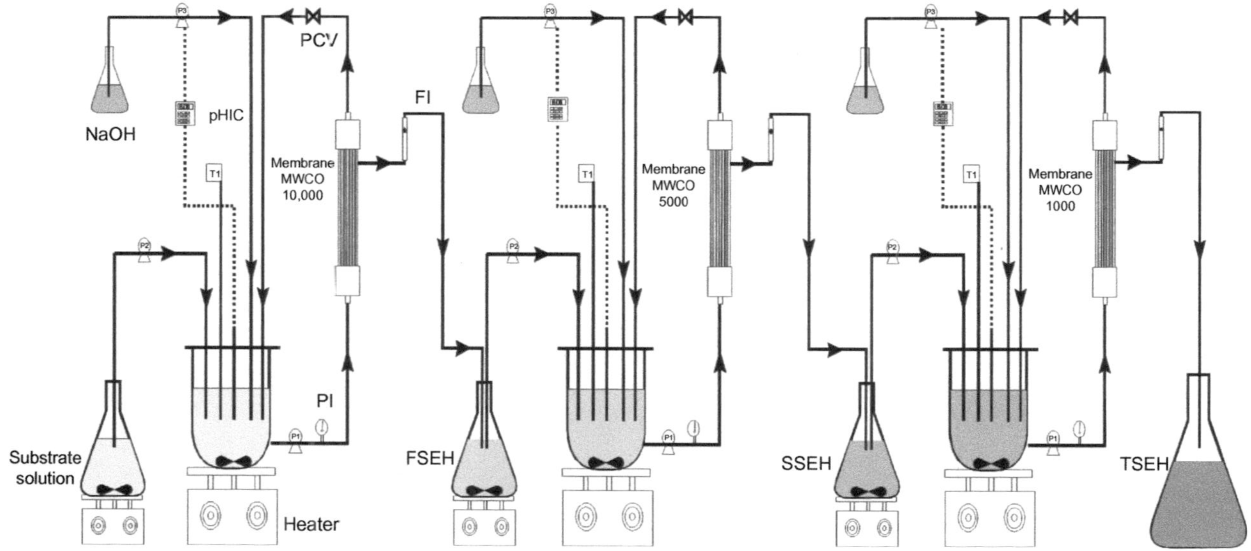

FIGURE 4.1 Schematic diagram of the three-step recycling membrane reactor for production and separation of enzymatic hydrolysates. PI, pressure indicator; FI, flow indicator; pHIC, pH indicator controller; FSEH, first hydrolysates; SSEH, second hydrolysates; TSEH, third hydrolysate.

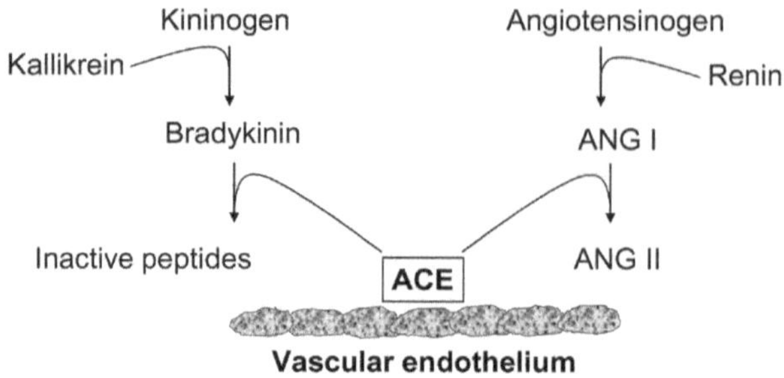

FIGURE 4.2 Schematic diagram of the renin–angiotensin system and kallikrein–kinin system. The angiotensin-converting enzyme is strategically poised to regulate the balance between ang II and bradykinin. ACE, angiotensin-converting enzyme; ang, angiotensin (Borer, 2007).

B. Shellfish

As ACE-inhibiting peptides have already been mentioned with regard to various vertebrate muscles, and as the structure of vertebrate and invertebrate muscle is very similar, invertebrates may be new sources of ACE-inhibiting peptides. However, while many ACE-inhibitor peptides have been derived from vertebrate muscle, very few studies on the subject have been conducted.

Two studies reporting on ACE-inhibitory peptides detected in them in hydrolysates of oyster. Among hydrolysis of oyster proteins with 11 proteases, hydrolysis with denazyme AP (from *Aspergillus oryzae*) (Matsumoto *et al.*, 1994) resulted in Table 4.1. The hydrolysate showed an inhibitory activity against ACE with an IC_{50} of 550 μg/ml. After separation by column chromatography, the most active fraction had an IC_{50} of 85 μg/ml. The peptide isolated from this fraction was identified as Leu-Phe (IC_{50}=126 μM). Katano *et al.* (2003) hydrolyzed pearl oyster meat with an alkaline protease. The hydrolysate was orally administrated to spontaneously hypertensive rats (SHRs) and resulted in a significant decrease of systolic blood pressure (SBP). Four active peptides were isolated and identified as Phe-Tyr, Ala-Trp, Val-Trp, and Gly-Trp. Further, the oyster, *Crassostrea talienwhanensis Crosse* (Wang *et al.*, 2008a), was known as a source of the ACE-inhibitory peptide. In these peptides, the amino acid sequence of *C. talienwhanensis* contains ACE-inhibitory peptides that are not present in the other amino acid sequences: Val-Val-Tyr-Pro-Trp-Try-Glu-Arg-Phe (Table 4.1). Particularly, for the invertebrate *C. talienwhanensis*, unique ACE-inhibitory peptides were found that were different from sardine muscle (Table 4.1) (Matsufuji *et al.*, 1994). Another study of oysters by Huang *et al.* (2011) separated the highest ACE-inhibitory peptide and the calculated the relative MW distributed between 1842 and 1396 Da. Moreover,

TABLE 4.1 ACE-inhibitory peptides derived from shellfish: origin, amino acid sequence, enzyme used for hydrolysis, and IC_{50} value

Origin	Amino acid sequence	Enzyme	IC_{50} (μM)	References
Oyster, *Pinctada fucata martencii*	Leu-Phe	Denazyme AP	126	Matsumoto *et al.* (1994)
Oyster, *Crassostrea talienwhanensis Crosse*	Val-Val-Tyr-Pro-Trp-Tyr-Glu-Arg-Phe	Pepsin	66	Wang *et al.* (2008a,b)
Freshwater clam, *Corbicula fluminea*	Val-Lys-Lys Val-Lys-Pro	Protamex	1,045 3.7	Tsai *et al.* (2006)
Fermented oyster sauce, *Crassostrea gigas*	Val-Lys-Lys	–	1,045	(Je and Kim, 2005)
Fermented blue mussel, *Mytilus edulis*	Glu-Val-Met-Ala-Gly-Asn-Leu-Tyr-Pro-Gly	–	81.91	Je *et al.* (2005a)

Shiozaki *et al.* (2010) hydrolyzed oyster (*Crassostrea gigas*) muscle, and the purified ACE-inhibitory activity peptide that resulted due to synergism was the hypertension-affecting peptide. One potent ACE-inhibitory peptide, Asp-Leu-Thr-Asp-Tyr. This raises the research objective to evaluate invertebrate species, such as mollusks, as potential sources of ACE-inhibitory peptides. In hydrolysates of the protein of the freshwater clam, *Corbicula fluminea*, hydrolyzed with Protamex, ACE-inhibitory activity was observed. A gastrointestinal digestion resulted in the highest ACE-inhibitory activity (Tsai *et al.*, 2006). This freshwater clam hydrolysate (peptide concentration 5mg/ml) was used as a drink administered to SHRs for 8 weeks. The SBP and diastolic blood pressure of the SHRs were significantly reduced by 22.0 and 13.2mmHg, respectively. Further, the antihypertensive activity of the purified peptide present in fermented oyster sauce, *C. gigas* (Je and Kim, 2005), was evaluated by measuring the change of SBP at 1, 2, 3, 6, and 9h after oral administration of 10mg/kg of body weight. There was no change in SBP in the controlled group investigated. *C. gigas*, with a recorded SBP reduction of 19mmHg at 3h after administration, was observed, and the activity was maintained for 6h. Captopril lowered SBP significantly from 1 to 6h after administration of the drug. The other fermented blue mussel, *Mytilus edulis*, was highest in ACE-inhibitoryactivity (Je *et al.*, 2005b). The IC_{50} value of the purified ACE-inhibitory peptide was 19.34μg/ml, and the 10 amino acid residues of the N-terminal sequence were Glu-Val-Met-Ala-

Gly-Asn-Leu-Tyr-Pro-Gly (Table 4.1). The purified peptide was evaluated for antihypertensive effects in SHRs following oral administration.

C. Marine zooplankton

Marine zooplankton (interstitial animals <1mm) is recognized as one of Earth's most diverse communities, and yet is one of the least known. Generally, larval fish have a high demand because of their high dietary protein. Hence, their food requires high protein content and a rapid growth rate. Rotifers have a lot of amino acid pools for producing high content proteins (Rønnestad *et al.*, 2003).

The ACE-inhibitory peptide was isolated from the marine zooplankton rotifer (*Brachionus rotundiformis*). The ACE-inhibitory peptide from the rotifer was hydrolyzed using various commercial enzymes. Among the various hydrolysates, the Alcalase hydrolysate had the highest ACE-inhibitory activity; the IC_{50} value of Alcalase hydrolysate for ACE-inhibitory activity was 0.63mg/ml. The ACE-inhibitory peptide from *B. rotundiformis* was fractionated using Sephadex G-25 and octadecyl silica column (Lee *et al.*, 2009). After a three-step purification process, the purified ACE-inhibitory peptide was identified as being 14 amino acid residues of Asp-Asp-Thr-Gly-His-Asp-Phe-Glu-Asp-Thr-Gly-Glu-Ala-Met, with a MW of 1538Da (IC_{50}=9.64μM). This peptide effectively lowered blood pressure after oral administration in SHRs at a dose of 50mg/body weight (kg) (Fig. 4.3). This peptide exhibited maximal blood pressure reduction 3h after oral administration and was more effective than oral administration of captopril. This peptide showed long-lasting antihypertensive activity since it contains ACE-inhibitory peptides with a variety of maximally effective times. In another study, the ACE-inhibitory peptide was purified and characterized from the enzymatic hydrolysate of a freshwater rotifer (*B. calyciflorus*) fractionation through gel chromatography and HPLC system (Lee *et al.*, 2010a). The ACE-inhibitory peptide was identified as being seven amino acid residues of Ala-Gln-Gly-Glu-Arg-His-Arg by N-terminal amino acid sequence analysis. The IC_{50} value of the purified ACE-inhibitory peptide was 47.1μM.

D. Crustaceans

ACE-inhibitory peptides from invertebrates such as crustaceans have been reported. The sequential hydrolysis of defatted Antarctic krill muscle via pepsin and trypsin resulted in an ACE-inhibitory extract. The active peptide isolated from the extract was found to be Lys-Leu-Lys-Phe-Val, showing an IC_{50} value of 30μM (Kawamura *et al.*, 1992). The other crustacean, *Acetes chinensis*, used the protease from *Bacillus* sp.

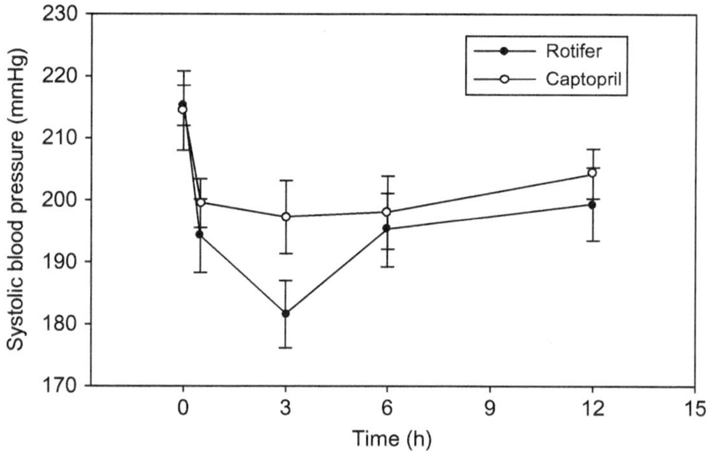

FIGURE 4.3 Antihypertensive activities of ACE-inhibitory peptides after single oral administration in SHRs. SHRs were administered captopril (○), the ACE-inhibitory peptide from rotifer (●) at a dose of 50 mg/kg. Changes in systolic blood pressure were expressed as mean±SE (n=5). Statistical analyses were done by Student t-test ($P<0.05$) (Skeggs *et al.*, 1957).

SM98011 to digest this kind of shrimp and found that the oligopeptide-enriched hydrolysate possessed antioxidant activity and high ACE-inhibitory activity with an IC_{50} value of 0.97 mg/ml (He *et al.*, 2006). The five peptides with high ACE-inhibitory activity were purified from the shrimp hydrolysates, and their sequences were identified by amino acid composition analysis and MW analysis. Three of them, Phe-Cys-Val-Leu-Arg-Pro, Ile-Phe-Val-Pro-Ala-Phe, and Lys-Pro-Pro-Glu-Try-Val, were novel ACE-inhibitory peptides (Table 4.2). Their IC_{50} values were 12.3, 3.4, and 24.1 μM, respectively. To test the ACE-inhibitory activity of the peptides after they were digested by digestive enzymes *in vivo*, derived peptides from Phe-Cys-Val-Leu-Arg-Pro and Ile-Phe-Val-Pro-Ala-Phe were synthesized based on their amino acid sequences and the cleavage sites of digestive enzymes. The IC_{50} values of the derived peptides were determined, and the result showed that except for Val-Pro-Ala-Phe, Phe-Cys and Phe-Cys-Val-Leu, the ACE-inhibitory activity of the other nine derived peptides did not significantly change when compared with their original peptides. Surprisingly, five peptides had lower IC_{50} values than their original peptides, particularly for Arg-Pro (IC_{50} value=0.39 μ M), which was about 30 times lower than its original peptide and nearly the lowest IC_{50} value overall for ACE-inhibitory peptides. It was found that the inhibitory activity of the peptide was intensified by 3.5 times from IC_{50} 15.9 to IC_{50} 4.5 μM after incubation with gastrointestinal proteases. The ACE-inhibitory peptide from *Acaudina molpadioidea* showed a clear

TABLE 4.2 ACE-inhibitory peptides derived from invertebrates: origin, amino acid sequence, enzyme used for hydrolysis, and IC_{50} value

Origin	Amino acid sequence	Enzyme	IC_{50} (μM)	References
Antarctic krill	Lys-Leu-Lys-Phe-Val	Pepsin+trypsin	30	Kawamura *et al.* (1992)
Shrimp, *Acetes chinensis*	Phe-Cys-Val-Leu-Pro	Protease from *Bacillus* sp. SM98011	12.3	He *et al.* (2006)
	Ile-Phe-Val-Pro-Ala-Phe		3.4	
	Lys-Pro-Pro-Gln-Try-Val		24.1	
	Tyr-Leu-Leu-Phe		172	
	Ala-Phe-Leu		65.2	
Shrimp, *Pandalopsis dispar*	Rich Tyr, Phe, Leu, Ile, Val, Lys	Alcalase	100–200 μg/ml	Cheung and Li-Chan (2010)
		Protamex	70 μg/ml	
Fresh sea bream	Val-Ile-Tyr	Alkaline protease	7.5	Fahmi *et al.* (2004)
	Val-Tyr		1.6	
Sea cucumber, *Acaudina molpadioidea*	Met-Glu-Gly-Ala-Gln-Glu-Ala-Gln-Gly-Asp	Bromelain+Alcalase	15.9	Zhao *et al.* (2009)

antihypertensive effect in SHRs at a dosage of 3 μM/kg. Cheung and Li-Chan (2010) produced hydrolysates of shrimp, *Pandalopsis dispar*, by Alcalase and Protamex, which possessed strong ACE-inhibitory activity (IC_{50}=100–200 μg/ml and 70 μg/ml, respectively). Further, ACE inhibition was positively correlated (r^2=0.87) with bitterness of the hydrolysates. Fractionation by size-exclusion chromatography revealed that the bitter substances, which also showed strong ACE inhibition, were <3 kDa in size and contained many hydrophobic residues, including Tyr, Phe, Leu, Ile, Val, and Lys. Despite the bitterness, these hydrolysates may have potential health benefits arising from their potent ACE-inhibitory activity.

E. Other invertebrates

Fahmi *et al.* (2004) produced the hydrolysate of collagen obtained from a sea bream scale and evaluated it with respect to ACE-inhibitory activity. The scales were hydrolyzed using an alkaline protease treatment by which 92% of the peptides were degraded to form hydrolysates. In addition, using SHRs, the oral administration of 300 mg of the peptides (kg of body weight)$^{-1}$ d^{-1} was shown to decrease blood pressure significantly ($P<0.05$). Four peptides that demonstrated high ACE-inhibitory activities were isolated from the hydrolysate of the scales using chromatographic methods. The amino acid sequences of inhibitory peptides were determined to be Gly-Tyr, Val-Tyr, Gly-Phe, and Val-Ile-Tyr. Among the purified peptides, Val-Ile-Tyr had the highest ACE-inhibitory activity (IC_{50}=7.5 μM) followed by Val-Tyr (IC_{50}=16 μM). The sea cucumber, *A. molpadioidea*, was a novel ACE-inhibitory peptide, showing very low similarities to other ACE-inhibitory peptide sequences and was sequenced as Met-Glu-Gly-Ala-Gln-Glu-Ala-Gln-Gly-Asp (Zhao *et al.*, 2009).

Byun and Kim (2001) reported that the aromatic amino acids at the C-terminal and aliphatic amino acids at the N-terminal were designated peptide binding sites for ACE as a competitive inhibitor. However, Maria *et al.* (2009) reported that the casein hydrolysate had a hydrophobic character and low MW. More specifically, they were rich in the amino acids Pro, Val, and Phe, which contributes to the correct location of the peptide in the active site of ACE. This is most likely due to the rigid structure of this residue. In addition, their study suggested that the Pro, Val, and Phe at the C-terminus may contribute to ACE-inhibiting activity. In this chapter, the purified ACE-inhibitory peptides revealed were Pro, Val, and Phe at the C-terminal, where they exhibited strong ACE-inhibitory activity. The tripeptides with Pro, Val, and Phe hydrophobic amino acids at the C-terminal have high ACE-inhibitory activity because of the interaction between three subsites at the active site of ACE (Pihlanto-Leppala, 2000). ACE-inhibitory activity correlation studies have indicated that ACE binding is strongly affected by the C-terminal tripeptide sequence of the substrate, and that the tripeptide could interact with ACE subsites S1, S′1, and S′2 (Pihlanto-Leppala, 2000). Therefore, marine invertebrate-derived ACE-inhibitory peptides have a potential use as functional ingredients in nutraceuticals and pharmaceuticals due to their effectiveness in both prevention and treatment of hypertension in addition to nutritive value. Some antihypertensive synthetic commercial drugs are known to produce side effects such as an abnormal elevation of the blood pressure after administration; however, marine invertebrate-derived bioactive peptides are well tolerated by the body and are not expected to cause any harmful side effects. In addition, cost-effective safe drugs can be produced from marine bioactive peptides, but further studies are needed with clinical trials for these marine-derived antihypertensive peptides.

IV. ANTI-ALZHEIMER'S ACTIVITY

A. Pathogenic mechanism of AD: Amyloid cascade hypothesis

According to recent epidemiologic studies, AD, a protein misfolding disease, is recognized as a major problem in public health in industrialized countries. This progressive neurodegenerative disease generally occurs after the age of 65 with prevalence around 2–4% at age 70 and somewhere between 30% and 50% by age 90 (Holtzman, 2002). It begins with short-term memory loss and continues with more widespread cognitive dysfunctions. The ultimate cause of the disease is unknown. Macroscopic changes found in brains with AD include shrinkage of the gyri, widening of the sulci, and enlargement of the ventricles along with two major microscopic lesions: extracellular amyloid plagues and intracellular neurofibrillary tangles (Giorgio *et al.*, 2006). The main constituent of the accumulated amyloidic plaques in the brain is β-amyloid peptide (Aβ). After the cloning of the amyloid precursor protein (APP), it was revealed that Aβ must be excised from the middle of its larger precursor protein (Kang *et al.*, 1987). The two necessary proteolytic cleavage events, one at the N-terminus by an enzyme termed "β-secretase" and one at the C-terminus by an enzyme termed "γ-secretase" have attracted a lot of attention. This is understandable because Aβ formation is the initial step in the hypothetical amyloid cascade (Hardy and Allsop, 1991) and is thus supposed to be ultimately responsible for the pathology of AD. Moreover, the necessity of proteolytic cleavage for Aβ generation immediately suggests the existence of two potential therapeutic intervention targets which could be addressed using standard protease inhibition approaches (John *et al.*, 2003). Consequently, APP processing and Aβ generation have been

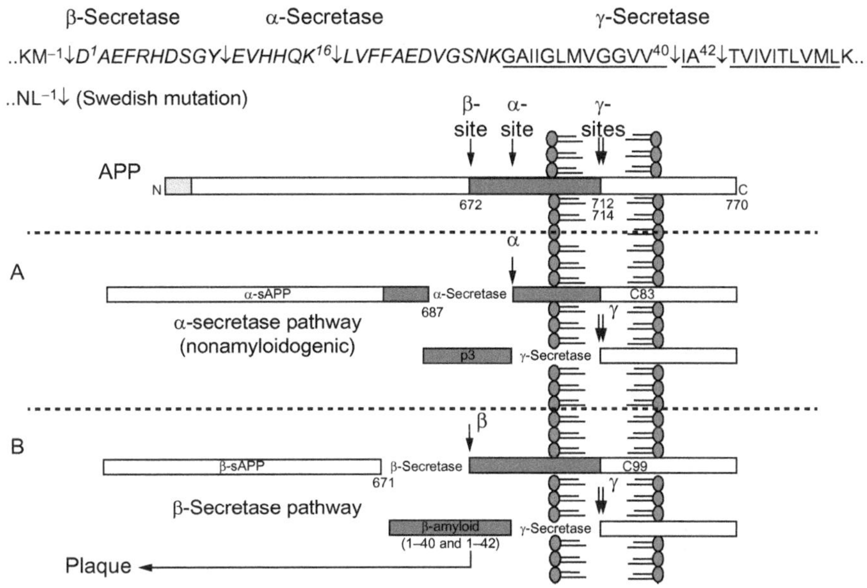

FIGURE 4.4 Schematic overview of APP processing by α-, β-, and γ-secretases (John *et al.*, 2003). The top panel shows the amino acid sequence of APP upstream of the transmembrane segment (underlined, bold) and encompasses the sequences of Aβ1-40 and Aβ1-42. The β-secretase cleaves at Asp^1 and Tyr^{10}. The α-secretase cleaves at Lys^{16}, and the γ-secretase cleaves at Val^{40} and/or Ala^{42}. Below the sequence is a representation of APP emphasizing its membrane localization and the residue numbers of interest in β- and γ-secretase processing. (A) The nonamyloidogenic α-secretase pathway in which sAPPα and C83 are generated. Subsequent hydrolysis by the γ-secretase produces a p3 peptide that does not form amyloid deposits. (B) The amyloidogenic pathway in which cleavage of APP by the β-secretase to liberate sAPPβ and C99 is followed by γ-secretase processing to release β-amyloid peptides (Aβ1-40 and Aβ1-42) found in plaque deposits. (For color version of this figure, the reader is referred to the Web version of this chapter.)

studied in a variety of systems by many investigators and their results are summarized in Fig. 4.4. At least three distinct protease activities are involved in processing the membrane protein APP along two major pathways: the α-secretase and the amyloid-forming β-secretase pathways.

A relatively small minority of APP molecules enter the β-secretase pathway in which β-secretase cleaves APP and releases a soluble fragment, sAPPβ. The C-terminal membrane-bound C99 peptide is then cleaved by γ-secretase within the transmembrane domain, and two major isoforms of 40 and 42 amino acid lengths with different C-termini, Aβ40 and Aβ42, are generated. Based on the amino acid sequence, β-secretase is predicted to be a type I transmembrane protein with the active site on the lumenal side of the membrane where β-secretase cleaves APP. β-Secretase is an attractive treatment for the AD target (John *et al.*, 2003).

B. Marine zooplankton

For the β-secretase inhibitor, many groups have focused on the identification of inhibitors using high-throughput screening of compound collections and natural product extracts. The peptidic β-secretase (aspartic protease, memapsin-2) inhibitor OM 99-1 and other aspartic protease inhibitors have been developed (Ghosh *et al.*, 2000). OM 99-2, an eight-residue transition state inhibitor (Hu *et al.*, 2006), and OM 00-3, a more potent eight-residue transition state inhibitor (Turner *et al.*, 2001), have also been developed. Chitosan derivatives from crab shells exhibited weak β-secretase inhibition (Byun *et al.*, 2005; Je and Kim, 2005). Catechins from green tea (Jeon, *et al.*, 2003), ellagic acid from pomegranate (Kwak *et al.*, 2005), hispidin from mycelial cultures of *Phellinus linteus* (Park *et al.*, 2004), and several compounds isolated from *Sanguisorbae radix* (Lee *et al.*, 2005) have all been studied as β-secretase inhibitors. In contrast, efforts to discover naturally occurring β-secretase inhibitors have been relatively limited. Several hydroxyl containing inhibitors have been reported (Cumming *et al.*, 2004). Byun *et al.* (2009a) studied β-secretase activity of six hydrolysates from marine zooplankton and rotifers with different enzymes (Alcalase, α-chymotrypsin, Neutrase, papain, pepsin, and trypsin). In their study of the β-secretase-inhibitory activity, the highest IC_{50} value was exhibited by tryptic hydrolysate at 0.21 mg/ml (Fig. 4.5). The IC_{50} value of the purified β-secretase-inhibitory activity peptide was 183 μM. The β-secretase-inhibitory peptide was identified as a sequence of

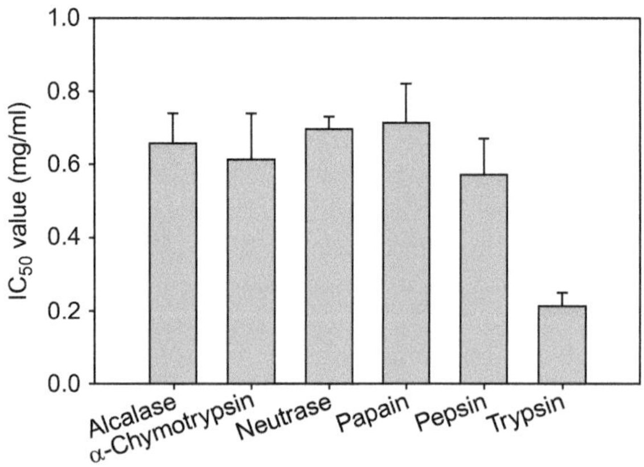

FIGURE 4.5 β-Secretase-inhibitory activity of various enzymatic hydrolysates from marine zooplankton, rotifer (Byun *et al.*, 2009a).

four amino acids, Gly-Arg-Gln-Lys by Q-TOF MS/MS. In addition, β-secretase-inhibitory peptides Gly-Arg-Gln-Lys, Gly-Arg-Gln, Gly-Arg, Arg-Gln-Lys, and Gln-Lys were synthesized by the solid-phase method. Among the five synthesized peptides, Arg-Gln-Lys had the highest β-secretase-inhibitory activity for 150.7 μM.

V. ANTIOXIDANT ACTIVITY

A. Oxidative stress

Oxidation is a vital process in all living organisms even though its side effect is the production of free radicals. Free radicals arise naturally during metabolism and during respiration in aerobic organisms, being by-products of normal reactions such as the production of calories, the degradation of lipids, the catecholamine response due to stress, and inflammatory processes (Wang *et al.*, 2008b). During normal body reactions, such as respiration, reactive oxygen species such as superoxide anion radicals (O^2), hydroxyl radicals (OH•), and radical species such as hydrogen peroxide (H_2O_2) and singlet oxygen ($^1O_2$) are formed (Gülçin, 2009, 2010). When free radicals are produced in excess or are produced and not eradicated, they can attack the closest molecules, subtracting electrons and starting a chain reaction in which the new molecule with a missing electron attacks other molecules (Kaur and Kapoor, 2001). It is well documented that free radical formation is implicated in many human diseases, such as health disease, stroke, arteriosclerosis, diabetes, and cancer (Dávalos *et al.*, 2004). In the human body, a variety of enzymatic antioxidants, such as catalase, superoxide dismutase, and glutathione peroxidase, helps maintain some control over oxidation processes (Dimitrios, 2006). However, when an excess of free radicals is formed, the body's inherent protective antioxidant enzymatic system may become overpowered and it can cause destructive and lethal cellular effects by oxidizing lipids, proteins, DNA, and enzymes (Philanto, 2006).

B. Shellfish

As antioxidant peptides are rarely present in marine invertebrates, they must be released from the parent protein by hydrolysis with enzymes. Various enzymes have been used to release peptides from muscle proteins. To date, different muscle proteins have been extracted, hydrolysed, and their antioxidant activities studied, which is among all invertebrate muscles the most similar to vertebrate skeletal muscle. Various studies have been conducted to investigate the antioxidant properties of hydrolysates or bioactive peptides from marine invertebrate sources like oysters

(*C. gigas*; Qian *et al.*, 2008b), fermented mussel sauce (*M. edulis*; Jung *et al.*, 2005; Rajapakse *et al.*, 2005), and blue mussel (*Mytilus coruscus*; Jung *et al.*, 2007). Table 4.3 presents a list of studies on the antioxidant activities as well as structural properties of peptides and hydrolysates.

Qian *et al.* (2008b) hydrolyzed gastrointestinal digestions of oysters by gastrointestinal proteases and the purified peptide with the amino acid sequence Leu-Lys-Gln-Glu-Leu-Glu-Asp-Leu-Leu-Glu-Lys-Gln-Glu (MW 1.60 kDa) and exhibited higher activity against polyunsaturated fatty acid (PUFA) peroxidation than that of the native antioxidant, α-tocopherol. The free radical scavenging assay conducted using electron spin resonance (ESR) spectroscopy clearly exhibited that it scavenged hydroxyl radicals and superoxide radicals at IC_{50} values of 28.76 and 78.97 μM, respectively. Jung *et al.* (2005) investigated the use of fermented blue mussel sauce (FBMS). The antioxidative activities of FBMSs were investigated and compared with that of a natural antioxidant with α-tocopherol standing as a reference. Using consecutive chromatographic methods, the antioxidative peptide with a molecular mass of 620 Da was purified from a 6-month-fermented sauce, and its sequence of the peptide was Phe-Gly-His-Pro-Tyr. In addition, 64.8 mM of the purified peptide could scavenge 89.5% of hydroxyl radicals in radical scavenging assay using ESR spectroscopy. Rajapakse *et al.* (2005) used fermented mussel, which was identical with Jung *et al.*'s (2005) FMBs. It and the hepta-peptide sequence, His-Phe-Gly-Asp-Pro-Phe-His (MW 962 Da), were found to be highly effective for radical scavenging. FMBs could scavenge superoxide, hydroxyl, carbon-centered, and 1,1-diphenyl-2-picrylhydrazyl (DPPH) radicals for 21, 34, 52, and 96 μM by IC_{50} values, respectively. Moreover, FMBs exhibited a strong lipid peroxidation inhibition at 54 μM concentration, and it was higher than α-tocopherol. Jung *et al.* (2007) also gastrointestinally digested blue mussel (*M. coruscus*) muscle by pepsin, trypsin, and α-chymotrypsin. The potent antioxidant peptide, which was identified as Leu-Val-Gly-Asp-Glu-Gln-Ala-Val-Pro-Ala-Val-Cys-Val-Pro (MW 1.59 kDa), exhibited higher protective activity against PUFA peroxidation than the native antioxidants, ascorbic acid, and α-tocopherol.

C. Marine zooplankton

In another study conducted on marine zooplankton (Byun *et al.*, 2009b), antioxidant activity was measured for the DPPH radical of hydrolysates produced by Alcalase, α-chymotrypsin, Neutrase, papain, pepsin, and trypsin. To identify antioxidant peptides, peptic hydrolysate was purified using consecutive chromatographic methods, and antioxidant peptides were identified to be Leu-Leu-Gly-Pro-Gly-Leu-Thr-Asn-His-Ala (MW 1076 Da) and Asp-Leu-Gly-Leu-Gly-Leu-Pro-Gly-Ala-His (MW 1033 Da) by Q-TOF ESI mass spectroscopy. IC_{50} values of purified peptides were

TABLE 4.3 Antioxidant activity peptides derived from invertebrates: origin, amino acid sequence, enzyme used for hydrolysis, and IC_{50} value

Origin	Amino acid sequence	Radical	Scavenging activity, IC_{50} (μM)	References
Oyster, *Crassostrea gigas*	Leu-Lys-Gln-Glu-Leu-Glu-Asp-Leu-Leu-Glu-Lys-Gln-Glu	Hydroxyl	28.76	Qian *et al.* (2008b)
		Superoxide	78.97	
Fermented blue mussel, *Mytilus edulis*	Phe-Gly-His-Pro-Tyr	Hydrl	64.8 (RAS at 89.5%)	Jung *et al.* (2005)
	His-Phe-Gly-Asp-Pro-Phe-His		34	Rajapakse *et al.* (2005)
			21	
		Hydroxyl	96	
Gastrointestinal digests of *Mytilus coruscus*	Leu-Val-Gly-Asp-Glu-Gln-Ala-Val-Pro-Ala-Val-Cys-Val-Pro	Superoxide	31.44 (RAS at 75.04%)	Jung *et al.* (2007)
Rotifer, *Brachionus rotundiformis*	Leu-Leu-Gly-Pro-Gly-Leu-Thr-Asn-His-Ala	DPPH	189.8	Byun *et al.* (2009b)
	Asp-Leu-Gly-Leu-Gly-Leu-Pro-Gly-Ala-His	DPPH	167.7	
Rotifer, *Brachionus calyciflorus*	Gly-His-Asp-Gly-Tyr-Glu-Pro-Leu-Ser-Ser	DPPH	100.8	Lee *et al.* (2010b)

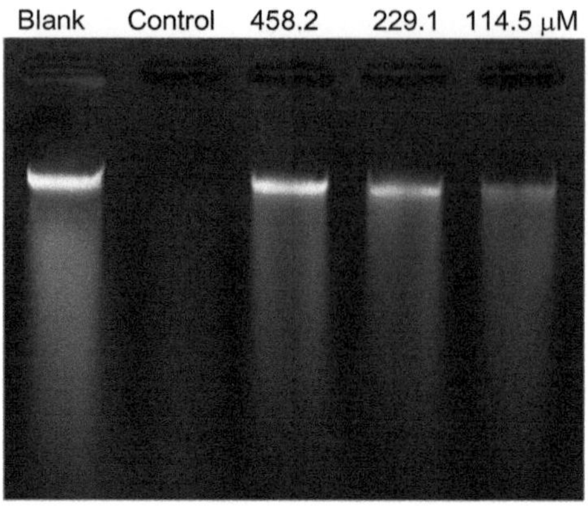

FIGURE 4.6 Protective effect on oxidative DNA damage of purified peptide from freshwater rotifer protein at various concentrations. Blank: untreated sample and H_2O_2, Control: distilled water instead of sample (Lee *et al.*, 2010b).

189.8 and 167.7 mM, respectively. In freshwater rotifers (Lee *et al.*, 2010b), the IC_{50} value of the purified antioxidant peptide was 100.8 μM. The antioxidant peptide was identified as a sequence of 10 amino acids, Gly-His-Asp-Gly-Tyr-Glu-Pro-Leu-Ser-Ser (MW 1091 Da), by N-terminal amino acid sequence analysis. The purified peptide exhibited an inhibitory effect against induced DNA oxidation (Fig. 4.6). Hydroxyl radicals have been known to cause oxidative damage in DNA structures by relaxing the DNA or denaturing it.

The results of this study strongly suggest that purified peptides can prevent oxidative damage to DNA when the DNA is exposed to OH• generated by the Fenton reaction. With the purified peptide of protective activity against DNA oxidation, a clear dose-dependent effect was observed. In this result, the effect of purified peptide was to protect hydroxyl radical-induced DNA damage.

D. Other invertebrates

Purple sea urchin (*Strongylocentrotus nudus*) gonads were treated separately with neutral protease, papain, pepsin, and trypsin (Qin *et al.*, 2011). The resultant hydrolysates were fractionated using a series of ultrafiltration membranes (MWCO 10, 5, 3, and 1 kDa). Five fractions were prepared from each hydrolysate, and the corresponding MW ranges were below 10 kDa, 5–10 kDa, 3–5 kDa, 1–3 kDa, and below 1 kDa. The peptide fractions were evaluated for antioxidant activity by using the DPPH assay and reducing power assay. Results indicated that all peptide fractions possessed DPPH radical scavenging capacities and reduced power in a dose-dependent manner. For all four hydrolysates, fractions below 1 kDa exhibited the highest DPPH radical scavenging capacity. The fractions below 1 kDa were prepared with neutral protease, papain, and pepsin, and the 1–3-kDa fraction prepared with trypsin showed the highest reducing capacity among corresponding hydrolysates.

Antioxidative properties of the peptides are more related to their composition, structure, and hydrophobicity (Chen *et al.*, 1998). Tyr, Trp, Met, Lys, Cys, and His are examples of amino acids that cause antioxidant activity (Wang and De Mejia, 2005). Amino acids with aromatic residues can donate protons to electron-deficient radicals. This property improves the radical-scavenging properties of the amino acid residues (Rajapakse *et al.*, 2005). It is proposed that the antioxidative activity of His-containing peptides is in relation with the hydrogen-donating, lipid peroxidation, radical trapping, and/or the metal ion-chelating ability of the imidazole group (Rajapakse *et al.*, 2005). On the other hand, the SH group in cysteine has an independently crucial antioxidant action due to its direct interaction with radicals (Qian *et al.*, 2008a). In addition to the presence of proper amino acids, their correct positioning in the peptide sequence plays an important role in antioxidant activity of peptides (Rajapakse *et al.*, 2005).

Rajapakse *et al.* (2005) purified antioxidant peptides following the structure of His-Phe-Gly-Asp-Pro-Phe-His from the digestion of fermented blue mussel *M. edulis*. It was revealed that the antioxidant activity of a peptide was more dependent on the His segment in the N-Pro-Phe-His domain. According to the results from the same study, the His sequence displayed the greatest antioxidative activity with rotifer *B. rotundiformis* (Byun *et al.*, 2009b). Table 4.3 provides information regarding the effect of amino acid compositions and their correct positioning in peptide sequences.

VI. ANTIMICROBIAL ACTIVITY

Antimicrobial peptides (AMPs) are a group of molecules exhibiting antimicrobial activity *in vitro*. In nature, they constitute an important part of the innate immune system in animals where they participate in the neutralization and elimination of intruding microorganisms (Zasloff, 2002). This chapter aims to discuss AMPs from marine invertebrates, mainly emphasizing the challenges and perspectives of purifying such peptides. In addition, different aspects concerning their antimicrobial potential are discussed. For a more general description of these molecules, recent review articles are available (Otero-Gonzalez *et al.*, 2010; Smith *et al.*, 2010). In this chapter, AMPs are considered as peptides/proteins between 1 and 12 kDa, exhibiting a profound activity against microorganisms *in vitro*, and which are coded by single genes and ribosomally synthesized. Many low MW, extensively modified, and cyclic peptides have been characterized from sponges and tunicates. These peptides are presumably produced nonribosomally or by associated microorganisms and will therefore not be implemented in this chapter. Peptides with antimicrobial activity but obtained from larger proteins or classified with other functions are also omitted. However, the distinction between AMPs and some other peptides showing antimicrobial activity is rather unclear.

Marine-derived AMPs are well described in the hemolymph of the many marine invertebrates (Tincu and Taylor, 2004), including spider crabs (Stensvag *et al.*, 2008), oysters (Liu *et al.*, 2008), American lobsters (Battison *et al.*, 2008), shrimp (Bartlett et al., 2002), and green sea urchins (Li *et al.*, 2008). Liu *et al.* (2008) isolated a novel peptide *CgPep33* with high inhibitory activity against bacteria such as *Escherichia coli*, *Pseudomonas aeruginosa*, *Bacillus subtilis*, and fungi such as *Botrytis cinerea* and *Penicillium expansum* from the Pacific oyster (*C. gigas*) via enzymatic hydrolysis, and purified with DEAE Sephadex A-25 ion exchange, Sephadex G-25 gel filtration, and HPLC. *CgPep33* inhibited the *in vitro* growth of *B. cinerea* by 50% at 20–40 μg/ml and by 100% at 120 μg/ml. More interestingly, the IC_{50} values of this peptide against the above bacteria and fungi were 18.6–48.2 μg/ml (Liu *et al.*, 2007, 2008). Bartlett *et al.* (2002) reported that arthropod AMPs, while characterized primarily from insects, also have been isolated from crustaceans. Expressed sequence tag analysis of hemocyte complementary DNA libraries from two species of shrimp, *Litopenaeus vannamei* and *Litopenaeus setiferus*, revealed transcripts with strong sequence similarity to an 11.5-kDa antibacterial peptide which identified Pro-Arg-Pro amino acid sequence. Further, an active peptide, Arg-Arg-Trp-Trp-Cys-Arg-X (X is an amino acid or an amino acid analog), against the herpes virus has been isolated from the enzymatic hydrolysate of oyster *C. gigas*. This peptide showed high inhibitory activity on the herpes virus (Zeng *et al.*, 2008). Zeng *et al.* (2008) obtained four fractions of hydrolysates according to MW <1, 1–5, 5–10, and higher than 10 kDa. Among them, the peptide purified from 5- to 10-kDa fractions had a higher biological activity than other fractions. The need to discover new antimicrobial substances is important due to the progressive development of resistance by pathogenic microorganisms against conventional antibiotics (Kim and Wijesekara, 2010). Marine invertebrates, representing an enormous genetic and biological diversity, have proven to be a rich source for discovering potent AMPs with novel and unique structural motifs. Therefore, it might be suggested that these AMPs have potent capacities for new antibiotic development in the pharmaceutical and food industries as novel antimicrobial agents.

VII. OTHER BIOLOGICAL ACTIVITY

The countless skin pathologies are correlated with exposure to UV light. Skin exposure to UV radiation induces critical effects for photodamage, which is characterized by distinct alterations in the composition of the dermal extracellular matrix resulting in wrinkles, laxity, coarseness, a mottled pigmentation, and histological changes that include increased epidermal thickness and connective tissue alteration (Kondo, 2000; Rittie and Fisher, 2002). The connective tissue of skin is made up of dozens of biomolecules including collagens, proteoglycans, and glycoproteins. Moreover, the breakdown of balance of the combination between these components leads to the detrimental effect in dermal fibroblasts like photoaging. Several investigations have described the deficiency of collagen in photoaged skin due to the inhibition of synthesis and the degradation mediated by matrix metalloproteinases (MMPs).

Kwon *et al.* (2007) revealed the antiwrinkle effects of peptides derived from collagens isolated from the starfish *Asterias amurensis*. The purified collagen peptides were fractions (F1: 116kDa; F2: 100kDa; F3: 58kDa; F4: 43kDa; F5: 24kDa), and these peptides were reduced to 34.8% for MMP-1 expression of UVA-induced human normal fibroblasts at 1.0mg/ml. The starfish *A. amurensis*' antiwrinkle collagen peptides had superior antiwrinkle effects of approximately 20kDa. Barnacle (Mannello *et al.*, 2003) (*Balanus amphitrite*) larvae MMPs showed biochemical characteristics different from those of vertebrate MMPs but common to other gelatinases from marine invertebrates; they were unaffected by several protease inhibitors and insensitive to specific activators' inhibitors of vertebrate MMPs.

VIII. CONCLUSION

In recent years, the marine environment has been shown to provide extremely rich biological active compounds. One of the hydrolysates derived from the marine invertebrate seems to possess many biological activities. Nowadays, major public health problems (including hypertensive, antioxidant, AD, and antibacterial) are in most cases treated by medicines containing synthetic drugs, which can cause serious side effects. Naturally occurring bioactivity peptides derived from marine organism proteins that are consumed daily can be used as components for functional foods or nutraceuticals. In this chapter, more particularly, marine invertebrates as a source of bioactive peptides were considered. Marine invertebrate muscle itself can act as a biologically active food through digestion with appropriate enzymes.

This suggests that products with bioactive peptides derived from marine invertebrates can meet the needs of marine organism-derived products due to health and/or religious reasons. From such a viewpoint, hydrolysates or bioactive peptides from marine invertebrates can be interesting sources of bioactivity peptides in the treatment of chronic diseases.

REFERENCES

Bartlett, T. C., Cuthbertson, B. J., Shepard, E. F., Chapman, R. W., Grops, P. S., and Warr, G. W. (2002). Crustins, homologues of an 11.5-kDa antibacterial peptide, from two species of penaeid shrimp, Litopenaeus vannamei and Litopenaeus setiferus. *Mar. Biotechnol.* **4**, 278–293.

Battison, A. L., Summerfield, R., and Patrzykat, A. (2008). Isolation and characterization of two antimicrobial peptides from haemocytes of the American lobster *Homarus americanus*. *Fish Shellfish Immunol.* **25**, 181–187.

Borer, J. S. (2007). Angiotensin-converting enzyme inhibition: A landmark advance in treatment for cardiovascular diseases. *Eur. Heart J. Suppl.* **9**, 2–9.

Byun, H. G. and Kim, S. K. (2001). Purification and characterization of angiotensin I converting enzyme (ACE) inhibitory peptides from Alaska Pollack (*Theragra chalcogramma*) skin. *Process Biochem.* **36**, 1155–1162.

Byun, H. G., Kim, Y. T., Park, P. J., Lin, X., and Kim, S. K. (2005). Chitooligosaccharides as a novel β-secretase inhibitor. *Carbohydr. Polym.* **61**, 198–202.

Byun, H. G., Jeon, J. K., Kim, S. K., Park, H. G., Lee, M. S., and Lee, J. K. (2009a). Composition extracted from zooplankton for inhibiting or preventing Alzheimer's disease and process thereof. patent 1020090108254.

Byun, H. G., Lee, J. K., Park, H. G., Jeon, J. K., and Kim, S. K. (2009b). Antioxidant peptides isolated from the marine rotifer, *Brachionus rotundiformis*. *Process Biochem.* **44**, 842–846.

Chen, H. M., Muramoto, K., Yamauchi, F., Fujimoto, K., and Nokihara, K. (1998). Antioxidative properties of histidine-containing peptides designed from peptide fragments found in the digests of a soybean protein. *J. Agric. Food Chem.* **46**, 49–53.

Cheung, I. W. Y. and Li-Chan, E. C. Y. (2010). Angiotensin-I-converting enzyme inhibitory activity and bitterness of enzymatically-produced hydrolysates of shrimp (*Pandalopsis dispar*) processing byproducts investigated by Taguchi design. *Food Chem.* **122**, 1003–1012.

Cumming, J. N., Iserloh, U., and Kennedy, M. E. (2004). Design and development of β-secretase inhibitors. *Curr. Opin. Drug. Discov. Devel.* **7**, 536–556.

Dávalos, A., Miguel, M., Bartolomé, B., and Lopez-Fandino, R. (2004). Antioxidant activity of peptides derived from egg white proteins by enzymatica hydrolysis. *J. Food Prot.* **67**, 1939–1944.

Deeslie, W. D. and Cheryan, M. (1981). Continuous enzymatic modification of proteins in an ultrafiltration reactor. *J. Food Sci.* **46**, 1035–1042.

Dimitrios, B. (2006). Sources of natural phenolic antioxidants. *Trends Food Sci. Technol.* **17**, 505–512.

Erdmann, K., Cheung, B. W. Y., and Schroder, H. (2008). The possible roles of food-derived bioactive peptides in reducing the risk of cardiovascular disease. *J. Nutr. Biochem.* **19**, 643–654.

Fahmi, A., Morimura, S., Gou, H. C., Shinematsu, T., Kida, K., and Uemura, Y. (2004). Production of angiotensin I converting enzyme inhibitory peptides from sea bream scales. *Process Biochem.* **39**, 1195–1200.

Ghosh, A. K., Shin, D., Koelsch, G., Lin, X., Ermolieff, J., and Tang, J. (2000). Design of potent inhibitors for human brain memapsin 2 (β-secretase). *J. Am. Chem. Soc.* **122**, 3522–3523.

Giorgio, S., Alessia, L., LUcilla, P., Daniele, T., and Francesso, A. (2006). Treatment of Alzheimer's disease: From pharmacology to a better understanding of disease pathophysiology. *Mech. Ageing Dev.* **127**, 148–157.

Gülçin, İ. (2009). Antioxidant activity of l-adrenaline: A structure-activity insight. *Chem. Biol. Interact.* **179**, 71–80.

Gülçin, İ. (2010). Antioxidant properties of resveratrol: A structure-insight. *Innov. Food Sci. Emerg. Technol.* **11**, 210–218.

Hardy, J. and Allsop, D. (1991). Amyloid deposition as the central event in the aetiology of Alzheimer's disease. *Trends Pharmacol. Sci.* **12**, 383–388.

He, H. L., Chen, X. L., Sun, S. Y., Zhang, Y. Z., and Zhou, B. C. (2006). Analysis of novel angiotensin-I-converting enzyme inhibitory peptides from protease-hydrolyzed marine shrimp *Acetes chinensis*. *J. Pept. Sci.* **12**, 726–733.

Holtzman, D. M. (2002). Aβ conformational change is central to Alzheimer's disease. *Neurobiol. Aging* **23**, 1085–1088.

Hu, B., Xiong, B., Qiu, B. Y., Li, X., Yu, H. P., Xiao, K., Wang, X., Li, J., and Shen, J. K. (2006). Construction of a small peptide library related to inhibitor OM99-2 and its structure-activity relationship to beta-secretase. *Acta Pharmacol. Sin.* **12**, 1586–1593.

Huang, R., Li, N., Zeng, Q., Deng, Q., and Chen, B. (2011). Extraction and separation of angiotensin-converting enzyme inhibitor from enzymatic hydrolysate of oyster. *Adv. Mat. Res.* **236**, 2610–2614.

Je, J. Y. and Kim, S. K. (2005). Water-soluble chitosan derivatives as a β-secretase inhibitor. *Bioorg. Med. Chem. Lett.* **13**, 6551–6555.

Je, J. Y., Park, J. Y., Jung, W. K., Park, P. J., and Kim, S. K. (2005a). Isolation of angiotensin I converting enzyme (ACE) inhibitor from fermented oyster sauce, Crassostrea gigas. *Food Chem.* **90**, 809–814.

Je, J. Y., Park, P. J., Byun, H. G., Jung, W. K., and Kim, S. K. (2005b). Angiotensin I converting enzyme (ACE) inhibitory peptide derived from the sauce of fermented blue mussel, *Mytilus edulis*. *Bioresour. Technol.* **96**, 1624–1629.

Jeon, Y. J., Byun, H. G., and Kim, S. K. (1999). Improvement of functional properties of cod frame protein hydrolysates using ultrafiltration membranes. *Process Biochem.* **35**, 471–478.

Jeon, S. Y., Bae, K. H., Seong, Y. H., and Song, K. S. (2003). Green tea catechins as a β-secretase (BACE1) inhibitor. *Bioorg. Med. Chem. Lett.* **13**, 3905–3908.

John, V., Beck, J. P., Bienkowski, M. J., Sinha, S., and Heinrikson, R. L. (2003). Human β-secretase (BACE) and BACE inhibitors. *J. Med. Chem.* **46**, 4625–4630.

Jung, W. K., Rajapakse, N., and Kim, S. K. (2005). Antioxidative activity of a low molecular weight peptide derived from the sauce of fermented blue mussel, *Mytilus edulis*. *Eur. Food Res. Technol.* **220**, 535–539.

Jung, W. K., Qian, Z. J., Lee, S. H., Choi, S. Y., Sung, N. J., Byun, H. G., and Kim, S. K. (2007). Free radical scavenging activity of a novel antioxidative peptide isolated from in vitro gastrointestinal digests of *Mytilus coruscus*. *J. Med. Food* **10**, 197–202.

Kang, J., Lemaire, H. G., Unterbeck, A., Salbaum, J. M., Msters, C. L., Grzeschik, K. H., Multhaup, G., Beyreuther, K., and Muller-Hill, B. (1987). The precursor of Alzheimer's disease amyloid A4 protein resembles a cell-surface receptor. *Nature* **325**, 733–736.

Katano, S., Oki, T., Matsuo, Y., Yoshihira, K., Nara, Y., Miki, T., Matsui, T., and Matsumoto, K. (2003). Antihypertensive effect of alkaline protease hydrolysate of the pearl oyster Pinctada fucata martencii & separation and identification of angiotensin-I converting enzyme inhibitory peptides. *Nippon Suisan Gakk.* **69**, 975–980.

Kaur, C. and Kapoor, H. C. (2001). Antioxidant in fruits and vegetables—The millennium's health. *Int. J. Food Sci. Technol.* **36**, 703–725.

Kawamura, Y., Takane, T., Stake, M., and Sugimoto, T. (1992). Physiologically active peptide motif in proteins: Peptide inhibitors of ACE from the hydrolysates of Antarctic krill muscle protein. *Jpn. Agric. Res. Quart.* **26**, 210–213.

Kim, S. K. and Mendis, E. (2006). Bioactive compounds from marine processing byproducts—A review. *Food Res. Int.* **39**, 383–393.

Kim, S. K. and Rajapakse, N. (2005). Enzymatic production and biological activities of chitosan oligosaccharides (COS): A review. *Carbohydr. Polym.* **62**, 357–368.

Kim, S. K. and Wijesekara, I. (2010). Development and biological activities of marine-derived bioactive peptides: A review. *J. Funct. Foods* **2**, 1–9.

Kim, S. K., Kim, Y. T., Byun, H. G., Park, P. J., and Ito, H. (2001). Purification and characterization of antioxidative peptides from bovine skin. *J. Biochem. Mol. Biol.* **34**, 214–219.

Kim, S. K., Nghiep, N. D., and Rajapakse, N. (2006). Therapeutic prospectives of chitin, chitosan, and their derivatives. *J. Chitin Chitosan* **11**, 1–10.

Kondo, S. (2000). The roles of cytokines in photoaging. *J. Dermatol. Sci.* **23**, 30–36.

Kwak, H. M., Jeon, S. Y., Sohng, B. H., Kim, J. G., Lee, J. M., Lee, K. B., Jeong, H. H., Hur, J. M., Kang, Y. H., and Song, K. S. (2005). β-secretase (BACE1) inhibitors from pomegranate (*Punica granatum*) husk. *Arch. Pharm. Res.* **28**, 1328–1332.

Kwon, M. C., Kim, C. H., Kim, H. S., Syed, A. Q., Hwang, B. Y., and Lee, H. Y. (2007). Antiwrinkle activity of low molecular weight peptides derived from the collagen isolated from *Asterias amurensis*. *Korean J. Food Sci. Technol.* **39**, 625–629.

Lee, H. J., Seong, Y. H., Bae, K. H., Kwon, S. H., Kwak, H. M., Nho, S. K., Kim, K. A., Hur, J. M., Lee, K. B., Kang, Y. H., and Song, K. S. (2005). Beta-secretase (BACE1) inhibitors from *Sanguisorbae radix*. *Arch. Pharm. Res.* **28**, 799–803.

Lee, J. K., Hong, S., Jeon, J. K., Kim, S. K., and Byun, H. G. (2009). Purification and characterization of angiotensin I converting enzyme inhibitory peptides from the rotifer, *Brachionus rotundiformis*. *Bioresour. Technol.* **100**, 5255–5259.

Lee, J. K., Lee, M. S., Park, H. G., Kim, S. K., and Byun, H. G. (2010a). Angiotensin I converting enzyme inhibitory peptide extracted from freshwater zooplankton. *J. Med. Food* **13**, 357–363.

Lee, J. K., Yun, J. H., Jeon, J. K., Kim, S. K., and Byun, H. G. (2010b). Effect of antioxidant peptide isolated from *Brachionus calyciflorus*. *J. Korean Soc. Appl. Biol. Chem.* **53**, 192–197.

Lee, J. K., Jeon, J. K., and Byun, H. G. (2011). Effect of angiotensin I converting enzyme inhibitory peptide purified from skate skin hydrolysate. *Food Chem.* **125**, 495–499.

Li, C., Haug, T., Styrvold, O. B., Jorgensen, T. O., and Stensvag, K. (2008). Strongylocins, novel antimicrobial peptides from the green sea urchin. *Dev. Comp. Immunol.* **32**, 1430–1440.

Liu, Z., Zeng, M., Dong, S., Xu, J., Song, H., and Zhao, Y. (2007). Effect of an antifungal peptide from oyster enzymatic hydrolysates for control of gray mold (*Botrytis cinerea*) on harvested strawberries. *Postharvest Biol. Technol.* **46**, 95–98.

Liu, Z., Dong, S., Xu, J., Zeng, M., Song, H., and Zhao, Y. (2008). Production of cysteine-rich antimicrobial peptide by digestion of oyster (*Crassostrea gigas*) with alcalase and bromelin. *Food Control* **19**, 231–235.

Mannello, F., Canesi, L., Faimali, M., Piazza, V., Gallo, G., and Geraci, S. (2003). Characterization of metalloproteinase-like activities in barnacle (*Balanus amphitrite*) nauplii. *Comp. Biochem. Physiol. B Biochem. Mol. Biol.* **135**, 17–24.

Maria, M. C., Rosalia, C., Maria, J. M., Mercedes, R., and Isidra, R. (2009). Novel casein-derived peptides with antihypertensive activity. *Int. Dairy J.* **19**, 566–573.

Matsufuji, H., Matsui, T., Seki, E., Osajima, K., Nakashima, M., and Osajima, Y. (1994). Angiotensin I-converting enzyme inhibitory peptides in an alkaline protease hydrolyzate derived from sardine muscle. *Biosci. Biotechnol. Biochem.* **58**, 2244–2245.

Matsumoto, K., Ogikubo, A., Yoshino, T., Matsui, T., and Osajima, Y. (1994). Separation and purification of angiotensin-I converting enzyme inhibitory peptide in peptic hydrolyzate of oyster. *J. Jpn. Soc. Food Sci. Technol.* **41**, 589–594.

Mendis, E., Rajapakse, N., and Kim, S. K. (2005). Antioxidant properties of a radicals scavenging peptide purified from enzymatically prepared fish skin gelatin hydrolysate. *J. Agric. Food Chem.* **53**, 581–587.

Oota, S. and Saitou, N. (1999). Phylogenetic relationship of muscle tissues deduced from superimposition of gene trees. *Mol. Biol. Evol.* **16**, 856–867.

Otero-Gonzalez, A. J., Magalhaes, B. S., Garcia-Villarino, M., Lopez-Abarrategui, C., Sousa, D. A., Dias, S. C., and Franco, O. L. (2010). Antimicrobial peptides from marine invertebrates as a new frontier for microbial infection control. *FASEB J.* **24**, 1320–1334.

Park, I. H., Jeon, S. Y., Lee, H. J., Kim, S. I., and Song, K. S. (2004). A β-secretase (BACE1) inhibitor hispidin from the mycelial cultures of *Phellinus linteus*. *Planta Med.* **70**, 143–146.

Philanto, A. (2006). Antioxidative peptides derived from milk proteins. *Int. Dairy J.* **16**, 1306–1314.

Pihlanto-Leppala, A. (2000). Bioactive peptides derived from bovine whey proteins: Opioid and ace-inhibitory. *Trends Food Sci. Technol.* **11**, 347–356.

Qian, Z. J., Jung, W. K., and Kim, S. K. (2008a). Free radical scavenging activity of a novel antioxidative peptide purified from hydrolysate of bullfrog skin, Rana catesbeiana Shaw. *Bioresour. Technol.* **99**, 1690–1698.

Qian, Z. J., Jung, W. K., Byun, H. G., and Kim, S. K. (2008b). Protective effect of an antioxidative peptide purified from gastrointestinal digests of oyster, *Crassostrea gigas* against free radical induced DNA damage. *Bioresour. Technol.* **99**, 3365–3371.

Qin, L., Zhu, B. W., Zhou, D. Y., Wu, H. T., Tan, H., Yang, J. F., Li, D. M., Dong, X. P., and Murata, Y. (2011). Preparation and antioxidant activity of enzymatic hydrolysates from purple sea urchin (*Strongylocentrotus nudus*) gonad. *LWT Food Sci. Technol.* **44**, 1113–1119.

Rajapakse, N., Mendis, E., Jung, W. K., Je, J. Y., and Kim, S. K. (2005). Purification of a radical scavenging peptide from fermented mussel sauce and its antioxidant properties. *Food Res. Int.* **38**, 175–182.

Rittie, L. and Fisher, G. J. (2002). UV-light-induced signal cascades and skin aging. *Ageing Res. Rev.* **1**, 705–720.

Rønnestad, I., Tonheim, S. K., Fyhn, H. J., Rojas-Garcia, C. R., Kamisaka, Y., Koven, W., Finn, R. N., Terjesen, B. F., Barr, Y., and Conceição, L. E. C. (2003). The supply of amino acids during stages of marine fish larvae: A review of recent findings. *Aquaculture* **227**, 147–164.

Shiozaki, K., Shiozaki, M., Masuda, J., Yamauchi, A., Ohwada, S., Nakano, T., Yamaguchi, T., Saito, T., Muranoto, K., and Sato, M. (2010). Identification of oyster-derived hypotensive peptide acting as angiotensin-I-converting enzyme inhibitor. *Fish Sci.* **76**, 865–872.

Simpson, B. K., Nayeri, G., Yaylayan, V., and Ashie, I. N. A. (1998). Enzymatic hydrolysis of shrimp meat. *Food Chem.* **61**, 131–138.

Skeggs, L. T., Kahn, J. E., and Shumway, N. P. (1957). The preparation and function of the angiotensin-converting enzyme. *J. Exp. Med.* **13**, 295–299.

Smith, V. J., Desbois, A. P., and Dyrynda, E. A. (2010). Conventional and unconventional antimicrobials from fish, marine invertebrates and micro-algae. *Mar. Drugs* **8**, 1213–1262.

Stensvag, K., Haug, T., Sperstad, S. V., Rekdal, O., Indrevoll, B., and Styrvold, O. B. (2008). Arasin 1, a proline-arginine-rich antimicrobial peptide isolated from the spider crab, Hyasaraneus. *Dev. Comp. Immunol.* **32**, 275–285.

Tincu, J. A. and Taylor, S. W. (2004). Antimicrobial peptides from marine invertebrates. *Antimicrob. Agents Chemother.* **48**, 3645–3654.

Tsai, J. S., Lin, T. C., Chen, J. L., and Pan, B. S. (2006). The inhibitory effects of freshwater clam (*Corbicula fluminea,* Muller) muscle protein hydrolysate on angiotensin I converting enzyme. *Process Biochem.* **41**, 2276–2281.

Turner, R. T., Koelsch, G., Hong, L., Castanheira, P., Ermolieff, J., Ghosh, A. K., and Tang, J. (2001). Subsite specificity of memapsin 2 (β-secretase): Implications for inhibitor design. *Biochemistry* **40**, 10001–10006.

Wang, W. Y. and De Mejia, E. G. (2005). A new frontier in soy bioactive peptides that may prevent age-related chronic diseases. *Compr. Rev. Food Sci. Food. Saf.* **4**, 63–78.

Wang, D., Wang, L. J., Zhu, F. X., Chen, X. D., Zou, L., Saito, M., and Li, L. T. (2008a). In vivo and in vitro studies on the antioxidant activities of the aqueous extracts of Douchi (a traditional Chinese salt-fermented soybean food). *Food Chem.* **107**, 1421–1428.

Wang, J., Hu, J., Cui, Z., Bai, X., and Du, Y. (2008b). Purification and identification of a ACE inhibitory peptide from oyster proteins hydrolysate and the antihypertensive effect of hydrolysate in spontaneously hypertensive rats. *Food Chem.* **111**, 302–308.

Zasloff, M. (2002). Antimicrobial peptides of multicellular organisms. *Nature* **415**, 389–395.

Zeng, M., Cui, W., Zhao, Y., Liu, Z., Dong, S., and Guo, Y. (2008). Antiviral active peptide from oyster. *Chin. J. Oceanol. Limnol.* **26**, 307–312.

Zhao, Y., Dong, S., Liu, Z., Zhao, X., Wang, J., and Zeng, M. (2009). A novel ACE inhibitory peptide isolated from *Acaudina molpadioidea* hydrolysate. *Peptides* **30**, 1028–1033.

CHAPTER **5**

Bioactive Marine Peptides: Nutraceutical Value and Novel Approaches

Anupam Giri and **Toshiaki Ohshima**[1]

Contents

Abstract

Marine organisms represent a valuable source of nutraceuticals and functional compounds. The biodiversity of the marine environment and the associated chemical diversity constitute a practically unlimited resource of novel active substances for the development of bioactive products. Recently, a great deal of interest has been expressed in marine-derived bioactive peptides because of their numerous beneficial health effects. Moreover, several studies have reported that marine bioactive peptides can be used as

Department of Food Science and Technology, Tokyo University of Marine Science and Technology, Tokyo, Japan
[1] Corresponding author: Toshiaki Ohshima, *E-mail address:* tohshima@kaiyodai.ac.jp

Advances in Food and Nutrition Research, Volume 65
ISSN 1043-4526, DOI: 10.1016/B978-0-12-416003-3.00005-6

antihypertensive, antioxidative, anticoagulant, and antimicrobial components in functional foods or nutraceuticals and pharmaceuticals due to their therapeutic potential in the treatment or prevention of disease. In this chapter, we provide an overview of bioactive peptides derived from marine organisms as well as information about their biological properties and mechanisms of action with potential applications in different areas.

I. INTRODUCTION

With marine species comprising ~50% of total global biodiversity, the sea represents an enormous resource for novel compounds (de Vries and Beart, 1995), and it has been classified as the largest remaining reservoir of natural molecules to be evaluated for drug activity (Gerwick, 1987). The importance of marine organisms as a source of novel bioactive substances is growing rapidly. Moreover, much attention has been paid to unraveling the structural, compositional, and sequential properties of bioactive peptides. The discovery of the bioregulatory role of different endogenous peptides in organisms as well as the characterization of the underlying molecular mechanisms of action of some novel natural bioactive peptides on specific cellular targets have contributed to the hypothesis that peptides could represent promising lead drug candidates. Recently, marine-derived peptides have opened a new area for the development of pharmaceutical compounds (Miljanich, 1997).

Marine-derived bioactive peptides have been shown to possess many physiological functions, including angiotensin I-converting enzyme (ACE) inhibition and antihypertensive (Byun and Kim, 2001; Je *et al.*, 2005a), antioxidant (Kim *et al.*, 2007; Mendis *et al.*, 2005a), anticoagulant (Jo *et al.*, 2008; Rajapakse *et al.*, 2005a), and antimicrobial (Liu *et al.*, 2008; Stensvag *et al.*, 2008) activity. Moreover, some of these bioactive peptides may have potential benefits for the promotion of human health and the reduction of disease risk (Defelice, 1995; Shahidi and Zhong, 2008); indeed, a role for food-derived bioactive peptides in the reduction of risk for cardiovascular disease has been well demonstrated (Erdmann *et al.*, 2008). Further, increasing consumer awareness about the link between diet and health has raised the demand for functional food ingredients and nutraceuticals. Bioactive peptides derived from marine organisms and the by-products from the processing of marine fish can potentially be used for the development of functional foods (Ariyoshi, 1993; Jung *et al.*, 2006a; Shahidi, 2007; Yamamoto, 1997).

Marine capture fisheries contribute to more than 50% of the total global production of fish, and >70% of this output is utilized for processing (FAOSTAT, 2001). As a result, a considerable amount of the yearly total catch is discarded as processing leftovers, which include trimmings, fins, frames, heads, skin, and viscera. In addition to fish processing, a large quantity of the by-products from marine bioprocessing plants consists of shells from crustaceans and shellfish. Therefore, there is a great potential for the marine bioprocess industry to convert and utilize more of these by-products as valuable products. Recent studies have identified a number of bioactive compounds in the discarded fish muscle proteins, collagen, gelatin, oil, bone, and internal organs and in shellfish and crustacean shells (Je *et al.*, 2005b; Jeon and Kim, 2000; Kim *et al.*, 2001). Generally, the production of foodstuffs for human consumption generates high levels of profitability; however, even higher levels of profit are expected to be derived from the production of bioactive compounds. Therefore, the development of new technology to identify and isolate novel bioactive compounds from marine processing by-products will increase the value of what is currently considered a waste product, and this represents a unique challenge and opportunity for the seafood industry.

This chapter provides an overview of the diversity of marine bioactive peptides, their biological activity, and their potential use as nutraceutical and pharmaceutical products.

II. EFFECT OF THE STRUCTURAL PROPERTIES OF PEPTIDES ON BIOACTIVITY

Bioactive peptides are inactive in the context of their parent protein, but can be released by enzymatic hydrolysis (Kim *et al.*, 1999; Lahl and Braun, 1994). Depending on their amino acid sequence, they may have roles in various biological processes such as opioid agonists or antagonists and demonstrate antihypertensive immunomodulatory, antithrombotic, antioxidant, anticancer, and antimicrobial activity, in addition to their use as nutritional compounds (Clare and Swaisgood, 2000; Elias *et al.*, 2008). On the basis of their structure and other factors, including hydrophobicity, charge, and microelement-binding properties, some bioactive peptides exhibit multifunctional activity (Cho *et al.*, 2008; Korhonen and Pihlanto-Leppala, 2003).

Bioactive peptides usually contain 3–20 amino acid residues (Pihlanto-Leppala, 2001), and the lower their molecular weight (MW), the higher their chance to cross the intestinal barrier and exert biological effects (Roberts *et al.*, 1999). Previous work on antioxidative peptides has shown that peptides with 5–16 amino acid residues can inhibit the autoxidation of linoleic acid (Chen *et al.*, 1995a). Lipid peroxidation is thought to proceed via the radical-mediated abstraction of hydrogen atoms from methylene carbons in polyunsaturated fatty acids (Rajapakse *et al.*, 2005b). Since the hydrophobicity of antioxidants is important for their accessibility to hydrophobic targets (Chen *et al.*, 1996), it is presumed that the presence of hydrophobic amino acids in the purified peptide might contribute to their lipid peroxidation inhibitory activity by increasing the solubility of peptides in lipids, thereby facilitating their improved interaction with radical species. In addition, the presence of a hydrophobic amino acid at the N-terminus of peptide sequences is thought to be important for their antioxidative activity (Chen *et al.*, 1995b; Ranathunga *et al.*, 2006) because it is assumed that such amino acids can increase the interaction between the peptides and fatty acids.

The specific amino acid composition of a peptide is a critical factor for its ACE-inhibitory activity. Glu, Asp, Pro, Gly, and Ala are observed in many ACE-inhibitory peptides derived from food proteins, for example, Arg-Ala-Asp-His-Pro-Phe from albumin (He *et al.*, 2004), Tyr-Gly-Leu from whey (He *et al.*, 2004), Ala-Glu-Leu from alga (He *et al.*, 2004), Asp-Leu-Pro from soy (He *et al.*, 2004), Val-Pro-Pro from skimmed milk (Pan *et al.*, 2005), Leu-Ala-Ile-Pro-Val-Asn-Lys-Pro from soybean protein (Kuba *et al.*, 2005), and Gly-Pro-Pro from buckwheat (Ma *et al.*, 2006). As reported by Cushman *et al.* (1981), the active sites of two domains in ACE are structurally and functionally homologous to a dipeptidyl carboxypeptidase, and the zinc coordination geometry is critical for their hydrolytic action. However, these two catalytic sites are differentially activated by chloride ions, and the physiological substrate angiotensin-I preferentially binds to the catalytic site of the C domain. The substrate also contributes to the chloride-mediated activation of the active site. Therefore, these differences indicate that, despite the high level of primary sequence homology, there are structural and functional differences between the active sites of the C and N domains. Three subsites, S1 (antepenultimate), S0 1 (penultimate), and S0 2 (ultimate), with distinct characteristics for the binding of the carboxy-terminal amino acids of the substrate or inhibitor, are located within two homologous active sites. For the binding and interaction of the inhibitor and enzyme, these three main subsites bind with the substrate. The binding of the inhibitor or the natural substrate to the enzyme predominantly takes place via the carboxy-terminal tripeptide residues of the inhibitor and substrate. Certain C-terminus amino acid residues of the inhibitor, for example, Pro and Phe, are reported to be favorable for ACE inhibition; Ala and Val are more favorable at the antepenultimate position (S1); while Pro and Leu in the substrate sequence are most favorable at the ultimate position (S0 2) with regard to its affinity for ACE (Cushman *et al.*, 1981; Jung and Kim, 2007).

The antioxidant activity of peptides has been shown to be a result of the cooperative effect of their entire amino acid sequence; however, there is still little information concerning the structural characteristics of antioxidative peptides. At present, the main strategy has been to identify and characterize antioxidative peptides from the hydrolysates of proteins;

however, novel methods describing the relationship between the structure of the peptide and its antioxidative activity are needed to predict the antioxidant potential of food protein hydrolysates. The Muramoto research group investigated the residue–activity relationship of antioxidative peptides by combinatorial chemistry (Chen *et al.*, 1996, 1998; Saito *et al.*, 2003). They isolated six antioxidative peptides from soybean protein. On the basis of the smallest peptide, Leu-Leu-Pro-His-His, 28 synthetic peptides were constructed and their antioxidant activity was compared. The results indicated that the N-terminal Leu had no effect on their antioxidant activity, while His and Pro played important roles in this activity. Further analysis of 22 synthetic peptides containing His residues demonstrated that His-containing peptides can act as metal-ion chelators, active-oxygen quenchers, and hydroxyl radical scavengers (Chen *et al.*, 1998). Subsequently, the Muramoto group constructed two series of tripeptide libraries (totaling 222 peptides) using combinatorial chemistry to explore the antioxidative properties of these peptides. Among the tested peptides, Tyr-(His, Lys, Arg)-Tyr was found to have the highest antioxidant activity.

According to the results from structure–activity relationship modeling by Li *et al.* (2011), the amino acid residue next to the C-terminus and the N-terminus amino acid residue are more important for antioxidant activity than the C-terminus. An amino acid residue next to the C-terminus with the ability to form hydrogen bonds and with low hydrophobicity might have higher antioxidant activity. A basic amino acid at the N-terminus with low hydrophobicity and a high hydrogen bond value was also predicted to generate high antioxidant activity. An amino acid residue next to the C-terminus with the ability to form hydrogen bonds also has an impressive effect on antioxidant activity. Saito *et al.* (2003) concluded that peptides containing Tyr-(His, Lys, Arg)-Tyr showed the highest antioxidant activity. For peptides composed of four or more amino acids, all of the four properties of the amino acid residue next to the C-terminus contribute to the antioxidant activity. In addition, the larger the hydrogen bonding and steric properties of this residue, the lower its hydrophobicity and the higher its activity. Therefore, the basic and acidic amino acids (Arg, Lys, His, Glu, and Asp) and the other hydrophilic amino acids (Thr, Ser, Asn, and Gln) have either a high value of hydrogen bonding or a high steric property with low hydrophobicity. In this position, they would have higher activity than the other amino acids. The hydrophobicity of the N-terminal amino acid residue is also very important for antioxidant activity, that is, the larger the hydrophobicity, the higher is the activity. Many other studies also speculated that hydrophobic amino acids on the N-terminus (Ala, Val, Leu, etc.) play an important role in antioxidant activity (Chen *et al.*, 1995a; Tsuge *et al.*, 1991).

The electronic property of the amino acid on the C-terminus also has an effect on antioxidant activity (Li *et al.*, 2011), that is, the larger the electronic property, the higher is the activity. The C-terminus is a polar position that is thus affected by its electrostatic potential, to some extent; therefore, the amino acids Trp, Glu, Leu, Ile, Met, Val, Tyr, etc. are suitable at the C-terminus. Some researchers have speculated that the identity of the amino acid on the C-terminus would play an important role in its activity. Suetsuna (2000) separated and identified a radical scavenging peptide, Tyr-Phe-Tyr-Pro-Glu-Leu, from casein hydrolysate, and it was confirmed that the Glu-Leu on the C-terminus mainly contributed to its antioxidant activity. Kim *et al.* (2009) speculated that the hydrophobic property of the amino acid on the C-terminus, for example, Val and Leu, had a distinct effect on the activity, as determined from the analysis of antioxidative peptides derived from venison hydrolysate.

Chen *et al.* (1996) investigated 28 synthetic peptides that were based on the antioxidative peptide Leu-Leu-Pro-His-His. They found that the deletion of the C-terminus His decreased the activity, whereas the deletion of the N-terminus Leu had no effect; therefore, it seemed that the C-terminus was related to the antioxidative activity of the peptide, while the N-terminus had no effect. Saito *et al.* (2003) pointed out that two Tyr-containing tripeptides (Tyr-X-Tyr) showed higher activity than did two His-containing tripeptides (His-X-His). According to the structural characteristics, the hydrophobicity of an amino acid on the N-terminus is the most important property for activity with respect to the same amino acid in the central position; in addition, Tyr on the N-terminus has stronger hydrophobicity than His. Therefore, it can be predicted that Tyr-X-Tyr has higher activity than His-X-His.

In addition, Cacciuttolo *et al.* (1993) reported that Tyr, Trp, and Phe, which have aromatic residues, can stabilize reactive oxygen species through electron transfer. Davalos *et al.* (2004) reported that Trp, Tyr, and Met have the highest antioxidant activity, followed by Cys, His, and Phe. The remaining amino acids (Arg, Asn, Gln, Asp, Pro, Ala, Val, Lys, Ile, Thr, Leu, Glu, and Gly) did not exhibit antioxidant activity. Therefore, several amino acids, for example, Tyr, Met, His, Cys, and Trp, are generally accepted as antioxidants that contribute to the activity of identified peptides (Chen *et al.*, 1995a; Hernandez-Ledesma *et al.*, 2005; Tsuge *et al.*, 1991). The antioxidative activity of peptides containing His has been attributed to the His residue because of the proton-donation ability of the His imidazole group.

In summary, the structural characteristics of peptides with high antioxidant activity are as follows: a hydrogen bonding and hydrophilic amino acid residue in the position next to the C-terminus, a hydrophobic amino acid residue at the N-terminus, and an electronic amino acid residue at the C-terminus.

III. BIOACTIVE PEPTIDES DERIVED FROM MARINE FISH

Peptides derived from fish proteins exert potent antioxidative activity in different oxidative systems (Jun *et al.*, 2004; Kim *et al.*, 2000; Rajapakse *et al.*, 2005a). There is increasing interest in identifying natural antioxidative substances that do not have side effects, and these compounds have the potential to address complications that arise from the oxidation of biomolecules. Recently, a number of studies have demonstrated that peptides derived from different marine protein hydrolysates act as potential antioxidants, for example, from marine fish such as hoki (Je *et al.*, 2005a; Kim *et al.*, 2007; Mendis *et al.*, 2005a), tuna (Je *et al.*, 2007, 2008), cod (Slizyte *et al.*, 2009), Pacific hake (Samaranayaka and Li-Chan, 2008), capelin (Amarowicz and Shahidi, 1997), scad (Thiansilakul *et al.*, 2007), Alaska pollock (Cho *et al.*, 2008; Je *et al.*, 2005b), conger eel (Ranathunga *et al.*, 2006), yellowfin sole (Jun *et al.*, 2004), and yellow stripe trevally (Klompong *et al.*, 2009; Table 5.1). The beneficial effects of marine-derived bioactive peptides include scavenging for free radicals and reactive oxygen species or preventing oxidative damage by interrupting the radical chain reaction of lipid peroxidation (Mendis *et al.*, 2005b; Qian *et al.*, 2008; Rajapakse *et al.*, 2005b; Ranathunga *et al.*, 2006). Kim *et al.* (2000) reported that some peptides derived from fish showed stronger antihypertensive activity by inhibiting the action of ACE than many other natural peptides. These peptides exhibited *in vivo* activity by lowering the blood pressure in spontaneously hypertensive rats (SHRs; Fujita and Yoshikawa, 1999; Je *et al.*, 2005a). Enzymatically hydrolyzed fish muscle peptides have also demonstrated anticoagulant and antiplatelet properties *in vitro*, and these results suggest the capability of fish peptides to inhibit the intrinsic coagulation pathway (Rajapakse *et al.*, 2005b). In addition, marine anticoagulant proteins have been purified from blood ark shell (Jung *et al.*, 2001), starfish (Koyama *et al.*, 1998), and yellowfin sole (Rajapakse *et al.*, 2005a).

Kim *et al.* (1999) and Jung *et al.* (2005b) reported that fish peptides can accelerate calcium absorption. Under many conditions, dietary calcium becomes unavailable for absorption due to the formation of insoluble compounds inside the dietary tract, and inadequate dietary calcium is associated with a number of common and chronic diseases including osteoporosis, osteoarthritis, cardiovascular disease (hypertension and stroke), diabetes, obesity, and cancer (Anderson and Garner, 1996). Further, fish protein hydrolysates contain hormone-like peptides and growth factors that accelerate calcium absorption (Fouchereau-Peron *et al.*, 1999). These peptides are capable of binding to the cell surface receptors on osteoclasts and have a role in calcium metabolism by decreasing the number of osteoclasts. Therefore, these peptides could be used in the treatment of osteoporosis and Paget's disease. Further,

TABLE 5.1 Bioactive peptides derived from marine organisms

Marine resources	Amino acid sequence/major amino acid in bioactive peptide	Activity	Reference
	Fish		
Big eye tuna (muscle)	Trp-Pro-Glu-Ala-Ala-Glu-Leu-Met-Met-Glu-Val-Asp-Pro	ACE inhibitor	Qian *et al.* (2007)
	H-Leu-Asn-Leu-Pro-Thr-Ala-Val-Tyr-Met-Val-Thr-OH	Antioxidant	Je *et al.* (2008)
	Val-Lys-Ala-Gly-Phe-Ala-Trp-Thr-Ala-Asn-Glu-Glu-Leu-Ser	Antioxidant	Je *et al.* (2007)
Big eye tuna (frame)	Gly-Asp-Leu-Gly-Lys-Thr-Thr-Thr-Val-Ser-Asn-Trp-Ser-Pro-Pro-Lys-Try-Lys-Asp-Thr-Pro	Antihypertensive	Lee *et al.* (2010)
Alaska pollock	Gly-Pro-Leu	ACE inhibitor	Byun and Kim (2001)
	Val-Leu-Ser-Gly-Gly-Thr-Thr-Met-Ala-Met-Tyr-Thr-Leu-Val	Antioxidant	Jung *et al.* (2006a)
Sea bream	Val-Ile-Tyr	ACE inhibitor	Fahmi *et al.* (2004)
Yellowfin sole	Met-Ile-Phe-Pro-Gly-Ala-Gly-Gly-Pro-Glu-Leu	ACE inhibitor	Jung *et al.* (2006a)
	Arg-Pro-Asp-Phe-Pro-Leu-Glu-Pro-Pro-Tyr	Antioxidant	Jun *et al.* (2004)
Horse mackerel (skin)	Asn-His-Arg-Tyr-Asp-Arg	Antioxidant	Sampath *et al.* (2011)
Croaker (skin)	Gly-Asn-Arg-Gly-Phe-Ala-Cys-Arg-His-Ala	Antioxidant	Sampath *et al.* (2011)
Conger eel	Leu-Gly-Leu-Asn-Gly-Asp-Asp-Val-Asn	Antioxidant	Ranathunga *et al.* (2006)
Hoki fish (skin)	His-Gly-Pro-Leu-Gly-Pro-Leu	Antioxidant	Mendis *et al.* (2005a)
Seaweed pipefish (muscle)	Thr-Phe-Pro-His-Gly-Pro	Antihypertensive	Wijesekara *et al.* (2011)
	His-Trp-Thr-Thr-Gln-Arg	Antihypertensive	Wijesekara *et al.* (2011)
Squid	Phe-Asp-Ser-Gly-Pro-Ala-Gly-Val-Leu	Antioxidant	Mendis *et al.* (2005a)
	Asn-Gly-Pro-Leu-Gln-Ala-Gly-Gln-Pro-Gly-Glu-Arg	Antioxidant	Mendis *et al.* (2005a)

TABLE 5.1 *(continued)*

Marine resources	Amino acid sequence/major amino acid in bioactive peptide	Activity	Reference
	Gly-Pro-Leu-Gly-Leu-Leu-Gly-Phe-Leu-Gly-Pro-Leu-Gly-Leu-Ser	ACE inhibitor	Alemán *et al.* (2011)
Oyster	Cys, Leu, Glu, Asp, Phe, Tyr, Ile, and Gly	Antimicrobial	Liu *et al.* (2008)
	Leu-Lys-Gln-Glu-Leu-Glu-Asp-Leu-Leu-Glu-Lys-Gln-Glu	Antioxidant	Qian *et al.* (2008)
	Leu-Leu-Glu-Tyr-Ser-Ile	Inhibitor of HIV-1 protease	Lee and Maruyama (1998)
	Leu-Leu-Glu-Tyr-Ser-Leu	Inhibitor of HIV-1 protease	Lee and Maruyama (1998)
	Arg-Arg-Trp-Trp-Cys-Arg-X	Inhibitory activity on herpes virus	Zeng *et al.* (2008)
Mussel	Glu-Ala-Asp-Ile-Asp-Gly-Asp-Gly-Gln-Val-Asn-Tyr-Glu-Glu-Phe-Val-Ala-Met-Met-Thr-Ser-Lys	Anticoagulant	Jung and Kim (2009)
	Leu-Val-Gly-Asp-Glu-Gln-Ala-Val-Pro-Ala-Val-Cys-Val-Pro	Antioxidant	Jung *et al.* (2007)
	Phe-Gly-His-Pro-Tyr	Antioxidant	Jung *et al.* (2005b)
American lobster	Gln-Tyr-Gly-Asn-Leu-Leu-Ser-Leu-Leu-Asn-Gly-Tyr-Arg	Antimicrobial	Battison *et al.* (2008)
Shrimp	Pro-Arg-Pro	Antimicrobial	Bartlett *et al.* (2002)
	Ile-Phe-Val-Pro-Ala-Phe	ACE inhibitor	Lun *et al.* (2006)
Sea urchin	Cys	Antimicrobial	Li *et al.* (2008)
Clam	Tyr-Asn	ACE inhibitor	Tsai *et al.* (2008)
	Val-Glu-Val	ACE inhibitor	Tsai *et al.* (2006)
	Val-Lys-Pro	ACE inhibitor	Tsai *et al.* (2006)
Sea cucumber	Met-Glu-Gly-Ala-Gln-Glu-Ala-Gln-Gly-Asp	ACE inhibitor	Zhao *et al.* (2009)
	Glu, Asp, Pro, Gly, and Ala	Antihypertensive	Zhao *et al.* (2007)

acidic peptide fractions from Atlantic cod hydrolysate demonstrate strong immunostimulatory effects, and treatment with these peptides generated an oxidative burst in Atlantic salmon leucocytes (Gildberg *et al.*, 1996). Basically, immunomodulators enhance the production of oxygen metabolites in macrophages, which are the primary source of these oxygen metabolites, and thus determine the oxidative burst. Oxidative burst reactions are of a major importance for the bactericidal power of phagocytes.

Working on Alaska pollock (*Theragra chalcogramma*) backbones discarded from industrial processing as a source of soluble calcium, Jung *et al.* (2006a) identified a backbone peptide, Val-Leu-Ser-Gly-Gly-Thr-Thr-Met-Ala-Met-Tyr-Thr-Leu-Val (MW: 1442 Da), exhibiting an affinity for calcium ions on the surface of hydroxyapatite crystals. This *in vitro* calcium-binding assay also demonstrated that this peptide can solubilize a similar amount of calcium as casein phosphopeptide; thus, it is possible to provide a novel nutraceutical with a high bioavailability for calcium through further studies with *in vivo* assays. Fish protein hydrolysates obtained from cod backbones also enhanced product stability by preventing oxidative deterioration (Slizyte *et al.*, 2009). The scavenging activity of 2,2-diphenyl-1-picrylhydrazyl (DPPH) showed that the antioxidative activity of hydrolysates could be due to their ability to scavenge lipid radicals. An increased degree of hydrolysis resulted in a slight increase of DPPH radical scavenging activity. These results also show that it is possible to obtain bioactive molecules from cod backbones by protein hydrolysis (gastrin/cholecystokinin- and calcitonin gene-related peptide-like peptides) and that the incorporation of cod hydrolysates into foods could be beneficial.

According to Lee *et al.* (2010), the peptide Gly-Asp-Leu-Gly-Lys-Thr-Thr-Thr-Val-Ser-Asn-Trp-Ser-Pro-Pro-Lys-Try-Lys-Asp-Thr-Pro, derived from big-eye tuna frame protein hydrolysate, showed a strong suppressive effect on the systolic blood pressure of SHRs, while its antihypertensive activity was similar to that of captopril, a commercially available antihypertensive drug. Further, they reported no side effects after the administration of this antihypertensive peptide to rats. In addition, these marine antihypertensive peptides exhibit stronger antihypertensive activity *in vivo* than *in vitro*. The exact mechanisms underlying this phenomenon have yet to be identified; however, it was suggested that bioactive peptides have a higher tissue affinity and are eliminated more slowly than captopril (Fujita and Yoshikawa, 1999).

To evaluate the immunomodulatory effects of marine oligopeptide preparation (MOP) from chum salmon (*Oncorhynchus keta*), Yang *et al.* (2009) demonstrated that MOP has the potential to enhance the immune function of mice, including cellular immunity, humoral immunity, and natural killer cell function, but has no significant effect on the activity of macrophages. Moreover, this positive immunomodulation by MOP is most likely attributed to the stimulation of T helper (Th) cells, causing the secretion of Th1 and Th2 cytokines. Therefore, as a novel food source, MOP may have a beneficial effect on the immune function of consumers, and consequently help to prevent disease. Further studies have been recommended to determine and evaluate the clinical significance of these findings, particularly in those with impaired immune function and/or with cancer.

ACE-inhibitory peptides have been isolated from enzymatic hydrolysates of various sources of fish waste, for example, Alaska pollock skin (Byun and Kim, 2001), sea bream scales (Fahmi *et al.*, 2004), and yellowfin sole frame protein (Jung *et al.*, 2006b). Enzymatically hydrolyzed fish skin gelatin has higher antioxidant and antihypertensive activity than peptides derived from fish muscle protein (Byun and Kim, 2001; Kim *et al.*, 2001; Mendis *et al.*, 2005a,b). Gelatin peptides contain unique repeated Gly-Pro-Ala sequences, and it is presumed that the antioxidative and antihypertensive properties of gelatin peptides are associated with their unique amino acid composition. In addition, these peptides can accelerate the absorption of dietary calcium in animal models, thereby increasing calcium bioavailability (Kim *et al.*, 1998).

IV. BIOACTIVE PEPTIDES DERIVED FROM LOBSTER, SHRIMP, AND CRABS

Marine-derived bioactive peptides are well described in the hemolymph of many marine invertebrates (Tincu and Taylor, 2004), including spider crab (Stensvag *et al.*, 2008), American lobster (Battison *et al.*, 2008), and shrimp (Bartlett *et al.*, 2002). Antimicrobial peptides (AMPs) in blue crab hemolymph (*Callinectes sapidus*) were highly inhibitory to Gram-negative bacteria (Edward *et al.*, 1996). Although there are several reports on the antibacterial activity of seminal plasma, few antibacterial peptides have been identified in the mud crab (*Scylla serrata*; Jayasankar and Subramonium, 1999). An AMP derived from American lobster (*Homarus americanus*) exhibited bacteriostatic activity against some Gram-negative bacteria and demonstrated protozoastatic and protozoacidal activity against two scuticociliate parasites (*Mesanophrys chesapeakensis* and *Anophryoides haemophilia*), the latter is a significant pathogen of *H. americanus* (Battison *et al.*, 2008). In addition, an AMP (arasin 1) derived from spider crab (*Hyas araneus*) inhibited the growth of *Corynebacterium glutamicum* (Stensvag *et al.*, 2008).

V. BIOACTIVE PEPTIDES DERIVED FROM SQUID, CLAMS, AND SEA URCHINS

Peptides derived from a tryptic hydrolysate of jumbo squid (*Dosidicus gigas*) skin gelatin exhibited strong inhibition of lipid peroxidation that was much higher than that of the natural antioxidant α-tocopherol (Mendis *et al.*, 2005a). Two representative peptides with comparatively higher antioxidant potency were purified and characterized as Phe-Asp-Ser-Gly-Pro-Ala-Gly-Val-Leu (880.18 Da) and Asn-Gly-Pro-Leu-Gln-Ala-Gly-Gln-Pro-Gly-Glu-Arg (1241.59 Da). The viability of radical-mediated oxidation-induced human lung fibroblasts was also enhanced following treatment with these peptides. However, they did not exhibit substantial ion chelation, and it was presumed that the observed radical scavenging potency of these peptides plays a vital role in their strong antioxidant activity. It was suggested that the hydrophobic amino acids of these peptides contributed greatly to their observed antioxidant activity.

Several ACE-inhibitory peptides were also prepared from clams, including a peptic digest of short-necked clam (Val-Glu-Val, $IC_{50}=8.7$ mM) and clam hydrolysate (Val-Lys-Pro, $IC_{50}=2.6$ mM; Tsai *et al.*, 2006). Tsai *et al.* (2008) reported that hard clam (*Meretrix lusoria*) residual meat extract can be recovered as a value-added by-product. One of the peptides isolated from hard clam residual meat hydrolysate (Tyr-Asn) showed high ACE-inhibitory activity with an IC_{50} of 51 mM (or 0.015 mg/mL).

Sea urchins possess an innate immune system and are regarded as a potential source for the discovery of new AMPs. Li *et al.* (2008) isolated two novel antibacterial peptides (strongylocin 1 and 2; 5.6 and 5.8 kDa, respectively) from coelomocyte extracts of the green sea urchin (*Strongylocentrotus droebachiensis*). These two peptides have putative isoforms (1b and 2b), similar to two putative proteins from the purple sea urchin (*Strongylocentrotus purpuratus*). The native strongylocin peptides are cationic, defensin-like peptides (cysteine-rich), but show no similarity to other known AMPs with respect to their cysteine distribution pattern. Strongylocin 1 consists of 83 amino acids that include a prepro sequence of 35 amino acids, whereas strongylocin 2a and 2b are composed of 89 and 90 amino acids, respectively, in which 38 amino acids represent a prepro sequence. The native peptides display potent activity against Gram-negative and -positive bacteria.

VI. BIOACTIVE PEPTIDES DERIVED FROM MOLLUSKS AND OYSTERS

Cytotoxic cyclic peptides have also been found in mollusks. Dolastatin peptides are a group of cyclic and linear peptides with prominent cell growth-suppressing activity that were isolated from the marine mollusk

(*Dolabella auricularia*). Another prominent family of peptides has also been isolated from mollusks; these highly compact and stable linear peptides, known as conotoxins, exhibit specific actions on the ion channels and membrane receptors of excitable cells.

D. auricularia is a shell-less mollusk that, therefore, initially appears to lack defenses against predators; however, this is only a preliminary supposition. Accumulated evidence supports the fact that Opisthobranchia mollusks have developed very powerful chemical defenses (Pettit *et al.*, 1989). Pettit *et al.* (1981) were the first to isolate some of these compounds; the pentapeptide dolastatin 10 (Fig. 5.1) was reported to be the most active natural anticancer substance at that time with an ED_{50} of 4.6×10^{-5} μg/mL against the P388 cell line (Pettit *et al.*, 2008). Dolastatin 10 was also shown to inhibit tubulin polymerization and tubulin-dependent GTP hydrolysis (Bai *et al.*, 1990).

Dolastatin 10 has been evaluated with promising results in a phase I clinical study in patients with solid tumors. Subsequently, its noticeable antitumor activity was well documented in various *in vitro* and *in vivo* tumor models (Madden *et al.*, 2000). More than a dozen dolastatin peptides have been isolated to date. Recent studies have shown, for example, that the depsipeptide dolastatin 11 arrests cells at cytokinesis by causing a rapid and massive rearrangement of the cellular actin filament network and induces the hyperpolymerization of purified actin (Bai *et al.*, 2001). The effects of dolastatin 11 were similar to those of the sponge-derived depsipeptide jasplakinolide; however, dolastatin 11 exhibited ~threefold more cytotoxicity than jasplakinolide in the cells studied.

Keenamide A is another novel cytotoxic cyclic hexapeptide isolated from a marine mollusk (*Pleurobranchus forskalii;* Fig. 5.2; Wesson and Hamann, 1996). Keenamide A exhibited significant activity against the P-388, A-549, MEL-20, and HT-29 tumor cell lines. New classes of anticancer drug candidates isolated from marine organisms have been shown to possess powerful cytotoxic activity against multiple tumor types.

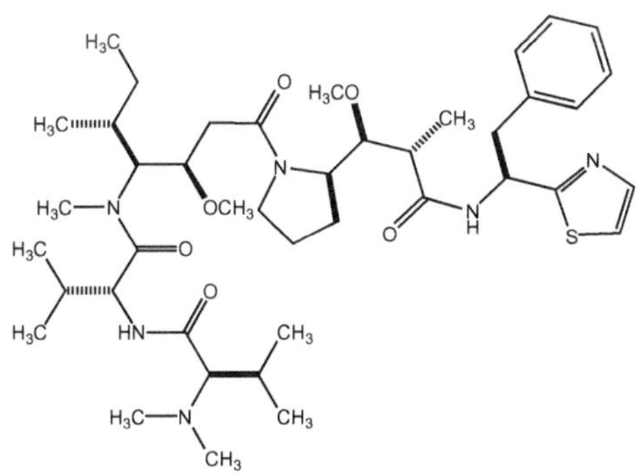

FIGURE 5.1 Structure of dolastatin 10 derived from the sea hare (mollusk) *Dolabella auricularia.*

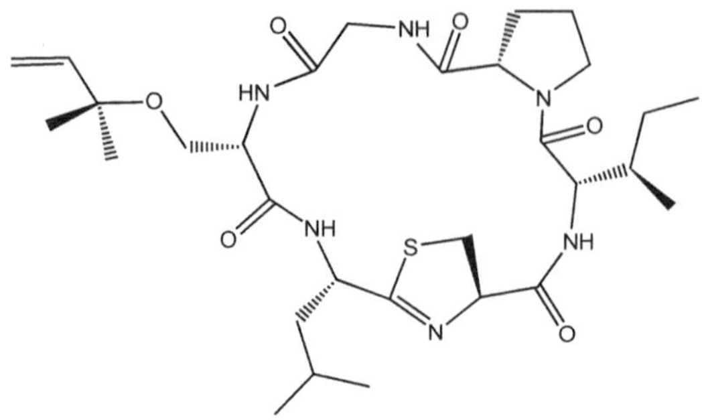

FIGURE 5.2 Structure of the cyclic peptide keenamide A derived from the notaspidean mollusk *Pleurobranchus forskalii.*

Peptides derived from marine organisms, including the oyster (Qian *et al.*, 2008) and blue mussel (Jung *et al.*, 2005b and Rajapakse *et al.*, 2005b), have been reported to exhibit functional properties. The beneficial effects of marine bioactive peptides are well known in scavenging for free radicals and reactive oxygen species or in preventing oxidative damage by interrupting the radical chain reaction of lipid peroxidation (Mendis *et al.*, 2005b; Qian *et al.*, 2008; Rajapakse *et al.*, 2005a; Ranathunga *et al.*, 2006). It has been shown that their antioxidant potency is mostly due to the presence of hydrophobic amino acids (Mendis *et al.*, 2005b). In addition, the bioactive antioxidant peptide Leu-Lys-Gln-Glu-Leu-Glu-Asp-Leu-Leu-Glu-Lys-Gln-Glu, isolated from the oyster (*Crassostrea gigas*), exhibited higher activity against polyunsaturated fatty acid peroxidation than α-tocopherol (Qian *et al.*, 2008). ACE-inhibitory peptides were also isolated from oyster protein digest (Leu-Phe; IC50=126mM; Matsumoto *et al.*, 1994) and pearl oyster meat hydrolysate (Leu-Val-Glu; IC_{50}=14.2 mM; Suetsuna, 2002). Liu *et al.* (2008) also reported the presence of AMPs in the oyster.

Moreover, the anticoagulant peptide Glu-Ala-Asp-Ile-Asp-Gly-Asp-Gly-Gln-Val-Asn-Tyr-Glu-Glu-Phe-Val-Ala-Met-Met-Thr-Ser-Lys, derived from the blue mussel, prolonged the blood clotting time (Jung and Kim, 2009). In addition, a protein derived from the blood arch shell prolonged the clotting time more than did heparin, a commercially available anticoagulant (Jung *et al.*, 2001). However, these marine-derived anticoagulant peptides are noncytotoxic and have the potential to be used as functional ingredients in nutraceutical or pharmaceutical products.

Further, the AMP CgPep33, derived from oyster (*C. gigas*), demonstrated antimicrobial activity and growth inhibition of bacteria, for example, *Escherichia coli*, *Pseudomonas aeruginosa*, and *Bacillus subtilis*; and fungi, for example, *Botrytis cinerea* and *Penicillium expansum* (Liu *et al.*, 2008). More interestingly, the IC_{50} values of this peptide against these bacteria and fungi ranged from 18.6 to 48.2μg/mL. In addition, antiviral bioactive peptides have also been identified. For example, Achour *et al.* (1997) found that an oyster protein extract could enhance the proliferation of immunocytes in human immunodeficiency virus 1 (HIV-1)-infected patients. In addition, Lee and Maruyama (1998) identified peptides that inhibit HIV-1 protease from the hydrolysate of oyster (*C. gigas*) proteins prepared with thermolysin. The amino acid sequences of these peptides were determined as Leu-Leu-Glu-Tyr-Ser-Ile and Leu-Leu-Glu-Tyr-Ser-Leu. These sequences are found in some proteins of various major viruses or human cytomegalovirus. Chemically synthesized Leu-Leu-Glu-Tyr-Ser-Ile and Leu-Leu-Glu-Tyr-Ser-Leu showed IC_{50} values of 20 and 15 nM, respectively and behaved as competitive inhibitors of HIV-1 protease with K_i values of 13 and 10nM, respectively. These peptides were more potent HIV-1 protease inhibitors than pepstatin A. Moreover, an active peptide, Arg-Arg-Trp-Trp-Cys-Arg-X (where X is an amino acid or an amino acid analog), against the herpes virus was isolated from the enzymatic hydrolysate of the oyster (*C. gigas*), and this peptide demonstrated high inhibitory activity against the herpes virus (Zeng *et al.*, 2008). Zeng *et al.* (2008) obtained four fractions of hydrolysates according to their MW (<1, 1–5, 5–10, and >10kDa). A peptide purified from the 5- to 10-kDa fractions had a higher biological activity than the other fractions.

VII. FERMENTED MARINE PEPTIDES AND NOVEL APPROACHES

Fermentation, one of the oldest food preservation techniques specifically practiced in South East Asian countries such as China, Japan, and Korea, is believed to enhance the nutraceutical value of fermented foods in addition to their long-term storage. The breakdown of food proteins by microbial proteases to produce bioactive peptides may be a possible reason for the development of such properties during fermentation. Therefore, interest has developed in identifying the biological activity of fermented foods, including fish and shellfish (Ichimura *et al.*, 2003; Kim *et al.*, 2004; Wong and Mine, 2004). Health-related functional properties such as antioxidative activity and radical scavenging capacity may present as promising biological benefits of these fermented foods. Fish sauce, being a fermented food, is assumed to contain many substances, including peptides, with ACE-inhibitory activity that is produced during the fermentation of fish proteins. As they are made by the fermentation of various kinds of fish, they are assumed to be rich in degradation products, for example, small peptides and amino acids. The ACE-inhibitory activity of fermented fish sauces made from salmon, sardine, or anchovy was reported (Okamoto *et al.*, 1995a), and three fermented salmon sauce-derived ACE-inhibitory peptides (Gly-Trp, Ile-Trp, and Val-Trp) containing Trp at the C-terminal position were further identified by Okamoto *et al.* (1995b). Earlier, Cheung *et al.* (1980) reported that Trp at the COOH terminus and Val and Ile at the NH_2 terminus of a dipeptide are the most favorable sequences for ACE-inhibitory activity. The dipeptide Gly-Trp, which was shown to be a considerable ACE inhibitor, has been isolated from sardine muscle hydrolysate by Seki *et al.* (1993). In addition, two ACE-inhibitory peptides, Gly-Pro-Pro and Val-Pro, were isolated from salted and fermented anchovy by Lee (1996). Ichimura *et al.* (2003) reported ACE-inhibitory activity in fermented fish sauce that was traditionally made from anchovy, sardine, or bonito, and purified the ACE-inhibitory peptides contained in the sauce (Table 5.2). ACE-inhibitory peptides such as Ala-Pro, Lys-Pro, Arg-Pro, Gly-Pro, Glu-Pro, Thr-Pro, Val-Pro, Gly-Ile, and Asp-Phe were isolated from fermented anchovy sauce. The ACE-inhibitory peptides Ala-Pro, Gly-Pro, Thr-Pro, Val-Pro,

TABLE 5.2 ACE-inhibitory activity of peptides derived from fermented fish sauce

Peptides	Origin	IC50[a] (μM) or % inhibition	Concentration of peptides[b] (μM)
Ala-Pro	Anchovy, sardine, bonito	29	31 (An), 29 (Sa), 1 (Bo)
Lys-Pro	Anchovy	22	28 (An)
Arg-Pro	Anchovy, bonito	21	42 (An), 2 (Bo)
Gly-Pro	Anchovy, sardine, bonito	360	1 (An), 140 (Sa), 6 (Bo)
Glu-Pro	Anchovy	1200	7 (An)
Thr-Pro	Anchovy, sardine	290	9 (An), 18 (Sa)
Val-Pro	Anchovy, sardine	570	37 (An), 8 (Sa)
Gly-Ile	Anchovy	1300	12 (An)
Asp-Phe	Anchovy	360	1 (An)
Asn-Pro	Sardine	2300	22 (Sa)
Asp-Met	Sardine	600	1 (Sa)
Asp-Leu	Sardine	2000	7 (Sa)
Ala-Val	Sardine	15% at 800 μM	8 (Sa)
Gly-Val	Sardine	17% at 800 μM	57 (Sa)
Ala-Gly-Pro	Bonito	560	4 (Bo)

Source: Ichimura *et al.* (2003). Permission has been obtained for use of copyrighted material from Elsevier B.V.
[a] IC50, concentration of the peptide required for 50% inhibition of ACE activity.
[b] Concentration of the peptide contained in the fermented fish sauce. As values were calculated from the quantity of finally purified peptide, the actual concentration of each peptide may be higher than the concentration indicated. An, anchovy; Sa, sardine; Bo, bonito.

Asn-Pro, Asp-Met, Asp-Leu, Ala-Val, and Gly-Val were isolated from fermented sardine sauce; further, the ACE-inhibitory peptides Ala-Pro, Arg-Pro, Gly-Pro, and Ala-Gly-Pro were isolated from fermented bonito sauce. Val-Pro was also identified in salted and fermented anchovy by Lee (1996). Among the peptides identified by Ichimura *et al.* (2003), Ala-Pro, Lys-Pro, and Arg-Pro showed strong and similar inhibitory activity. Ichimura *et al.* (2003) also isolated nine types of peptides containing Pro residues in their carboxy terminals. Due to the unique structure of Pro as an imino acid, peptide bonds containing Pro residues are often resistant to hydrolysis by common peptidases. This may be the reason why these Pro-containing dipeptides survived after long-term fermentation. Among these peptides, Lys-Pro was further evaluated *in vivo* in male SHRs (Charles River Japan, Yokohama) by oral administration. As shown in Fig. 5.3, orally administered Lys-Pro shows a tendency to lower the blood pressure of SHRs.

Salt-fermented anchovy sauce (FAS), a fermented fish product of Southeast and Far East Asia, made by salting anchovies, also contains anticoagulation agents (Kim *et al.*, 2004). The authors reported that the degree of fibrin-clotting inhibition increased with increasing amounts of desalted FAS (Fig. 5.4), indicating the presence of a bioactive peptide that acts as an anticoagulant against fibrin clotting. It has been suggested that the active peptide has a molecular mass much smaller than nattokinase, a known fibrinolytic enzyme from fermented foods; thus, it may be more useful for oral use in the prevention of thrombotic disease. For its wider

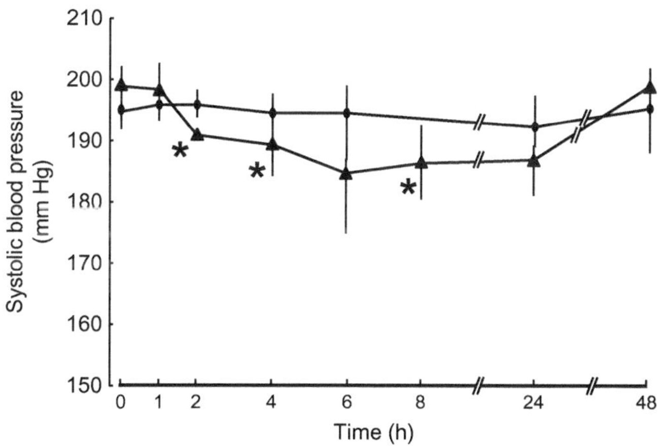

FIGURE 5.3 Change in systolic blood pressure after the oral administration of Lys-Pro. Values are the mean ± standard deviation (SD) ($n=6$). Circles, control group (distilled water); triangles, Lys-Pro group (50 mg/kg). *$P<0.05$ (paired *t*-test) compared to 0 h in the Lys-Pro group. *Source*: Ichimura *et al.* (2003). Permission has been obtained for the use of copyrighted material from Elsevier B.V.

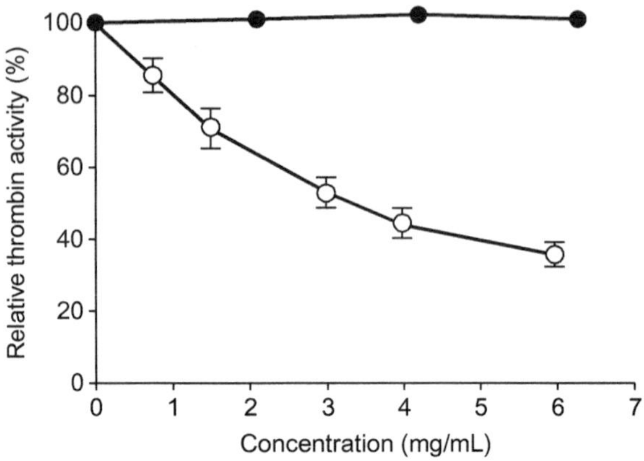

FIGURE 5.4 Relative fibrin-clotting and amidolytic activity of thrombin as a function of the concentration of desalted fermented anchovy sauce (dFAS). Anticoagulation activity was assessed by the increase in the fibrin clotting time (empty circles) and the release of *p*-nitroaniline (closed circles). Data are presented as the mean ± SD ($n = 3$). *Source*: Kim *et al.* (2004). Permission has been obtained for the use of copyrighted material from Elsevier B.V.

application, additional characteristics, for example, high thermal and pH stability and safety, are necessary for its long-term intake. The capability of this substance to control the risk of thrombosis by incorporating it into the daily diet without any adverse side effects on human health is most desirable for its application as a functional foodstuff with the intention of decreasing the medical cost of circulatory diseases.

Rajapakse *et al.* (2005b) extracted radical scavenging peptides from fermented marine blue mussel sauce. They found that the purified peptides from mussel sauce exerted strong scavenging effects on the tested radicals, although with different capacities. In addition, the mussel sauce-derived radical scavenging peptides exhibited different antioxidative mechanisms, indicating their protective action against free radicals. A cell viability assay clearly indicated the ability of these peptides to protect cells against oxidative stress-mediated injuries in a dose-dependent manner, which was similar to the observed activity of α-tocopherol (Fig. 5.5). Jung *et al.* (2005b) reported that blue mussel sauce that had been fermented for 6 months showed the highest antioxidative activity, exhibiting ~54.1% inhibition of linoleic acid peroxidation. The amino acid sequence of a purified peptide with a molecular mass of 620 Da was determined as Phe-Gly-His-Pro-Tyr. The activity of this peptide may be attributed to the chelating and lipid radical trapping ability of its imidazole ring. In addition, this peptide contained a Tyr residue, which is a potent hydrogen donor. This bioactive peptide showed antioxidative activity in lipid or fatty acid autoxidation systems and radical scavenging activity; thus, it is

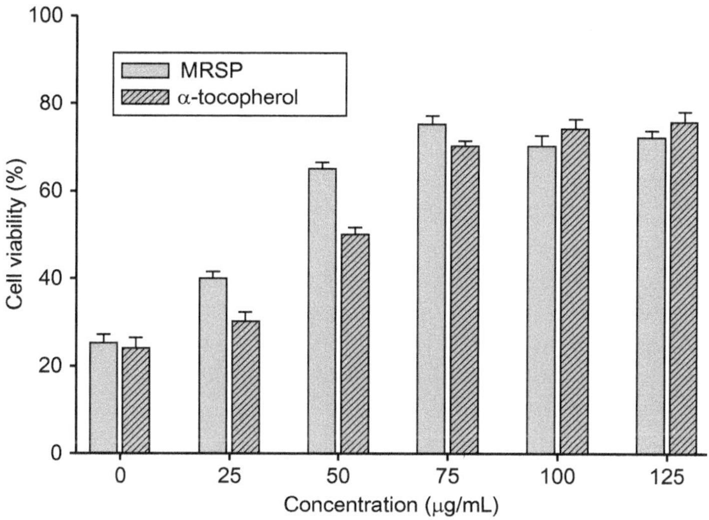

FIGURE 5.5 Effect of fermented mussel-derived radical scavenging peptides (MRSPs) on cell viability. Human lung fibroblasts were cultured in Dulbecco's modified Eagle's medium without fetal bovine serum and incubated with the purified peptide (MRSP) at various concentrations for 16h. Oxidative stress was artificially induced in cultured cells by incubation with 200 μM *t*-BHP for 2h. Viable cells remaining after the oxidative stress was measured as the percentage of viable cells compared to control cells using the MTT assay. Values are given as the mean ± standard error. *Source*: Rajapakse *et al.* (2005c). Permission has been obtained for the use of copyrighted material from Elsevier B.V.

possible to provide a natural antioxidant and nutraceutical food through further studies of its mechanism of action and *in vivo* assays.

The proteolytic activation of bioactive sequences by lactic acid bacteria has been debated recently due to the great advantage of using food-grade microorganisms to enrich foods with bioactive substances (Gobbetti *et al.*, 2002); however, limited applications are reported for fermented marine proteins. Wang *et al.* (2008) attempted to ferment *Acetes chinensis*, an underutilized shrimp species thriving in the Bohai Gulf of China, with lactic acid bacteria to produce a fermented shrimp sauce with high ACE-inhibitory activity. Three peptides with high ACE-inhibitory activity were isolated from the fermented shrimp sauce (Asp-Pro, Gly-Thr-Gly, and Ser-Thr). Previous studies have shown that di-, tri-, and tetra-peptides are more rapidly absorbed and reach a higher concentration in the blood than free amino acids; further, these short peptides are less susceptible to proteolytic enzymes (Chun *et al.*, 1996; Craft *et al.*, 1968; Matthews and Payne, 1980). Hence, it can be expected that Asp-Pro, Gly-Thr-Gly, and Ser-Thr will be readily absorbed by the alimentary canal, thereby reaching the peripheral target sites and producing their effect *in vivo*. The production of *A. chinensis* fermented by *L. fermentum* SM 605 may be useful as an antihypertensive component of functional foods or nutraceuticals.

The antioxidant activity of shrimp protein hydrolysate has been demonstrated (Suetsuna, 2000). Using the water extract from Mun Goong, a paste extracted from the cephalothorax of white shrimp (*Litopenaeus vannamei*), Binsan *et al.* (2008) revealed that the water extract contained antioxidative peptides with mass ranges of *m/z* 400–1000. A shrimp (*A. chinensis*) hydrolysate prepared with a crude protease from *Bacillus* sp. SM98011, containing oligopeptides with a molecular mass <3 kDa, exhibited antioxidant and ACE-inhibitory activity (He *et al.*, 2006).

Shrimp waste, containing 35–50% crude protein, is also an important source of bioactive molecules (Sachindra *et al.*, 2005); its antioxidative property was found to be improved by ensilaging with lactic acid (Sachindra and Bhaskar, 2008). This protein-rich liquor has beneficial biological functions due to the inherent protein hydrolysis that occurs during fermentation. The potential for producing functional bioactive peptides through enzymatic hydrolysis of shrimp by-products has also been suggested by several recent reports, for example, antioxidant peptides (Guerard *et al.*, 2007), ACE-inhibitory peptides (Cheung and Li-Chan, 2010; He *et al.*, 2006), and AMPs (Bartlett *et al.*, 2002).

A recent report by Huang *et al.* (2010) demonstrated the generation of an iron-binding peptide by the enzymatic hydrolysis of shrimp processing by-product. The MW of the hydrolysate ranged from 1 to 6 kDa, and the MW of the metal-binding peptides isolated from other proteins belong to this range, for example, a 1561-Da protein from hoki frame following pepsinolytic hydrolysis (Jung and Kim, 2007), a 3500-Da protein from hoki bone following carnivorous intestine crude proteinase hydrolysis (Jung *et al.*, 2005a), and a 1442-Da protein from Alaska pollock (*T. chalcogramma*) backbone following pepsinolytic hydrolysis (Jung *et al.*, 2006a). The infrared spectra of the iron-binding peptides and the iron–peptide complex shown in Fig. 5.6 suggest that the principal site of iron binding primarily corresponds to the carboxylate groups and, to a lesser extent, the peptide bonds. These attempts might facilitate the development of strong iron-binding peptides as natural functional additives in the food industry.

Meat-based products have been fermented with *Aspergillus* in an attempt to identify novel properties. Yin *et al.* (2005) reported that *Aspergillus oryzae* produces multiple enzymes and can hydrolyze minced mackerel. The present authors (Giri *et al.*, 2009a,b) also developed a marine fish meat-based functional paste by utilizing the traditional Japanese koji fermentation technique with improved food functionality and aroma attributes. Several trash fish, including horse mackerel, spotted mackerel, lizard fish, and squid meat, were utilized to produce a functional paste

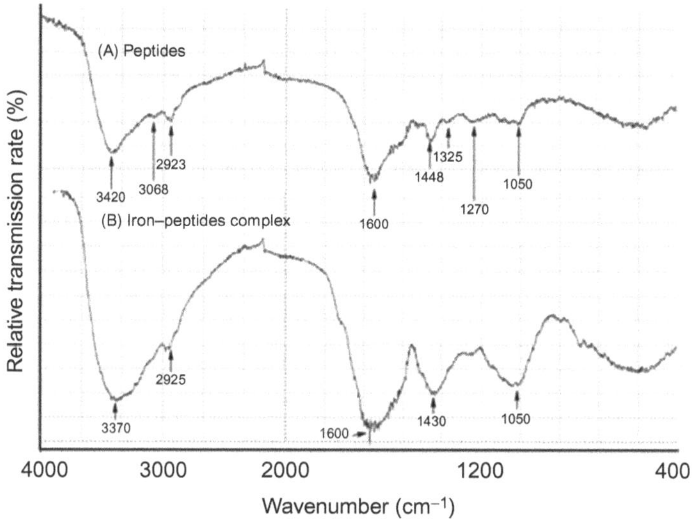

FIGURE 5.6 Infrared spectra of iron-binding peptides (A) and iron–peptide complex (B). *Source*: Huang *et al.* (2010). Permission has been obtained for the use of copyrighted material from Springer. (For color version of this figure, the reader is referred to the Web version of this chapter.)

from *A. oryzae*-inoculated koji. Analysis of several physicochemical parameters of the finished products, including free amino acid, oligopeptide, organic acid, and mineral content, revealed the potential utility of marine fish meat for the production of miso-like fermented fish pastes (Giri *et al.*, 2009a,b, 2010a,b). There are extensive reports on the nutritional value, taste, and aroma (Giri *et al.*, 2009a,b, 2010b) as well as the antioxidative properties (Giri *et al.*, 2011a) of fish miso. A fish miso preparation of rice malt koji inoculated with *A. oryzae* provides several proteolytic, lipolytic, and amylolytic enzymes; thus, hydrolyzed protein and carbohydrate substrates were efficiently produced from marine fish meat (Giri *et al.*, 2011b). Interestingly, the scavenging activity of aqueous fish miso extracts against DPPH, hydroxyl, NO, and carbon-centered radicals, estimated through ESR, increased with the prolongation of fermentation (Fig. 5.7). These time course observations indicated that the substrate responsible for radical scavenging developed during the process of fermentation. Using the online high-performance liquid chromatography-DPPH method, the production of peptides with radical-scavenging activity and their molecular mass distributions were estimated (Fig. 5.8). As fermentation proceeded (>60 days), peptides with a low molecular mass (1.45 kDa) and radical-scavenging ability developed, indicating the involvement of those peptides in the improved antioxidative properties of fish miso. Scavenging activity increased remarkably after 270 days due to the

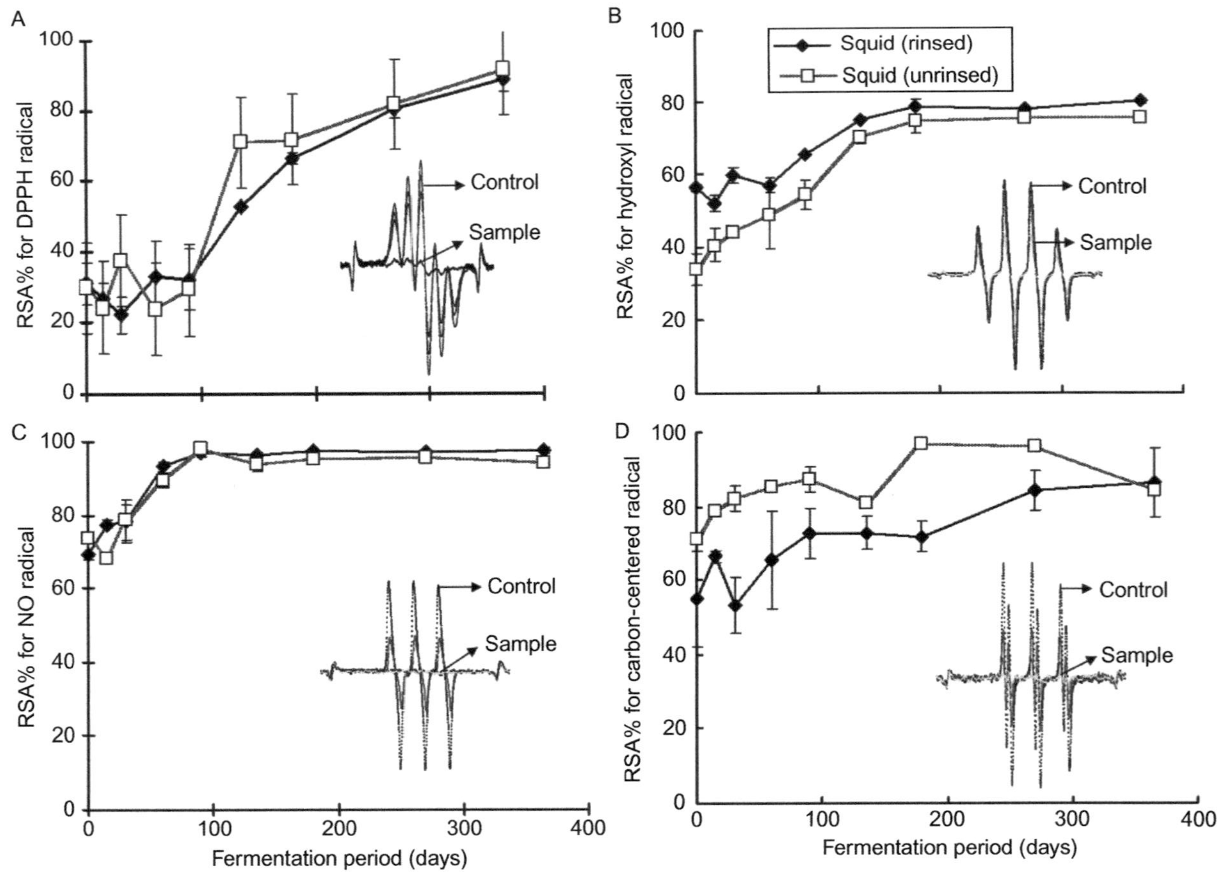

FIGURE 5.7 Changes in the levels of DPPH (A), hydroxyl (B), NO (C), and carbon-centered (D) radical-scavenging activity of miso prepared from rinsed and unrinsed squid meat at different points of the fermentation period (the ESR signal patterns for the control and sample are inset for all types of radicals). *Source*: Giri *et al.* (2011a). Permission has been obtained for the use of copyrighted material from Elsevier B.V.

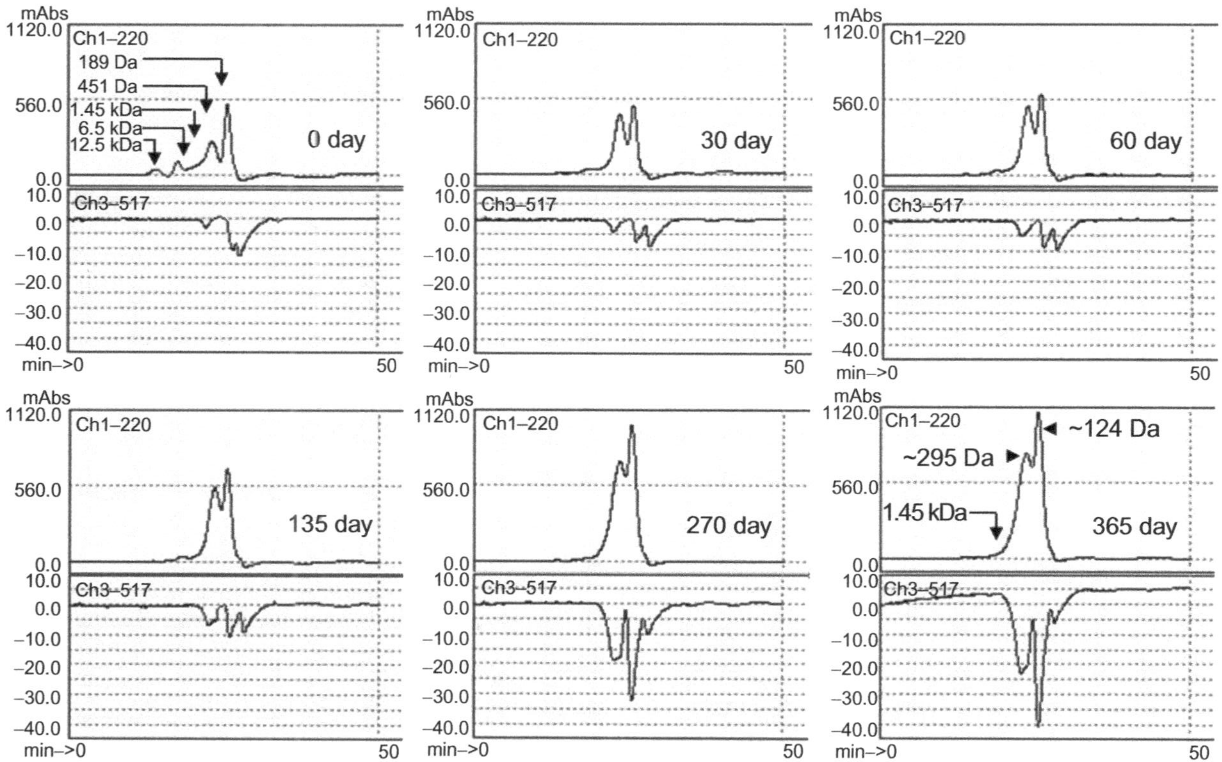

FIGURE 5.8 Changes in the molecular mass distribution of peptides and the radical-scavenging capacity of miso extracts prepared from unrinsed squid meat at different points in the fermentation period measured with an online HPLC-DPPH system. *Source*: Giri *et al.* (2011a). Permission has been obtained for the use of copyrighted material from Elsevier B.V. (For color version of this figure, the reader is referred to the Web version of this chapter.)

development of peptides with molecular masses below ~500 Da. Changes in the free amino acid content of fish miso indicated that the levels of free amino acids, including Asp/Asn, Glu/Gln, Ala, and Leu, increased rapidly over the 270 days of fermentation (Table 5.3). The levels of oligopeptide amino acids (Table 5.4) also increased during this time period, demonstrating the development of low MW peptides as fermentation proceeded. The oligopeptides were mainly comprised of Glu/Gln, Pro, Gly, Ala, Val, Lys, and Leu. The presence of peptides with enhanced hydrophobicity (derived from many protein sources) is linked to antioxidative properties (Chen *et al.*, 1995a). An increase in hydrophobicity increases the affinity of peptides to lipids and, therefore, enhances their antioxidative properties (Rajapakse *et al.*, 2005a), suggesting that the hydrophobic amino acids, for example, Leu, Val, and Ala, in fish miso peptides might have promoted their radical-scavenging properties. These amino acids are effective as inhibitors of linoleic acid oxidation in the oil/water emulsion model system (Marcuse, 1962). In addition, the amino acid sequences of peptides strongly affect their antioxidative activity; the correct positioning of Glu/Gln, Leu, and His in antioxidative peptide sequences improves their radical-scavenging activity (Chen *et al.*, 1996). Decomposition of the imidazole ring in His is responsible for the strong radical-scavenging activity of this amino acid; thus, it was assumed that the radical-scavenging activity of fish miso resulted from the biophysical properties of some of the peptides that developed during fermentation.

VIII. CONCLUDING REMARKS

The marine-derived peptidic compounds discussed here were obtained from very different marine organisms, exhibited different chemical structures, and displayed a large variety of pharmacological effects on human health and nutrition. The possibilities of designing new functional foods and pharmaceuticals to support the reduction or regulation of diet-related chronic malfunctions are highly promising. Further, the processing of fish by-products as food proteins can be easily utilized for the production of bioactive peptides. This review suggests that due to their valuable biological functions with beneficial health effects, marine-derived bioactive peptides have the potential to be used as active ingredients for the preparation of various functional foods or nutraceutical and pharmaceutical products. Considering that the majority of the studies discussed in this chapter were conducted *in vitro* or in mouse model systems, further studies are needed to investigate the bioactivity of functional peptides in human subjects.

TABLE 5.3 Free amino acids and related cocompounds (mg/g dry extract) and mol% (in parenthesis) in squid *miso* during fermentation period

Amino acids	0 day	30 day	60 day	135 day	270 day	365 day
Taurine	38.05^{a} (25.74)	27.56^{b} (10.42)	24.08^{b} (8.66)	22.34^{b} (7.64)	20.99^{c} (7.35)	17.49^{c} (6.23)
Aspartic acid	4.07^{d,e} (2.59)	21.70^{b} (7.71)	30.67^{a,b} (10.37)	39.80^{a} (12.80)	42.63^{a} (14.03)	45.07^{a} (15.09)
Threonine	2.25^{d,e} (1.59)	7.18^{c,d} (2.85)	7.73^{d} (2.92)	8.16^{d} (2.93)	7.93^{d} (2.91)	7.31^{d} (2.73)
Serine	3.51^{d,e} (2.83)	12.25^{c} (5.51)	13.26^{c} (5.68)	14.15^{c,d} (5.76)	12.58^{c,d} (5.24)	10.41^{c,d} (4.41)
Glutamic acid	3.52^{d,e} (2.02)	24.50^{b} (7.87)	29.58^{a,b} (9.05)	36.88^{a} (10.73)	34.91^{b} (10.39)	27.53^{b} (8.34)
Proline	14.50^{c} (10.66)	14.07^{c} (5.78)	15.57^{c} (6.09)	17.73^{c} (6.59)	18.18^{b,c} (6.92)	14.68^{c} (5.68)
Glycine	3.17^{d,e} (3.58)	6.24^{d} (3.93)	7.22^{d} (4.33)	8.11^{d} (4.62)	8.02^{d} (4.68)	7.16^{d} (4.25)
Alanine	7.84^{d} (7.45)	22.31^{b,c} (11.84)	25.21^{b} (12.74)	27.56^{b} (13.24)	27.30^{b} (13.42)	25.15^{b} (12.58)
Valine	3.49^{d,e} (2.51)	10.34^{c} (4.17)	11.17^{c,d} (4.29)	12.06^{c,d} (4.40)	13.10^{c,d} (4.90)	12.35^{c,d} (4.70)
Cysteine	4.27^{d,e} (2.98)	2.86^{d,e} (1.11)	2.72^{e} (1.01)	2.68^{e} (0.94)	3.58^{e} (1.29)	15.72^{c} (5.78)
Methionine	2.52^{d,e} (1.42)	6.53^{d} (2.07)	5.83^{e} (1.75)	5.24^{e} (1.50)	5.94^{d,e} (1.74)	9.52^{c,d} (2.84)
Isoleucine	2.28^{d,e} (1.46)	6.65^{d} (2.39)	7.02^{d,e} (2.41)	7.25^{d,e} (2.36)	7.95^{d,e} (2.65)	9.47^{c,d} (3.21)
Leucine	28.00^{b} (18.01)	40.89^{a} (14.69)	38.05^{a,b} (13.01)	34.94^{a} (11.36)	25.42^{b,c} (8.45)	24.55^{b} (8.31)
Tyrosine	3.65^{d,e} (1.70)	6.73^{d} (1.75)	7.00^{d} (1.74)	7.17^{d} (1.69)	7.64^{d} (1.84)	8.00^{d} (1.96)
Phenylalanine	3.01^{d,e} (1.54)	6.38^{d} (1.82)	6.23^{d} (1.69)	6.44^{d} (1.67)	7.58^{d} (2.01)	6.65^{d,e} (1.79)
β-Alanine	0.13^{e} (0.12)	0.13^{e} (0.07)	0.12^{f} (0.06)	0.11^{f} (0.05)	0.09^{f} (0.04)	0.09^{f} (0.04)
4-Aminobutyric acid	2.27^{d,e} (1.86)	1.74^{d,e} (0.79)	1.54^{e,f} (0.67)	1.43^{e,f} (0.59)	1.47^{e,f} (0.62)	1.21^{e} (0.52)
Histidine	1.63^{d,e} (0.88)	2.27^{d,e} (0.69)	2.11^{e,f} (0.61)	1.85^{e,f} (0.51)	1.48^{e,f} (0.41)	1.39^{e} (0.40)
Ornithine	0.33^{e} (0.21)	0.35^{e} (0.12)	0.29^{f} (0.09)	0.84^{f} (0.27)	1.98^{e,f} (0.65)	4.05^{d,e} (1.36)
Lysine	8.06^{c} (4.15)	24.39^{b} (7.03)	24.35^{b} (6.68)	21.95^{b,c} (5.72)	16.06^{c} (4.28)	12.83^{c,d} (3.48)
Arginine	13.67^{c} (6.64)	26.85^{b} (7.29)	23.53^{b} (6.08)	18.51^{c} (4.55)	24.35^{b,c} (6.12)	24.32^{b} (6.22)

Different letters (*a–f*) represent significant differences at $P<0.05$.
Source: Giri *et al.* (2011a). Permission has been obtained for use of copyrighted material from Elsevier B.V.

TABLE 5.4 Amino acids and related compounds in oligopeptides (mg/g dry extract) and mol% (in parenthesis) in squid *miso* during fermentation period

Amino acids	0 day	30 day	60 day	135 day	270 day	365 day
Taurine	7.24^{c} (3.78)	10.03^{d} (3.32)	13.33^{d} (3.97)	13.52^{d} (4.01)	$9.60^{d,e}$ (2.60)	12.66^{e} (3.17)
Aspartic acid/asparagine	13.69^{b} (6.73)	$17.08^{c,d}$ (5.33)	14.91^{d} (4.18)	$9.71^{d,e}$ (2.70)	$10.49^{d,e}$ (2.67)	10.56^{e} (2.49)
Threonine	6.26^{c} (3.43)	10.46^{d} (3.64)	12.50^{d} (3.91)	13.39^{d} (4.17)	13.45^{d} (3.83)	14.05^{e} (3.70)
Serine	$8.16^{b,c}$ (5.08)	9.23^{d} (3.65)	$10.68^{d,e}$ (3.79)	$11.23^{d,e}$ (3.96)	$11.12^{d,e}$ (3.59)	12.89^{e} (3.85)
Glutamic acid/glutamine	38.53^{a} (17.13)	64.25^{a} (18.14)	72.05^{a} (18.27)	72.70^{a} (18.34)	77.17^{a} (17.81)	88.60^{a} (18.90)
Proline	$8.63^{b,c}$ (4.90)	15.10^{c} (5.45)	12.33^{d} (3.99)	17.17^{d} (5.53)	27.82^{c} (8.20)	29.19^{c} (7.96)
Glycine	16.45^{b} (14.34)	$21.65^{b,c}$ (11.98)	22.69^{c} (11.28)	$23.19^{c,d}$ (11.46)	$20.88^{c,d}$ (9.44)	22.18^{d} (9.27)
Alanine	$10.63^{b,c}$ (7.80)	13.86^{d} (6.46)	15.10^{d} (6.32)	15.14^{d} (6.31)	14.76^{d} (5.62)	$18.44^{d,e}$ (6.49)
Valine	6.72^{c} (3.75)	12.72^{d} (4.51)	15.30^{d} (4.87)	16.86^{d} (5.34)	18.37^{d} (5.32)	$21.19^{d,e}$ (5.67)
Cysteine	5.48^{c} (2.95)	15.53^{d} (5.32)	14.78^{d} (4.55)	$4.02^{e,f}$ (1.23)	$12.85^{d,e}$ (3.60)	2.99^{f} (0.77)
Methionine	1.55^{d} (0.68)	$6.98^{e,f}$ (1.94)	$11.77^{d,e}$ (2.94)	$4.61^{e,f}$ (1.14)	1.76^{f} (0.40)	5.54^{f} (1.16)
Isoleucine	5.77^{c} (2.88)	12.18^{d} (3.85)	15.16^{d} (4.31)	17.33^{d} (4.90)	19.70^{d} (5.10)	20.31^{d} (4.86)
Leucine	16.02^{b} (7.96)	$19.24^{b,c}$ (6.06)	5.98^{e} (1.69)	$12.01^{d,e}$ (3.38)	$11.30^{d,e}$ (2.91)	12.16^{e} (2.89)
Tyrosine	$8.61^{b,c}$ (3.11)	13.58^{d} (3.11)	27.40^{c} (5.64)	29.04^{c} (5.95)	32.81^{c} (6.14)	46.34^{b} (8.02)
Phenylalanine	6.33^{c} (2.50)	1.86^{e} (0.46)	$10.27^{d,e}$ (2.32)	8.40^{e} (1.88)	$10.97^{d,e}$ (2.25)	11.50^{e} (2.18)
β-Alanine	$3.55^{c,d}$ (2.60)	$7.30^{d,e}$ (3.40)	8.47^{e} (3.55)	$9.54^{d,e}$ (3.97)	$10.86^{d,e}$ (4.13)	11.48^{e} (4.04)
4-Aminobutyric acid	0.31^{e} (0.19)	0.56^{f} (0.22)	0.99^{f} (0.35)	1.39^{f} (0.50)	2.22^{f} (0.73)	3.11^{f} (0.94)
Histidine	$3.22^{c,d}$ (1.35)	$6.73^{e,f}$ (1.80)	7.31^{e} (1.75)	7.39^{d} (1.76)	$5.93^{e,f}$ (1.29)	5.20^{f} (1.05)
Ornithine	$3.94^{c,d}$ (1.94)	$4.19^{e,f}$ (1.31)	$4.42^{e,f}$ (1.24)	$4.78^{d,e}$ (1.34)	$11.65^{d,e}$ (2.99)	$7.66^{e,f}$ (1.81)
Lysine	14.52^{b} (5.78)	31.72^{b} (8.02)	37.61^{b} (8.55)	40.71^{b} (9.20)	45.75^{b} (9.46)	47.81^{b} (9.14)
Arginine	$2.81^{c,d}$ (1.05)	8.29^{d} (1.97)	$11.50^{d,e}$ (2.46)	13.33^{d} (2.84)	$9.42^{d,e}$ (1.83)	8.61^{e} (1.55)

Different letters (*a–f*) represent significant differences at $P<0.05$.
Source: Giri *et al.* (2011a). Permission has been obtained for use of copyrighted material from Elsevier B.V.

REFERENCES

Achour, A., Lachgar, A., and Astgen, A. (1997). Potentialization of IL-2 effect on immune cells by oyster extract (JCOE) in normal and HIV-infected individuals. *Biomed. Pharmacother.* **51**, 427–429.

Alemán, A., Giménez, B., Pérez-Santin, E., Gómez-Guillén, M. C., and Monter, P. (2011). Contribution of Leu and Hyp residues to antioxidant and ACE-inhibitory activities of peptide sequences isolated from squid gelatin hydrolysate. *Food Chem.* **125**, 334–341.

Amarowicz, R. and Shahidi, F. (1997). Antioxidant activity of peptide fractions of capelin protein hydrolysates. *Food Chem.* **58**, 355–359.

Anderson, J. J. B. and Garner, S. C. (1996). Calcium and phosphorous nutrition in health and disease: Introduction. *In* "Calcium and Phosphorous in Health and Disease", (J. J. B. Anderson and S. C. Garner, Eds), pp. 1–5. CRC Press, New York.

Ariyoshi, Y. (1993). Angiotensin-converting enzyme inhibitors derived from food proteins. *Trends Food Sci. Technol.* **4**, 139–144.

Bai, R., Pettit, G. R., and Hamel, E. (1990). Binding of dolastatin 10 to tubulin at exchangeable nucleotide and Vinca alkaloid sites. *J. Biol. Chem.* **265**, 17141–17149.

Bai, R., Verdier-Pinard, P., Gangwar, S., Stessman, C. C., McKlure, K. J., Sausville, E. A., Pettit, G. R., Bates, R. B., and Hamel, E. (2001). Dolastatin 11, a marine depsipeptide, A arrests cells at cytokinesis and induces hyperpolymerization of purified actin. *Mol. Pharmacol.* **59**, 462–469.

Bartlett, T. C., Cuthbertson, B. J., Shepard, E. F., Chapman, R. W., Grops, P. S., and Warr, G. W. (2002). Crustins, homologues of an 11.5-kDa antibacterial peptide, from two species of penaeid shrimp, *Litopenaeus vannamei* and *Litopenaeus setiferus*. *Mar. Biotechnol.* **4**, 278–293.

Battison, A. L., Summerfield, R., and Patrzykat, A. (2008). Isolation and characterization of two antimicrobial peptides from haemocytes of the American lobster *Homarus americanus*. *Fish Shellfish Immunol.* **25**, 181–187.

Binsan, W., Benjakul, S., Visessanguan, W., Roytrakul, S., Tanaka, M., and Kishimura, H. (2008). Antioxidative activity of Mungoong, an extract paste, from the cephalothorax of white shrimp (*Litopenaeus vannamei*). *Food Chem.* **106**, 185–193.

Byun, H. G. and Kim, S. K. (2001). Purification and characterization of angiotensin I converting enzyme (ACE) inhibitory peptides from Alaska Pollack (*Theragra chalcogramma*) skin. *Process Biochem.* **36**, 1155–1162.

Cacciuttolo, M. A., Trinh, L., Lumpkin, J. A., and Rao, G. (1993). Hyperoxia induces DNA damage in mammalian cells. *Free Radic. Biol. Med.* **14**, 267–276.

Chen, H. M., Muramoto, K., and Yamauchi, F. (1995a). Structural analysis of antioxidative peptides from soybean β-conglycinin. *J. Agric. Food Chem.* **43**, 574–578.

Chen, J., Suetsuna, K., and Yamauchi, F. (1995b). Isolation and characterization of immunostimulative peptides from soybean. *J. Nutr. Biochem.* **6**, 310–313.

Chen, H., Muramoto, K., Yamauchi, F., and Nokihara, K. (1996). Antioxidant activity of designed peptides based on the antioxidative peptide isolated from digests of a soybean protein. *J. Agric. Food Chem.* **44**, 2619–2623.

Chen, H., Muramoto, K., Yamauchi, F., Fujimoto, K., and Nokihara, K. (1998). Antioxidative properties of histidine-containing peptides designed from peptide fragments found in the digests of a soybean protein. *J. Agric. Food Chem.* **46**, 49–53.

Cheung, I. W. Y. and Li-Chan, E. C. Y. (2010). Angiotensin-I-converting enzyme inhibitory activity and bitterness of enzymatically-produced hydrolysates of shrimp (*Pandalopsis dispar*) processing byproducts investigated by Taguchi design. *Food Chem.* **122**, 1003–1012.

Cheung, H. S., Wang, F. L., Ondetti, M. A., Sabo, E. F., and Cushman, D. W. (1980). Binding of peptide substrates and inhibitors of angiotensin-converting enzyme. *J. Biol. Chem.* **255**, 401–407.

Cho, S. S., Lee, H. K., Yu, C. Y., Kim, M. J., Seong, E. S., Ghimire, B. K., Son, E. H., Choung, M. G., and Lim, J. D. (2008). Isolation and characterization of bioactive peptides from Hwangtae (yellowish dried Alaska pollack) protein hydrolysate. *J. Food Sci. Nutr.* **13**, 196–203.

Chun, H., Sasaki, M., Fujiyama, Y., and Bamba, T. (1996). Effect of peptide chain length on absorption and intact transport of hydrolyzed soybean peptide in rat intestinal everted sac. *J. Clin. Biochem. Nutr.* **21**, 131–140.

Clare, D. A. and Swaisgood, H. E. (2000). Bioactive milk peptides: A prospectus. *J. Dairy Sci.* **83**, 1187–1195.

Craft, I. L., Geddes, D., Hyde, C. W., Wise, I. J., and Matthews, D. M. (1968). Absorption and malabsorption of glycine and glycine peptides in man. *Gut* **9**, 425–437.

Cushman, D. W., Cheung, H. S., Sabo, E. F., and Ondetti, M. A. (1981). Angiotensin-converting enzyme inhibitors: Evaluation of a new class of antihypertensive drugs. *In* "Angiotensin-Converting Enzyme Inhibitors: Mechanism of Action and Clinical Implications", (Z. P. Horovitz, Ed.), pp. 1–25. Urban and Schwarzenberg, Baltimore and Munich.

Davalos, A., Miguel, M., Bartolomé, B., and López-Fandiño, R. (2004). Antioxidant activity of peptides derived from egg white proteins by enzymatic hydrolysis. *J. Food Prot.* **67**, 1939–1944.

de Vries, D. J. and Beart, P. M. (1995). Fishing for drugs from the sea: Status and strategies. *Trends Pharmacol. Sci.* **16**, 275–279.

Defelice, S. L. (1995). The nutritional revolution: Its impact on food industry R&D. *Trends Food Sci. Technol.* **6**, 59–61.

Edward, N. J., Arroll, T. A., and Fan, Z. (1996). Specificity and some physicochemical characteristics of the antibacterial activity from blue crab *Callinectus sapidus*. *Fish Shellfish Immunol.* **6**, 403–413.

Elias, R. J., Kellerby, S. S., and Decker, E. A. (2008). Antioxidant activity of proteins and peptides. *Crit. Rev. Food Sci. Nutr.* **48**, 430–441.

Erdmann, K., Cheung, B. W. Y., and Schroder, H. (2008). The possible roles of food-derived bioactive peptides in reducing the risk of cardiovascular disease. *J. Nutr. Biochem.* **19**, 643–654.

Fahmi, A., Morimura, S., Guo, H. C., Shigematsu, T., Kida, K., and Uemura, Y. (2004). Production of angiotensin I converting enzyme inhibitory peptides from sea bream scales. *Process Biochem.* **39**, 1195–1200.

FAOSTAT, FAO Statistical Databases, Fisheries Data (2001). Food and Agriculture Organization of the United Nations, Rome, Italy. http://www.fao.org. Available from Internet, URL.

Fouchereau-Peron, M., Duvail, L., Michel, C., Gildberg, A., Batista, I., and Gal, Y. I. (1999). Isolation of an acid fraction from a fish protein hydrolysate with a calcitonin–gene-related-peptide-like biological activity. *Biotechnol. Appl. Biochem.* **29**, 87–92.

Fujita, H. and Yoshikawa, M. (1999). LKPNM: A prodrug-type ACE-inhibitory peptide derived from fish protein. *Immunopharmacology* **44**, 123–127.

Gerwick, W. H. (1987). Drugs from the Sea—The Search Continues. *J. Pharm. Tech.* (July/August), 136–141.

Gildberg, A., Bogwald, J., Johansen, A., and Stenberg, E. (1996). Isolation of acid peptide fractions from a fish protein hydrolysate with strong stimulatory effect on Atlantic salmon (*Salmo salar*) head kidney leucocytes. *Comp. Biochem. Physiol.* **11**, 97–101.

Giri, A., Osako, K., and Ohshima, T. (2009a). Effect of raw materials on the extractive component sand taste aspects of fermented fish paste: Sakana miso. *Fish. Sci.* **75**, 785–796.

Giri, A., Osako, K., and Ohshima, T. (2009b). Extractive components and taste aspects of fermented fish pastes and bean pastes prepared using different koji molds as starters. *Fish. Sci.* **75**, 481–489.

Giri, A., Osako, K., and Ohshima, T. (2010a). Identification and characterization of headspace volatiles of fish miso, a Japanese fish meat based fermented paste, with special emphasis on effect of fish species and meat washing. *Food Chem.* **120**, 621–631.

Giri, A., Osako, K., Okamoto, A., and Ohshima, T. (2010b). Olfactometric characterization of aroma active compounds in fermented fish paste in comparison with fish sauce, fermented soy paste and sauce products. *Food Res. Int.* **43**, 1027–1040.

Giri, A., Osako, K., Okamoto, A., Okazaki, E., and Ohshima, T. (2011a). Antioxidative properties of aqueous and aroma extracts of squid miso prepared with *Aspergillus oryzae*-inoculated koji. *Food Res. Int.* **44**, 317–325.

Giri, A., Osako, K., Okamoto, A., Okazaki, E., and Ohshima, T. (2011b). Effect of meat washing on the development of impact odorants in fish miso prepared from spotted mackerel. *J. Sci. Food Agric.* **91**, 850–859.

Gobbetti, M., Stepaniak, L., Angelis, M. D., Corsetti, A., and Cagno, R. D. (2002). Latent bioactive peptides in milk proteins: Proteolytic activation and significance in dairy processing. *Crit. Rev. Food Sci. Nutr.* **42**, 223–239.

Guerard, F., Sumaya-Martinez, M. T., Laroque, D., Chabeaud, A., and Dufosse, L. (2007). Optimization of free radical scavenging activity by response surface methodology in the hydrolysis of shrimp processing discards. *Process Biochem.* **42**, 1486–1491.

He, H., Chen, X., Sun, C., and Zhang, Y. (2004). Research progress in inhibitory peptides of angiotensin-converting enzyme. *China Biotechnol.* **24**, 7–11.

He, H., Chen, X., Sun, C., Zhang, Y., and Gao, P. (2006). Preparation and functional evaluation of oligopeptide-enriched hydrolysate from shrimp (*Acetes chinensis*) treated with crude protease from *Bacillus* sp SM98011. *Bioresour. Technol.* **97**, 385–390.

Hernandez-Ledesma, B., D'avalos, A., Bartolom'e, B., and Amigo, L. (2005). Preparation of antioxidant enzymatic hydrolysates from α-lactalbumin and β-lactoglobulin. Identification of active peptides by HPLC-MSMS. *J. Agric. Food Chem.* **53**, 588–593.

Huang, G., Ren, Z., and Jiang, J. (2010). Separation of iron-binding peptides from shrimp processing by-products hydrolysates. *Food Bioprocess Technol.* **4**, 1527–1532.

Ichimura, T., Hu, J., Aita, D. Q., and Maruyama, S. (2003). Angiotensin I-converting inhibitory activity and insulin secretion stimulative activity of fermented fish sauce. *J. Biosci. Bioeng.* **5**, 496–499.

Jayasankar, V. and Subramonium, T. (1999). Antibacterial activity of seminal plasma of the mud crab *Scylla serrata* (forskal). *J. Exp. Mar. Biol. Ecol.* **236**, 253–259.

Je, J. Y., Park, J. Y., Jung, W. K., Park, P. J., and Kim, S. K. (2005a). Isolation of angiotensin I converting enzyme (ACE) inhibitor from fermented oyster sauce, *Crassostrea gigas*. *Food Chem.* **90**, 809–814.

Je, J. Y., Park, P. J., Byun, H. K., Jung, W. K., and Kim, S. K. (2005b). Angiotensin I converting enzyme (ACE) inhibitory peptide derived from the sauce of fermented blue mussel, *Mytilus edulis*. *Bioresour. Technol.* **96**, 1624–1629.

Je, J. Y., Qian, Z. J., Byun, H. G., and Kim, S. K. (2007). Purification and characterization of an antioxidant peptide obtained from tuna backbone protein by enzymatic hydrolysis. *Process Biochem.* **42**, 840–846.

Je, J. Y., Qian, Z. J., Lee, S. H., Byun, H. G., and Kim, S. K. (2008). Purification and antioxidant properties of bigeye tuna (*Thunnus obesus*) dark muscle peptide on free radical-mediated oxidation systems. *J. Med. Food* **11**, 629–637.

Jeon, Y. J. and Kim, S. K. (2000). Continuous production of chitooligosaccharides using a dual reactor system. *Process Biochem.* **35**, 623–632.

Jo, H. Y., Jung, W. K., and Kim, S. K. (2008). Purification and characterization of a novel anticoagulant peptide from marine echiuroid worm, *Urechis unicinctus*. *Process Biochem.* **43**, 179–184.

Jun, S. Y., Park, P. J., Jung, W. K., and Kim, S. K. (2004). Purification and characterization of an antioxidative peptide from enzymatic hydrolysates of yellowfin sole (*Limanda aspera*) frame protein. *Eur. Food Res. Technol.* **219**, 20–26.

Jung, W. K. and Kim, S. K. (2007). Calcium-binding peptide derived from pepsinolytic hydrolysates of hoki (*Johnius belengerii*) frame. *Eur. Food Res. Technol.* **224**, 763–767.

Jung, W. K. and Kim, S. K. (2009). Isolation and characterization of an anticoagulant oligopeptide from blue mussel, *Mytilus edulis*. *Food Chem.* **117**, 687–692.

Jung, W. K., Je, J. Y., and Kim, S. K. (2001). A novel anticoagulant protein from *Scapharca broughtonii*. *J. Biochem. Mol. Biol.* **35**, 199–205.

Jung, W. K., Park, P. J., Byun, H. G., Moon, S. H., and Kim, S. K. (2005a). Preparation of hoki (*Johnius belengerii*) bone oligophosphopeptide with a high affinity to calcium by carnivorous intestine crude proteinase. *Food Chem.* **91**, 333–340.

Jung, W. K., Rajapakse, N., and Kim, S. K. (2005b). Antioxidative activity of a low molecular weight peptide derived from the sauce of fermented blue mussel, *Mytillus edulis*. *Eur. Food Res. Technol.* **220**, 535–539.

Jung, W. K., Karawita, R., Heo, S. J., Lee, B. J., Kim, S. K., and Jeon, Y. J. (2006a). Recovery of a novel Ca-binding peptide from Alaska pollack (*Theragra chalcogramma*) backbone by pepsinolytic hydrolysis. *Process Biochem.* **41**, 2097–2100.

Jung, W. K., Mendis, E., Je, J. Y., Park, P. J., Son, B. W., Kim, H. C., Choi, Y. K., and Kim, S. K. (2006b). Angiotensin-I-converting enzyme inhibitory peptide from yellowfin sole (*Limanda aspera*) frame protein and its antihypertensive effect in spontaneously hypertensive rats. *Food Chem.* **94**, 26–32.

Jung, W. K., Qian, Z. J., Lee, S. H., Choi, S. Y., Sung, N. J., Byun, H. G., and Kim, S. K. (2007). Free radical scavenging activity of a novel antioxidative peptide isolated from in vitro gastrointestinal digests of *Mytilus coruscus*. *J. Med. Food* **10**, 197–202.

Kim, G. H., Jeon, Y. J., Byun, H. G., Lee, Y. S., and Kim, S. K. (1998). Effect of calcium compounds from oyster shell bound fish skin gelatine peptide in calcium deficient rats. *J. Korean Fish. Soc.* **31**, 149–159.

Kim, S. K., Jeon, Y. J., Byun, H. G., and Park, P. J. (1999). Calcium absorption acceleration effect on phosphorylated and nonphosphorylated peptides from hoki (*Johnius belengeri*) frame. *J. Korean Fish. Soc.* **32**, 713–717.

Kim, S. K., Choi, Y. R., Park, P. J., Choi, J. H., and Moon, S. H. (2000). Screening of biofunctional peptides from cod processing wastes. *J. Korean Soc. Agric. Chem. Biotechnol.* **43**, 225–227.

Kim, S. K., Kim, Y. T., Byun, H. G., Nam, K. S., Joo, D. S., and Shahidi, F. (2001). Isolation and characterization of antioxidative peptides from gelatin hydrolysate of Allaska Pollack skin. *J. Agric. Food Chem.* **49**, 1984–1989.

Kim, D. C., Chae, H. J., and In, M. J. (2004). Existence of stable fibrin-clotting inhibitor in salt fermented anchovy sauce. *J. Food Compost. Anal.* **17**, 113–118.

Kim, S. Y., Je, J. Y., and Kim, S. K. (2007). Purification and characterization of antioxidant peptide from hoki (*Johnius balengerii*) frame protein by gastrointestinal digestion. *J. Nutr. Biochem.* **18**, 31–38.

Kim, E., Lee, S., Jeon, B., Moon, S., Kim, B., Park, T., Han, J., and Park, P. (2009). Purification and characterisation of antioxidative peptides from enzymatic hydrolysates of venison protein. *Food Chem.* **114**, 1365–1370.

Klompong, V., Benjakul, S., Yachai, M., Visessanguan, W., Shahidi, F., and Hayes, K. D. (2009). Amino acid composition and antioxidative peptides from protein hydrolysates of yellow stripe trevally (*Selaroides leptolepsis*). *J. Food Sci.* **74**, 126–133.

Korhonen, H. and Pihlanto-Leppala, A. (2003). Food-derived bioactive peptides: Opportunities for designing future foods. *Curr. Pharm. Des.* **9**, 1297–1308.

Koyama, T., Noguchi, K., Aniya, Y., and Sakanashi, M. (1998). Analysis for sites of anticoagulant action of plancinin, a new anticoagulant peptide isolated from the starfish *Acanthaster planci*, in the blood coagulation cascade. *Gen. Pharmacol.* **31**, 277–282.

Kuba, M., Tana, C., Tawata, S., and Yasuda, M. (2005). Production of angiotensin-I converting enzyme inhibitory peptides from soybean protein with *Monascus purpureus* acid proteinase. *Process Biochem.* **40**, 2191–2196.

Lahl, W. J. and Braun, S. D. (1994). Enzymatic production of protein by hydrolysates for food use. *Food Technol.* **48**, 68–71.

Lee, T. G. (1996). Functional peptides from hydrolysate of marine oyster proteins. Ph.D. thesis, Pukyong National University, Korea.

Lee, T. G. and Maruyama, S. (1998). Isolation of HIV-1 protease-inhibiting peptides from thermolysin hydrolysate of oyster proteins. *Biochem. Biophys. Res. Commun.* **253**, 604–608.

Lee, S. H., Qian, Z. J., and Kim, S. K. (2010). A novel angiotensin I converting enzyme inhibitory peptide from tuna frame protein hydrolysate and its antihypertensive effect in spontaneously hypertensive rats. *Food Chem.* **118**, 96–102.

Li, C., Haug, T., Styrvold, O. B., Jørgensen, T.Ø., and Stensvag, K. (2008). Strongylocins, novel antimicrobial peptides from the green sea urchin, *Strongylocentrotus droebachiensis*. *Dev. Comp. Immunol.* **32**, 1430–1440.

Li, Y. W., Li, B., He, J., and Qian, P. (2011). Structure–activity relationship study of antioxidative peptides by QSAR modeling: The amino acid next to C-terminus affects the activity. *J. Pept. Sci.* **17**, 454–462.

Liu, Z., Dong, S., Xu, J., Zeng, M., Song, H., and Zhao, Y. (2008). Production of cysteine-rich antimicrobial peptide by digestion of oyster (*Crassostrea gigas*) with alcalase and bromelin. *Food Control* **19**, 231–235.

Lun, H. H., Lan, C. X., Yun, S. C., Zhong, Z. Y., and Cheng, Z. B. (2006). Analysis of novel angiotensin-I-converting enzyme inhibitory peptides from protease-hydrolyzed marine shrimp *Acetes chinensis*. *J. Pept. Sci.* **12**, 726–733.

Ma, M. S., Bae, I. Y., Lee, H. G., and Yang, C. B. (2006). Purification and identification of angiotensin-I-converting enzyme inhibitory peptide from buckwheat. *Food Chem.* **96**, 36–42.

Madden, T., Tran, H. T., Beck, D., Huie, R., Newman, R. A., Pusztai, L., Wright, J. J., and Abbruzzese, J. L. (2000). Novel marine-derived anticancer agents: A phase I clinical, pharmacological, and pharmacodynamic study of dolastatin 10 (NSC 376128) in patients with advanced solid tumors. *Clin. Cancer Res.* **6**, 1293–1301.

Marcuse, R. (1962). The effect of some amino acids on the oxidation of linoleic acid and its methyl ester. *J. Am. Oil Chem. Soc.* **39**, 97–107.

Matsumoto, K., Ogikubo, A., Yoshino, T., Matsui, T., and Osajima, Y. (1994). Separation and purification of angiotensin I converting enzyme inhibitory peptide in peptic hydrolysate of oyster. *Nippon Shokuhin Kogyo Gakkaishi* **41**, 589–594. (in Japanese).

Matthews, D. M. and Payne, J. W. (1980). Transmembrane transport of small peptides. *J. Curr. Top. Membr. Transp.* **14**, 331–425.

Mendis, E., Rajapakse, N., and Kim, S. K. (2005a). Antioxidant properties of a radical-scavenging peptide purified from enzymatically prepared fish skin gelatin hydrolysate. *J. Agric. Food Chem.* **53**, 581 587.

Mendis, E., Rajapakse, N., Byun, H. G., and Kim, S. K. (2005b). Investigation of jumbo squid (*Dosidicus gigas*) skin gelatin peptides for their in vitro antioxidant effects. *Life Sci.* **77**, 2166–2178.

Miljanich, G. P. (1997). Venom peptides as human pharmaceuticals. *Sci. Med.* (Sept/Oct), 6–15.

Okamoto, A., Hanagata, H., Matsumoto, E., Kawamura, Y., Koizumi, Y., and Yanagida, F. (1995a). Angiotensin I converting enzyme inhibitory activities of various fermented foods. *Biosci. Biotech. Biochem.* **59**, 1147–1149.

Okamoto, A., Matsumoto, E., Iwashita, A., Yasuhara, T., Kawamura, Y., Koizumi, Y., and Yanagida, F. (1995b). Angiotensin I-converting enzyme inhibitory action of fish sauce. *Food Sci. Technol. Int.* **1**, 101–106.

Pan, D. D., Luo, Y. K., and Tanokura, M. (2005). Antihypertensive peptides from skimmed milk hydrolysate digested by cell-free extract of *Lactobacillus helveticus*. *Food Chem.* **91**, 123–129.

Pettit, G. R., Kamano, Y., Fuji, Y., Herald, C. L., Inoue, M., Brown, P., Gust, D., Kitahara, K., Shmidt, J. M., Doubek, D. L., and Michel, C. (1981). Marine animal biosynthetic constituents for cancer chemotherapy. *J. Nat. Prod.* **44**, 482–485.

Pettit, G. R., Kamano, Y., Herald, C. L., Dufresne, C., Cerny, R. L., Herald, D. L., Schmidt, J. M., and Kizu, H. (1989). Isolation and structure of the cytostatic depsipeptide dolastatin 13 from the sea hare *Dolabella auricularia*. *J. Am. Chem. Soc.* **111**, 5015–5017.

Petitt, R., Hogan, F., Xu, J. P., Tan, R., Nogawa, T., Cichacz, Z., Pettit, R. K., Du, J., Ye, Q. H., Cragg, G. M., Herald, C. L., *et al.* (2008). Antineoplastic Agents. 536. New Sources of Naturally Occurring Cancer Cell Growth Inhibitors from Marine Organisms, Terrestrial Plants, and Microorganisms. *J. Nat. Prod.* **71**(3), 438–444.

Pihlanto-Leppala, A. (2001). Bioactive peptides derived from bovine proteins: Opioid and ACE-inhibitory peptides. *Trends Food Sci. Technol.* **11**, 347–356.

Qian, Z. J., Je, J. Y., and Kim, S. K. (2007). Antihypertensive effect of angiotensin I converting enzyme-inhibitory peptide from hydrolysates of bigeye tuna dark muscle, Thunnus obesus. *J. Agric. Food Chem.* **55**, 8398–8403.

Qian, Z. J., Jung, W. K., Byun, H. G., and Kim, S. K. (2008). Protective effect of an antioxidative peptide purified from gastrointestinal digests of oyster, *Crassostrea gigas* against free radical induced DNA damage. *Bioresour. Technol.* **99**, 3365–3371.

Rajapakse, N., Jung, W. K., Mendis, E., Moon, S. H., and Kim, S. K. (2005a). A novel anticoagulant purified from fish protein hydrolysate inhibits factor XIIa and platelet aggregation. *Life Sci.* **76**, 2607–2619.

Rajapakse, N., Mendis, E., Byun, H. G., and Kim, S. K. (2005b). Purification and *in vitro* antioxidative effects of giant squid muscle peptides on free radical-mediated oxidative systems. *J. Nutr. Biochem.* **16**, 562–569.

Rajapakse, N., Mendis, E., Jung, W. K., Je, J. Y., and Kim, S. K. (2005c). Purification of a radical scavenging peptide from fermented mussel sauce and its antioxidant properties. *Food Res. Int.* **38**, 175–182.

Ranathunga, S., Rajapakse, N., and Kim, S. K. (2006). Purification and characterization of antioxidantative peptide derived from muscle of conger eel (*Conger myriaster*). *Eur. Food Res. Technol.* **222**, 310–315.

Roberts, P. R., Burney, J. D., Black, K. W., and Zaloga, G. P. (1999). Effect of chain length on absorption of biologically active peptides from the gastrointestinal tract. *Digestion* **60**, 332–337.

Sachindra, N. M. and Bhaskar, N. (2008). *In vitro* antioxidant activity of liquor from fermented shrimp biowaste. *Bioresour. Technol.* **99**, 9013–9016.

Sachindra, N. M., Bhaskar, N., and Mahendrakar, N. S. (2005). Carotenoids in different body components of Indian shrimps. *J. Sci. Food Agric.* **85**, 167–172.

Saito, K., Jin, D., Ogawa, T., Muramoto, K., Hatakeyama, E., Yasuhara, T., and Nokihara, K. (2003). Antioxidative properties of tripeptide libraries prepared by the combinatorial chemistry. *J. Agric. Food Chem.* **51**, 3668–3674.

Samaranayaka, A. G. P. and Li-Chan, E. C. Y. (2008). Autolysis-assisted production of fish protein hydrolysates with antioxidant properties from Pacific hake (*Merluccius productus*). *Food Chem.* **107**, 768–776.

Sampath, K. N. S., Nazeer, R. A., and Jaiganesh, R. (2011). Purification and identification of antioxidant peptides from the skin protein hydrolysate of two marine fishes, horse mackerel (*Magalaspis cordyla*) and croaker (*Otolithes ruber*). *Amino Acids.* doi: 10.1007/s00726-011-0858-6.

Seki, E., Osajima, K., Matsui, T., and Osajima, Y. (1993). Separation and purification of angiotensin I converting enzyme inhibitory peptides from heated sardine meat by treatment with alkaline protease. *Nippon Shokuhin Kogyo Gakkaishi* **40**, 783–791. (in Japanese).

Shahidi, F. (2007). Antioxidants from marine by-products. *In* ''Maximising the Value of Marine By-products'', (F. Shahidi, Ed.), pp. 397–412. CRC Press, USA.

Shahidi, F. and Zhong, Y. (2008). Bioactive peptides. *J. AOAC Int.* **91**, 914–931.

Slizyte, R., Mozuraityte, R., Martı'nez-Alvarez, O., Falch, E., Fouchereau-Peron, M., and Rustad, T. (2009). Functional, bioactive and antioxidative properties of hydrolysates obtained from cod (*Gadus morhua*) backbones. *Process Biochem.* **44**, 668–677.

Stensvag, K., Haug, T., Sperstad, S. V., Rekdal, O., Indrevoll, B., and Styrvold, O. B. (2008). Arasin 1, a proline-arginine-rich antimicrobial peptide isolated from the spider crab, *Hyas araneus*. *Dev. Comp. Immunol.* **32**, 275–285.

Suetsuna, K. (2000). Antioxidant peptides from the protease digest of prawn (*Penaeus Japonicus*) muscle. *Mar. Biotechnol.* **2**, 5–10.

Suetsuna, K. (2002). Identification of antihypertensive peptides from peptic digest of the short-necked clam tapes philippinarum and the pearl oyster pinctada fucata martensii. *Fish Sci.* **68**, 233–235.

Thiansilakul, Y., Benjakul, S., and Shahidi, F. (2007). Antioxidative activity of protein hydrolysate from round scad muscle using alcalase and flavourzyme. *J. Food Biochem.* **31**, 266–287.

Tincu, J. A. and Taylor, S. W. (2004). Antimicrobial peptides from marine invertebrates. *Antimicrob. Agents Chemother.* **48**, 3645–3654.

Tsai, J. S., Lin, T. C., Chen, J. L., and Pan, B. S. (2006). The inhibitory effects of freshwater clam (*Corbicula fluminea*, Muller) muscle protein hydrolysates on angiotensin I converting enzyme. *Process Biochem.* **41**, 2276–2281.

Tsai, J. S., Chen, J. L., and Pan, B. S. (2008). ACE-inhibitory peptides identified from the muscle protein hydrolysate of hard clam (*Meretrix lusoria*). *Process Biochem.* **43**, 743–747.

Tsuge, N., Eikawa, Y., Nomura, Y., Yamamoto, M., and Sugisawa, K. (1991). Antioxidative activity of peptides prepared by enzymatic hydrolysis of egg-white albumin. *Nippon Nogeikagaku Kaishi* **65**, 1635–1641.

Wang, Y. K., He, H. L., Chen, X. L., Sun, C. Y., Zhang, Y. Z., and Zhou, B. C. (2008). Production of novel angiotensin I-converting enzyme inhibitory peptides by fermentation of marine shrimp *Acetes chinensis* with *Lactobacillus fermentum* SM 605. *Appl. Microbiol. Biotechnol.* **79**, 785–791.

Wesson, K. J. and Hamann, M. T. (1996). Keenamide A, a bioactive cyclic peptide from the marine mollusk *Pleurobranchus forskalii*. *J. Nat. Prod.* **59**, 629–631.

Wijesekara, I., Qian, Z. J., Ryu, B. M., Ngo, D. H., and Kim, S. K. (2011). Purification and identification of antihypertensive peptides from seaweed pipefish (Syngnathus schlegeli) muscle protein hydrolysate. *Food Res. Int.* **44**(3), 703–707.

Wong, A. H. K. and Mine, Y. (2004). Novel fibrinolytic enzyme in fermented shrimp paste, a traditional Asian fermented seasoning. *J. Agric. Food Chem.* **52**, 980–986.

Yamamoto, N. (1997). Antihypertensive peptides derived from food proteins. *Biopolymers* **43**, 129 131.

Yang, R., Zhang, Z., Pei, X., Han, X., Wang, J., Wang, L., Long, Z., Shen, X., and Li, Y. (2009). Immunomodulatory effects of marine oligopeptide preparation from chum salmon (*Oncorhynchus keta*) in mice. *Food Chem.* **113**, 464–470.

Yin, L. J., Tong, Y. L., and Jiang, S. T. (2005). Effect of combining proteolysis and lactic acid bacterial fermentation on the characteristics of minced mackerel. *J. Food Sci.* **70**, 186–192.

Zeng, M., Cui, W., Zhao, Y., Liu, Z., Dong, S., and Guo, Y. (2008). Antiviral active peptide from oyster. *Chin. J. Oceanol. Limnol.* **26**, 307–312.

Zhao, Y., Li, B., Liu, Z., Dong, S., Zhao, X., and Zeng, M. (2007). Antihypertensive effect and purification of an ACE inhibitory peptide from sea cucumber gelatin hydrolysate. *Process Biochem.* **42**, 1586–1591.

Zhao, Y., Bafang, L., Dong, S., Liu, Z., Zhao, X., Wang, J., and Zeng, M. (2009). A novel ACE inhibitory peptide isolated from *Acaudina molpadioidea* hydrolysate. *Peptides* **30**, 1028–1033.

CHAPTER **6**

Chitosan and Fish Collagen as Biomaterials for Regenerative Medicine

Yoshihiko Hayashi,[*,1] **Shizuka Yamada,**[*] **Kajiro Yanagi Guchi,**[†] **Zenya Koyama,**[†] and **Takeshi Ikeda**[†]

Contents

Abstract

This chapter focuses and reviews on the characteristics and biomedical application of chitosan and collagen from marine products and advantages and disadvantages of regeneration medicine. The understanding of the production processes of chitosan and collagen and the conformation of these biomaterials are indispensable for promoting the theoretical and practical availability. The initial inflammatory reactions associated with chitosan application to

[*] Department of Cariology, Graduate School of Biomedical Sciences, Nagasaki University, Nagasaki, Japan
[†] Department of Cariology, Nagasaki University Hospital, Nagasaki, Japan
[1] Corresponding author: Yoshihiko Hayashi, *E-mail address:* hayashi@nagasaki-u.ac.jp

Advances in Food and Nutrition Research, Volume 65
ISSN 1043-4526, DOI: 10.1016/B978-0-12-416003-3.00006-8

hard and soft tissues need to be controlled before it can be considered for clinical application as scaffold. Further, as chitosan takes too long for biodegradation *in vivo*, generally it is not suitable for the scaffold for degenerative medicine in especially dental pulp tissue. The collagen extract from the scales of tropical fish has been reported to have a degeneration temperature of 35 °C. The properties of biocompatibility and biodegradation of fish atelocollagen are suitable for the scaffold in regenerative medicine.

I. INTRODUCTION

The regenerative medicine consists of three components: cell, nutrient, and scaffold. The combinatory usage of these components is important. For the scaffold manufacturing, bioactive natural organic materials originated from marine products are indispensable because the severe inflection problems such as bovine spongiform encephalopathy, avian and swine influenzas, and tooth-and-mouth disease in bovine, pig, and buffalo occur all over the world.

Chitin is mainly contained in the shells of crabs and shrimps. Chitosan is produced by the deacetylation of chitin, and it has numerous pharmacological activities, such as immunopotentiation, antihypertensive, serum cholesterol-lowering, antibacterial, and wound-healing-promoting properties (Asaoka, 1996; Koide, 1998). These biological activities would be favorable for scaffold fabrication. Further, marine collagen from fish scales, skin, and bone has been widely investigated to apply as a scaffold and a carrier because of its bioactive properties such as its excellent biocompatibility, low antigenicity, high biodegradability, and cell growth potential (Dillow and Lowman, 2002; Yang *et al.*, 2001).

This chapter describes the characteristics and biomedical application of chitosan and collagen from marine resources with the advantages and disadvantages in relation to regenerative medicine. The understandings of the production of those materials and their biological activities are very important for promoting them as potential source in regenerative medicine.

II. GENERAL PROPERTIES OF SCAFFOLD FOR REGENERATIVE MEDICINE

The basic principle of tissue engineering is that cells, genes, and proteins are delivered via a degradable material, termed as scaffold, to regenerate tissue. This concept was first elucidated by Langer, Vacanti, Griffith, and their colleagues (Cima *et al.*, 1991, Langer and Vacanti, 1993, 1999; Langer *et al.*, 1990). They laid out the basic requirements for the scaffold, as (1) choosing a material for a support matrix that was biocompatible and could be readily processed into desired shapes, (2) characterizing cell interaction with the material based on the tissue structural and metabolic demands, and (3) evaluating the performances of the matrices *in vitro* and *in vivo* through quantitative molecular and histological assays. These principles laid the foundation for tissue engineering scaffold research and development.

The scaffold functions to provide structural integrity, define a potential space for the engineered tissue, guide the restructuring that occurs through the proliferation of the donor cells and in-growth of the host tissue, maintain distances between the parenchymal cells that permit diffusion of the gas and nutrients and possibly the in-growth of vasculature from the host bed, and transmit the tissue-specific mechanical forces to cue the behavior of the cells within it (Marler *et al.*, 1998). Based on these functions, the sponge form is most suitable and reasonable for the scaffold structure (Madihally and Matthew, 1999).

Beyond knowing what parameters can influence tissue regeneration, it is difficult to know what quantitative measures can be used to characterize these regeneration-enhancing parameters. Three scaffold-design parameters are accepted as influencing tissue regeneration: (i) modification of scaffold surfaces to enhance cell interaction, (ii) controlled release of growth factors from scaffolds, and (iii) scaffold mass transport (Hollister, 2009).

Enhancing tissue regeneration by controlling cell-scaffold interaction and the necessity to accommodate cellular metabolic demands through scaffold diffusivity were two fundamental scaffold-design requirements enunciated in the early 1990s (Cima *et al.*, 1991; Langer *et al.*, 1990).

Scaffold mass transport can be characterized by scaffold diffusivity and permeability. As with mechanical properties, native tissue diffusivity and permeability can be regarded as a starting point for defining scaffold-transport design targets (Hollister, 2009). One of the major effects of designed diffusivity and permeability is to affect oxygen diffusion to cells and regenerative tissues. Partial oxygen pressure is a factor clearly affected by scaffold mass-transport characteristics that can affect cell differentiation. Most studies on differentiation of progenitor cells or behavior of fully differentiated cells reflect required permeability and diffusivity values (Domm *et al.*, 2002; Malda *et al.*, 2004).

III. CHEMICAL AND PHYSICAL PROPERTIES OF SCAFFOLD

A. Chitosan

1. Molecular weight and degree of deacetylation

The term chitosan describes a series of chitosan polymers with different molecular weight (MW), viscosity, and degree of deacetylation (DD) (40–98%). It is a linear polyamine with a number of amino groups that are readily available for chemical reaction and salt formation with acids.

Important characteristics of chitosan are its MW, viscosity, DD (Bodek, 1994; Ferreira *et al.*, 1994a,b), crystallinity index, number of monomeric units, water retention value, pKa, and energy of hydration (Kas, 1997). Chitosan has a high charge density, adheres to negatively charged surfaces, and chelates metal ions.

The MW of chitosan is affected by deproteinization conditions used for the isolation of the chitinous substrate. It is difficult to prepare a chitosan with a DD higher than 90% without significant degradation of polysaccharide molecules. It was also reported that the relationship of attachment and growth of any cells with a percentage of acetylation of chitosan followed a general trend with the higher deacetylated chitosan supporting attachment and subsequent growth of the cultured cells (Prasitsilp *et al.*, 2000). The DD and MW of chitosan influence the properties important for many applications, such as solubility of the product in dilute acids, viscosity of the obtained solutions, and their biological activity.

Generally, the MW of chitosan exerts a major influence on its biological and physicochemical characteristics. The potential of chitosan (MW <5000–10,000; DD 55.3–65.4%) in gene delivery was investigated by Richardson *et al.* (1999). Chitosan molecular mass fraction was observed to readily complex DNA even down to a chitosan:DNA charge ratio of 1:0.1, which also resulted in a significant decrease in the degradation by DNase II with no degradation being apparent at a charge ratio of 1:1. Gene delivery studies have shown that siRNA transfection efficiency can be modulated by the MW of chitosan, and since the MW affects polymer chain entanglement, this in turn influences its complexing ability with negatively charged siRNA. For instance, high MW chitosan will entangle siRNA more readily than low MW chitosan, which results in binding siRNA more efficiently and protecting the condensed siRNA from enzymatic degradation and serum components.

Liu *et al.* (2007) found that size, zeta potential, morphology, and complex stability as well as *in vitro* gene silencing of chitosan/siRNA nanoparticles were dependent on MW and DD. High MW and high DD samples produced stable nanoparticles, while those prepared with low MW (10kDa) and an nitrogen/phosphorus (N/P) charge ratio of 50 showed almost no knockdown of endogenous EGFP in H1299 human lung carcinoma cells. On the other hand, nanoparticles prepared from MWs in the range of 65–170kDa and a DD of 80% showed a gene-silencing efficiency between 45% and 65%. The highest gene-silencing efficiency of 80% was achieved when using an N/P ration of 150 for MWs of 114 and 170kDa having a DD of 84%.

Fernandez-Urrusuno *et al.* (1999) investigated the potential of chitosan (MW <50,000–130,000; DD 70–87%) nanoparticles as a system for improving the systemic absorption of insulin following nasal instillation. Nanoparticles prepared by ionotropic gelation with tripolyphosphate enhanced the nasal absorption of the peptide to a greater extent than that of aqueous solution of chitosan in a conscious rabbit model by monitoring the plasma glucose levels. The amount and MW of chitosan did not have a significant effect on insulin release.

LeHoux and Grondin (1993) investigated the effects of chitosan on plasma and liver cholesterol levels, liver weight, and 3-hydroxy-3-methylglutaryl coenzyme A reductase in rats fed on a sterol diet (1% cholesterol and 0.2% cholic acid). Chitosan at a level of 5% lowered plasma and liver cholesterol levels by 54% and 64%, respectively. High MW chitosan (>750 kDa) had less hypocholesterolemic potential than that of 70 kDa.

2. Cross-linking with bioactive agent

Growth factors including platelet-derived growth factor-BB, insulin-like growth factor, and transforming growth factor-β function as modulators to promote wound healing, cell proliferation, and bone regeneration (Hollinger, 1993; Park *et al.*, 1998). Typical cross-linker binding chitosan polymer with these growth factors is tripolyphosphate pentasodium at 5% (Lee *et al.*, 2000; Park *et al.*, 2000). This chemical cross-linking procedure brings the development of drug delivery system using chitosan scaffold.

3. Mechanical strength

Mechanical properties of scaffold can be evaluated similar to those of biomaterials in medicine and dentistry.

i. Tensile test Measurement is conducted after swelling with phosphate-buffered saline (PBS) as described by Tomihata and Ikada (1997) and Chen *et al.* (2009). The scaffold is subjected to tensile test using an Autograph (cross-head speed: 10 mm/min). The tensile strength is calculated as the breaking load divided by the initial cross-sectional area. A 250–400 g/mm^2 of tensile strength is reported in chitosan with a DD higher than 60%.

ii. Compression test Scaffold is rehydrated in deionized water before the test. The test is performed on a special machine as described by Subramanian and Lin (2005). The cross-head speed is 2 mm/min and a 50-kgf load cell is used. The load (kgf)–displacement (mm) data are recorded by the computer software and converted to stress–strain curves to obtain elastic modulus (kPa). The cross-linked scaffolds have about 2–5 times higher elastic modulus (7.4–19.9 kPa) compared to uncross-linked scaffolds (3.8 kPa).

iii. Load–displacement test The scaffold is evaluated in the dry state by the depth-sensing indentation approach using a nanoindenter (Depan *et al.*, 2011). A maximum load of 0.15 mN is set, and 15 indents are made at 35-μm intervals for each sample. The maximum indentation depth is set to 1000 nm. The load–displacement data are recorded continuously through one complete cycle of loading and unloading. Young's modulus is 0.06–0.1 GPa in chitosan and derivatives.

B. Fish collagen

1. Amino acids composition

Biochemical composition of marine collagen is thought to be different from that of mammalian collagen. For biochemical analyses, the strict condition for sample preservation is important and indispensable before collagen extraction. This means that the hydroxyproline content in relation to collagen stability strongly depends on these sampling procedures (Swatschek *et al.*, 2002). Several works showed that amino acid composition of fish collagens was almost similar to those of mammalian collagens (Bae *et al.*, 2008; Kimura, 1983; Kimura *et al.*, 1988; Nagai *et al.*, 2001, 2004). Glycine was the most abundant amino acid and accounted for more than 30% of all amino acids. Further, the degree of hydroxylation of proline was calculated to be 40–48%, which was also similar level to that of the mammalian (about 45%). The linear relationship between collagen stability and hydroxyproline content was demonstrated by that the degree of hydroxylation (Table 6.1) of proline in the fish collagen peptides was calculated to be about 35% (personal communication from the laboratory of Prof. M. Yamauchi, Oral Health Institute, University of North Carolina). Further, it is very interesting that the degree of hydroxylation of proline of fish in cold sea, for example, chum salmon, was reported to be low (35–37%) (Kimura *et al.*, 1988; Matsui *et al.*, 1991) compared to that of fish in relatively warm sea, which is related to the denaturation temperature (Table 6.1).

2. Degeneration temperature

Fish collagen fibrillar gels have not been studied, with the exception of shark collagen (Nomura *et al.*, 2000a,b), probably due to their low denaturation temperature (T_d), which renders these materials difficult to handle. The T_d of shark collagen solution is approximately 30 °C (Nomura *et al.*, 1995), which results in the dissolution of the fibrillar gel of this collagen at 37 °C (Nomura *et al.*, 2000a). This indicates that the gel could not be practically used at the actual physical temperature of human medical application. The T_d of chum salmon is approximately 19 °C (Kimura *et al.*, 1988; Matsui *et al.*, 1991), which is the main reason to be unstable at the actual physical temperature of human body. As the

TABLE 6.1 Degree of hydroxylation

Fish	%
Squid	47.8
Carp	43.3
Eel	40.2
Common mackerel	41.1
Saury	40.5
Chum salmon	38.0
Tilapia	43.0
Tiger puffer	34.5
Dusky spinefoot	37.6
Sea chubs	40.4
Eagle ray	41.6
Red stingray	46.9
Yantai stingray	40.6

denaturation temperature of fish collagen is lower than the mammalian body temperature, fish collagen melts when placed in contact with the human body for a clinical application. T_d of collagen extracted from ray skin or the scales of a tropical fish (tilapia) has been reported between 33–34 °C (Bae *et al.*, 2008) and 35 °C (Ikoma *et al.*, 2003), respectively. Further, the improvement can be achieved by chemical cross-linking *in vitro* collagen fibrillogenesis. This method brings T_d of salmon collagen (SC) to 55 °C, and its biocompatible properties have been demonstrated by several studies (Nagai *et al.*, 2004, 2007).

3. Cross-linking for stability

Numerous attempts have been recently made to use type I collagen for biomaterials. The cross-linking methods for stabilization of collagen are divided into physical treatment such as ultraviolet irradiation (Weadock *et al.*, 1995), dehydrothermal (DHT) treatment (Gorham *et al.*, 1992; Koide *et al.*, 1993; Pieper *et al.*, 1999; Wang *et al.*, 1994; Weadock *et al.*, 1995), and treatments involving chemical such as glutaraldehyde (White *et al.*, 1973), carbodiimide (Pieper *et al.*, 1999), and 1-ethyl-3-(3-dimethyl-aminopropyl)-carbodiimide (Yunoki *et al.*, 2004). Chemical treatments confer remarkably high strength and stability to the collagen matrix, but may result in potential cytotoxicity or poor biocompatibility (Huang-Lee *et al.*, 1990), while physical treatments have no cytotoxicity and can provide sufficient stability (Koide *et al.*, 1993; Yunoki *et al.*, 2003).

4. Mechanical strength

Compression test is a typical method for measuring mechanical strength of collagen (Lee *et al.*, 2001; Lyons *et al.*, 2010; Sugiura *et al.*, 2009). The collagen from sponges is soaked in PBS, immediately degassed, and subjected to test. The specimens are compressed using a cylindrical probe at a cross-head speed of 0.2 mm/s to achieve a strain of 50% and immediately returned to the original position. The compressive modulus is calculated from the slope of the stress–strain curve in the linear region (strain below 15%). Modules are different (6.7–30.9 kPa) depending on the cross-linked conditions.

IV. BIOCOMPATIBILITY AND ALLERGY

A. Chitosan

Chitosan shows potential proinflammatory properties through the release of chemical mediators (Bianco *et al.*, 2000; Usami *et al.*, 1998). Histological findings indicate that this material stimulates the migration of polymorphonuclear leukocytes and macrophages (Minami *et al.*, 1997; Peluso *et al.*, 1994; Usami *et al.*, 1994), and promotes angiogenesis, reorganization of the extracellular matrix, and granulation tissue formation (Minami *et al.*, 1993; Okamoto *et al.*, 1993).

Crustaceans are consumed in many coastal countries. In Japan, large amounts of shrimp, lobster, spiny lobster, and crab are imported from Asian countries and many other regions, and are processed as materials for commercial foods. Several clinical cases reported that crustaceans are well-known allergens (Becker *et al.*, 2004; Lehrer *et al.*, 2003). It is known that crustacean allergy generally presents as skin and respiratory tract symptoms. Further, anaphylaxis can be induced in sensitive patients by the intake of trace amounts of crustacean (Department of Food Safety, 2002; Tomikawa *et al.*, 2006). Although the wound dressing material originated from crab shell has been clinically used over 25 years in Japan, fortunately, no allergic incidents or side effects were reported. This should be brought through the good product management including the process of deproteinization.

B. Collagen

An IgE-reactive protein was clinically observed in surimi from walleye pollack, one of fish foods, by ELISA using one patient serum positive to the high MW allergen suggesting it to be collagen (Hamada *et al.*, 2000). It is provided that the evidence that the high MW allergen recognized by plural patient sera is collagen, and in competitive ELISA inhibition experiments, the big eye tuna collagen almost completely inhibited the

IgE reactivity to the heated extracts from five species of fish (Hamada *et al.*, 2001). Some fish-sensitive patients possessed IgE antibody to fish gelatin. Fish gelatin (type I collagen) might be an allergen in subjects with fish allergy (Sakaguchi *et al.*, 2000). Atelocollagen is a processed natural biomaterial produced from bovine type I collagen. It inherits useful biomaterial characteristics from collagen, such as rare inflammatory responses, a high biocompatibility, and a high biodegradability (Hanai *et al.*, 2006; Miyata *et al.*, 1992). The parts of collagen that are attributed to its immunogenicity, namely telopeptides, are eliminated in the process of atelocollagen production. Therefore, atelocollagen possesses little immunogenicity (Sano *et al.*, 2003). If substantial amounts of collagen could be obtained from fish wastes (scale, skin, and bone), they would provide an alternative to bovine collagen in food, cosmetics, and biomedical materials.

Jellyfish collagen scaffolds had a highly porous and interconnected pore structure, which is useful for a high-density cell seeding, an efficient nutrient, and oxygen supply to the cells cultured in the three-dimensional matrices. To determine whether jellyfish collagen evokes any specific inflammatory response compared to that induced by bovine collagen or gelatin, the levels of proinflammatory cytokines and antibody secretions were measured and the population changes of immune cells after *in vivo* implantation were monitored. Jellyfish collagen was demonstrated to induce an immune response at least comparable to those caused by bovine collagen and gelatin (Song *et al.*, 2006).

Elastic SC vascular grafts were prepared by incubating a mixture of acidic SC solution and a fibrillogenesis-inducing buffer containing a cross-linking agent, water-soluble carbodiimide (WSC). Subsequently, re-cross-linking in ethanol solution containing WSC was performed. Upon placement in rat subcutaneous pouches, the SC grafts brought little inflammatory reaction (Nagai *et al.*, 2008).

Collagen sponges with microporous structures from tilapia were fabricated reconstituted collagen fibrils using freeze-drying and cross-linked by DHT treatment or additional treatment with WSC treatment. The pellet implantation tests into the paravertebral muscle of rabbits demonstrated that tilapia collagen caused rare inflammatory responses at 1- and 4-week implantation, statistically similar to those of porcine collagen and a high-density polyethylene as a negative control (Sugiura *et al.*, 2009).

V. BIODEGRADATION

A. Chitosan

The cotton-like chitosan (MW: about 200 kDa; 35, 70, and 100% DD) was implanted into the alveolar bone cavities. The histopathological examination was carried out at 1, 3, 6, 9, and 12 months after the implantation.

All the various types of chitosans were degraded with time, in conjunction with the bone regeneration (Ikeda *et al.*, 2002). However, it takes about 9months after the implantation for the almost complete disappearance of chitosan in the bone tissue. Only the monomer type of chitosan, D-glucosamine which is effective to relieve the signs of osteoarthritis, is easy to completely dissolve immediately *in vitro* and *in vivo*.

B. Collagen

Upon placement in rat subcutaneous pouches, the SC grafts were gradually and slowly biodegraded. At 1 month after implantation, fibroblasts and macrophages started penetrating the surface of the graft without exhibiting any signs of necrosis (Nagai *et al.*, 2008). The biodegradation rates of both the collagen implants were similar, except for the DHT-treated tilapia collagen sponges at 1-week implantation. Various types of treated collagens did not disappear in the tissue even at 4-week implantation (Sugiura *et al.*, 2009).

VI. CONCLUSIONS

The initial inflammatory reactions associated with chitosan application to hard and soft tissues need to be controlled before it can be considered for clinical application as scaffold. Further, as it takes too long for biodegradation of implanted chitosan *in vivo*, generally chitosan is concluded to be not suitable for the scaffold for degenerative medicine in especially dental pulp tissue surrounding hard tissue.

The properties of biocompatibility and biodegradation of fish atelocollagen are suitable for the scaffold in regenerative medicine. However, these phenomena strongly depend on the procedures for cross-linking.

REFERENCES

Asaoka, K. (1996). Chitin-Chitosan: The Choice Food Supplement for Over 10000 Physicians in Japan. Vantage Press, New York.

Bae, I., Osatomi, K., Yoshida, A., Osako, K., Yamaguchi, A., and Hara, K. (2008). Biochemical properties of acid-soluble collagen extracted from the skins of underutilized fishes. *Food Chem.* **108**, 49–54.

Becker, W., Brasseur, D., Bresson, J-L., Flynn, A., Jackson, A. A., Lagiou, P., Mingrone, G., Moseley, B., Palou, A., Przyrembel, H., Salminen, S., Strobel, S., and van Loveren, H. (2004). Opinion of the scientific panel on dietetic products, nutrition and allergies on a request from the commission relating to the evaluation of allergic foods for labeling purpose. EFSA J. **32**, 1–197 Request No. EFSA-Q-2003–016.

Bianco, I. D., Balsinde, J., Beltramo, D. M., Castagna, L. F., Landa, C. A., and Dennis, E. A. (2000). Chitosan-induced phospholipase A_2 activation and arachidonic acid mobilization in P388D_1 macrophages. *FEBS Lett.* **466**, 292–294.

Bodek, K. H. (1994). Potentiometric method for determination of the degree of acetylation of chitosan. *In* "Chitin World", (Z. S. Karnicki, M. M. Breziski, P. J. Bykowsky, and A. Wojtasz-Pajak, Eds), pp. 456–461. Wirtschaftsverlag NW-Verlag, Germany.

Chen, F., Su, Y., Mo, X., He, C., Wang, H., and Ikada, Y. (2009). Biocompatibility, alignment degree and mechanical properties of an electrospun chitosan-P(LLA-CL) fibrous scaffold. *J. Biomater. Sci. Polym. Ed.* **20**, 2117–2128.

Cima, L. G., Vacanti, J. P., Vacant, C., Ingber, D., Mooney, D., and Langer, R. (1991). Tissue engineering by cell transplantation using degradable polymer substrates. *J. Biomech. Eng.* **113**, 143–151.

Depan, D., Venkata Surya, P. K. C., Girrase, B., and Misra, R. D. K. (2011). Organic/inorganic hybrid network structure nanocomposite scaffolds based on grafted chitosan for tissue engineering. *Acta Biomater.* **7**, 2163–2175.

Department of Food Safety (2002). The Ministry of Health, Labour and Welfare of Japan. Notification No. 1106001.

Dillow, A. K. and Lowman, A. M. (2002). Biomimetic Materials and Design: Biointerfacial Strategies, Tissue Engineering, and Targeted Drug Delivery. Marcel Dekker, Inc., New York.

Domm, C., Schunke, M., Christesen, K., and Kurz, B. (2002). Redifferentiation of dedifferentiated bovine articular chondrocytes in alginate culture under low oxygen tension. *Osteoarthritis Cartilage* **10**, 13–22.

Fernandez-Urrusuno, R., Calvo, P., Remunan-Lopez, C., Vila-Jato, J. L., and Alonso, M. J. (1999). Enhancement of nasal absorption of insulin using chitosan nanoparticles. *Pharm. Res.* **16**, 1576–1581.

Ferreira, M. C., Marvao, M. R., Duarte, M. L., Domard, A., Nunes, T., and Feio, G. (1994a). Chitosan degree of acetylation: Comparison of two spectroscopic methods (^{13}C CP/MAS NMR and dispersive IR). *In* "Chitin World", (Z. S. Karnicki, M. M. Breziski, P. J. Bykowsky, and A. Wojtasz-Pajak, Eds), pp. 476–479. Wirtschaftsverlag NW-Verlag, Germany.

Ferreira, M. C., Marvao, M. R., Duarte, M. L., Domard, A., Nunes, T., and Feio, G. (1994b). Optimisation of the measuring of chitin/chitosan degree of acetylation by FT-IR spectroscopy. *In* "Chitin World", (Z. S. Karnicki, M. M. Breziski, P. J. Bykowsky, and A. Wojtasz-Pajak, Eds), pp. 480–488. Wirtschaftsverlag NW-Verlag, Germany.

Gorham, S. D., Light, N. D., Diamond, A. M., Willins, M. J., Bailey, A. J., Wess, T. J., and Leslie, N. J. (1992). Effects of chemical modifications on the susceptibility of collagen to proteolysis. II. Dehydrothermal crosslinking. *Int. J. Biol. Macromol.* **14**, 129–138.

Hamada, Y., Genka, E., Ohira, M., Nagashima, Y., and Shiomi, K. (2000). Allergenicity of fish paste products and surimi from walleye Pollack. *Shokuhin Eiseigaku Zasshi* **41**, 38–43 (in Japanese).

Hamada, Y., Nagashima, Y., and Shiomi, K. (2001). Identification of collagen as a new fish allergen. *Biosci. Biotechnol. Biochem.* **65**, 285–291.

Hanai, K., Takeshita, F., Honma, K., Nagahara, S., Maeda, M., Minakuchi, Y., Sano, A., and Ochiya, T. (2006). Atelocollagen-mediated systemic DDS for nucleic acid medicines. *Ann. N. Y. Acad. Sci.* **1082**, 9–17.

Hollinger, J. (1993). Factors for osseous repair and delivery: Part I. *J. Craniofac. Surg.* **4**, 102–108.

Hollister, S. J. (2009). Scaffold design and manufacturing: From concept to clinic. *Adv. Mater.* **21**, 3330–3342.

Huang-Lee, L. L. H., Cheung, D. T., and Nimni, M. E. (1990). Biochemical changes and cytotoxicity associated with the degradation of polymeric glutaraldehyde derived crosslinks. *J. Biomed. Mater. Res.* **24**, 1185–1201.

Ikeda, T., Yanagiguchi, K., and Hayashi, Y. (2002). Application to dental medicine—In focus on dental caries and alveolar bone healing. *Bioindustry* **19**, 22–30 (in Japanese).

Ikoma, T., Kobayashi, H., Tanaka, J., Walsh, D., and Mann, S. (2003). Physical properties of type I collagen extracted from fish scales of Pagrus major and Oreochromis niloticas. *Int. J. Biol. Macromol.* **32**, 199–204.

Kas, H. S. (1997). Chitosan: Properties, preparation and application to microparticulate systems. *J Microencapsul.* **14**, 689–711.

Kimura, S. (1983). Vertebrate skin type I collagen: Comparison of bony fishes with lamprey and calf. *Comp. Biochem. Physiol. B* **74**, 525–528.

Kimura, S., Zhu, X.-P., Matsui, R., Shinjoh, M., and Takamizawa, S. (1988). Characterization of fish muscle type I collagen. *J. Food Sci.* **53**, 1315–1318.

Koide, S. S. (1998). Chitin-chitosan: Properties, benefits and risks. *Nutr. Res.* **18**, 1091–1101.

Koide, M., Osali, K., Konishi, J., Oyamada, K., Katakura, T., Takahashi, A., and Yoshizato, K. (1993). A new type of biomaterial for artificial skin: Dehydrothermally cross-linked composites of fibrillar and denatured collagen. *J. Biomed. Mater. Res.* **27**, 79–84.

Langer, R. and Vacanti, J. P. (1993). Tissue engineering. *Science* **260**, 920–926.

Langer, R. and Vacanti, J. P. (1999). Tissue engineering: The challenges ahead. *Sci. Am.* **280**, 86–89.

Langer, R., Cima, L. G., Tamada, J. A., and Wintermantel, E. (1990). Future directions in biomaterials. *Biomaterials* **11**, 738–745.

Lee, Y. M., Park, Y. J., Lee, S. J., Ku, Y., Han, S. B., Klokkevold, P. R., and Chung, C. P. (2000). The bone regenerative effect of platelet-derived growth factor-BB delivery with a chitosan/tricalcium phosphate sponge carrier. *J. Periodontol.* **71**, 418–424.

Lee, C. R., Grodzinsky, A. J., and Pector, M. (2001). The effects of cross-linking of collagen-glycosaminoglycan scaffolds on compressive stiffness, chondrocyte-mediated contraction, proliferation and biosynthesis. *Biomaterials* **22**, 3145–3154.

LeHoux, J. G. and Grondin, F. (1993). Some effects of chitosan on liver function in the rat. *Endocrinology* **132**, 1078–1084.

Lehrer, S. B., Ayuso, R., and Reese, G. (2003). Seafood allergy and allergen: A review. *Marine Biotechnol.* **5**, 339–348.

Liu, X. D., Howard, K. A., Dong, M. D., Anderson, M. O., Rahbek, U. L., Johnsen, M. G., Hansen, O. C., Besenbacher, F., and Kjems, J. (2007). The influence of polymeric properties on chitosan/siRNA nanoparticle formation and gene silencing. *Biomaterials* **28**, 1280–1288.

Lyons, F. G., Al-Munajjed, A. A., Kieran, S. M., Toner, M. E., Murphy, C. M., Duffy, G. P., and O'Brien, F. J. (2010). The healing of bony defects by cell-free collagen-based scaffolds compared to stem cell-seeded tissue engineered constructs. *Biomaterials* **31**, 9232–9245.

Madihally, S. V. and Matthew, W. T. (1999). Porous chitosan scaffolds for tissue engineering. *Biomaterials* **20**, 1133–1142.

Malda, J., van Blitterswijk, C. A., van Geffen, M., Martens, D. E., Tramper, J., and Riesle, J. (2004). Low oxygen tension stimulates the redifferentiation of dedifferentiated adult human nasal chondrocytes. *Osteoarthritis Cartilage* **12**, 306–313.

Marler, J. J., Upton, J., Langer, R., and Vacanti, J. P. (1998). Translation of cells in matrices for tissue regeneration. *Adv. Drug Deliv. Rev.* **33**, 165–182.

Matsui, R., Ishida, M., and Kimura, S. (1991). Characterization of an 3 chain from the skin type I collagen of chum salmon (Oncorhynchus keta). *Comp. Biochem. Physiol. B* **99**, 171–174.

Minami, S., Okamoto, Y., Tanioka, S., Sashiwa, H., Saimoto, H., Matsuhashi, A., and Shigemasa, Y. (1993). Effects of chitosan on wound healing. *In* "Carbohydrates and Carbohydrate Polymers", (M. Yalpani, Ed.), pp. 141–152. ALT Press, Illinois.

Minami, S., Masuda, M., Suzuki, H., Okamoto, Y., Matsuhashi, A., Kato, K., and Shigemasa, Y. (1997). Subcutaneous injected chitosan induces systemic activation in dogs. *Carbohydr. Polym.* **33**, 285–294.

Miyata, T., Taira, T., and Noishiki, Y. (1992). Collagen engineering for biomaterial use. *Clin. Mater.* **9**, 139–174.

Nagai, T., Yamashita, E., Taniguchi, K., Kanamori, N., and Suzuki, N. (2001). Isolation and characterization of collagen from the outer skin waste material of cuttlefish (Sepia lycidas). *Food Chem.* **78**, 173–177.

Nagai, N., Yuniki, S., Suzuki, T., Sakata, M., Tajima, K., and Munekata, M. (2004). Application of cross-linked salmon atelocollagen to the scaffold of human periodontal ligament cells. *J. Biosci. Bioeng.* **101**, 511–514.

Nagai, N., Mori, K., Satoh, Y., Takahashi, N., Yunoki, S., Tajima, K., and Munekata, M. (2007). In vitro growth and differentiated activities of human periodontal ligament fibroblasts cultured on salmon collagen gel. *J. Biomed. Mater. Res.* **82A**, 395–402.

Nagai, N., Nakayama, Y., Zhou, Y. M., Takamizawa, K., Mori, K., and Munekata, M. (2008). Development of salmon collagen vascular graft: Mechanical and biological properties and preliminary implantation study. *J. Biomed. Mater. Res.* **87B**, 432–439.

Nomura, Y., Yamano, M., and Shirai, K. (1995). Renaturation of alpha 1 chains from shark skin collagen type I. *J. Food Sci.* **60**, 1233–1236.

Nomura, Y., Toki, S., Ishii, Y., and Shirai, K. (2000a). The physicochemical property of shark type I collagen gel and membrane. *J. Agric. Food Chem.* **48**, 2028–2032.

Nomura, Y., Toki, S., Ishii, Y., and Shirai, K. (2000b). Improvement of the material property of shark type I collagen by comparison with pig type I collagen. *J. Agric. Food Chem.* **48**, 6332–6336.

Okamoto, Y., Minami, S., Matsuhashi, A., Sashiwa, H., Saimoto, H., Shigemasa, Y., Tanigawa, T., Tanaka, Y., and Tokura, S. (1993). Application of polymeric N-acetyl-D-glucosamine (chitin) to veterinary practice. *J. Vet. Med. Sci.* **55**, 742–747.

Park, Y. J., Ku, Y., Chung, C. P., and Lee, S. J. (1998). Controlled release of platelet-derived growth factor from porous poly(L-lactide) membranes for guided tissue regeneration. *J. Control. Release* **51**, 201–211.

Park, Y. J., Lee, Y. M., Park, S. N., Sheen, S. Y., Chung, C. P., and Lee, S. J. (2000). Platelet derived growth factor releasing chitosan sponge for periodontal bone regeneration. *Biomaterials* **21**, 153–159.

Peluso, G., Perillo, O., Ranieri, M., Santi, M., Ambrosio, L., Calabro, D., Avallone, B., and Balsammo, G. (1994). Chitosan-mediated stimulation of macrophage function. *Biomaterials* **15**, 1215–1220.

Pieper, J. S., Oosterhof, A., Dijkstra, P. J., Veerkamp, J. H., and van Kuppevelt, T. H. (1999). Preparation and characterization of porous crosslinked collagenous matrices containing bioavailable chondroitin sulphate. *Biomaterials* **20**, 847–858.

Prasitsilp, M., Jenwithisuk, R., Kongsuwan, K., Damrongchai, N., and Watts, P. (2000). Cellular responses to chitosan in vitro: The importance of deacetylation. *J. Mater. Sci. Mater. Med.* **11**, 773–778.

Richardson, S. C. W., Kolbe, H. V. J., and Duncan, R. (1999). Potential of low molecular mass chitosan as a DNA delivery system. Biocompatibility, body distribution and ability to complex and protect DNA. *Int. J. Pharm.* **178**, 231–243.

Sakaguchi, M., Toda, M., Ebihara, T., Irie, S., Hori, H., Imai, A., Yanagida, M., Miyazawa, H., Ohsuna, H., Ikezawa, Z., and Inoue, S. (2000). IgE antibody to fish gelatin (type I collagen) in patients with fish allergy. *J. Allergy Clin. Immunol.* **106**, 579–584.

Sano, A., Maeda, M., Nagahara, S., Ochiya, T., Honma, K., Itoh, H., Miyata, T., and Fujioka, K. (2003). Atelocollagen for protein and gene delivery. *Adv. Drug Deliv. Rev.* **55**, 1651–1677.

Song, E., Kim, S. Y., Chun, T., Byun, H. J., and Lee, Y. M. (2006). Collagen scaffolds derived from a marine source and their biocompatibility. *Biomaterials* **27**, 2951–2961.

Subramanian, A. and Lin, H. Y. (2005). Crosslinked chitosan: Its physical properties and the effects of matrix stiffness on chondrocyte cell morphology and proliferation. *J. Biomed. Mater. Res.* **75A**, 742–753.

Sugiura, H., Yunoki, S., Kondo, E., Ikoma, T., Tanaka, J., and Yasuda, K. (2009). In vivo biological responses and bioresorption of tilapia scale collagen as a potential biomaterial. *J. Biomater. Sci. Polym. Ed.* **20**, 1353–1368.

Swatschek, D., Schatton, W., Kellermann, J., Mullaer, W. E. G., and Kreuter, J. (2002). Marine sponge collagen: Isolation, characterization and effects on the skin parameters surface-pH, moisture and sebum. *Eur. J. Pharm. Biopharm.* **53**, 107–113.

Tomihata, K. and Ikada, Y. (1997). In vitro and in vivo degradation of films of chitin and its deacetylated derivatives. *Biomaterials* **18**, 567–575.

Tomikawa, M., Suzuki, N., Urist, A., Tsuburai, T., Ito, S., Shibata, R., Ito, K., and Ebisawa, M. (2006). Characteristics of shrimp allergy from childhood to adulthood in Japan. *Arerugi* **55**, 1536–1542, (in Japanese).

Usami, Y., Okamoto, Y., Minami, S., Matsuhashi, A., Kumazawa, N. H., Tanaka, S., and Shigemasa, Y. (1994). Chitin and chitosan induce migration of bovine polymorphonuclear cells. *J. Vet. Med. Sci.* **56**, 761–762.

Usami, Y., Okamoto, Y., Takayama, T., Shigemasa, Y., and Minami, S. (1998). Chitin and chitosan stimulate canine polymorphonuclear cells to release leukotriene B_4 and prostaglandin E_2. *J. Biomed. Mater. Res.* **42**, 517–522.

Wang, M. C., Pins, G. D., and Silver, F. H. (1994). Collagen fibers with improved strength for the repair of soft tissue injuries. *Biomaterials* **15**, 507–512.

Weadock, K. S., Miller, F. J., Bellincampi, L. D., Zawadsky, J. P., and Dunn, M. G. (1995). Physical crosslinking of collagen fibers: Comparison of ultraviolet irradiation and dehydrothermal treatment. *J. Biomed. Mater. Res.* **29**, 1373–1379.

White, M. J., Kohno, I., Rubin, A. I., Stenzel, K. H., and Miyata, T. (1973). Collagen films: Effect of cross-linking on physical and biological properties. *Biomater. Med. Devices Artif. Organs* **1**, 703–715.

Yang, A. F., Leong, K. F., Du, Z. H., and Chua, C. K. (2001). The design of scaffolds for use in tissue engineering. Part 1. Traditional factors. *Tissue Eng.* **7**, 679–689.

Yunoki, S., Suzuki, T., and Takai, M. (2003). Stabilization of low denaturation temperature collagen from fish by physical cross-linking methods. *J. Biosci. Bioeng.* **96**, 575–577.

Yunoki, S., Nagai, N., Suzuki, T., and Munekata, M. (2004). Novel biomaterial from reinforced salmon collagen gel prepared by fibril formation and cross-linking. *J. Biosci. Bioeng.* **98**, 40–47.

CHAPTER 7

Chitosan as Potential Marine Nutraceutical

Jae-Young Je[*,1] and **Se-Kwon Kim**[†]

Contents

Abstract

Chitosan, the most abundant marine mucopolysaccharide, is derived from chitin by alkaline deacetylation, and possesses versatile biological properties such as biocompatibility, biodegradability, and a non-toxic nature. Due to these characteristics, considerable attention has been given to its industrial applications in the food, pharmaceutical, agricultural, and environmental industries. Currently, chitosan can be considered as a potential marine nutraceutical because its remarkable biological activities have been investigated and reported, in order to exploit its nutraceutical

[*] School of Food Technology and Nutrition, Chonnam National University, Yeosu, Republic of Korea
[†] Department of Chemistry and Marine Bioprocess Research Center, Pukyong National University, Busan, Republic of Korea
[1] Corresponding author: Jae-Young Je, *E-mail address:* jjy1915@jnu.ac.kr

Advances in Food and Nutrition Research, Volume 65
ISSN 1043-4526, DOI: 10.1016/B978-0-12-416003-3.00007-X

properties. This chapter, therefore, reviews the biological activities of chitosan including antioxidant, hypocholesterolemic, antimicrobial, and anti-inflammatory activities, as well as its mode of action in each activity.

I. INTRODUCTION

Chitin, a naturally occurring biopolymer, consists of 2-acetamido-2-deoxy-(1-4)-β-D-glucopyranose residues (*N*-acetyl-D-glucosamine units) and is a well-known mucopolysaccharide and distributed in the shell of crustaceans, in the cuticle of insects, and in the cell wall of some fungi and microorganisms. Chitin has a strong intra- and intermolecular hydrogen bonding network, which provides the water-insoluble property of chitin. Chitin production is estimated as approximately 1×10^9 metric tons annually, and it is the second most abundant natural biopolymer. A great attention has been paid to its biomedical applications such as a drug carrier, an antitumor agent, and a wound-healing agent because of its useful biological properties including nontoxicity, biocompatibility, biodegradability, hemostatic activity, and wound healing (Muzzarelli, 1994).

Chitosan is also a natural cationic polymer derived by deacetylation of chitin under alkaline conditions. Chitosan is a linear polysaccharide consisting of 2-amino-2-deoxy-(1-4)-β-D-glucopyranose residues (D-glucosamine units). Due to the deacetylation of chitin, chitosan can be dissolved in water by making salt with various acids on the amino group of D-glucosamine units. Further, chitosan can be dissolved in water alone as it has a partially acetylated degree of about 50% D-glucosamine units (Aiba, 1989).

Major applications of chitosan were previously focused on sludge dewatering, food processing, and metal ion chelation until the mid-1980s. Further, it has received considerable attention for its commercial applications in agriculture, chemistry, biomedical, food, cosmetic, and biotechnology industries (Alasalvar *et al.*, 2002; Knorr, 1984; Kurita, 2006). In recent years, however, its numerous biological activities were documented such as antioxidant (Park *et al.*, 2004a), antibacterial (Allan and Hadwiger, 1979; Hirano and Nagao, 1989; Jeon and Kim, 2000; Jeon *et al.*, 2001), hypocholesterolemic (Sugano *et al.*, 1992), and immune-stimulating (Jeon and Kim, 2001), and due to these biological properties, chitosan has received much attention for development as a potential marine nutraceutical. This chapter, therefore, deals with producing chitosan from chitin, and will overview the biological properties of chitosan and its derivatives as potential marine nutraceuticals.

II. PREPARATION OF CHITOSAN

Due to strong hydrogen bonds, chitin is insoluble in water as well as in most solvents, making it difficult to apply in industry. Chitin can be converted into chitosan by enzymatic means or alkali deacetylation, this being the most utilized method. Chitosan is soluble in some organic acids such as lactic acid and acetic acid, as well as in inorganic acids. In addition, chitosan salts can be water-soluble depending on factors such as pH and degree of acetylation (Illum, 1998).

A. Chitosan from crustacean chitin

Several sources have been used for the production of chitosan, and the most exploited sources of chitin are the processing waste of shellfish and marine crustaceans, especially crab, lobster, shrimp, krill, oysters, and squid (Rasmussen and Morrissey, 2007). Generally, marine crustacean shells contain around 15–40% chitin (dry weight) with proteins and calcium carbonate (Kurita, 2006). To isolate chitin from natural resources, other components such as cellulose, mannan, polygalactosamine, glucan as well as proteins, pigments, and minerals (especially calcium carbonate) must be removed during extraction procedures. Generally, decalcification of samples is conducted by using dilute hydrochloric acid followed by deproteination in dilute aqueous sodium hydroxide solution. Then the decalcified and deproteinated product undergoes decolorization using 0.5% $KMnO_4$ and oxalic acid. Finally, chitosan can be obtained from chitin followed by deacetylation of the chitin using a hot concentrated sodium hydroxide solution (Fig. 7.1).

B. Chitosan from other sources

Shell wastes from shrimp, crab, and lobster processing industries are the traditional source of chitin. However, commercial production of chitosan by deacetylation of crustacean chitin with a strong alkali appears to have limited potential for industrial acceptance because of seasonal and limited supply, difficulties in processing, particularly with the large amount of waste of concentrated alkaline solution causing environmental pollution, and inconsistent physico-chemical properties (Chatterjee *et al.*, 2005).

Because the biological properties of chitosan are dependent on its physico-chemical properties such as molecular weight (MW) and degree of deacetylation (DD), uniform physico-chemical properties are a prerequisite to specific industrial applications. As an alternative method, fermentation technology for chitosan preparation from fungal cell walls has received much attention as an eco-friendly pathway. Rane and

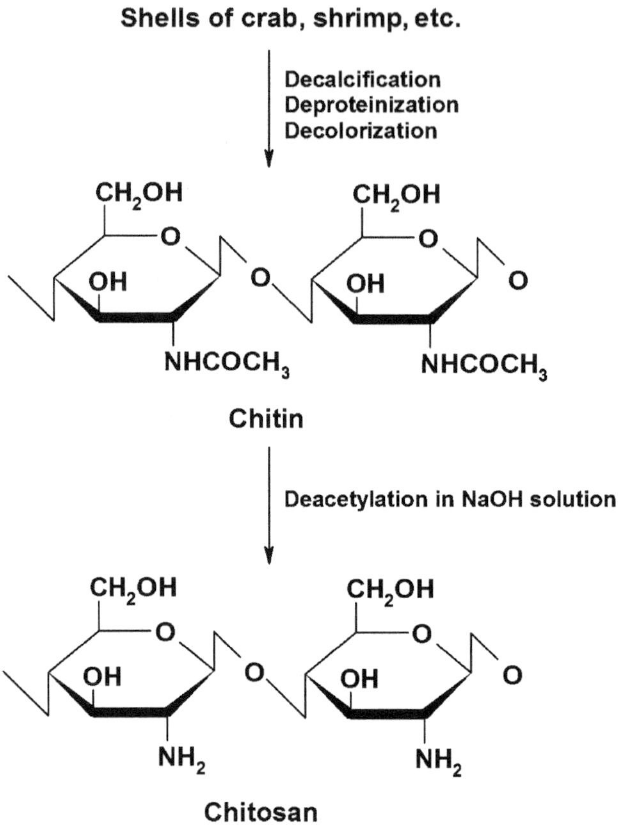

FIGURE 7.1 Preparation of chitosan from chitin.

Hoover (1993) described the production of chitosan from mycelia of *Absidia coerulea, Mucor rouxii, Gongronella butieri, Phycomyces blakesieeanus,* and *Absidia blakesleeana. A. coerulea* ATCC 14076 and NRRL 1315 were the best producers of chitosan among the strains examined. Crestini *et al.* (1996) also examined the production and isolation of chitosan through submerged and solid fermentation of *Lentinus edodes.* Chatterjee *et al.* (2005) determined the optimum conditions for the production of chitosan from *M. rouxii,* and investigated the MW, protein content, and DD of the isolated chitosan. Suntornsuk *et al.* (2002) utilized food processing by-products for the growth of fungal strains including *Aspergillus niger* TISTR3245, *Rhizopus oryzae* TISTR3189, *Zygosaccharomyces rouxii* TISTR5058, and *Candida albicans* TISTR5239, and their chitosan production yields were in the range of 0.4–4.3g/kg of soybean residue and 0.5–1.6g/kg of mungbean residue. These fungal approaches have benefits of easy handling, harvesting, and control in producing high quality chitosan, as well as different types of chitosans produced by diverse strains with individual environmental and nutritional conditions (Tajdini *et al.*, 2010; Wang *et al.*, 2008).

III. HEALTH BENEFITS OF CHITOSAN AND ITS DERIVATIVES

Because it has a variety of biological activities such as antioxidant, antimicrobial, anti-inflammatory, anticancer, and immune-stimulating effects, chitosan offers a number of uses in food, cosmetic, biomedical, and pharmaceutical industries. In this section, the nutraceutical properties of chitosan and its derivatives are discussed.

A. Antioxidant activity

Reactive oxygen species (ROS) are produced by abnormal metabolic processes as well as sunlight, ultraviolet light, and chemical reactions, and can change the structure of DNA, membrane lipids, and protein, which may result in diseases such as cancer, aging, inflammation, and atherosclerosis (Moskovitz *et al.*, 2002). Generally, ROS are effectively eliminated by antioxidant enzymes and compounds in our body. However, over production of ROS can cause considerable damage to biomacromolecules by ROS-mediated oxidative stress. The term antioxidant is defined as any substance, that when present at low concentrations compared to those of an oxidizable substrate, significantly delays or inhibits oxidation of that substrate (Park *et al.*, 2001). Therefore, supplementation of antioxidants, which can overcome oxidation-mediated problems, may prevent the body from a set of diseases by directly quenched ROS. There has been a growing interest to develop natural antioxidant compounds from many sources to overcome the ROS-mediated deleterious effects in biological systems. Numerous natural antioxidants such as tocopherols, ascorbic acid, polyphenols, and peptides are marketed around the world.

A number of studies have examined the antioxidant activities of chitosan from various sources. Park *et al.* (2004a) prepared three kinds of partially deacetylated hetero-chitosans such as 90% deacetylated, 75% deacetylated, and 50% deacetylated chitosan from crab chitin, and their antioxidant properties were measured using electron spin resonance spectrometry. Park and coworkers found that their antioxidant activities were dependent on the DD, and the 90% deacetylated chitosan showed the highest free radical scavenging activities. Yen *et al.* (2008) also found that a sample with more amino groups at the C-2 position showed the highest antioxidant activity. Tomida *et al.* (2009) examined the protective effects of seven different MW chitosans on plasma protein from oxidation by peroxyl radicals. In the ability to protect plasma protein from oxidation, it was found that the low MW chitosan was more effective in preventing the formation of carbonyl groups in plasma protein exposed to peroxyl radicals.

In vivo studies were also conducted by several researchers. Anraku *et al.* (2009) examined the antioxidant effects of water-soluble chitosan in normal subjects by measuring the reduction of indices of oxidative stress. Treatment with chitosan for 4weeks produced a significant decrease in levels of plasma glucose and the atherogenic index, and led to an increase in high-density lipoprotein cholesterol (HDL-C). Chitosan treatment also lowered the ratio of oxidized to reduced albumin and increased total plasma antioxidant activity. Further, Anraku *et al.* (2011) proved the antioxidant effects of high MW chitosan in normal volunteers, and the obtained results were consistent with previous results observed by Anraku *et al.* (2009).

B. Hypocholesterolemic effects

Obesity and high cholesterol levels are major factors influencing human health and can lead to increased risk of cardiovascular disease, type II diabetes, and some cancers. The hypocholesterolemic action of chitosan can be explained by decreased cholesterol absorption through entrapment of bile acids and interference with bile acid absorption. It is well-known that bile acids are produced from cholesterol, which is a major route of cholesterol metabolism, and about half of the synthesized cholesterol in our body is used for bile acid synthesis. About 90% of excreted bile acids are reabsorbed and recycled in enterohepatic circulation that moves bile acids to the liver and the gallbladder. Due to the negatively charged form of bile acids, chitosan can form complexes with bile acids due to its positively charged form, thereby bile acids are subsequently removed from reabsorption and recycling systems, and fewer bile acids are returned to the liver. This increases the synthesis of bile acids in the liver from the oxidation of cholesterol, thereby decreasing the total blood cholesterol level (Lee *et al.*, 1999).

A number of studies have revealed that chitosan can reduce plasma and liver triacylglycerol (TG) and total cholesterol levels *in vitro* and *in vivo*. Recently, the effects of the DD and the viscosity-average molecular weight (Mv) of chitosan samples on their fat-binding and cholesterol- and bile acid-binding capacities *in vitro* were evaluated (Liu *et al.*, 2008a,b; Zhou *et al.*, 2006). The fat-binding capacity of chitosan increased with increasing DD and Mv values, whereas cholesterol-binding ability was not affected by the values of DD and Mv. In addition, high Mv chitosan samples showed the greatest bile acid-binding ability but were not affected by the DD value.

In vivo studies also showed that chitosan has hypocholesterolemic and hypolipidemic activities (Nagyvary *et al.*, 1979; Sugano *et al.*, 1978, 1980).

It was found that chitosan with a relatively low MW (5–120kDa) and 79% DD had the greatest cholesterol-lowering effects, and short oligomers seemed to be ineffective (Sugano *et al.*, 1988; Trautwein *et al.*, 1997; Ylitalo *et al.*, 2002). Rats fed diets with high DD chitosan exhibited significantly lowered plasma triglyceride, total cholesterol, and low-density lipoprotein cholesterol (LDL-C) levels as well as elevated high-density lipoprotein cholesterol (HDL-C) levels. Moreover, food-efficiency ratio decreased with increasing DD (Li *et al.*, 2007; Liu *et al.*, 2008b). In addition, liver hepatic and lipoprotein lipase activities were reduced by chitosan, which indicated that it could regulate lipase activity; chitosan did not only prevent hyperlipidemia induced by long-term administration of a high-fat diet but also reduced serum lipid levels and liver-fat accumulation in hyperlipidemic rats. Xu *et al.* (2007) also investigated the effects of chitosan on lipid metabolism in hyperlipidemic rats. Chitosan lowered total cholesterol levels, LDL-C in plasma, and total cholesterol and total triglycerides in the liver, and increased fecal bile acid excretion, but did not change levels of triglyceride and HDL-C in plasma. It was also found that hepatic LDL receptor mRNA levels were increased by chitosan.

Bokura and Kobayashi (2003) suggested that in the digestive system, chitosan acts by forming a gel in the intestinal tract, which entraps cholesterol and lipids, and ingested chitosan develops an HCl layer in the stomach. Chitosan particles form agglomerates with cholesterol and fatty acids in the duodenum, which resulted in inhibiting cholesterol adsorption from the gastrointestinal tract. In a randomized, double-blind, and placebo-controlled clinical study, oral administration of chitosan reduced serum total cholesterol, especially in elderly women, while chitosan exhibited mild effects of reducing total and LDL cholesterol after 8weeks (Bokura and Kobayashi, 2003).

Several researchers suggested a possible mechanism with regard to the hypocholesterolemic activity of chitosan as follows. Tai *et al.* (2000) explained that the possible mechanism of the hypocholesterolemic activity of chitosan is similar to dietary fiber. Chitosan has high positive charge densities at low pH (6.5); therefore, chitosan is positively charged in gastric acid, which allows it to make complexes between positively charged chitosan and negatively charged cholesterol, fatty acids, and bile acids. The complexes then move into the intestines under an alkaline condition and change into an insoluble gel form. When the complexes change into the gel form, they cannot be hydrolyzed by pancreatic lipase and are excreted in the feces. A number of studies had also reported that chitosan increased fecal excretion of cholesterol and bile acids in animals and humans (Imonek and Bartonova, 2005; Lengsfeld *et al.*, 1999; Yao and Chiang, 2006). Due to the electrostatic interaction between chitosan and negatively charged cholesterol, fatty acids, and bile acids, higher DD chitosan shows the highest hypocholesterolemic activity but viscosity is not the major factor influencing the hypocholesterolemic effects of chitosan in the upper gastrointestinal tract.

C. Antimicrobial activity

Many studies have been conducted on the antimicrobial activity of chitosan since Allan and Hadwiger (1979) first reported that chitosan and its derivatives had broad-spectrum antimicrobial effects. The group of researches proved that chitosans are capable of inhibiting the growth of some microorganisms including bacteria, yeasts, and fungi.

Jeon *et al.* (2001) prepared 89% deacetylated chitosan, and the MIC values of chitosan against Gram-negative and Gram-positive bacteria were less than 0.06%. No *et al.* (2002) also examined the antibacterial activities of six chitosans and six chitosan oligomers with different MWs against four Gram-negative (*Escherichia coli, Pseudomonas fluorescens, Salmonella typhimurium,* and *Vibrio parahaemolyticus*) and seven Gram-positive bacteria (*Listeria monocytogenes, Bacillus megaterium, Bacillus cereus, Staphylococcus aureus, Lactobacillus plantarum, Lactobacillus brevis,* and *Lactobacillus bulgaricus*). The MICs of the chitosans ranged from 0.05% to more than 0.1% depending on the bacterial species and the MW of each chitosan, and the activity was higher than that of chitosan oligomers. Seo *et al.* (1992) tested the effects of chitosan on the growth of 11 different bacteria and found that the MIC values of the chitosan ranged from 10 to 1000ppm. Among the organisms tested, the growth of *E. coli, P. fluorescens, B. cereus,* and *S. aureus* was inhibited by chitosan concentrations of 20, 500, 500, and 1000ppm (i.e., 0.002, 0.05, 0.05, and 0.1%, respectively), respectively. Uchida *et al.* (1989) previously reported that the MIC values of chitosan for *E. coli* and *S. aureus* were 0.025% and 0.5%, respectively. According to the literature, the antimicrobial activity of chitosan is influenced by a number of factors such as the type of chitosan, molecular mass, and some of its other physico-chemical properties (Shibasaki *et al.*, 1994; Uchida *et al.*, 1989). In recent years, the antimicrobial activity of chitosan against methicillin-resistant *S. aureus* (MRSA) and vancomycin-resistant *S. aureus* (VRSA) was investigated by several researchers. Lee *et al.* (2009) prepared β-chitosan (685kDa, 20cps) from arrow squid pens and evaluated it against MRSA and VRSA, and found a MIC value of 0.625mg/mL against MRSA and VRSA, respectively. Park *et al.* (2009) also prepared chitosan (P6-MMC) with an average MW of 110 kDa from high MW chitosan by enzymatic hydrolysis, and evaluated its antimicrobial activity against MRSA. It was observed that P6-MMC exhibited superior anti-MRSA activity in a dose-dependent manner compared to native chitosan.

So far, a tremendous amount of literature supports the essential importance of the polycationic structure of chitosan in antimicrobial activity. A higher positive charge leads to electrostatic interaction between chitosan and negatively charged cell walls. A high DD shows a higher positive charge in chitosan. Park *et al.* (2004b) proved that high DD chitosan revealed a higher antimicrobial activity than low DD chitosan. Takahashi *et al.* (2008) also provided evidence that a higher DD with more positive charge was especially successful in inhibiting the growth of *S. aureus*, suggesting the antibacterial activity of chitosan toward *S. aureus* was enhanced with increasing DD. These results suggest that a free amino group plays a pivotal role in the antimicrobial activity. However, numerous studies on the bactericidal activity of chitosan have generated ambiguous results concerning the correlation between bactericidal activity and chitosan MW. Several reports revealed that high MW chitosan showed higher antimicrobial activity than low MW chitosan because high MW chitosan has more positive charge. On the other hand, low MW chitosan showed higher inhibitory activities against *Fusarium oxysporum*, *Phomopsis fukushi*, and *Alternaria alternata* than high MW chitosan (Hirano and Nagao, 1989). Jeon and Kim (2000) also showed that low MW chitosan with MW of 5000–10,000 Da had stronger antimicrobial activity than high MW chitosan. Chitosan is known to have chelating ability for various trace metal ions including Ni^{2+}, Zn^{2+}, Co^{2+}, Fe^{2+}, Mg^{2+}, and Cu^{2+} in acid conditions. It is well known that the cell walls of microorganisms, especially the outer membrane of Gram-negative bacteria, are stabilized by interactions between metal ions and cell wall molecules such as lipopolysaccharides. Therefore, the chelating of chitosan toward trace metal ions disrupts the integrity of the outer membrane, resulting in loss of the barrier function or blocking of the nutrient flow with concomitant bacterial death due to depletion of nutrients.

Currently, chitosan has been approved as a food additive in Korea and Japan since 1995 and 1983, respectively (KFDA, 1995; Weiner, 1992), and is mostly applied as a food additive or preservative and as a component of packaging material. In addition, the antimicrobial characteristics of chitosan present a profitable potential for developing functional food ingredients; however, they may prove to be useful in nutraceuticals for the ability to stimulate host defenses against a variety of bacteria, fungi, and yeast (Lim and Hudson, 2003).

D. Anti-inflammatory activity

Inflammation involving a wide variety of physiological and pathological processes is a kind of defense mechanism in an organism and is initiated by the invasion of pathogens or by tissue injury caused by biological, chemical, or physical damage. Inflammation is largely divided into acute and chronic inflammation. Acute inflammation is a short-term normal response, whereas chronic inflammation may progress from acute inflammation if the stimuli persist, causing individual tissue damage (Drayton *et al.*, 2006). Usually, leukocytes including macrophages, neutrophils, and eosinophils lead to tissue repair by the production of a variety of inflammatory mediators in order to remove the cause of inflammation. Among the cellular responses, the activation of macrophages is essential to the initiation and continuation of defensive reactions. Once they are stimulated by stimuli, macrophages released proinflammatory cytokines such as tumor necrosis factor-α (TNF-α), interleukin-1β (IL-1β), and IL-6, as well as inflammatory mediators including nitric oxide and prostaglandin E_2 (PGE_2), which are synthesized by inducible nitric oxide and cyclooxygenase-2 (COX-2), in order to enhance the defense capacity. Secretion of the inflammatory mediators against invading pathogens is essential for innate immunity and promotes the actions of the specific immune system; however, overproduction of macrophage-derived mediators are involved in the causation of many human diseases including rheumatoid arthritis, asthma, atherosclerosis, and endotoxin-induced multiple organ injury (Guslandi, 1998; Ritchlin *et al.*, 2003; Simons *et al.*, 1996). Therefore, anti-inflammatory agents reduce the magnitude of the released mediators by inflammatory response, and several nutraceuticals such as ω-3 PUFA, soy isoflavones, plants, and spices are being discovered.

Although a number of studies regarding the biological properties of chitosan have been performed, a few on its anti-inflammatory activities have recently been published. Ueno *et al.* (2001) described that chitosan promotes phagocytosis and the production of osteopontin and leukotriene B by polymorphonuclear leukocytes; activities such as phagocytosis; the production of IL-1, transforming growth factor b1, and platelet-derived growth factor by macrophages; and the production of IL-8 by fibroblasts, enhancing immune responses. Chitosan was also able to prevent pulmonary inflammation by inhibiting type 2 helper T cells and reducing levels of IL-4 and IL-5, and water-soluble chitosan inhibited the secretion of both IL-8 and TNF-α from mast cells (Hall *et al.*, 2002; Kim *et al.*, 2004). Since mast cells are necessary for allergic reactions and have been implicated in a number of neuroinflammatory diseases, these results suggest that chitosan has the potential to reduce the allergic inflammatory response. Chou *et al.* (2003) also found that chitosan inhibited PGE_2 formation and COX-2 induction in LPS-stimulated RAW264.7 macrophages. A high amount of PGE_2 derived from COX-2 induced by many proinflammatory mediators including TNF-α, IL-1β, and LPS has been implicated in the pathogenesis of sepsis and inflammation (Futaki *et al.*, 1997; Hammond *et al.*, 1999). In addition, treatment of a selective COX-2 inhibitor suppressed PGE_2 as well as normalized proinflammatory cytokine formation and subsequently improved survival after trauma, suggesting that inhibition of PGE_2 overproduction by suppressing COX-2 expression may be a potential pharmacologic target to treat inflammatory diseases (Mark Strong *et al.*, 2001). In another study, Kim *et al.* (2002) investigated the anti-inflammatory effect of water-soluble chitosan in human astrocytoma cells activated by amyloid β peptide and IL-1β, because a chronic inflammatory response associated with β-amyloid and IL-1β is responsible for the pathology of Alzheimer's disease. After stimulation by β-amyloid and IL-1β, the production of proinflammatory cytokines was amplified; however, the production of TNF-α and IL-6 was significantly inhibited by pretreatment of water-soluble chitosan in the human astrocytoma cells. As described above, chitosan may be able to modulate some inflammatory cytokines as well as inflammatory gene expression; therefore, chitosan could be used as an ingredient in functional foods for the prevention or alleviation of inflammatory diseases.

E. Other biological properties

Chitosan has also been reported to prevent increases in blood pressure. A high level of sodium chloride in the diet has been shown to increase blood pressure in human and animal models of hypertension (Boon and Aronson, 1985). In addition, the ingestion of nonchloride sodium salt did not increase blood pressure, but an elevation of blood pressure was observed with the ingestion of chloride salts. These results suggest that chloride ions play an important role in the elevation of blood pressure in human and animal models of hypertension. Okuda *et al.* (1997) proved that chitosan could reduce blood pressure with the administration of a high-salt diet containing chitosan, and serum angiotensin-converting enzyme was also significantly reduced by chitosan in stroke-prone spontaneously hypertensive rats.

The immunostimulating activity of chitosan has also been reported. A 70% DD chitosan showed immunostimulating effects by activating macrophages and natural killer cells in rats when infected with *E. coli* and *Sendai virus* (Nishimura *et al.*, 1984). Chang *et al.* (2004) also reported the immune-enhancing effects of chitosan as a novel adjuvant to an inactivated influenza vaccine, and the antibody content in serum remarkably increased the antiviral defenses of mice.

Further, the antiulcer effect of chitosan was investigated. Anandan *et al.* (2004) induced ulcers in rats by HCl–ethanol, and investigated the effects of ulcers in the presence of chitosan. Significant increases were observed in the volume and acidity of gastric fluid and in the level of lipid peroxidation, along with significant decreases in glutathione levels, activity of antioxidant enzymes, and peptic activity in the gastric mucosa. However, the ulcerogenic effects of HCl–ethanol were decreased in the presence of chitosan to normal levels of gastric activity and peptic activity.

IV. FUTURE PROSPECTS

Chitosan is a deacetylated form of chitin from the exoskeletons of marine crustaceans, shellfish, and some fungi. Due to its biocompatible, biodegradable, and nontoxic nature, considerable interest and attention have been focused on chitosan to develop nutraceuticals. According to past research, chitosans exhibit a variety of biological activities including antioxidant, hypocholesterolemic, antimicrobial, immune-stimulating, and anti-inflammatory activities that give them the potential to be highly effective nutraceuticals. However, it should be noted that the described biological activities of chitosan were dependent on its MW and DD, and not all of these biological activities are displayed in the same chitosan. Prior to potential application of chitosan as a nutraceutical, special types of chitosan displaying special bioactivities should be developed. Moreover, high MW chitosans are water-insoluble, and are difficult to absorb in the intestine, which is a major limitation for applications. In addition, the exact mechanism of each biological activity of chitosan should be understood in molecular-level detail. Although a tremendous amount of research on chitosan with regard to exploiting it as a nutraceutical has been conducted, some problems still remain unsolved; therefore, solutions to these problems may help accelerate its future applications.

REFERENCES

Aiba, S. (1989). Studies on chitosan: 2. Solution stability and reactivity of partially N-acetylated chitosan derivatives in aqueous media. *Int. J. Biol. Macromol.* **11**, 249–252.

Alasalvar, C., Shahidi, F., and Quantick, P. (2002). Food and health applications of marine nutraceuticals: A review. *In* "Seafoods: Quality, Technology, and Nutraceutical Applications", (C. Alasalvar and T. Taylor, Eds), pp. 175–204. Springer-Verlag, Berlin.

Allan, G. R. and Hadwiger, L. A. (1979). The fungicidal effect of chitosan on fungi of varying cell wall composition. *Exp. Mycol.* **3**, 285–287.

Anandan, R., Mathew, S., and Viswanathan, N. P. G. (2004). Antiulcerogenic effects of chitin and chitosan on mucosal antioxidant defense system in HCl-ethanol induced ulcer in rats. *J. Pharm. Pharmacol.* **56**, 265–269.

Anraku, M., Fujii, T., Furutani, N., Kadowaki, D., Maruyama, T., Otagiri, M., Gebiciki, J. M., and Tomida, H. (2009). Antioxidant effects of a dietary supplement: Reduction of indices of oxidative stress in normal subjects by water-soluble chitosan. *Food Chem. Toxicol.* **47**, 104–109.

Anraku, M., Fujii, T., Kono, T., Kojima, E., Hata, T., Tabuchi, N., Tsuchiya, D., Goromaru, T., Tsutsumi, H., Kadowaki, D., Maruyama, T., Otagiri, M., *et al.* (2011). Antioxidant properties of high molecular weight dietary chitosan in vitro and in vivo. *Carbohydr. Polym.* **83**, 501–505.

Bokura, H. and Kobayashi, S. (2003). Chitosan decreases total cholesterol in women. A randomized, double-blind, placebo-controlled trial. *Eur. J. Clin. Nutr.* **57**, 721–725.

Boon, N. A. and Aronson, J. K. (1985). Dietary salt and hypertension: Treatment and prevention. *Br. Med. J.* **290**, 949–950.

Chang, H. Y., Chen, J. J., Fang, F., and Chen, Z. (2004). Enhancement of antibody response by chitosan, a novel adjuvant of inactivated influenza vaccine. *Chin. J. Biol.* **17**, 21–24.

Chatterjee, S., Adhya, M., Guha, A. K., and Chatterjee, B. P. (2005). Chitosan from *Mucor rouxii*: Production and physico-chemical characterization. *Process Biochem.* **40**, 395–400.

Chou, T. C., Fu, E., and Shen, E. C. (2003). Chitosan inhibits prostaglandin E2 formation and cyclooxygenase-2 induction in lipopolysaccharide-treated RAW 264.7 macrophages. *Biochem. Biophys. Res. Commun.* **308**, 403–407.

Crestini, C., Kovac, B., and Sermanni, G. G. (1996). Production and isolation of chitosan from submerged and solid state fermentation from *Lentinus edodes*. *Biotechnol. Bioeng.* **50**, 207–210.

Drayton, D. L., Liao, S., Mounzer, R. H., and Ruddle, N. H. (2006). Lymphoid organ development: From ontogeny to neogenesis. *Nat. Immunol.* **7**, 344–353.

Futaki, N., Tajahashi, S., Kitagawa, T., Yamakawa, Y., Tanka, M., and Higuchi, S. (1997). Selective inhibition of cyclooxygenase-2 by NS-398 in endotoxin shock rats in vivo. *Inflamm. Res.* **46**, 496–502.

Guslandi, M. (1998). Nitric oxide and inflammatory bowel diseases. *Eur. J. Clin. Invest.* **28**, 904–907.

Hall, G., Lund, L., Lamb, J. R., and Jarman, E. R. (2002). Kinetics and mode of peptide delivery via the respiratory mucosa determine the outcome of activation versus TH2 immunity in allergic inflammation of the airways. *J. Allergy Clin. Immunol.* **110**, 883–890.

Hammond, R. A., Hannon, R., Frean, S. P., Armstrong, S. J., Flower, R. J., and Bryant, C. E. (1999). Endotoxin induction of nitric oxide synthase and cyclooxygenase-2 in equine alveolar macrophages. *Am. J. Vet. Res.* **60**, 426–431.

Hirano, S. and Nagao, N. (1989). Effects of chitosan, pectic acid, lysozyme and chitinase on the growth of several phytopathogens. *Agric. Biol. Chem.* **53**, 3065–3066.

Illum, L. (1998). Chitosan and its use as a pharmaceutical excipient. *Pharm. Res.* **15**, 1326–1331.

Imonek, J. and Bartonova, H. (2005). Effect of dietary chitin and chitosan on cholesterolemia of rats. *Acta Vet. Brno* **74**, 491–499.

Jeon, Y. J. and Kim, S. K. (2000). Production of chitooligosaccharides using an ultrafiltration membrane reactor and their antibacterial activity. *Carbohydr. Polym.* **41**, 133–141.

Jeon, Y. J. and Kim, S. K. (2001). Potential immuno-stimulating effect of antitumoral fraction of chitosan oligosaccharides. *J. Chitin Chitosan* **6**, 163–167.

Jeon, Y. J., Park, P. J., and Kim, S. K. (2001). Antimicrobial effect of chitooligosaccharides produced by bioreactor. *Carbohydr. Polym.* **44**, 71–76.

KFDA (1995). Food Additives Code. Korea Food and Drug Administration, Seoul p. 449.

Kim, M. S., Sung, M. J., Seo, S. B., Yoo, S. J., Lim, W. K., and Kim, H. M. (2002). Water-soluble chitosan inhibits the production of pro-inflammatory cytokine in human astrocytoma cells activated by amyloid β peptide and interleukin-1β. *Neurosci. Lett.* **321**, 105–109.

Kim, M. S., You, H. J., You, M. K., Kim, N. S., Shim, B. S., and Kim, H. M. (2004). Inhibitory effect of water-soluble chitosan on TNF-α and IL-8 secretion from HMC-1 cells. *Immunopharmacol. Immunotoxicol.* **26**, 401–409.

Knorr, D. (1984). Use of chitinous polymers in food-a challenge for food research and development. *Food Technol.* **38**, 85–97.

Kurita, K. (2006). Chitin and chitosan: Functional biopolymers from marine crustaceans. *Mar. Biotechnol.* **8**, 203–226.

Lee, J. K., Kim, S. U., and Kim, J. H. (1999). Modification of chitosan to improve its hypocholesterolemic capacity. *Biosci. Biotechnol. Biochem.* **63**(5), 833–839.

Lee, S. J., Kim, E. K., Moon, S. H., Jeon, B. T., Park, P. J., and Song, M. D. (2009). Antimicrobial activity of β-chitosan against VRSA and MRSA. *J. Chitin Chitosan* **14**, 130–133.

Lengsfeld, H., Fleury, A., Nolte, M., Piquerez, J. C., Hadvary, P., and Beglinger, C. (1999). Effect of orlistat and chitosan on faecal fat excretion in young healthy volunteers. *Obes. Res.* **7**(Suppl. 1), 50S.

Li, L., Zhang, J. L., Liu, J. N., and Xia, W. S. (2007). Effects of chitosan on serum lipid and fat liver. *Chin. J. Mar. Drugs* **26**, 7–9.

Lim, S. H. and Hudson, S. M. (2003). Review of chitosan and its derivatives as antimicrobial agents and their uses as textile chemicals. *J. Macromol. Sci.* **43**, 223–269.

Liu, J. N., Xia, W. S., and Zhang, J. L. (2008a). Effects of chitosans physico-chemical properties on binding capacities of lipid and bile salts in vitro. *Chin. Food Sci.* **29**, 45–49.

Liu, J. N., Zhang, J. L., and Xia, W. S. (2008b). Hypocholesterolemic effects of different chitosan samples in vitro and in vivo. *Food Chem.* **107**, 419–425.

Mark Strong, V. E., Mackrell, P. J., Concannon, E. M., Mestre, J. R., Smyth, G. P., Schaefr, P. A., Stapleton, P. P., and Daly, J. M. (2001). NS-398 treatment after trauma modifies NF-kappa B activation and improves survival. *J. Surg. Res.* **98**, 40–46.

Moskovitz, L., Yim, M. B., and Chock, P. B. (2002). Free radicals and disease. *Arch. Biochem. Biophys.* **397**, 354–359.

Muzzarelli, R. (1994). *In vivo* biochemical significance of chitin-based medical items. *In* "Polymeric Biomaterials", (S. Dumitriu, Ed.), pp. 179–197. Marcel Dekker, New York.

Nagyvary, J. J., Falk, J. D., Hill, M. L., Schmidt, M. L., Wilkins, A. K., and Bradbury, E. L. (1979). The hypolipidemic activity of chitosan and other polysaccharides in rats. *Nutr. Rep. Int.* **20**, 677–684.

Nishimura, K., Nishimura, S., Nishi, N., Saiki, I., Tokura, S., and Azuma, I. (1984). Immunological activity of chitin and its derivatives. *Vaccine* **2**, 93–99.

No, H. K., Park, N. Y., Lee, S. H., and Meyers, S. P. (2002). Antibacterial activity of chitosans and chitosan oligomers with different molecular weights. *Int. J. Food Microbiol.* **74**, 65–72.

Okuda, H., Kato, H., and Tsujita, T. (1997). Antihypertensive and antihyperlipidemic actions of chitosan. *J. Chitin Chitosan* **2**, 49–59.

Park, P. J., Jung, W. K., Nam, K. S., Shahidi, F., and Kim, S. K. (2001). Purification and characterization of antioxidative peptides from protein hydrolysate of lecithin-free egg yolk. *J. Am. Oil Chem. Soc.* **78**, 651–656.

Park, P. J., Je, J. Y., and Kim, S. K. (2004a). Free radical scavenging activities of differently deacetylated chitosans using an ESR spectrometer. *Carbohydr. Polym.* **55**, 17–22.

Park, P. J., Je, J. Y., Byun, H. G., Moon, S. H., and Kim, S. K. (2004b). Antimicrobial activity of hetero-chitosans and their oligosaccharides with different molecular weights. *J. Microbiol. Biotechnol.* **14**, 317–323.

Park, J. K., Moon, H. R., Kim, W. J., Kim, S. M., and Park, Y. I. (2009). Preparation and antimicrobial activity of medium-molecular weight chitosan. *J. Chitin Chitosan* **14**, 138–142.

Rane, K. D. and Hoover, D. G. (1993). Production of chitosan by fungi. *Food Biotechnol.* **7**, 11–33.

Rasmussen, R. S. and Morrissey, M. T. (2007). Chitin and chitosan. *In* "Marine Nutraceuticals and Functional Foods", (F. Shahidi and C. Barrow, Eds), pp. 155–182. CRC Press, New York.

Ritchlin, C. T., Haas-Smith, S. A., Li, P., Hicks, D. G., and Schwarz, E. M. (2003). Mechanisms of TNF-α and RANKL-mediated osteoclastogenesis and bone resorption in psoriatic arthritis. *J. Clin. Invest.* **111**, 821–831.

Seo, H., Mitsuhashi, K., and Tanibe, H. (1992). Antibacterial and antifungal fiber blended by chitosan. *In* "Advances in Chitin and Chitosan", (C. J. Brine, P. A. Sanford, and J. P. Zikakis, Eds), pp. 34–40. Elsevier Applied Science, London and New York.

Shibasaki, K., Sano, H., Matsukubo, T., and Takaesu, Y. (1994). Effects of low molecular chitosan and pH changes in human dental plague. *Bull. Tokyo Dent. Coll.* **35**, 33–39.

Simons, R. K., Junger, W. G., Loomis, W. H., and Hoyt, D. B. (1996). Acute lung injury in endotoxemic rats is associated with sustained circulating IL-6 levels and intrapulmonary CINC activity and neutrophil recruitment-role of circulating TNF-α and IL-β? *Shock* **6**, 39–45.

Sugano, M., Fujikawa, T., Hiratsuji, Y., and Hasegawa, Y. (1978). Hypocholesterolemic effects of chitosan in cholesterol-fed rats. *Nutr. Rep. Int.* **18**, 531–537.

Sugano, M., Fujikawa, T., Hiratsuji, Y., Nakashima, K., Fukuda, N., and Hasegawa, Y. (1980). A novel use of chitosan as a hypocholesterolemic agent in rat. *Am. J. Clin. Nutr.* **33**, 787–793.

Sugano, M., Watanabe, S., Kishi, A., Izumi, M., and Ohtakara, A. (1988). Hypocholesterolemic action of chitosans with different viscosity in rats. *Lipids* **23**, 187–191.

Sugano, M., Yoshida, K., Hashimoto, H., Enomoto, K., and Hirano, S. (1992). Hypocholesterolemic activity of partially hydrolyzed chitosan in rats. *In* "Advances in Chitin and Chitosan", (C. J. Brine, P. A. Sanford, and J. P. Zikakis, Eds), pp. 472–478. Elsevier Applied Science, London and New York.

Suntornsuk, W., Pochanavanich, P., and Suntornsuk, L. (2002). Fungal chitosan production on food processing by-products. *Proc. Biochem.* **37**, 727–729.

Tai, T. S., Sheu, W., Lee, W. J., Yao, H. T., and Chiang, M. T. (2000). Effect of chitosan on plasma lipoprotein concentrations in type 2 diabetic subjects with hypercholesterolemia. *Diabetes Care* **23**, 1703–1704.

Tajdini, F., Ali Amini, M., Nafissi-Varcheh, N., and Ali Faramarzi, M. (2010). Production, physiochemical and antimicrobial properties of fungal chitosan from *Rhizomucor miehei* and *Mucor racemosus*. *Int. J. Biol. Macromol.* **47**, 180–183.

Takahashi, T., Imaia, M., Suzukia, I., and Sawai, J. (2008). Growth inhibitory effect on bacteria of chitosan membranes regulated by the deacetylation degree. *Biochem. Eng. J.* **40**, 485–491.

Tomida, H., Fujii, T., Furutani, N., Michihara, A., Yasufuku, T., Akasaki, K., Maruyama, T., Otagiri, M., Gebicki, J. M., and Anraku, M. (2009). Antioxidant properties of some different molecular weight chitosans. *Carbohydr. Polym.* **344**, 1690–1696.

Trautwein, E. A., Jurgensen, U., and Erbersdobler, H. F. (1997). Cholesterol-lowering and gallstone-preventing action of chitosans with different degrees of deacetylation in hamsters fed cholesterol-rich diets. *Nutr. Res.* **17**, 1053–1065.

Uchida, Y., Izume, M., and Ohtakara, A. (1989). Preparation of chitosan oligomers with purified chitosanase and its application. *In* "Chitin and Chitosan", (G. Skjak-Braek, T. Anthonsen, and P. Standford, Eds), pp. 373–382. Elsevier Applied Science, United Kingdom.

Ueno, H., Mori, T., and Fujinaga, T. (2001). Topical formulations and wound healing applications of chitosan. *Adv. Drug Deliv. Rev.* **52**, 105–115.

Wang, W., Du, Y., Qiu, Y., Wang, X., Hu, Y., Yang, J., Cai, J., and Kennedy, J. F. (2008). A new green technology for direct production of low molecular weight chitosan. *Carbohydr. Polym.* **74**, 127–132.

Weiner, M. L. (1992). An overview of the regulatory status and of the safety of chitin and chitosan as food and pharmaceutical ingredients. *In* "Advances in Chitin and Chitosan", (C. J. Brine, P. A. Sanford, and J. P. Zikakis, Eds), pp. 663–670. Elsevier Applied Science, London and New York.

Xu, G., Huang, X., Qin, L., Wu, J., and Hu, Y. (2007). Mechanism study of chitosan on lipid metabolism in hyperlipidemic rats. *Asia Pac. J. Clin. Nutr.* **16**, 313–317.

Yao, H. T. and Chiang, M. T. (2006). Effect of chitosan on plasma lipids, hepatic lipid and fecal bile acid in Hamsters. *J. Food Drug Anal.* **14**, 183–189.

Yen, M. T., Yang, J. H., and Mau, J. L. (2008). Antioxidant properties of chitosan from crab shells. *Carbohydr. Polym.* **74**, 840–844.

Ylitalo, R., Lehtinen, S., Wuolijoki, E., Ylitalo, P., and Lehtimaki, T. (2002). Cholesterol-lowering properties and safety of chitosan. *Arzneimittelforschung* **52**, 1–7.

Zhou, K., Xia, W., Zhang, C., and Yu, L. (2006). In vitro binding of bile acids and triglycerides by selected chitosan preparations and their physicochemical properties. *LWT Food Sci. Technol.* **39**, 1087–1092.

CHAPTER 8

Bioactive Compounds from Marine Sponges and Their Symbiotic Microbes: A Potential Source of Nutraceuticals

Se-Kwon Kim[*,†,1] and **Pradeep Dewapriya**[*]

Contents

Abstract

Sponges are considered as the chemical factory in marine environment because of its immense production of chemically diverse compounds. Other than the chemical diversity, these compounds possess remarkable bioactivities. This great potential has aroused applications of sponge-derived compounds as therapeutics and at present, a number of promising compounds are in clinical and preclinical trials. Recently, nutraceuticals have received considerable interest among the health conscious community because of its multiple therapeutic effects. Natural health-promoting substances

[*] Marine Biochemistry Laboratory, Department of Chemistry, Pukyong National University, Busan, Republic of Korea
[†] Marine Bioprocess Research Center, Pukyong National University, Busan, Republic of Korea
[1] Corresponding author: Se-Kwon Kim, *E-mail address:* sknkim@pknu.ac.kr

Advances in Food and Nutrition Research, Volume 65
ISSN 1043-4526, DOI: 10.1016/B978-0-12-416003-3.00008-1

gain continuous popularity as nutraceuticals due to its reduced risk of side effects. This overview discusses the potentials of marine sponge-derived bioactivities as natural health-promoting compounds.

I. INTRODUCTION

Marine sponges are one the most prolific and important source of new bioactive compounds. Hundreds of interesting new compounds have been discovered from sponges. Hence, marine sponges have been considered as a gold mine to scientists (Baby and Sujatha, 2010). Chemical and physical conditions in the marine environment and intensive pressure from competitors, predators, and pathogens have resulted in producing wide array of chemically diverse secondary metabolites. There have been several reviews which discuss different classes of secondary metabolites of marine sponges and their bioactivities (John *et al.*, 2010; Keyzers and Coleman, 2005; Taylor *et al.*, 2007; Webster and Taylor, 2011). Medicinal applications of sponges go back to ancient times, and sponges saturated with solutions had been used for various diseases such as heartaches, detoxifications of bites of poisonous animals, and wound healing (Sipkema *et al.*, 2005a). Discovery of an unusual nucleoside and development of commercially important anticancer drug Ara-C (Fig. 8.1) and

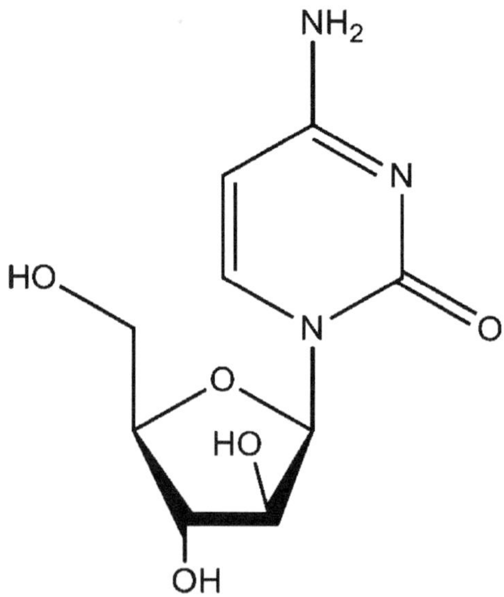

FIGURE 8.1 Ara-C (cytarabine).

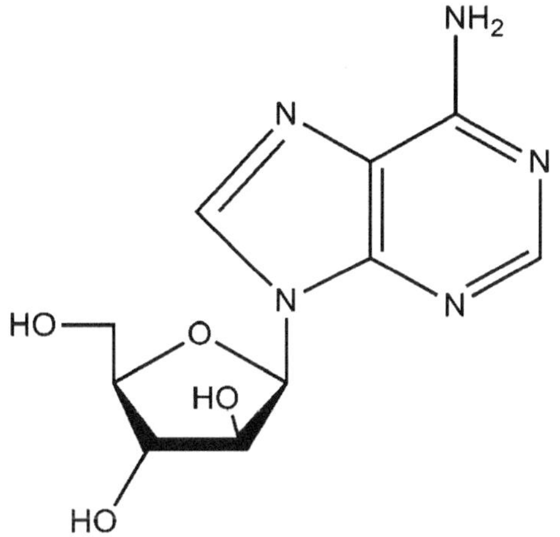

FIGURE 8.2 Ara-A (spongothymidine).

antiviral drug Ara-A (Fig. 8.2) aroused pharmaceutical interest of marine sponge (Proksch *et al.*, 2002). A number of secondary metabolites that have been isolated from sponges are in clinical trials for use as drugs. Further, health-promoting abilities of many more compounds have been proved (David and Gordon, 2004).

There have been continuous search for new substances that can improve biological functions and possibility of applying these compounds to promote well-being of human. This has resulted in development of products such as vitamins, dietary supplements, functional foods, nutraceuticals, phytochemicals, bio-chemopreventatives, and designer foods (Zeisel, 1999). Nutraceuticals, accounted for an estimated 5% market growth annually, are the fastest growing segment. The popularity has gained through the belief that nutraceuticals prevent chronic diseases, promote human health, and delay the onset of aging (Le and Pathak, 2011). The term nutraceutical is a hybrid of two terms "nutrition" and "pharmaceutical." A number of definitions have been given based on countries' regulations. Barrow and Shahidi (2008) define marine nutraceuticals as marine-derived substances that can be used as dietary supplements or food ingredients that provide medicinal or health benefits beyond nutrition. Among the commercially available nutraceuticals, products derived from marine sources are considered as a rich source of health-promoting substances (Shahidi, 2009). Long-chain ω-3 fatty acids and glucosamine are well-established marine-derived nutraceuticals. However, other than common health benefits of nutraceuticals, innovative developments are required to combat with emerging "lifestyle diseases" (Dharti *et al.*, 2010).

Thousands of bioactive compounds that have fascinating potentials to develop as nutraceuticals have been identified from marine sources. This chapter investigates the possibility of developing nutraceuticals from bioactive compounds, identified from marine sponges and symbiotic microorganisms.

II. MARINE SPONGES AND THEIR SYMBIOTIC MICROBES

Sponges, sessile metazoans consist of a gelatinous material—mesophyl, are the simplest form of multicellular animals. The hollow pitcher-like body is reinforced by spicules (a needle-like silica or calcium structure) and spongia (collagen fibers) (Uriz *et al.*, 2003). This efficient filter feeder obtained microbes and microparticles in the water as nutrients through the unique water circulation system (Lafi *et al.*, 2005). Basically, marine sponges consist of three sublineages such as Hexactinellida, Calcarea, and Demospongiae. True sponging is found in the class of demospongiae and is considered as the most economically important group of sponges. Wide range of chemically diverse compounds, alkaloids, terpenoids, polyketides, macrolides, polyphenolic compounds, peptides, and sterols, exhibit the potential of marine sponge (Thoms and Schupp, 2007).

Association of sponges and microbes has been highlighted in several reviews (Hentschel *et al.*, 2006; Taylor *et al.*, 2007; Thomas *et al.*, 2010; Webster and Taylor, 2011). Mesophyl matrix is a good shelter of extracellular heterotrophic and autotrophic microbes while the outer layer is occupied by photosynthetic microbes. Novel molecular biological techniques such as 16S rRNA gene library, fluorescence *in situ* hybridization, denaturing gradient gel electrophoresis, and lipid biomarkers enable identification of microbial community associated with marine sponges (Hentschel *et al.*, 2006). Interestingly, microbial density in some marine sponges may account up to 60% of sponge biomass. Microbial association of sponge spans a wide range of microbial groups starting from prokaryotic archaea to filamentous fungi. The symbiotic microbial community contributes to a number of physiological processes such as photosynthesis, nitrogen and carbon cycles, and dehalogenation. Bright colors of sponges are usually a result of symbiotic cyanobacteria. Despite the fact that sponge harbors large number of microorganism, it has been proved that symbiotic microbes are responsible for production of biologically important chemical compounds (Table 8.1). This has given a hope of producing active metabolites in large scale to address the so-called supply problem (Taylor *et al.*, 2007).

TABLE 8.1 Secondary metabolites of marine sponges associated microorganisms.

Group of organisms	Compound class	Bioactivity	Reference
Fungi	Polyketides, cyclohexane, terpenes, amino acid derivatives	Cytotoxic, antioxidant, antimicrobial, kinase inhibitors	Li *et al.* (2011); Bugni and Ireland (2004)
Bacteria	Proteins, peptides, propionic acid, indole, PAFA	Antimicrobial	Boobathy *et al.* (2009); Taylor *et al.* (2007); Devi *et al.* (2010); Patnayak and Sree (2005)
Cyanobacteria	Ether, chlorinated metabolites, pigments		Hentschel *et al.* (2006)
Actinobacteria	Amino acid, polyketides, glycosides, methyl esters	Anti-inflammation, antimicrobial, antioxidant, hypocholesterolemic	Schneemann *et al.* (2010)
Actinomycetes	Carotenoids	Antioxidant	Dharmaraj *et al.* (2009a)
Yeast	Indole derivatives	Radical scavengers	Sugiyama *et al.* (2009)

III. BIOACTIVE COMPOUNDS

A. Anti-inflammatory compounds

Classically, inflammation is a protective reaction of the body in response to some physical, chemical, or microbial injury and insult of the cells. Acute inflammation, rapid onset and shorter duration, is considered as a healthy response. However, when inflammation continues for prolonged period of time, it becomes detrimental and may raise the first step of a chronic disease (Medzhitov, 2008). Arachidonic acid/COX and nuclear factor-κB (NF-κB) are well known inflammatory pathways which induce production of inflammatory mediators such as prostaglandins, thromboxanes, leukotrienes, and cytokines. Most commonly accepted mechanism of anti-inflammation is inhibition of cyclooxygenase (COX) activity. There are two kinds of cyclooxygenases: COX-1 is natural protective enzyme of intestinal mucosa while COX-2 is induced by tissue damage as an inflammatory mediator (Maroon *et al.*, 2006). NF-κB is a recently identified complex inflammatory pathway which produces inflammatory cytokines and proinflammatory enzymes. Curcumin, γ-linolenic acid, epigallocatechin gallate, eugenol, [6]-gingerol, and many more flavanoids are well known natural anti-inflammatory agents (Aggarwal *et al.*, 2009). However, commonly used non-steroidal anti-inflammatory drugs (NSAIDs) have been creating some complications like gastrointestinal hemorrhage, myocardial infarction, and stroke due to lack of selectivity (Maroon *et al.*, 2006). Thus, natural and selective inhibitors of the inflammatory cascade have gained much attention.

Sponge-derived sesterterpenoids are potent anti-inflammatory metabolites which inhibit phospholipase A_2 (Sipkema *et al.*, 2005a). In arachidonic acid pathway of inflammatory response (Fig. 8.3), phospholipase A_2 catalyzes the release of membrane-bound phospholipids to produce inflammatory mediators once stimulated by tissue injury. Manoalide (Fig. 8.4) is a best known sponge-derived anti-inflammatory sesterterpenoid. Dysidotronic acid (Fig. 8.5), non-complex manoalide analogue, has been identified as an anti-inflammatory metabolite of the sponge *Dysidea* sp. The mechanism by which dysidotronic acid inhibit inflammation is much more selective and potent than manoalide. This sesquiterpenoid

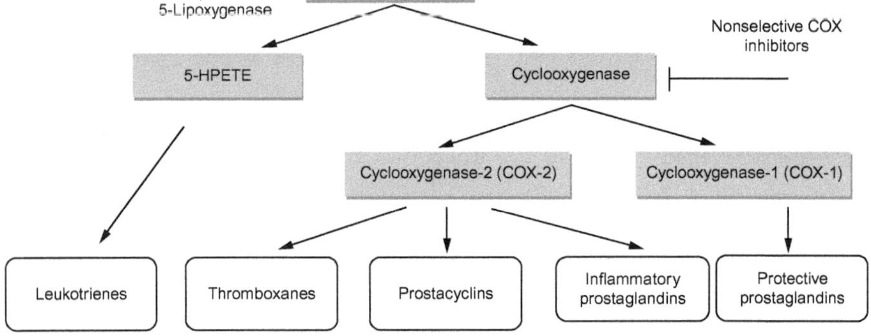

FIGURE 8.3 Schematic diagram of arachidonic acid pathway of inflammatory response. In tissue damage or insult, inflammatory mediators leukotrienes, thromboxanes, and prostaglandins are produced as a result of releasing membrane-bound arachidonic acid by phospholipase A_2. 5-Lipoxygenase and cyclooxygenase-2 (COX-2) mediate the conversion of free arachidonic acid into inflammatory mediates. Sponge metabolites inhibit phospholipase A_2 and 5-lipoxygenase followed by inhibition of production of inflammatory mediators.

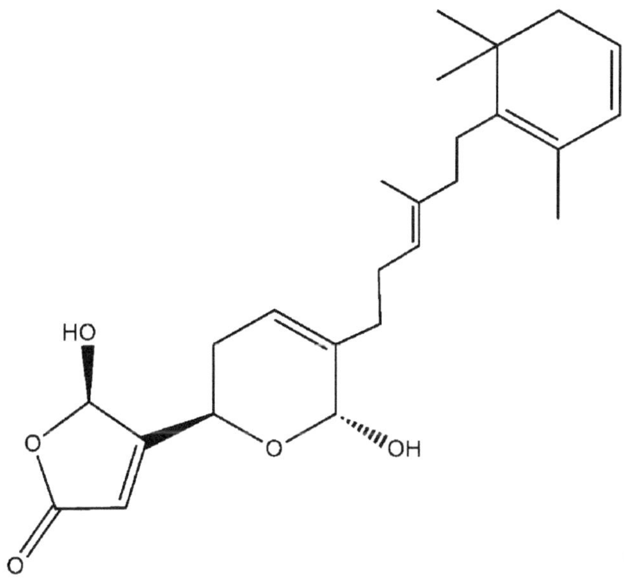

FIGURE 8.4 Manoalide.

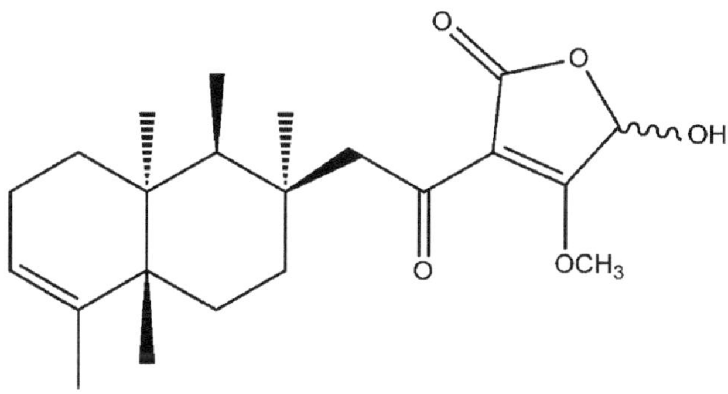

FIGURE 8.5 Dysidotronic acid.

inhibits cytokine production, consequently decreasing the production of inducible nitric oxide synthase and prostaglandin E_2 (Posadas *et al.*, 2001). Carroll *et al.* (2001) revealed that bioactive metabolites of *Jaspis splendens* and *Suberea* sp. inhibit lipoxygenase, a group of enzymes that oxygenate polyenoic fatty acids followed by production of inflammatory mediators leukotrienes, lipoxins, and hepoxilins. In contrast to the NSAIDs, sponge-derived anti-inflammatory metabolites are selective inhibitors of enzymes. Therefore, these metabolites have a clear potential of overcoming complication associated with non-selective inhibitors.

B. Novel antioxidants

Environmental as well as lifestyle changes are responsible for increasing reactive oxygen species in the body. Overproduction of oxidative free radicals, when cell's natural controlling mechanism (superoxide dismutase and catalase) is not effective or enough, creates an oxidative stress due to imbalance of pro-oxidants and antioxidants (Kelsey *et al.*, 2010). The chronic oxidative stress leads to various kinds of degenerative diseases and aging process. In this scenario, antioxidant-rich sources and supplements are under spotlight to offer extra protection. Although many chemical substances have been proved for their antioxidant activity, there has been continuous search for new antioxidants due to poor bioavailability, cumulative effects inside the body, and adverse side effects of synthetic antioxidants (Jorge *et al.*, 2011). Among novel antioxidants, marine sponges are highly ranked source. A number of metabolites derived from marine sponges, indole derivatives, aromatic alkaloids, aromatic polyketides, and phenolic compounds have exhibited strong antioxidant effect compared to vitamin E, ascorbic acid, and butylated hydroxyanisole (Longeon *et al.*, 2011; Utkina, 2009; Utkina *et al.*, 2004). Aromatic polyketides isolated from marine sponge-derived fungus *Aspergillus versicolor* have shown significantly higher antioxidant capacity than that of butylated hydroxytoluene (Li *et al.*, 2011). Sugiyama *et al.* (2009) reported that marine-derived yeasts are capable of producing antioxidative indole derivatives. These findings were very attractive since these microbes are culturable, and there is a possibility of reproducing the compound in large scale.

C. Hypocholesterolemic compounds

Cholesterol, an essential steroid which is produced by human body, acts as a cell membrane modulator and facilitator of absorption in the intestine. In addition, it serves as a substrate for hormones, bile acid, and vitamin D (Chen *et al.*, 2008). Water-insoluble cholesterol is transported throughout the body by four lipoproteins in the blood. Among them, cholesterol carrier lipoprotein, low-density lipoprotein (LDL), and high-density lipoprotein are strictly regulated to maintain cholesterol hemostasis in blood (Deng, 2009). High total serum cholesterol level has been recognized as a highly ranked risk factor of world biggest killer, cardiovascular disease (Roth *et al.*, 2011). Inhibition of cholesterol biosynthesis to control serum cholesterol is a widely accepted hypothesis, and various natural and synthetic products have been employed at different steps of the biosynthesis pathway. Cholesterol biosynthesis enzyme HMG-CoA reductase is the target of most of the hypocholesterolemic drugs since it is the rate-limiting enzyme. However, myositis, gastrointestinal symptoms, pharyngitis, and pain are commonly associated side effects with these types of drugs. In this regard, recent studies have focused on identifying more specific inhibitors, especially later stage of the biosynthesis pathway (Korosec *et al.*, 2008).

Marine sponge *Spirastrella abata*-derived Lyso-PAF analogues and lysophosphatidylcholines have been reported as successful inhibitors of cholesterol biosynthesis (Fig. 8.6). An *in vitro* study shows that active compounds specifically blocked the conversion of lanosterol into cholesterol in the liver cells with more specific inhibition (Shin *et al.*, 1999). Zhao *et al.* (2003) have reported that marine sponges are a source of novel lysophosphatidylcholines and thereby aroused an interest of hypocholesterolemic compounds from marine sponge due to short lifespan of conventional lysophosphatidylcholines *in vivo*. Another interesting target of hypocholesterolemia is increasing bile acid biosynthesis in liver, as bile acids are the major metabolites of cholesterol (Chen *et al.*, 2008). Farnesoid X-activated receptor (FXR) antagonists from a marine sponge *Spongia* sp. have been discovered as an interesting molecular target of hypocholesterolemia. The FXR suppresses the production of cholesterol 7a-hydroxylase (CYP7A1), which is the rate-limiting enzyme of the bile acid biosynthetic pathway. This novel compound increases the biosynthesis of bile acid and

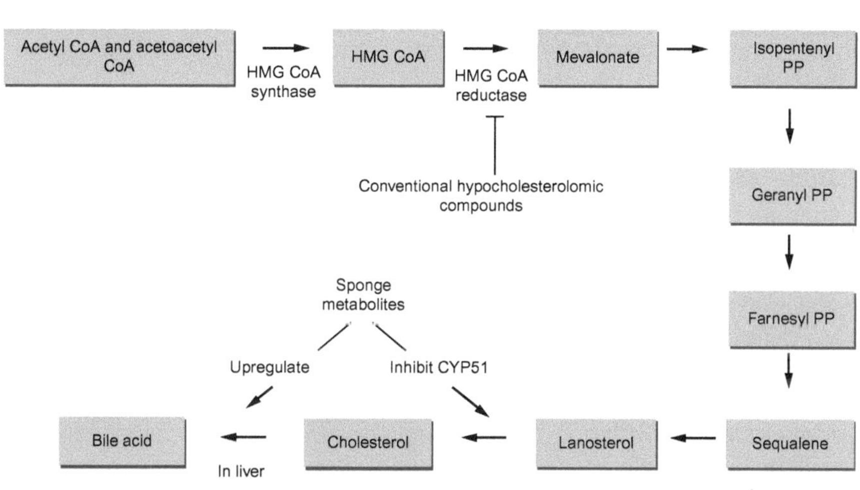

FIGURE 8.6 Cholesterol biosynthesis pathway. 3-Hydroxy-3-methyl-glutaryl coenzyme A (HMG-CoA) reductase is considered as the rate-limiting step of the pathway. Even though HMG-CoA is the main target of common hypocholesterolemic drugs, more specific inhibitors which act on later stages of the pathway claim lesser number of adverse side effects. Lyso-PAF analogues isolated from marine sponge inhibit lanosterol 14α-demethylase (CYP51) followed by inhibiting conversion of lanosterol into cholesterol. Farnesoid X-activated receptor (FXR) antagonists upregulate the cholesterol metabolism in the liver and lower the serum cholesterol.

lower the hepatic cholesterol followed by removal of cholesterol from the circulation and decreases LDL level in the blood (Nam *et al.*, 2006).

D. Sponge collagen

Collagen is the most abundant structural protein in animals, accounts up to 25–35% of total protein content. Unique triple-strand helical polypeptide chains are rich in proline and glycine. This natural polymer is widely used in pharmaceutics and cosmetics because of its biodegradability, biocompatibility, low toxicity, and immunogenicity. Biomaterials as a coating or a carrier of drugs, moisturizers, and health supplements are known applications of collagen (Olsen *et al.*, 2003; Swatschek *et al.*, 2002). Even though animal-derived collagen is the main source of collagen, transmissible diseases (bovine spongiform encephalopathy), antigenic response and its variations, and issues related to food laws limit its applicability and create a demand for alternative sources (Pallela *et al.*, 2011).

Biochemical similarity of amino acid composition and structural similarities to bovine collagen fibrils observed under TEM, SEM, X-ray diffraction, and atomic force microscopy have proved the possibility of substituting conventional collagen for sponge collagen (Heinemann *et al.*, 2007). Overcoming the difficulties associated with isolation and purification of marine sponge collagen, Swatschek *et al.* (2002) suggested a modified method yielding 30% of collagen. Among natural polymers which have been approved for drug coatings for sustained delivery of a dosage, sponge collagen shows distinguish characteristics owing to its stability against pepsin, trypsin, collagenases, and acid media. Nicklas *et al.* (2009) used these properties and developed delayed-release enteric coating from marine sponge *Chondrosia reniformis* Nardo (Demospongiae: Hadromerida: Chondrosiidae) with good mechanical properties and storage stability. In addition, collagen possesses a promising ability of delivering drugs into specific target sites overcoming undesirable interactions (Ruszczak and Friess, 2003). Improved ability of sponge microparticles can be used to deliver anti-inflammatory compounds as they have a tendency of accumulation on inflamed tissues. Moreover, sponge collagen is the best solution for susceptibility of degradation of calf collagen by intestinal pepsin (Berthold *et al.*, 1998; Swatschek *et al.*, 2002).

E. Natural pigments

Appearance, particularly color, is an important characteristic of a food or a supplement because the first impression is made on that. Thus, natural or synthetic colorants play a vital role in food and pharmaceutical industries. Certain legal aspects related to biological effects, purity of synthetic colorants, and poor stability of natural colorants under processing conditions demand alternatives (Mortensen, 2006).

Pigments from marine sponges have long been an interesting topic because of their brilliant color range. Sponge colors are either their own synthesis in pigment granules or a product of symbiotic microbes (Bandaranayake, 2005). Drumm *et al.* (1945) have isolated and characterized carotenoid pigments of marine sponge *Hymeniacidon sanguineum*. Carotenoid pigments are responsible for most bright color sponges, and more than 20 characteristic aryl carotenoids have been identified in sponges (Maoka, 2011). By contrast, carotenoid pigments are strong radical scavengers, are inhibitors of lipid peroxidation, and offer health benefits and higher stability for food and pharmaceuticals than other colorants (Hix *et al.*, 2004). In recent years, the interest of carotenoid production by microbial fermentation has increased instead of extracting from plant tissues or chemical synthesis. Marine sponge-derived actinomycetes sp. (Dharmaraj *et al.*, 2009a), *Streptomyces* sp. (Dharmaraj *et al.*, 2009b), and fungus *Paecilomyces lilacinus* (Elbandy *et al.*, 2009) have been proved for the production of carotenoid pigments and showed the potential as sources of natural colorants. In addition, mytiloxanthin derivatives, red color pigments of bright orange sponge *Phakellia stellidem*, have been identified as bioactive pigments (Higa *et al.*, 1994).

IV. SUSTAINABLE PRODUCTION OF SPONGE METABOLITE

Despite the chemical diversity and biological activities of marine sponge metabolites, large-scale commercial supply of these compounds remains as a challenge. This seems to be a reason why only a few sponge metabolites, out of thousands of novel compounds, are in clinical trials. Wild harvest of sponges for bioactive compounds is not feasible as it threatens the marine environment. Likewise, chemical synthesis of compounds is much more difficult since they are structurally complex (Thakur and Müller, 2004). Therefore, production of bioactive compounds in a sustainable manner became an interesting focus of sponge's research. Several attempts including sponge cell culture, culturing of symbiotic microorganism, or metagenomic approaches in which large gene fragments responsible for production of the bioactive compounds are identified and transferred into a suitable host and sponge aquaculture have been performed to overcome this problem.

Discovery of medicinally important metabolites from marine sponges renewed the interest of sponge aquaculture which has been studied early in the twentieth century for commercial supply of sponge metabolites. Sponge mariculture is considered to be the most cost-effective method (Duckworth and Battershill, 2003). Several drawbacks associated with the mariculture turn the way of sponge culturing into *ex situ* cultivation with controlled conditions. Optimal conditions for growth as well as metabolite production and continuous growth could be achieved with *ex situ* cultivation (Sipkema *et al.*, 2005b). This is an economically feasible option for the production of sponge metabolites, and optimal conditions close to *in situ* double the production of metabolites (Kloppel *et al.*, 2008). Sponge-associated microbes are a promising hope to obtain bioactive metabolites. In addition to culturing of symbiotic microbes, metagenomic approach offers biotechnological solution for unculturable microbial species. Sponge cell culture for production of bioactives was started a decade ago. Muller *et al.* (2000) revealed that proliferative cells of *Dysidea avara* can be used to produce avarol *in vitro*, and cell assemblies termed primmorphs are more capable of producing metabolites than single cells. With these interesting findings, several studies were carried out to optimize the condition of sponge cell culture (Caralt *et al.*, 2007; Schippers *et al.*, 2011; Zhao *et al.*, 2005). In recent future, sponge metabolites for therapeutic applications would be produced *in vitro* instead of wild harvesting.

REFERENCES

Aggarwal, B. B., Kuiken, M. E. V., Iyer, L. H., Harikumar, K. B., and Sung, B. (2009). Molecular targets of nutraceuticals derived from dietary spices: Potential role in suppression of inflammation and tumorigenesis. *Exp. Biol. Med.* **243**, 825–849.

Baby, J. and Sujatha, S. (2010). Pharmacologically important natural products from marine sponges. *J. Nat. Prod.* **4**, 5–12.

Bandaranayake, W. M. (2005). The nature and role of pigments of marine invertebrates. *Nat. Prod. Rep.* **23**, 223–255.

Barrow, C. and Shahidi, F. (2008). Marine Nutraceuticals and Functional Foods. CRC Press Publishing, New York.

Berthold, A., Cremer, K., and Kreuter, J. (1998). Collagen microparticles: Carriers for glucocorticosteroids. *Eur. J. Pharm. Biopharm.* **45**, 23–29.

Boobathy, S., Kumar, T. T. A., and Kathiresann, K. (2009). Isolation of symbiotic bacteria and bio active proteins from marine sponge, Callyspongia diffusa. *Indian J. Biotechnol.* **8**, 272–275.

Bugni, T. S. and Ireland, C. M. (2004). Marine derived fungi: A chemically and biologically diverse group of microorganism. *Nat. Prod. Rep.* **21**, 143–163.

Caralt, S. D., Uriz, M. J., and Wijffels, R. H. (2007). Cell culture from sponges: Pluripotency and immortality. *Trends Biotechnol.* **25**, 467–471.

Carroll, J., Jonsson, E. N., Ebel, R., Hartman, M. S., Holman, T. R., and Crews, P. (2001). Probing sponge-derived terpenoids for human 15-lipoxygenase inhibitors. *J. Org. Chem.* **66**, 6847–6851.

Chen, Z., Jiao, R., and Ma, K. Y. (2008). Cholesterol-lowering nutraceuticals and functional foods. *J. Agric. Food Chem.* **56**, 8761–8773.

David, J. N. and Gordon, M. C. (2004). Marine natural products and related compounds in clinical and advanced preclinical trials. *J. Nat. Prod.* **67**, 1216–1238.

Deng, R. (2009). Food and food supplements with hypocholesterolemic effects. *Recent Pat. Food. Nutr. Agric.* **1**, 15–24.

Devi, P., Wahidullah, S., Rodrigues, C., and Souza, L. D. (2010). The Sponge-associated bacterium bacillus licheniformis SAB1: A source of antimicrobial compounds. *Mar. Drugs* **8**, 1203–1212.

Dharmaraj, S., Ashokkumar, B., and Dhevendaran, K. (2009a). Fermentative production of carotenoids from marine actinomycetes. *Iran. J. Microbiol.* **1**, 36–41.

Dharmaraj, S., Ashokkumar, B., and Dhevendaran, K. (2009b). Food-grade pigments from Streptomyces sp. isolated from the marine sponge Callyspongia diffusa. *Food Res. Int.* **42**, 487–492.

Dharti, T. S., Gandhi, S., and Shah, M. (2010). Nutraceuticals—Portmanteau of science and nature. *Int. J. Pharm. Sci. Rev. Res.* **5**, 33–38.

Drumm, P. J., O'connor, W. F., and Renouf, L. P. (1945). The pigments of sponges. *Biochem. J.* **39**(2), 208–210.

Duckworth, A. and Battershill, C. (2003). Sponge aquaculture for the production of biologically active metabolites: The influence of farming protocols and environment. *Aquaculture* **221**, 311–329.

Elbandy, M., Shinde, P. B., Hong, J., Bae, K. S., Kim, M. A., Lee, S. M., and Jung, J. H. (2009). α-pyrones and yellow pigments from the sponge-derived fungus paecilomyces lilacinus. *Bull. Korean chem. Soc.* **30**, 188–192.

Heinemann, S., Ehrlich, H., Douglas, T., Heinemann, C., Worch, H., Schatton, W., and Hanke, T. (2007). Ultrastructural studies on the collagen of the marine sponge Chondrosia reniformis Nardo. *Biomacromolecules* **8**, 3452–3457.

Hentschel, U., Usher, K. M., and Taylor, M. W. (2006). Marine sponges as microbial fermenters. *FEMS Microbiol. Ecol.* **55**, 167–177.

Higa, T., Tanaka, J., Kitamura, A., Koyama, T., Takahashi, M., and Uchida, T. (1994). Bioactive compounds from marine sponges. *Pure Appl. Chem.* **66**, 2227–2230.

Hix, L. M., Lockwood, S. F., and Bertram, J. S. (2004). Bioactive carotenoids: Potent antioxidants and regulators of gene expression. *Redox Rep.* **9**, 181–191.

John, W. B., Brent, R. C., Murray, H. G. M., Peter, T. N., and Michele, R. P. (2010). Marine natural products. *Nat. Prod. Rep.* **27**, 165–237.

Jorge, A. T. S., Arroteia, K. F., Lago, J. C., Sa´-Rocha, V. M. D., Gesztesi, J., and Moreira, P. L. (2011). A new potent natural antioxidant mixture provides global protection against oxidative skin cell damage. *Int. J. Cosmet. Sci.* **33**, 113–119.

Kelsey, N. A., Wilkins, H. M., and Linseman, D. A. (2010). Nutraceutical antioxidants as novel neuroprotective agents. *Molecules* **15**, 7792–7814.

Keyzers, R. A. and Coleman, M. T. D. (2005). Anti-inflammatory metabolites from marine sponges. *Chem. Soc. Rev.* **34**, 355–365.

Kloppel, A., Pfannkuchen, M., Putz, A., Proksch, P., and Brummer, F. (2008). Ex situ cultivation of Aplysina aerophoba close to in situ conditions: Ecological, biochemical and histological aspects. *Mar. Ecol.* **29**, 259–272.

Korosec, T., Acimovic, J., Seliskar, M., Kocjan, D., Tacer, K. F., Rozman, D., and Urleb, U. (2008). Novel cholesterol biosynthesis inhibitors targeting human lanosterol 14a-demethylase (CYP51). *Bioorg. Med. Chem.* **16**, 209–221.

Lafi, F. F., Garson, M. J., and Fuerst, J. A. (2005). Culturable bacterial symbionts isolated from two distinct sponge species (Pseudoceratina clavata and Rhabdastrella globostellata) from the great barrier reef display similar phylogenetic diversity. *Microb. Ecol.* **50**, 213–220.

Le, U. and Pathak, Y. (2011). Nutraceuticals—Advancing in a right direction. *In* "Hand Book of Nutraceuticals", (Y. Pathak, Ed.), Vol. 2, pp. 1–14.

Li, J. L., Lee, Y. M., Hong, J., Bae, K. S., Choi, J. S., and Jung, J. H. (2011). A new antioxidant from the marine sponge-derived fungus *Aspergillus versicolor*. *Nat. Prod. Sci.* **17**(1), 14–18.

Longeon, A., Copp, B. R., Quévrain, E., Roué, M., Kientz, B., Cresteil, T., Petek, S., Debitus, C., and Kondracki, M. L. B. (2011). Bioactive indole derivatives from the south pacific marine sponges. *Mar. Drugs* **9**, 879–888.

Maoka, T. (2011). Carotenoids in marine animals. *Mar. Drugs* **9**, 278–293.

Maroon, J. C., Bost, J. W., Borden, M. K., Lorenz, K. M., and Ross, N. A. (2006). Natural anti-inflammatory agents for pain relief in athletes. *Neurosurg. Focus* **21**, 1–13.

Medzhitov, R. (2008). Origin and physiological roles of inflammation. *Nature* **454**, 428–435.

Mortensen, A. (2006). Carotenoids and other pigments as natural colorants. *Pure Appl. Chem.* **78**, 1477–1491.

Muller, W. E. G., Bohm, M., Batel, R., Rosa, S. D., Tommonaro, G., Muller, I. M., and Schroder, H. C. (2000). Application of cell culture for the production of bioactive compounds from sponges: Synthesis of avarol by primmorphs from Dysidea avara. *J. Nat. Prod.* **63**, 1077–1081.

Nam, S. J., Ko, H., Shin, M., Ham, J., Chin, J., Kim, Y., Kim, H., Shin, K., Choi, H., and Kang, H. (2006). Farnesoid X-activated receptor antagonists from a marine sponge Spongia sp. *Bioorg. Med. Chem. Lett.* **16**, 5398–5402.

Nicklas, M., Schatton, W., Heinemann, S., Hanke, T., and Kreuter, J. (2009). Enteric coating derived from marine sponge collagen. *Drug Dev. Ind. Pharm.* **35**, 1384–1388.

Olsen, D., Yang, C., Bodo, M., Chang, R., Leigh, S., Baez, J., Carmichael, D., Perala, M., Hamalainen, E. R., Jarvinen, M., and Polarek, J. (2003).*Adv. Drug Deliv. Rev.* **55**, 1547–1567.

Pallela, R., Bojja, S., and Janapala, V. R. (2011). Biochemical and biophysical characterization of collagens of marine sponge, Ircinia fusca (Porifera: Demospongiae: Irciniidae). *Int. J. Biol. Macromol.* **49**, 85–92.

Patnayak, S. and Sree, A. (2005). Screening of bacterial associates of marine sponges for single cell oil and PUFA. *Lett. Appl. Microbiol.* **40**, 358–363.

Posadas, I., Terencio, M. C., Giannini, C., D'Auria, M. V., and Paya, M. (2001). Dysidotronic acid, a new sesquiterpenoid, inhibits cytokine production and the expression of nitric oxide synthase. *Eur. J. Pharmacol.* **415**, 285–292.

Proksch, P., Edrada, R. A., and Ebel, R. (2002). Drugs from the seas—Current status and microbiological implications. *Appl. Microbiol. Biotechnol.* **59**, 125–134.

Roth, G. A., Fihn, S. D., Mokdad, A. H., Aekplakorn, W., Hasegawa, T., and Lim, S. S. (2011). High total serum cholesterol, medication coverage and therapeutic control: An analysis of national health examination survey data from eight countries. *Bull. World Health Organ.* **89**, 92–101.

Ruszczak, Z. and Friess, W. (2003). Collagen as a carrier for on-site delivery of antibacterial drugs. *Adv. Drug Deliv. Rev.* **55**, 1679–1698.

Schippers, K. J., Martens, D. E., Pomponi, S. A., and Wijffels, R. H. (2011). Cell cycle analysis of primary sponge cell cultures. *In Vitro Cell. Dev. Biol. Anim.* **47**, 302–311.

Schneemann, I., Nagel, K., Kajahn, I., Labes, A., Wiese, J., and Imhof, J. F. (2010). Comprehensive investigation of marine actinobacteria associated with the sponge Halichondria panacea. *Appl. Environ. Microbiol.* **76**, 3702–3714.

Shahidi, F. (2009). Nutraceuticals and functional foods: Whole versus processed foods. *Trends Food Sci. Technol.* **20**, 376–387.

Shin, B. A., Kim, Y. R., Lee, I. S., Sung, C. K., Hong, J., Sim, C. J., Im, K. S., and Jung, J. S. (1999). Lyso-PAF analogues and lysophosphatidylcholines from the marine sponge *Spirastrella abata* as inhibitors of cholesterol biosynthesis. *J. Nat. Prod.* **62**, 1554–1557.

Sipkema, D., Franssen, M. C. R., Osinga, R., Tamper, J., and Wijffels, R. H. (2005a). Marine sponges as pharmacy. *Mar. Biotechnol.* **7**, 142–162.

Sipkema, D., Osinga, R., Schatton, W., Mendola, D., Tramper, J., and Wijffels, R. H. (2005b). Large-scale production of pharmaceuticals by marine sponges: Sea, cell, or synthesis? *Biotechnol. Bioeng.* **90**, 202–222.

Sugiyama, Y., Ito, Y., Suzuki, M., and Hirota, A. (2009). Indole Derivatives from a marine sponge-derived yeast as DPPH radical scavengers. *J. Nat. Prod.* **72**, 2069–2071.

Swatschek, D., Schatton, W., Muller, W. E. G., and Kreuter, J. (2002). Microparticles derived from marine sponge collagen (SCMPs): Preparation, characterization and suitability for dermal delivery of all-trans retinol. *Eur. J. Pharm. Biopharm.* **54**, 125–133.

Taylor, M. W., Radax, R., Steger, D., and Wagner, M. (2007). Sponge-associated microorganisms: Evolution, ecology, and biotechnological potential. *Microbiol. Mol. Biol. Rev.* **71**, 295–347.

Thakur, N. L. and Müller, W. E. G. (2004). Biotechnological potential of marine sponges. *Curr. Sci.* **86**, 1506–1512.

Thomas, T. R. A., Kavlekar, D. P., and LokaBharathi, P. A. (2010). Marine drugs from sponge-microbe association—A review. *Mar. Drugs* **8**, 1417–1468.

Thoms, C. and Schupp, P. J. (2007). Chemical defense strategies in sponges: A review. *In* "Porifera Research: Biodiversity, Innovation and Sustainability", (M. R. Custódio, G. L. Hajdu, E. Hajdu, and G. Muricy, Eds), pp. 627–637. IMOS, Brazil.

Uriz, M., Turon, X., Becerro, M. A., and Agell, G. (2003). Siliceous spicules and skeleton frameworks in sponges: Origin, diversity, ultrastructural patterns, and biological functions. *Microsc. Res. Tech.* **62**, 279–299.

Utkina, N. K. (2009). Antioxidant activity of aromatic alkaloids from marine sponges Aaptos aaptos and Hyrtios sp. *Chem. Nat. Compd.* **6**, 849–853.

Utkina, N. K., Makarchenko, A. E., Shchelokova, O. V., and Virovaya, M. V. (2004). Antioxidant activity of phenolic metabolites from marine sponges. *Chem. Nat. Compd.* **40**, 373–377.

Webster, N. S. and Taylor, M. W. (2011). Marine sponges and their microbial symbionts: Love and other relationships. *Environ. Microbiol.* doi: 10.1111/j.1462-2920.2011.02460.x.

Zeisel, S. H. (1999). Regulation of nutraceuticals. *Science* **285**, 1835–1855.

Zhao, Q., Mansoor, T. A., Hong, J., Lee, C. O., Im, K. S., Lee, D. S., and Jung, J. H. (2003). New lysophosphatidylcholines and monoglycerides from the marine sponge Stelletta sp. *J. Nat. Prod.* **66**, 725–728.

Zhao, Q., Zhang, W., Jin, M., Yu, X., and Deng, M. (2005). Formulation of a basal medium for primary cell culture of the marine sponge Hymeniacidon perleve. *Biotechnol. Prog.* **21**, 1009–1012.

CHAPTER 9

Medicinal Benefits of Marine Invertebrates: Sources for Discovering Natural Drug Candidates

Mahanama De Zoysa[1]

Contents

Abstract

Marine invertebrates are one of the major groups of organisms, which could be diversified under the major taxonomic groups of Porifera, Cnidaria, Mollusca, Arthropoda, Echinodermata, and many other minor phyla. To date, range of medicinal benefits and a significant number of marine natural products (MNPs) have been discovered from marine invertebrates. Seafood diet from edible marine invertebrates such as mollusks and crustaceans has been linked with various medicinal benefits to improve human health. Among marine invertebrates, spongers from phylum Porifera is the most dominant group responsible for discovering large number of

College of Veterinary Medicine, Chungnam National University, Yuseong-gu, Daejeon, Republic of Korea
[1] Corresponding author: Mahanama De Zoysa, *E-mail address:* mahanama@cnu.ac.kr

Advances in Food and Nutrition Research, Volume 65
ISSN 1043-4526, DOI: 10.1016/B978-0-12-416003-3.00009-3

MNPs, which have been used as template to develop therapeutic drugs. MNPs isolated from invertebrates have shown wide range of therapeutic properties including antimicrobial, antioxidant, antihypertensive, anticoagulant, anticancer, anti-inflammatory, wound healing and immune modulator, and other medicinal effects. Therefore, marine invertebrates are rich sources of chemical diversity and health benefits for developing drug candidates, cosmetics, nutritional supplements, and molecular probes that can be supported to increase the healthy life span of human.

I. INTRODUCTION

Marine ecosystem has a higher diversity of living organisms compared to the land that provides numerous resources to human societies (Hill and Fenical, 2010). Marine invertebrates are a diverse group having habitats in all ocean ecosystems, ranging from intertidal zone to the deep sea environments. Animals belong to marine invertebrates are composed of different taxonomic groups, which can be classified into several major phyla, namely, Porifera (sponges), cnidaria (corals, sea anemones, hydrozoans, jellyfish), annelida (Polychaetes, marine worms), bryozoa (moss animals or sea mats), mollusca (oysters, abalones, clams, mussels, squids, cuttlefish, octopuses), arthropoda (horseshoe, lobsters, crabs, shrimps, prawns, crayfish), and echinodermata (sea stars, sea cucumbers, sea urchin) (Thorpe *et al.*, 2000). Historically, the dietary and medicinal uses of marine invertebrates were recorded in various literatures of different countries and communities (Gopal *et al.*, 2008; Lev, 2003; Yesilada, 2005). The most notably the "father of modern medicine," Hippocrates has described the use of various marine invertebrates, their ingredients, and therapeutic effects on human health (Grammaticos and Diamantis, 2008; Voultsiadou, 2010; Voultsiadou and Vafidis, 2007). Marine invertebrates account nearly 40% of the world fishery trade, and sea foods obtained from them are highly recognized for health benefits and wellbeing that have been attributed to higher amounts of polyunsaturated fatty acids, various peptides, minerals (selenium and iodine), and other bioactive compounds such as carotenoids and taurine (Borresen, 2009; Caddy, 1989; Erwin *et al.*, 2010; Harris *et al.*, 2009; Larsen *et al.*, 2011; Voultsiadou, 2010).

In 1951, the pioneering work related to nucleosides spongouridine and spongothymidine from the Caribbean sponge *Tethya crypta* (formerly *Cryptotethya crypta*) has introduced new era on drug discovery from marine invertebrates (Bergmann and Feeney, 1951). Thereafter, marine invertebrates are increasingly being selected for screening of bioactive marine natural products (MNPs) to develop new therapeutic agents (Hill and Fenical, 2010; Pawlik, 1993). A recent study conducted by Hu *et al.* (2011)

described that nearly 20,000 MNPs have been isolated from all marine organisms to date and around 75% of that are from marine invertebrates. The most common chemical classes of natural products are belonging to terpenoids, steroids, alkaloids, ethers, phenols, strigolactones, and peptides which all are greatly isolated from marine invertebrates compared to marine plants and microorganisms (Hu *et al.*, 2011). Compounds or extracts obtained from marine invertebrates have shown antiviral, antibacterial, antifungal, antiprotozoal, anthelmintic, anti-inflammatory, antihypertension, anticancer, antituberculosis, and immune modulatory properties, and these candidate molecules lead to enter research pipelines that targeted to develop novel therapeutic agents (Blunt *et al.*, 2004, 2007; Erwin *et al.*, 2010; Folmer *et al.*, 2008; Jha and Zi-rong, 2004; Mayer *et al.*, 2009; Newman and Cragg, 2004; Riguera, 1997; Rinehart *et al.*, 1981; Smith *et al.*, 2010). More importantly, MNPs isolated from marine invertebrates are being evaluating different levels of clinical trials at present and few of them may enter into commercial scale as marine-derived drugs. It is further reported that among the 13 clinically testing MNPs, 12 of them are new drug candidates from marine invertebrates (Thomas *et al.*, 2010). Also, recent trends in MNPs research have reached to set another important stage by developing MNPs database which contains over 12,000 novel compounds isolated from marine organisms (http:momdc.sysu.edu.cn). Additionally, present status of marine-derived compounds that are either been approved by Food Drug Administration (FDA) or are in different phases of clinical pipelines is available at http://marinepharmacology.midwestern.edu/clinPipeline.htm.

Sample collection is the first and one of the most important steps to be considered during natural product drug discovery programs. It is hypothesized that taxonomic diversity correlates to chemical diversity of natural products (Williams, 2008). Therefore, marine ecosystem offers unlimited sources of marine invertebrates to isolate new bioactive compounds under diversified chemical spectrum, which are potent templates for drug discovery. However, researchers need to have a sound understanding about the most potential marine invertebrates for developing health foods or therapeutic agents. Even though there have been a number of recent reviews illustrating MNPs related to their chemical structure, biological function, and applications (Bugni *et al.*, 2008; Erwin *et al.*, 2010; Hill and Fenical, 2010; Hu *et al.*, 2011; Mayer *et al.*, 2009; Voultsiadou, 2010), it is important to update the species of marine invertebrates, which are potential sources for medicinal foods and MNPs for selecting drug candidate species more systematically. However, the research effort on marine drug discovery has not been equally distributed among taxonomic groups. Moreover, discussing on medicinal benefits of these animals will be aimed to introduce new types of invertebrates, which can consume or otherwise use for commercial applications in future. This chapter aims to focus the medicinal benefits, bioactive MNPs, and therapeutic properties of some of the marine invertebrates' phyla such as Porifera, cnidaria, bryozoa, mollusca, arthropoda, and echinodermata. It is also discussing on marine invertebrates as potential sources for discovering natural drug candidates. This approach allows researchers to culture the most promising and demanded groups of marine invertebrates for the production of MNPs.

II. PHYLUM PORIFERA

Spongers from phylum Porifera are the most primitive, evolutionarily ancient metazoan animals. Although spongers are not using as human food, particular sponge species such as *Spongia officinalis* are roasted to produce syrups for homeopathic treatments (Sipkema *et al.*, 2005). Sponges are the major contributing species among invertebrates for discovering MNPs, as they are one of the richest animals with natural secondary metabolites (Belarbi *et al.*, 2003; Blunt *et al.*, 2007; Hu *et al.*, 2011). Hence, they have been regarded as a gold mine and future miracles of sea during the past 50 years (Sipkema *et al.*, 2005). Sponges are simple multicellular animals attached to a substrate and distributed mainly in tropical and temperate waters (Hentschel *et al.*, 2002). Nearly 15,000 sponges have been identified and majority of them belong to be in marine water. Sponges are grouped in three major classes, namely, Calcarea, Demospongiae, and Hexactinellida (Fieseler *et al.*, 2004). They are filter feeders and have unique feeding system, which contains numerous tiny pores on body surface. It facilitates higher chances to intake, and circulates the water with various ions, microorganisms that support for producing secondary metabolites in their body. However, they play an important role in the food chains at intermediate level, as sponges are eaten by other marine animals like chitons, snails, crustaceans, and fish (Meylan, 1990).

Since ancient time, spongers have been used for enhancing of blood coagulation, treating wounds, bone fractures, dropsy, stomach disorders, and other infectious diseases showing that they have being continuously applying in human medicine (Gopal *et al.*, 2008; Sipkema *et al.*, 2005). Large-scale screening of marine spongers for MNPs has been conducted massively. As few examples, it was reported that 37 unidentified sponge species from total of 71 species have antimicrobial properties, in which samples were isolated from west coast of Baja California and Gulf of California (Rinehart *et al.*, 1981). It was revealed that 28% of the sponge samples (from total 302 species) collected from New Zealand, Antarctica, and Western Samoa has shown antimicrobial activity (Munro *et al.*, 1989). The most reported therapeutic properties of MNPs from sponge include antiviral (Cutignano *et al.*, 2000; Martinez *et al.*, 2008), antibacterial (Amade *et al.*, 1987; Andersson *et al.*, 1983; Linington *et al.*, 2006; Lippert *et al.*, 2003; Tadesse *et al.*, 2008; Volk and Kock, 2004), antifungal (Phillipson and Rinehart, 1983), antiparasitic (Capon *et al.*, 2005), antitumor (Schwartsmann *et al.*, 2001), and anti-inflammatory (Fakhr *et al.*, 2006). Cytarabine drug used for treating leukemia and lymphoma patients was developed based on C-nucleosides isolated from Caribbean sponge *C. crypta* (Schwartsmann *et al.*, 2001). Sipkema *et al.* (2005) have summarized bioactive compounds and products from the marine spongers with relation to their various therapeutic properties which show the remarkable potency of sponge-derived medicinal drugs.

The important question is which factors associated with availability of higher MNPs particularly in sponges. Mainly, it is believed that sponges have been emerged for 700–800 million years ago, and early appearance has supported them to develop diversified, complex, and advance defense system against their pathogens and predators through these MNPs (Belarbi *et al.*, 2003; Sipkema *et al.*, 2005). Another important fact is that sponges maintain numerous symbiotic relationships between other marine organisms including bacteria, fungi, and microalgae, which associations are involved in the biosynthesis of range of MNPs (Lee *et al.*, 2001; Thomas *et al.*, 2010). Sponge-bacteria or fungal associations have yielded large number of biologically active compounds such as acetic acid-butyl-ester (antimicrobial), chloriolin B (antitumor), oleic acids (anti-inflammatory), sorbicillactone A (anti-HIV), and roridin A (antileukemic) (Thomas *et al.*, 2010). These finding are very promising, as identifying specific sponge species and related symbiosis relationships can enhance the production of these compounds through mariculture. Antiviral compounds Ara-A and anticancer Ara-C (cytarabine) are two marine sponge (*C. crypta*)-derived compounds currently using as commercial drugs (Newman and Cragg, 2007). As a summary, results discussed in large number of studies greatly proved that marine sponges are promising source for the discovery of therapeutic drugs and large number of sponge species have not been screened for their potential yet.

III. PHYLUM CNIDARIA

Invertebrate animals such as corals, jellyfish, and sea anemones are belonging to phylum Cnidaria. Corals are used as an ingredient in traditional medicine for treating various diseases such as pulmonary tuberculosis, asthma, chronic bronchitis, urinary diseases, and cancer (Gopal *et al.*, 2008). Corals contain mostly Ca and minor amounts of Mg, Fe, and P. Calcined corals or calcium extracted from marine corals have shown effective as therapeutic agent for controlling cancer, diabetes, and many other deadly diseases. Corals with high species diversity in the tropical oceans are often rich in bioactive molecules and preferential targets in drug discovery (Sammarco and Coll, 1992). In 1969, natural source of prostaglandins was first discovered in coral *Plexaura homomalla* (Weinheimer and Spraggins, 1969). The discovery of prostaglandins has greatly contributed to realize the potential and importance of corals as rich source of MNPs. Antimicrobial activities have been reported from different species of corals such as *Xenia macrospiculata* (Kelman *et al.*, 2006), *Alcyonium digitatum* (Andersson *et al.*, 1983, 2005; Tadesse *et al.*, 2008), and *Alcyonium paessleri* (Slattery *et al.*, 1995). Fluorescent proteins of seven Caribbean hard coral species have shown significantly higher H_2O_2 scavenging activity indicating their antioxidant potential to reduce the oxidative stress caused by reactive oxygen species (Palmer *et al.*, 2009). Corals of genus *Montipora* contains high levels of acetylenic compounds and proved to process antimicrobial and cytotoxic properties. More than 10 diacetylene compounds have been isolated from the stony coral *Montipora* spp. which are cytotoxic against human lung, ovarian, skin, and colon cancer cells (Alam *et al.*, 2002). Based on various other Cnidarians, derived MNPs indicate that they have open wide area to conduct further research for extracting new bioactive materials.

Jellyfish has been consuming as a food for more than 1000 years in the world. The processed jellyfish (dried) was first introduced by China for human consumption and recognized as one of Asia's top food with unique taste (Hsieh *et al.*, 2001). Analysis of chemical composition of jellyfish has shown that it contains low levels of protein and calories, with less than 0.35mg/100g of cholesterol (Hsieh *et al.*, 2001). Also, fresh jellyfish believed to be rich in minerals such as Na, Ca, K, and Mg (Hsieh *et al.*, 1996). Based on the nutritional facts of jellyfish, it is considered as low-fat and cholesterol-free marine food. Collagen is the main protein ingredient in jellyfish, which acts as a major connective and building component of tissues, cartilages, and bones. Moreover, jellyfish has been effectively used for obtaining medicinal benefits against various types of diseases such as arthritis, hypertension, ulcers and improving digestion, and skin condition. Also, powder of jellyfish has strong curing power of burning damage (Hsieh and Rudloe, 1994). However, the exact bioactive molecules associated with jellyfish medicinal effects are largely unknown, yet suggesting that further investigating of medicinal values and MNPs can have higher chances to discover natural drug candidates for these diseases.

IV. PHYLUM BRYOZOA

More than 5700 bryozoans species have been recorded and most of them are marine sessile species (Montoya-Cadavid *et al.*, 2007). Lesser number of MNPs have been identified from phylum Bryozoa compared to other main invertebrates (Jha and Zi-rong, 2004). Two novel nematocidal compounds, (brominated alkaloids) named as convolutindol A and convolutamine H, have been isolated from marine bryozoan *Amathia convoluta* extracts (Narkowicz *et al.*, 2002). Most of the MNPs isolated from bryozoans are belonging to group of alkaloid chemicals (Blunt *et al.*, 2004). Bryozoan *Bugula neritina* grows in upright and branching tufts up to 15 m, which is a rich source of anticancer compounds named as bryostatins (Davidson and Haygood, 1999). However later, it was reported that symbiosis bacteria such as *Candidatus Endobugula sertula* could be the main source of bryostatins. It was explained by growing *B. neritina* colonies in the laboratory conditions with antibiotic treatments (to reduce the symbiosis bacteria population) that finally resulted in the reduction of bryostatin production (Davidson *et al.*, 2001). Until today, 20 different bryostatins have been synthesized and some of them are under clinical trials for the treatments against cancer and Alzheimer's disease (Hale and Manaviazar, 2010). Moreover, in animal models, it has been acted as a memory enhancing agent (Kuzirian *et al.*, 2006). If these clinical trials are successful, bryostatin originated from Bryozoans will become a commercial drug for treating different diseases.

V. PHYLUM MOLLUSKA

Mollusks are the second largest animal phylum on earth, and approximately 50,000 living marine species with estimated animals of 100,000–200,000 have been reported (Bouchet, 2006; Pechenic, 2000). They are belonged to eight taxonomic classes including Aplacophora, Monoplacophora, Polyplachophora (chitons), Scaphopoda (tusk shells and tooth shells), Gastropoda (abalones, sea slugs, sea snails, and sea hares), Bivalves (scallops, mussels, oysters, and clams), and Cephalopoda (squids, octopuses, cuttlefish, and nautiluses). It is estimated that about 90% of molluscan species are belonged to one class, the Gastropoda next followed by Bivalves (Ponder and Lindberg, 2008). Among them, gastropods, Bivalves, and cephalopods play important role as food source. Shelled mollusk species are widely used in traditional medicine compared with other groups of mollusks (Benkendorff, 2010). Whole-body or ground shell powders of mollusks with different combinations of other natural sources are common approach in traditional medicine.

Several reports have described the application of oyster shells for treating wound and healing urinary diseases (Gopal *et al.*, 2008; Voultsiadou, 2010), osteoporosis (Fujita *et al.*, 1990), headache, dizziness, and uterine bleeding (Yeung, 1983). Also, dietary supplement containing natural taurine has developed from Pacific oyster *Crassostrea gigas* powder for treating liver disorders, arthritis, and skin problems

(Aroma New Zealand Ltd., 2007). Lipid extracts of oyster *C. gigas* have shown anticancer properties (Kim *et al.*, 2010). Antioxidant peptide purified from gastrointestinal digests of oyster *C. gigas* has shown protective effect of free radical-induced DNA damage (Qian *et al.*, 2008). Experiments conducted using other animal models also have shown health benefits of oyster-derived products. Prevention of gastric ulcers and liver damage has been reported in rat by treating with oyster shell materials (Nie *et al.*, 1994) and water extracts (Kimura *et al.*, 1998), respectively.

Abalone is another important marine gastropod mollusk with high medicinal effects, therefore recommended to physically weak or sick patients. There are number of evidences for beneficial effects or bioactivities of abalone extracts that include antioxidant (Kim *et al.*, 2006), anticancer (Lee *et al.*, 2010; Uchida *et al.*, 1987), antihypertensive (Kim *et al.*, 2006), and anticoagulant (Kim *et al.*, 2006). Immunomodulatory effects have been reported from abalone glycoprotein fraction separated from water extract (Uchida *et al.*, 1987). Marine bivalve clam species are good source of anticoagulants. Sulfated polysaccharides with anticoagulant activities have been isolated from several clam species *Callista chione* (Luppi *et al.*, 2005), *Anomalocardia brasiliana* and *Tivela mactroides* (Pejler *et al.*, 1987), *Mercenaria mercenaria* (Jordan and Marcum, 1986), and *Tapes philippinarum* (Cesaretti *et al.*, 2004). Also, two epidioxysterols isolated from hard clam *Meretrix lusoria* can induce the apoptosis in HL-60 cells and showed its potential for treating liver disease and hepatitis (Pan *et al.*, 2007). Low-molecular-weight (620 Da) peptide with antioxidant activity was discovered from the sauce of fermented blue mussel *Mytilus edulis* (Jung *et al.*, 2005). Marine invertebrates like oysters and mussels are fermented to make varieties of food sources, which contain considerably higher nutritional value and biological activities (Jung *et al.*, 2005). It indicates that fermentation of marine invertebrates can further enhanced the potential of developing health foods and different bioactive molecules. Moreover, mollusks have different feeding habits such as predatory, herbivorous, and scavenging and filter feeding. Most of the mollusks consume large amount of micro- and macroalgae (seaweed) such as *Ecklonia*, *Laminaria*, and *Undaria*. These algae are also rich sources for various MNPs such as polysaccharides, glycoproteins which exhibit potent antimicrobial, anticancer, and antioxidant like therapeutic properties. Therefore, a visceral extract of the mollusks may contain algae-derived bioactive products, thus producing these marine foods with added health benefits.

Corn shell is carnivorous snail belongs to the world's largest genus of marine invertebrates (Conus) under the phylum of mollusks (Bingham *et al.*, 2010). Corn shell species are able to secrete venom, which contains corn shell toxins (conotoxins) of bioactive peptides called as conopeptides (Layer and McIntosh, 2006). Different structural classes of conopeptides belong to six super-families (A, M, O, P, S, and T) that have been identified from various conus species such as *Conus amadis*, *C. geographus*, *C. imperialis*, and *C. magus*, in which data were summarized in a recent review by Bingham *et al.* (2010). Conopeptides are being widely used in clinical as well as in biomedical research for developing analgesic, antinoceptive, anticancer, and cardio- and neuroprotective drugs (Han *et al.*, 2008; Layer and McIntosh, 2006; Twede *et al.*, 2009). Conopeptide isolated from mollusk *C. geographus* has been used to develop drug called Ziconotide/Prialt® which represent the first corn shell drug from marine invertebrate snail (Olivera *et al.*, 1984). It is now in the list of FDA-approved drugs in the United States as well as in European Union for treating chronic pain. Moreover, there are few conotoxins candidates like conantokin-G, conantokin-T, and contulakin-G which are being under clinical phase of drug discovery (Bingham *et al.*, 2010; Skov *et al.*, 2007).

The sea hares are another potential mollusk group with reported pharmacologic substances (Kamiya *et al.*, 2006). The sea hare species Aplysia and Dolabella have been shown to produce antibacterial molecules and cytotoxins (Yamazaki, 1993). Mainly, these bioactive molecules are reported from purple gland or immune response tissues which are involved in the host defense of these invertebrates (Melo *et al.*, 1998; Rajaganapathi *et al.*, 2002). The 60-kDa protein named as bursatellanin-P purified from purple fluid of sea hare *Bursatella leachii* has shown anti-HIV activity (Rajaganapathi *et al.*, 2002). Dolastatins are cytotoxic peptides isolated from mollusk sea hare, *Dolabella auricularia*, which entered into phase II clinical trials as anticancer agents (Schwartsmann *et al.*, 2001). Even though large number of bioactive products has been isolated from mollusks, there are only few approved drugs derived from MNPs isolated from marine mollusks.

As in all invertebrates, mollusks only have an innate immunity and do not develop adaptive immune reactions against pathogens. With higher pathogenic exposure from the environment, marine mollusks require to have efficient defense system through various immune components like antimicrobial peptides (AMPs). These AMPs are cationic and amphipathic peptides with low molecular mass that generally expressed constitutively and induced upon stress and pathogenic challenge to maintain efficient innate immune responses (Smith *et al.*, 2008). Purified proteins or gene encoding for AMPs are being reported at significant phase due to rapid development of genomics in marine organisms. Researchers are continuously searching different AMPs, and their mechanisms of action using native proteins or purifying recombinant proteins since AMPs are alternatives to antibiotics. AMPs like defensins have been identified from various mollusks species including Pacific oyster, *C. gigas* defensins (Gonzales *et al.*, 2007), American oyster *Crassostrea virginica* (Seo *et al.*,

2005), abalone *Haliotis discus discus* (De Zoysa *et al.*, 2010), and Mediterranean mussel *Mytilus galloprovincialis* (Hubert *et al.*, 1996). Several types of AMPs that are identified from various marine invertebrates including mollusk are summarized by Sperstad *et al.* (2011). Even though mollusk AMPs have shown strong activities against serious pathogens, development of mollusk AMPs for using clinical purposes needs further research. With the growing interest on MNPs and rapid development of biotechnology, various species of mollusks have become sources for isolating bioactive products and studying their medicinal values. However, many mollusk species have not been screened for their medicinal benefits to date, and with utilization of these new species of Phylum, mollusk can have unlimited potential for discovering novel drug candidates.

VI. PHYLUM ARTHROPODA

Arthropoda is the largest phylum of all living organisms and having members with characteristic exoskeleton and segmented bodies. Class Crustacea is the largest and the most economically important group of marine arthropods. Among them, crabs, shrimps, and prawns are some of the most attracted species and giving attention in relation to their use as health foods and various types of MNPs (Smith *et al.*, 2008). New sources of marine crustacean such as krill species have been suggested for human consumption and to develop value added health food products due to presence of higher nutritional value (Tou *et al.*, 2007). Higher fat consumption mainly with saturated fatty acids is linked with risk of cardiovascular diseases. But omega 3 polyunsaturated fatty acids, particularly eicosapentaenoic acids (EPA) and decosahexaenoic acids (DHA), are associated with reducing cardiovascular diseases. Analysis of nutrient composition has shown that marine crustaceans like shrimps and krills contain lower total lipids and higher protein, vitamin A, EPA, DHA than some selected fish (Tou *et al.*, 2007).

Chitin is a polysaccharide primarily found in the exoskeleton of crustaceans. It is the main raw material of chitosan, which both have wide range of applications including dietary supplements, functional foods, drugs, cosmetics, antioxidants, and immune stimulants. Harish Prashanth and Tharanathan (2007) have reviewed the use of chitin and chitosan and their potential for medicinal benefits such as hemostasis, drug release, antitumor, anticoagulant, wound healing, antiulcer, antimicrobial, and immune modulatory agents. Biomedical and therapeutic importance of chitin and chitosan highlights the marine crustaceans as an important source of raw materials.

Crabs, lobsters, and shrimps are important sources for natural carotenoids, which are a group of fat soluble pigments such as astaxanthin, canthaxanthin, cryptoxanthin and lutein. Presence of astaxanthin as a major carotenoid and higher levels of unsaturated fatty acids have been found in different crab species (Sachindra *et al.*, 2005). Carotenoids have shown antioxidant properties, and therefore, edible crustaceans are candidates for developing functional food ingredients. It was reported that astaxanthin is potential antiaging compound, while canthaxanthin could be used for treating cancer, stroke, high cholesterol, and Parkinson's disease (Miyashita and Hosokawa, 2008).

Numerous types of AMPs from crustaceans are also isolated and evaluated for their ability to control pathogenic microorganisms. The first crustacean AMP of proline-rich cationic peptide was purified from the hemocytes of shore crab *Carcinus maenas* with activity against both Gram-negative and -positive bacteria (Schnapp *et al.*, 1996). As an example, over 50 different sequences of crustin type AMPs have been reported from crabs, crayfish, and white shrimp. These crustins have shown to inhibit the growth of pathogenic bacteria such as *Staphylococcus aureus*, *Streptococcus iniae*, and *Bacillus* spp. (Smith *et al.*, 2008). Hovingh and Linker (1982) have reported the presence of glucuronic acid-containing heparan sulfate in a species of Crustacea (lobsters) *Homarus americanus*. Numerous reports of marine crustaceans indicate their potential in health benefits and medical applications.

VII. LIMITATIONS OF MARINE INVERTEBRATES AS SOURCE FOR HEALTH FOODS AND DRUG CANDIDATES

There are several problems associated with medicinal foods and discovery of natural drug candidates from marine invertebrates. Even though large numbers of potential marine invertebrates are available as health foods, there is a high risk of biotoxins and environmental intoxicants on public health. Different toxic levels have been reported by consuming marine invertebrates including species of sea anemone, sea urchins, clam, mussels, oysters, abalone, squid, octopus, crab, and lobsters (Bagnis *et al.*, 1970). Therefore, proper quality assurance is necessary for introducing marine invertebrates as health foods for human. However, those toxic marine organisms are potential sources of new pharmaceutical products. Also, drug discovery from marine invertebrates has been slowed due to lack of sufficient amounts of biological materials (Folmer *et al.*, 2008). The sample collection from marine environment may be more difficult than terrestrial sources, although some animals are easily accessible (spongers) which can be collected by hand. The most commonly reported technical problem is the presence of large quantities of inorganic salts in the invertebrate extracts. Additionally, wide variation of chemical diversity within one type of source can exhibit different or opposite activities and presence of major active compounds can mask the activity of minor molecules in the crude extracts. So that could be affected with low detection of bioactive molecules from crude extract of marine invertebrates during drug discovery screening programs (Bugni *et al.*, 2008). Another limitation is that it required very sensitive analytical tools, as the recovery of bioactive material is significantly low. Most of the invertebrates are maintaining symbiotic relationship with other organisms such as fungi, bacteria, and microalgae that could cause difficulties to find out exact origin of the bioactive material (Riguera, 1997). There were some controversial suspicions that MNP isolated from invertebrates may be produced by symbiotic microorganisms. In general, several limiting factors are associated and make it more complicated in the search of natural drug screening programs from marine invertebrates. In many occasions, MNPs isolated from invertebrates have shown excellent activities as therapeutic compound but unable to use as medicinal drug due to high toxicity or other side effects.

VIII. CONCLUSION

As summary, marine invertebrates have a wide range of biological activities with unique chemical structures. Also, focus of medicinal foods and drug discovery from marine invertebrates mainly have touched the tropical and temperate environments whereas large extents of polar area are unexploded yet. Therefore, it is necessary to identify potential species with specific parts or organs of the animal, where the bioactive molecules are abundant. Based on the present status, we can assume that these marine invertebrates will become sources of MNPs than any other terrestrial animal group to provide health benefits for developing drug candidates, cosmetics, nutritional supplements, and molecular probes that can be supported to increase the healthy life span of human.

REFERENCES

Alam, N., Hong, J., Lee, C. O., Choi, J. S., Im, K. S., and Jung, J. H. (2002). Additional cytotoxic diacetylenes from the stony coral *Montipora* sp. *Chem. Pharm. Bull. (Tokoyo)* **50**, 661–662.

Amade, P., Charroin, C., Baby, C., and Vacelet, J. (1987). Antimicrobial activities of marine sponges from the Mediterranean Sea. *Mar. Biol.* **94**, 271–275.

Andersson, L., Lidgren, G., Bohlin, L., Magni, L., Ogren, S., and Afzelius, L. (1983). Studies of Swedish marine organisms I: Screening of biological activity. *Acta Pharm. Suec.* **20**, 401–414.

Aroma New Zealand Ltd (2007). Company Profile and Product Specification. www.aromanz.com.

Bagnis, R., Berglund, F., Elias, P. S., Van Esch, G. J., Halstead, B. W., and Kojima, K. (1970). Problems of toxicants in marine food products. *Bull. World Health Organ.* **42**, 69–88.

Belarbi, E. H., Gomez, A. C., Chisti, Y., Camacho, F. G., and Grima, E. M. (2003). Producing drugs from marine sponges. *Biotechnol. Adv.* **21**, 585–598.

Benkendorff, K. (2010). Molluscan biological and chemical diversity: Secondary metabolites and medicinal resources produced by marine mollusks. *Biol. Rev.* **85**, 757–775.

Bergmann, W. and Feeney, R. J. (1951). Contributions to the study of marine products XXXII. The nucleosides of sponges. *J. Org. Chem.* **16**, 981–987.

Bingham, J. P., Mitsunaga, E., and Bergeron, Z. L. (2010). Drugs from slugs-past, present and future perspectives of ώ-conotoxin research. *Chem. Biol. Interact.* **183**, 1–18.

Blunt, J. W., Copp, B. R., Munro, M. H. G., Northcote, P. T., and Prinsep, M. R. (2004). Marine natural products. *Nat. Prod. Rep.* **21**, 1–49.

Blunt, J. W., Copp, B. R., Hu, W. P., Munro, M. H. G., Northcote, P. T., and Prinsep, M. R. (2007). Marine natural products. *Nat. Prod. Rep.* **24**, 31–86.

Borresen, T. (2009). Seafood for improved health and wellbeing. *Food Technol.* **63**, 88.

Bouchet, P. (2006). The magnitude of marine biodiversity. *In* "The Exploration of Marine Biodiversity, Scientific and Technological Challenges", (C. M. Duarte, Ed.), pp. 33–64. Foundation BBVA, Francewww.fbbva.es.

Bugni, T. S., Harper, M. K., McCulloch, M. W. B., Reppart, J., and Ireland, C. M. (2008). Fractionated marine invertebrate extract libraries for drug discovery. *Molecules* **13**, 1372–1383.

Caddy, J. F. (1989). Marine Invertebrate Fisheries: Their Assessment and Management. John Wiley and Sons, New York, p. 752.

Capon, R. J., Vuong, D., Lacey, E., and Gill, J. H. (2005). Echinobetaine A: Isolation, structure elucidation, synthesis, and preliminary SAR studies on a new nematocidal betaine from a southern Australian marine sponge, *Echinodictyum* sp.. *J. Nat. Prod.* **68**, 179–182.

Cesaretti, M., Luppi, E., Maccari, M., and Volpi, N. (2004). Isolation and characterization of a heparin with high anticoagulant activity from the clam *Tapes philippinarum*. Evidence for the presence of a high content of antithrombin III-binding site. *Glycobiology* **14**, 1275–1284.

Cutignano, A., Bifulco, G., Bruno, I., Casapullo, A., Gomez-Paloma, L., and Riccio, R. (2000). Dragmacidin F: A new antiviral bromoindole alkaloid from the Mediterranean sponge *Halicortex* spp. *Tetrahedron* **56**, 3743–3748.

Davidson, S. K. and Haygood, M. G. (1999). Identification of sibling species of the Bryozoan *Bugula neritina* that produce different anticancer bryostatins and harbor distinct strains of the bacterial symbiont "*Candidatus Endobugula sertula*". *Biol. Bull.* **196**, 273–280.

Davidson, S. K., Allen, S. W., Lim, G. E., Anderson, C. M., and Haygood, M. G. (2001). Evidence for the biosynthesis of bryostatins by the bacterial symbiont "*Candidatus* Endobugula sertula" of the Bryozoan *Bugula neritina*. *Appl. Environ. Microbiol.* **67**, 4531–4537.

De Zoysa, M., Whang, I., Lee, Y., Lee, S., Lee, J. S., and Lee, J. (2010). Defensin from disk abalone *Haliotis discus discus*: Molecular cloning, sequence characterization and immune response against bacterial infection. *Fish Shellfish Immunol.* **28**, 261–266.

Erwin, P. M., Lopez-Legentil, S., and Schuhmann, P. W. (2010). The pharmaceutical value of marine biodiversity for anti-cancer drug discovery. *Ecol. Econ.* **70**, 445–451.

Fakhr, I., Hamdy, N., Radwan, M., El-Batran, S., and El Shabrawy, O. (2006). Studies on the anti-inflammatory and analgesic effects of extracts from marine sponges. *Nat. Prod. Sci.* **12**, 74–78.

Fieseler, L., Horn, M., Wagner, M., and Hentschel, U. (2004). Discovery of the novel candidate Phylum "*Poribacteria*" in marine sponges. *Appl. Environ. Microbiol.* **70**, 3724–3732.

Folmer, F., Jaspars, M., Dicato, M., and Diederich, M. (2008). Marine natural products as targeted modulators of the transcription factor NF-κB. *Biochem. Pharmacol.* **75**, 603–617.

Fujita, T., Fukase, M., Miyamoto, H., Matasumoto, T., and Ohue, T. (1990). Increase of bone mineral density by calcium supplement with oyster shell electrolysate. *Bone Miner.* **11**, 85–91.

Gonzales, M., Gueguen, Y., Desserre, G., de Lorgeril, J., Romestand, B., and Bachere, E. (2007). Molecular characterization of two isoforms of defensin from hemocytes of the oyster *Crassostrea gigas*. *Dev. Comp. Immunol.* **31**, 332–339.

Gopal, R., Vijayakumaran, M., Venkatesan, R., and Kathiroli, S. (2008). Marine organisms in Indian medicine and their future prospects. *Nat. Prod. Rad.* **7**, 139–145.

Grammaticos, P. C. and Diamantis, A. (2008). Useful known and unknown views of the father of modern medicine Hippocrates and his teacher Democritus. *Hell. J. Nucl. Med.* **11**, 2–4.

Hale, K. J. and Manaviazar, S. (2010). New approaches to the total synthesis of bryostatin antitumor macrolides. *Chem. Asian J.* **5**, 704–754.

Han, T. S., Teichert, R. W., Olivera, B. M., and Bulaj, G. (2008). Conus venoms-a rich source of peptide-based therapeutics. *Curr. Pharm. Des.* **14**, 2462–2479.

Harish Prashanth, K. V. and Tharanathan, R. N. (2007). Chitin/chitosan: Modifications and their unlimited application potential. *Trends Food Sci. Technol.* **18**, 117–131.

Harris, M., Bruhn, C., Schor, D., Kapsak, W. R., and Blakistone, B. (2009). Communicating the net benefits of seafood consumption. *Food Technol.* **63**, 38–44.

Hentschel, U., Hopke, J., Horn, M., Friedrich, A. B., Wagner, M., Hacker, J., and Moore, B. S. (2002). Molecular evidence for a uniform microbial community in sponges from different oceans. *Appl. Environ. Microbiol.* **68**, 4431–4440.

Hill, R. T. and Fenical, W. (2010). Pharmaceuticals from marine natural products: Surge or ebb? *Curr. Opin. Biotechnol.* **21**, 777–779.

Hovingh, P. and Linker, A. (1982). An unusual heparin sulfate isolated from lobsters (Homarus americanus). *J. Biol. Chem.* **16**, 9840–9844.

Hsieh, Y. H. P. and Rudloe, J. (1994). Potential of utilizing jellyfish as food in Western countries. *Trends Food Sci. Technol.* **5**, 225–229.

Hsieh, Y. H. P., Leong, F. M., and Barnes, K. W. (1996). Inorganic constituent in fresh and processed cannonball jellyfish (*Stomolophus meleagris*). *J. Agric. Food Chem.* **44**, 3117–3119.

Hsieh, Y. H. P., Leong, F. M., and Rudloe, J. (2001). Jellyfish as food. *Hydrobiologia* **451**, 11–17.

Hu, G. P., Yuan, J., Sun, L., She, Z. G., Wu, J. H., Lan, X. J., Zhu, X., Lin, Y. C., and Chen, S. P. (2011). Statistical research on marine natural products based on data obtained between 1985 and 2008. *Mar. Drugs* **9**, 514–525.

Hubert, F., Noel, T., and Roch, P. (1996). A member of the arthropod defensin family from edible Mediterranean mussels (*Mytilus galloprovincialis*). *Eur. J. Biochem.* **240**, 302–306.

Jha, R. K. and Zi-rong, X. (2004). Biomedical compounds from marine organisms. *Mar. Drugs* **2**, 123–146.

Jordan, R. E. and Marcum, J. A. (1986). Anticoagulant active heparin from clam (*Mercenaria mercenaria*). *Arch. Biochem. Biophys.* **248**, 690–695.

Jung, W. K., Rajapakse, N., and Kim, S. K. (2005). Antioxidative activity of a low molecular weight peptide derived from the sauce of fermented blue mussel, *Mytilus edulis*. *Eur. Food Res. Technol.* **220**, 535–539.

Kamiya, H., Sakai, R., and Jimbo, M. (2006). Bioactive molecules from sea hares. *Prog. Mol. Subcell. Biol.* **43**, 215–239.

Kelman, D., Kashman, Y., Rosenberg, E., Kushmaro, A., and Loya, Y. (2006). Antimicrobial activity of red sea corals. *Mar. Biol.* **149**, 357–363.

Kim, H. L., Kang, S. G., Kim, I. C., Kim, S. J., Kim, D. W., Ma, S. J., Gao, T., Li, H., Kim, M. U., and Lee, T. H. (2006). In vitro Anti-hypertensive, antioxidant and anticoagulant activities of extracts from *Haliotis discus hannai*. *J. Korean Soc. Food Sci. Nutr.* **35**, 835–840.

Kim, E. K., Kim, Y. S., Lee, S. J., Jeon, Y. J., Lee, J. D., Son, T. I., Ahn, C. B., Kim, Y. T., Moon, S. H., and Jeon, B. T. (2010). Anticancer effect of lipids partially purified from Pacific oyster, *Crassostrea gigas* on PC3 cells. *Food Sci. Biotechnol.* **19**, 213–217.

Kimura, Y., Ohminami, H., and Okuda, H. (1998). Effects of extract of oyster on lipid metabolism in rats. *J. Ethnopharmacol.* **59**, 117–123.

Kuzirian, A. M., Epstein, H. T., Gagliardi, C. J., Nelson, T. J., Sakakibara, M., Taylor, C., Scioletti, A. B., and Alkon, D. L. (2006). Bryostatin enhancement of memory in Hermissenda. *Biol. Bull.* **210**, 201–214.

Larsen, R., Eilertsen, K. E., and Elvevoll, E. O. (2011). Health benefits of marine foods and ingredients. *Biotechnol. Adv.* **29**, 508–518.

Layer, R. T. and McIntosh, J. M. (2006). Conotoxins: Therapeutic potential and application. *Mar. Drugs* **4**, 119–142.

Lee, Y. K., Lee, J. H., and Lee, H. K. (2001). Microbial symbiosis in marine sponges. *J. Microbiol.* **39**, 254–264.

Lee, C. G., Kwon, H. K., Ryu, J. H., Kang, S. J., Im, C. R., Kim, J. I., and Im, S. H. (2010). Abalone visceral extract inhibit tumor growth and metastasis by modulating Cox-2 levels and CD8+ T cell activity. *BMC Complement. Altern. Med.* **10**, 60.

Lev, E. (2003). Traditional healing with animals (Zootherapy): Medieval to present day Levantine practice. *J. Ethnopharmacol.* **86**, 107–118.

Linington, R. G., Robertson, M., Gauthier, A., Finlay, B. B., MacMillan, J. B., Molinski, T. F., van Soest, R., and Andersen, R. J. (2006). Caminosides B–D, antimicrobial glycolipids isolated from the marine sponge *Caminus sphaeroconia*. *J. Nat. Prod.* **69**, 173–177.

Lippert, H., Brinkmeyer, R., and Iken, K. (2003). Antimicrobial activity in sub-Arctic marine invertebrates. *Polar Biol.* **26**, 591–600.

Luppi, E., Cesaretti, M., and Volpi, N. (2005). Purification and characterization of heparin from the Italian clam *Callista chione*. *Biomacromolecules* **6**, 1672–1678.

Martinez, M. J. A., Olmo, L. M. B. D., and Benito, P. B. (2008). Natural marine antiviral products. *Bioact. Nat. Prod.* **35**, 101–134.

Mayer, A. M. S., Rodriguez, A. D., Berlinck, R. G. S., and Hamann, M. T. (2009). Marine pharmacology in 2005–6: Marine compounds with anthelmintic, antibacterial, anticoagulant, antifungal, antiinflammatory, antimalarial, antiprotozoal, antituberculosis, and antiviral activities; affecting the cardiovascular, immune and nervous systems, and other miscellaneous mechanisms of action. *Biochim. Biophys. Acta* **5**, 283–308.

Melo, V. M. M., Fonseca, A. M., Vasconcelos, I. M., and Carvalho, A. F. (1998). Purification of a novel antibacterial and haemagglutinating protein from the purple gland of the sea hare, *Aplysia dactylomela* Rang, 1828. *Toxicon* **38**, 1415–1427.

Meylan, A. (1990). Nutritional characteristics of the sponges in the diet of the hawksbill turtle *Eretmochelys imbricata*. *In* "New Perspectives in Sponge Biology", (K. Rutzler, Ed.), pp. 472–477. Smithsonian, Institution Press, Washington, DC.

Miyashita, K. and Hosokawa, M. (2008). Beneficial health effects of seaweed carotenoid, fucoxanthin. *In* "Marine Nutraceuticals and Functional Foods", (C. Barrow and F. Shahidi, Eds), pp. 297–320. CRC Press, Boca Raton, USA.

Montoya-Cadavid, E., Romero, P. F., and Winston, J. E. (2007). Check list of the marine Bryozoa of the Colombian Caribben. *Biota Colombiana* **8**, 159–184.

Munro, M. H. G., Blunt, J. W., Barns, G., Battershill, C. N., Lake, R. J., and Perry, N. B. (1989). Biological activity in New Zealand marine organisms. *Pure Appl. Chem.* **61**, 529–534.

Narkowicz, C. K., Blackman, A. J., Lacey, E., Gill, J. H., and Heiland, K. (2002). Convolutindole A and convolutamine H, new nematocidal brominated alkaloids from the marine bryozoan *Amathia convoluta*. *J. Nat. Prod.* **65**, 938–941.

Newman, D. and Cragg, G. M. (2004). Marine natural products and related compounds in clinical and advanced preclinical trials. *J. Nat. Prod.* **67**, 1216–1238.

Newman, D. and Cragg, G. M. (2007). Natural products as sources of new drugs over the last 25 years. *J. Nat. Prod.* **70**, 461–477.

Nie, S. Q., Li, T. L., Jiang, A. H., and Li, G. Q. (1994). A comparative study on anti-ulcer action of unprepared and calcined oyster shell. *Zhongguo Zhong Yao Za Zhi* **19**, 405–407.

Olivera, B. M., McIntosh, J. M., Lourdes, J. C., Luque, F. A., and Gray, W. R. (1984). Purification and sequence of a presynaptic peptide toxin from *Conus geographus* venom. *Biochemistry* **23**, 5087–5090.

Palmer, C. V., Modi, C. K., and Mydlarz, L. D. (2009). Coral fluorescent proteins as antioxidants. *PLoS One* **4**, e7298.

Pan, H. M., Huang, Y. T., Chang, C. I., Ho, C. T., and Pan, B. S. (2007). Apoptotic-inducing epidioxysterols identified in hard clam (*Meretrix lusoria*). *Food Chem.* **102**, 788–795.

Pawlik, J. R. (1993). Marine invertebrate chemical defenses. *Chem. Rev.* **93**, 1911–1922.

Pechenic, J. A. (2000). Biology of Invertebrates. 4th edn. McGraw Hill, New York.

Pejler, G., Danielsson, A., Bjork, I., and Lindahl, U. (1987). Structure and antithrombin-binding properties of heparin isolated from the clams *Anomalocardia brasiliana* and *Tivela mactroides*. *J. Biol. Chem.* **262**, 11413–11421.

Phillipson, D. W. and Rinehart, K. L., Jr. (1983). Antifungal peroxide-containing acids from two Caribbean sponges. *J. Am. Chem. Soc.* **105**, 7735–7736.

Ponder, W. F. and Lindberg, D. R. (2008). Phylogeny and Evolution of Mollusca. University of California Press, Berkley, p. 469.

Qian, Z. J., Jung, W. K., Byun, H. G., and Kim, S. K. (2008). Protective effect of an antioxidative peptide purified from gastrointestinal digests of oyster, *Crassostrea gigas* against free radical induced DNA damage. *Bioresour. Technol.* **99**, 3365–3371.

Rajaganapathi, J., Kathiresan, K., and Singh, T. P. (2002). Purification of Anti-HIV Protein from Purple fluid of the sea hare *Bursatella leachii* de Blainville. *Mar. Biotechnol.* **4**, 447–453.

Riguera, R. (1997). Isolating bioactive compounds from marine organisms. *J. Mar. Biotechnol.* **5**, 187–193.

Rinehart, K. L., Shaw, P. D., Shield, L. S., Gloer, J. B., Harbour, G. C., Koker, M. E. S., Samain, D., Schwartz, R. E., Tymiak, A. A., Weller, D. L., Carter, G. T., Munro, M. H. G., *et al.* (1981). Marine natural products as sources of antiviral, antimicrobial, and antineoplastic agents. *Pure Appl. Chem.* **59**, 795–817.

Sachindra, N. M., Bhaskar, N., and Mahendrakar, N. S. (2005). Carotenoids in crabs from marine and fresh waters of India. *Food Sci. Technol.* **38**, 221–225.

Sammarco, P. W. and Coll, J. C. (1992). Chemical adaptations in the Octocorallia: Evolutionary considerations. *Mar. Ecol. Prog. Ser.* **88**, 93–104.

Schnapp, D., Schnapp, D., Kemp, G. D., and Smith, V. J. (1996). Purification and characterization of a proline-rich antibacterial peptide, with sequence similarity to bactenecin 7, from the haemocytes of the shore crab, *Carcinus maenas*. *Eur. J. Biochem.* **240**, 532–539.

Schwartsmann, G., da Rocha, A. B., Berlinck, R. G., and Jimeno, J. (2001). Marine organisms as a source of new anticancer agents. *Lancet Oncol.* **2**, 221–225.

Seo, J. K., Crawford, J. M., Stone, K. L., and Noga, E. J. (2005). Purification of a novel arthropod defenisn from the American oyster, *Crassostrea virginica*. *Biochem. Biophys. Res. Commun.* **338**, 1998–2004.

Sipkema, D., Franssen, M. C. R., Osinga, R., Tramper, J., and Wijffels, R. H. (2005). Marine sponges as pharmacy. *Mar. Biotechnol.* **7**, 142–162.

Skov, M. J., Beck, J. C., de Kater, A. W., and Shopp, G. M. (2007). Nonclinical safety of ziconotide: An intrathecal analgesic of a new pharmaceutical class. *Int. J. Toxicol.* **26**, 411–421.

Slattery, M., McClintock, J. B., and Heine, J. N. (1995). Chemical defense in Antarctic soft corals: Evidence for antifouling compounds. *J. Exp. Mar. Biol. Ecol.* **190**, 61–71.

Smith, V. J., Fernandes, J. M. O., Kemp, G. D., and Hauston, C. (2008). Crustins: Enigmatic WAP domain-containing antibacterial proteins from crustaceans. *Dev. Comp. Immunol.* **32**, 758–772.

Smith, V. J., Desbois, A. P., and Dyrynda, E. A. (2010). Conventional and unconventional antimicrobials from fish, marine invertebrates and micro-algae. *Mar. Drugs* **8**, 1213–1262.

Sperstad, S. V., Haug, T., Blencke, H. M., Styrvol, O. B., Li, C., and Stensvag, K. (2011). Antimicrobial peptides from marine invertebrates: Challengers and perspectives in marine antimicrobial peptide discovery. *Biotechnol. Adv.* **29**, 519–530.

Tadesse, M., Gulliksen, B., Strøm, M. B., Styrvold, O. B., and Haug, T. (2008). Screening for antibacterial and antifungal activities in marine benthic invertebrates from northern Norway. *J. Invertebr. Pathol.* **99**, 286–293.

Thomas, T. R. A., Kavlekar, D. P., and LokaBharathi, P. A. (2010). Marine drugs from sponge-microbe association. *Mar. Drugs* **8**, 1417–1468.
Thorpe, J. P., Sole-Cava, A. M., and Watts, P. C. (2000). Exploited marine invertebrates: Genetics and fisheries. *Hydrobiologia* **420**, 165–184.
Tou, J. C., Jaczynski, J., and Chen, Y. C. (2007). Krill for human consumption: Nutritional value and potential health benefits. *Nutr. Rev.* **65**, 63–77.
Twede, V. D., Miljanich, G., Olivera, B. M., and Bulaj, G. (2009). Neuroprotective and cardioprotective conopeptides: An emerging class of drug leads. *Curr. Opin. Drug Discov. Devel.* **12**, 231–239.
Uchida, H., Sasaki, T., Uchida, N. A., Takasuka, N., Endo, Y., and Kamiya, H. (1987). Oncostatic and immunomodulatory effects of a glycoprotein fraction from water extract of abalone, *Haliotis discus hannai. Cancer Immunol. Immunother.* **24**, 207–212.
Volk, C. A. and Kock, M. (2004). Viscoline: New 3-alkyl pyridinium from the Arctic sponge *Haliclona viscose. Org. Biomol. Chem.* **2**, 1827–1830.
Voultsiadou, E. (2010). Therapeutic properties and uses of marine invertebrates in the ancient Greek world and early Byzantum. *J. Ethnopharmacol.* **130**, 237–247.
Voultsiadou, E. and Vafidis, D. (2007). Marine invertebrates diversity in Aristotle's zoology. *Contrib. Zool.* **76**, 103–120.
Weinheimer, A. J. and Spraggins, R. L. (1969). The occurrence of two new prostaglandin derivatives (15-epi-PGA2 and its acetate, methyl ester) in the gorgonian Plexaura homomalla, Chemistry of coelenterates. *Tetrahedron Lett.* 5185–5188.
Williams, P. G. (2008). Panning for chemical gold: Marine bacteria as a source of new therapeutics. *Trends Biotechnol.* **27**, 45–52.
Yamazaki, M. (1993). Antitumor and antimicrobial glycoproteins from sea hare. *Comp. Biochem. Physiol. C* **105**, 141–146.
Yesilada, E. (2005). Past and future contributions to traditional medicine in the health care system of the Middle-East. *J. Ethnopharmacol.* **100**, 135–137.
Yeung, H. C. (1983). Handbook of Chinese Herbal Formulas. Institute of Chinese Medicine, Rosemead, LA, USA.

CHAPTER 10

Sea-Originated Cytotoxic Substances

Norihito Maru and **Daisuke Uemura**[1]

Contents

Abstract

Cancer accounted for huge number of deaths, which represents about 13% of all deaths worldwide, and the number of the deaths due to cancer is increasing. Natural products and their synthetic analogs are widely used as antitumor drugs. As represented by these drugs, many anticancer drugs originated from cytotoxic compounds. Marine natural products are a gold mine of strong bioactive compounds with unique structures created in evolution of organisms over hundred million years. However, in the field of drug discovery, most studies have focused on plant essences and bacterial metabolites, and candidate compounds from marine origin are still remaining relatively unexplored.

Department of Chemistry, Faculty of Science, Kanagawa University, Hiratsuka, Japan
[1] Corresponding author: Daisuke Uemura, *E-mail address:* uemurad@kanagawa-u.ac.jp

Advances in Food and Nutrition Research, Volume 65
ISSN 1043-4526, DOI: 10.1016/B978-0-12-416003-3.00010-X

I. INTRODUCTION

Important anticancer drugs are not found without the tremendous effort of scientists. In 2008, cancer accounted for nearly 7.6 million deaths, which represents about 13% of all deaths worldwide (Ferlay *et al.*, 2010). Further, the number of the deaths due to cancer is increasing, and has been projected to reach to 11.5 million in 2030. Therefore, there is a constant need for new compounds that can serve as leads for new anticancer drugs. Currently, natural products and their synthetic analogs are widely used as antitumor drugs, including paclitaxel, vinblastine, and etoposide. As represented by these drugs, many anticancer drugs originated from cytotoxic compounds. Cytotoxicity is ambiguous concept that is also paraphrase of recasting of cell death (necrosis and apoptosis) or arrest in cell division. Mechanisms of action are huge diverse modes, such as membrane integrity losing, metabolic enzyme inhibition, signal-transfer defection, apoptosis induction, DNA replication inhibition, mitosis inhibition, and so on. Due to wide variety of mode of action, cytotoxic compounds could also contribute to the discovery of new type drugs.

Marine natural products are a gold mine of strong bioactive compounds with unique structures created in evolution of organisms over hundred million years. However, in the field of drug discovery, most studies have focused on plant essences and bacterial metabolites, and candidate compounds from marine origin are still remaining relatively unexplored. This review shows cytotoxic compound of an antitumor drug lead and compounds with potential for use for drug discovery that isolated from marine phenomena.

II. HALICHONDRINS, ANTITUMOR COMPOUNDS FROM *HALICHONDRIA OKADAI*

H. okadai Kadota (Fig. 10.1) is a sponge that is abundant in the tidal zone in the Pacific Ocean south of the Boso Peninsula in Japan. Sponges take seawater into their bodies, gather food by filtration, and coexist with symbiotic microorganisms, which account for about 40% of the sponge's volume (Vogel, 2008). Thus, sponges might be aggregates of marine microorganisms. Since sponges have existed for about 500 million years without being able to physically move, they must possess some hidden capability for endurance. After the oil crisis in the 1970s, Gorgonians were shown to contain large quantities of prostaglandins, and this increased the level of interest in natural marine resources. This represented a shift from mass chemicals to fine chemicals. Under these circumstances, the level of interest in marine organisms, and particularly sponges, increased,

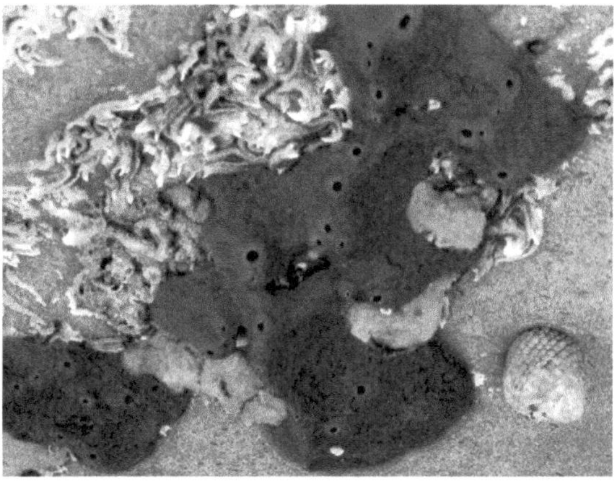

FIGURE 10.1 Black sponge *Halichondria okadai*. (For color version of this figure, the reader is referred to the Web version of this chapter.)

and research on antitumor compounds in *H. okadai* Kadota was started (Uemura, 2010). The inhibitory effects of the constituent compounds were closely examined based on the proliferation of B16 melanoma cells *in vivo*, and eight halichondrin homologues (Fig. 10.2) were identified (Hirata and Uemura, 1986).

Based on the name of this particular sponge, the antitumor substances were named halichondrins. However, 600kg of the sponge gave <15mg of the most active homolog, halichondrin B. Fortunately, useful experience was gained by determining the structure of palytoxin, and we concluded that its structure could not be determined based solely on detailed NMR data. In particular, halichondrin B was very fragile and was obtained in small quantities, which made our task very difficult. Thus, norhalichondrin A, which was present in greater quantities, was purified in the form of a *p*-bromophenacyl ester, and a high-quality crystal was obtained. X-ray crystal structure analyses were performed, and extremely complex and specific structures were seen (Uemura *et al.*, 1985). The 2,6,9-trioxatricyclo[3.3.2.$0^{3,7}$]decane ring system inside the molecule was the most complex. Both ends of the molecule were diverse: at one end, the number and length of the carbon chains varied, and at the other end, the number and oxidation state of the hydroxy groups varied.

While we suspected that there are a total of nine variations in halichondrin A, we have not identified all of them, mostly due to issues regarding limited quantities. The antitumor activities of halichondrin B have been clarified by *in vivo* experiments, and it appears to be a promising compound. However, toxicity has always been an issue. In a joint Japan–US science seminar on cancer research held in Hawaii in 1987, B. A. Chabner at NCI and G. R. Pettit at Arizona State University were

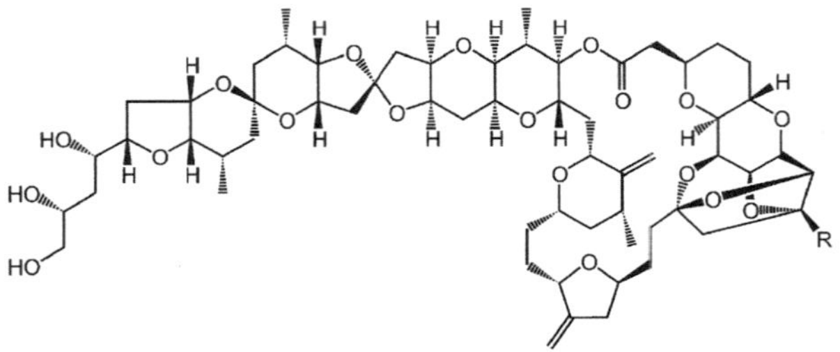

Halichondrin B: R = H
Halichondrin C: R = OH

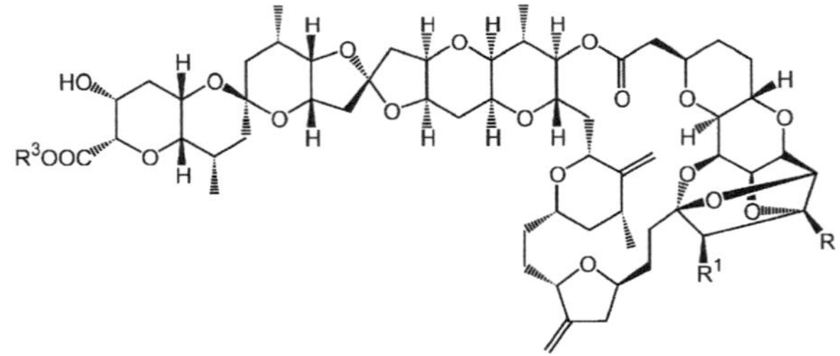

Norhalichondrin A: $R^1 = R^2 = OH$, $R^3 = H$
Norhalichondrin B: $R^1 = R^2 = R^3 = H$
Norhalichondrin C: $R^1 = R^3 = H$, $R^2 = OH$
p-Bromophenacyl derivative: $R^1 = R^2 = OH$, $R^3 = CH_2COC_6H_4Br$

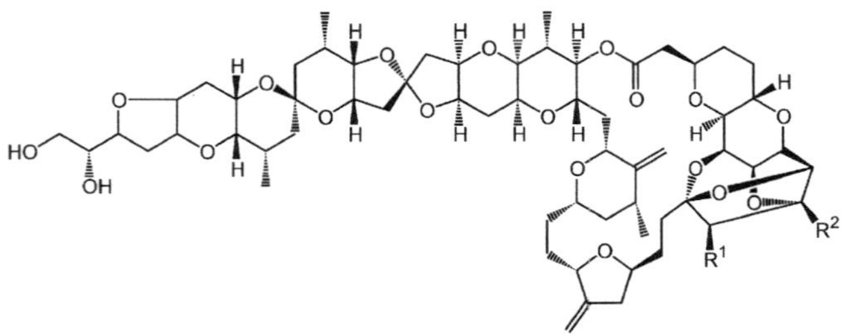

Homohalichondrin A: $R^1 = R^2 = OH$
Homohalichondrin B: $R^1 = R^2 = H$
Homohalichondrin C: $R^1 = H$, $R^2 = OH$

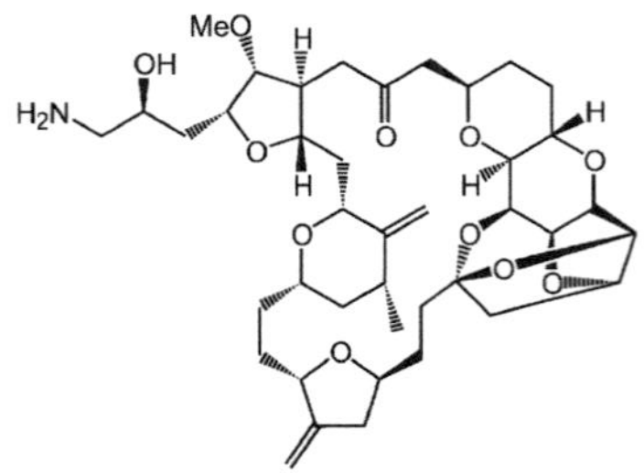

Eribulin

FIGURE 10.2 Halichondrins and eribulin.

very excited by our research. In fact, Pettit and M. Munro at Canterbury University in New Zealand had obtained halichondrin B from different sponges. However, they had not been able to determine its structure.

Once the framework of a complicated natural product has been elucidated, the overall structure can be determined using various analytical techniques. Therefore, it is extremely important to elucidate the skeleton, and only researchers can appreciate the joy at being the first to reveal this information. In 1992, Y. Kishi, a leading researcher in the total synthesis of complex natural products, who had already successfully synthesized palytoxin carboxylate (Armstrong *et al.*, 1989), synthesized halichondrin B (Aicher *et al.*, 1992). Based on these findings, the antitumor activities of the intermediates were tested at Eisai's US laboratory, and the results clarified that the right half of the molecule was active (Towle *et al.*, 2001). While these results were expected, they were very encouraging because this meant that such compounds could be produced at total synthesis laboratories. Subsequently, many researchers studied the molecule and fine-tuned its structure to develop eribulin mesylate (Newman, 2007). With the help of process chemistry, which is at the center of modern organic synthesis chemistry, they succeeded in mass production, and there are high expectations for its application in breast cancer therapy (Jarvis, 2007). Eribulin has been approved by the US Food and Drug Administration, the European Medicines Agency, and the Japan Ministry of Health, Labour, and Welfare for locally advanced or metastatic breast cancer.

It is not easy to obtain halichondrin B from living organisms. Munro's group planned and implemented sponge cultivation and has conducted studies, but the process has not always gone smoothly. As mentioned above, sponges coexist with many microorganisms and filter food from their environment. At this point, we do not yet know which organism actually produces halichondrins, although many laboratories have isolated and cultured microorganisms from inside sponges. However, only $<1\%$ of the microorganisms can be cultured (Rouhi, 1999). Subsequently, with the use of a metagenome technique, gene groups were viewed as compound sources, and fosmid libraries have been prepared. Overall, about 150,000 libraries have been made, and research is currently focused on compound production and genomic information (Abe *et al.*, unpublished work). In the near future, we believe that a new biosynthesis system will be elucidated.

III. CYTOTOXIC COMPOUNDS RELATED TO THE CORAL COMMUNITY

Coral reefs are rich resources as primary producers in tropical and subtropical areas. However, coral is being destroyed by many different external factors, including overgrowth by organisms that cover coral and feeding by coral predators. The discovery of new ecologically active compounds often triggers the development of basic scientific concepts in the field of biological sciences, since such compounds have direct physiological and behavioral effects on other living organisms (Uemura *et al.*, 2009).

Recently, we encountered a catastrophic change in the Nakijin coral reef in Okinawa Prefecture: the coral's surface was covered with the black sponge *Terpios hoshinota* (Fig. 10.3). This sponge emits compounds that kill corals and cover the dead bodies, which suggest that the sponge might be injecting a toxin into the corals. Guided by cytotoxicity assays, nakiterpiosin (IC_{50}=16nM; Teruya *et al.*, 2003), nakiterpiosinone (IC_{50}=16nM; Teruya *et al.*, 2004), and terpiodiene (Fig. 10.4; Teruya *et al.*, 2002) were isolated from the sponge *T. hoshinota*. Notably, nakiterpiosin and its analog have a unique highly oxidized ring that corresponds to the steroid A-ring and contain bromine and chlorine atoms. Its steroidal skeleton, consisting of C-nor and D-homo components, can be found on land, but this is the first example to be found in the marine environment. Nakiterpiosin and nakiterpiosinone were recently synthesized and their stereostructures have been revised (Gao *et al.*, 2009). The mode of action was also considered. Chen and coworkers found that synthetic nakiterpiosin arrests cell cycle in the G2/M phase. However, in *in vitro* tubulin polymerization assays, nakiterpiosin did not show any activity, in contrast to taxol and nocodazole (Gao et al., 2010a,b). On the other hand, another study by Seemann indicated that nakiterpiosin interacted with tubulin directly and inhibits microtubule polymerization in an *in vitro* assay, and suggested that it may decrease the polymer mass in cells (Wei and Seemann, 2010). In any event, nakiterpiosins are promising compounds that have potential for use as antitumor agents.

FIGURE 10.3 Corals overgrown by black sponge *Terpios hoshinota*. (For color version of this figure, the reader is referred to the Web version of this chapter.)

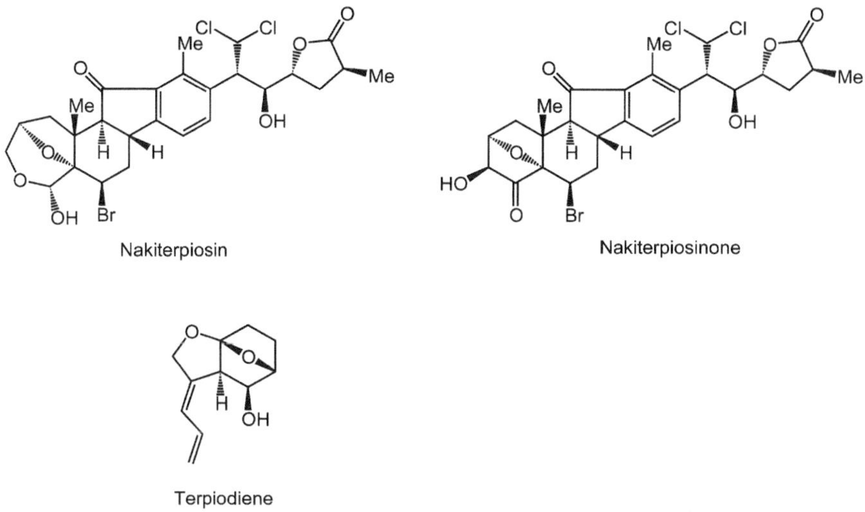

FIGURE 10.4 Cytotoxic compounds from *Terpios hoshinota*.

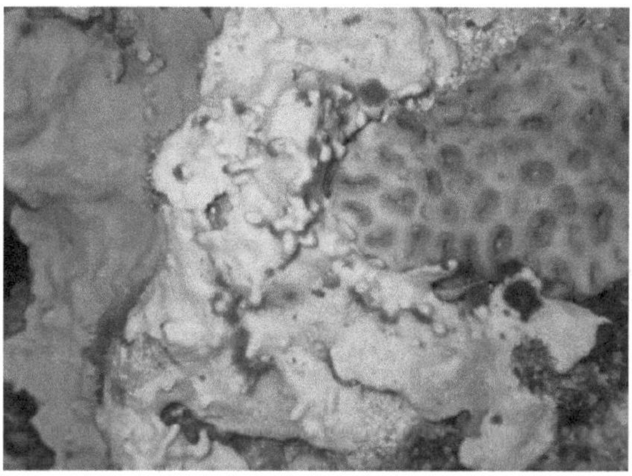

FIGURE 10.5 White sponge *Dysidea* sp.

In addition to the sponge Terpios, many other kinds of sponges, soft corals, and algae cover corals compete for survival, including the sponges *Dysidea* sp. and *Aplysina* sp.

White sponge, *Dysidea* sp. (Fig. 10.5), has been observed to overgrow members of the family Poritidae at Okinawa Island. Extracts from this sponge have been shown to contain brominated diphenyl ethers, which show potent cytotoxicity against P388 cells. Two known compounds, 3,6-dibromo-2-(2′,4′-dibromophenoxy) and 3,6-dibromo-2-(4′-dibromophenoxy)phenol (Fu and Schmitz, 1996; Fu *et al.*, 1995; Fig. 10.6), were

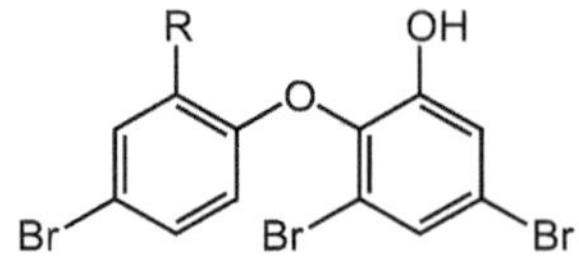

3,6-dibromo-2-(2',4'-dibromophenoxy)phenol: R = Br
3,6-dibromo-2-(2'-dibromophenoxy)phenol: R = H

FIGURE 10.6 Brominated diphenyl ether from *Dysidea* sp.

FIGURE 10.7 Yellow sponge *Aplysina* sp.

isolated from this extract. Intriguingly, these bromophenols showed strong toxicity against dinoflagellates that lives in an endosymbiotic relationship with corals (Maru *et al.*, unpublished work).

Similarly, *Aplysina* sp.(Fig. 10.7), which show a dramatic change in color from yellow to black when exposed to air, contain cytotoxic compounds including bromotyrosine-derivative alkaloids, purealidins (Ishibashi *et al.*, 1991; Yagi *et al.*, 1993), aplysamine 2 (Xynas and Capon, 1989), and purpuramin G (Fattorusso *et al.*, 1970; Fig. 10.8).

Bromotyrosine-derivative alkaloids are characteristic secondary metabolites from marine sponges of the order Verongida. The structures of these metabolites are typically based on one or two spirocyclohexadienyl isoxazole moieties that are connected to diverse side chains, as seen in aerothinin (Hernández-Guerrero *et al.*, 2007) and aerophobin-1 (Cimino *et al.*, 1983). Sunabedine is a new compound in this family that was isolated from an Okinawan sponge, order Verongida (Maru *et al.*, 2010a). The structure of sunabedine contains a 1,4-disubstituted imidazole ring that is linked to two bromotyrosines via methylene chains. The biological activity of sunabedine was examined with regard to cytotoxicity against B16 mouse melanoma cells and toxicity against brine shrimp

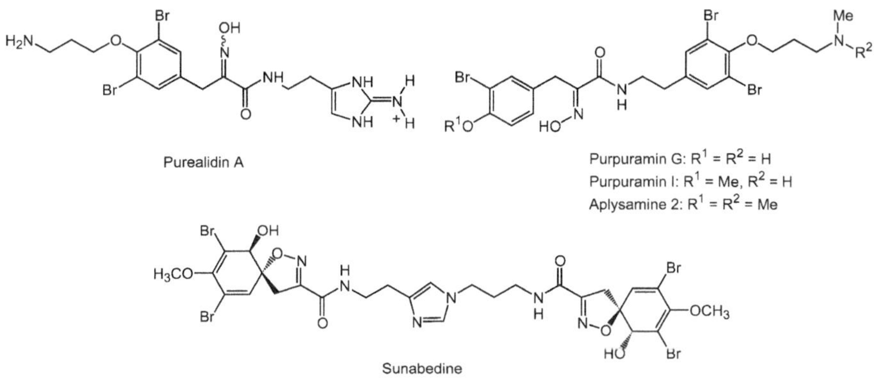

FIGURE 10.8 Bromotyrosine-derivative alkaloids.

FIGURE 10.9 Red algae *Laurencia papillosa.*

(genus *Artemia*). After incubation, sunabedine showed an IC_{50} of 39 μM against B16 cells and an LD_{50} of 110 μM against brine shrimp.

The red algae *Laurencia papillosa* (Fig. 10.9) generally grow in the intertidal zone in tropical and subtropical seas (Masuda *et al.*, 1997). Members of the genus *Laurencia* are widely distributed throughout tropical and temperate zones, and are known to be a rich source of secondary metabolites. However, there have been few phytochemical studies on *L. papillosa* (Wright *et al.*, 1996). We found that *L. papillosa* overgrew hermatypic corals of the families Acroporidae and Poritidae at Ishigaki Island, Okinawa Prefecture. The novel fatty acid amide papillamide (Fig. 10.10) was isolated

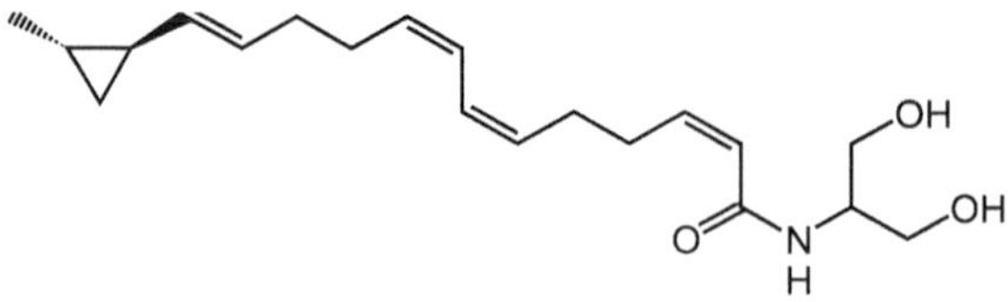

Papillamide

FIGURE 10.10 Papillamide from *Laurencia papillosa*.

from extract of this alga (Maru *et al.*, 2010b). NMR studies showed that the structure of papillamide contained a cyclopropane ring and serinol. Despite the strong cytotoxic activity of the extract, papillamide did not show definite cytotoxic activity (IC_{50}>100gm/L) toward P388 mouse leukemia cells or B16 mouse melanoma cells. This result shows that papillamide may have other biological activities that are not related to a toxic or predative effect, and such activities are now being investigated.

IV. CYTOTOXIC COMPOUNDS FROM CYANOBACTERIA

Cyanobacteria are photosynthetic prokaryotes that are widely distributed throughout marine and terrestrial environments. Members of the marine cyanobacteria genus *Lyngbya* are known to produce structurally interesting and biologically active secondary metabolites. Typically, linear/cyclic peptides and depsipeptides that include various nonproteinogenic amino acids are the major groups of these metabolites (Fig. 10.11), which can exhibit potent cytotoxicity.

Bisebromoamide is cytotoxic linear peptide isolated from the marine cyanobacterium *Lyngbya* sp. (Teruya *et al.*, 2009). Structure of bisebromoamide contains rich unusual amino acid derivatives with D-amino acids, N-methylated amino acids, a brominated tyrosine, a modified 4-methylproline, a 2-substituted thiazoline-4-methyl-4-carboxylic acid unit, and a rare 2-(1-oxopropyl)pyrrolidine moiety. Recently, stereochemistry at thiazoline ring moiety was reassigned by synthetic research (Gao et al., 2010a,b). Bisebromoamide showed strong cytotoxicity against HeLa S3 cells with an IC_{50} value of 39nM. In observation under the microscope, bisebromoamide induced nuclear protrusion type morphological changes in HeLa cells as a result of its mode of action, actin depolymerization (Sumiya *et al.*, 2011).

The cyanobacteria *Lyngbya* sp. was found to overgrow corals at Ishigaki Island. Apratoxin A (Luesch *et al.*, 2001) was isolated as the main toxic compound from this alga. Apratoxins were originally isolated in Guam and Palau in the *Lyngbya majuscule* known to cause swimmers' itch (Luesch *et al.*, 2002). Apratoxin A has a macrocyclic structure with parts of

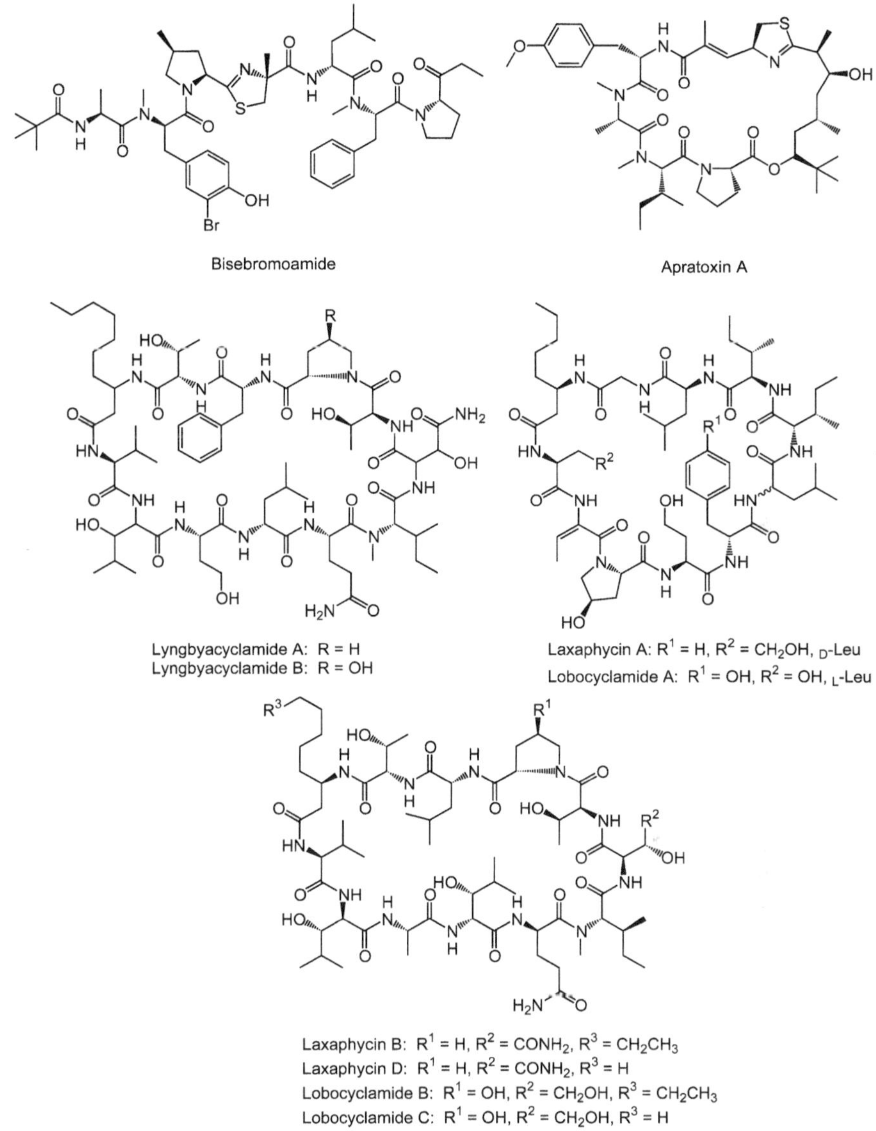

FIGURE 10.11 Cytotoxic peptide from cyanobacteria.

amino acids and a polyketide unit, and shows potent cytotoxicity against P388 cells with an IC_{50} value of 0.3nM. Another *Lyngbya* sp. was also collected at Ishigaki Island, and the aqueous ethanol extract of this alga showed strong cytotoxicity against tumor cells. Two new peptides, lyngbyacyclamides A and B, were isolated via a guided bioassay

(Maru *et al.*, 2010c). The structures of these cyclic peptides were determined by spectroscopic analyses and degradation reactions. The members of the amino acid residues of lyngbyacyclamide A were assigned with the proteinogenic L-amino acids, a Val, two Thr, a Pro, a Gln, and unusual amino acids, a D-Phe, a D-Leu, an N-methyl Ile, β-hydroxy leucine and asparagine, a homoserine, and a β-amino decanoic acid. The structure of lyngbyacyclamide B only differs at a hydroxyl group on the proline residue from lyngbyacyalamide A. The biological activities of lyngbyacyclamides were examined with regard to cytotoxicity against B16 cells and toxicity against brine shrimp. After incubation, lyngbyacyclamides A and B showed an IC_{50} of 0.7 μM against B16 cells. However, they did not show definite toxicity at 70 μM against brine shrimp. This result suggests that lyngbyacyclamides may have a unique mode of action and could contribute to the discovery of new drugs. The structures of lyngbyacyclamides resemble those of the natural products laxaphycins (Bonnard *et al.*, 1997, 2007) and lobocyclamides (MacMillan *et al.*, 2002). Laxaphycins were originally isolated as major products from antifungal extract of cyanobacterium *Anabaena laxa* and *L. majuscule*. Isolated laxaphycins did not show antifungal effect, but especially laxaphycin B shows strong cytotoxicity against the cell line of CCRF-CEM human leukemic lymphoblasts (IC_{50}=1.1 μM), CEM/VLB100 vinblastine-resistant subline (IC_{50}=1.0 μM), and CEM/VM-1 subline (IC_{50}=1.4 μM). Interestingly, laxaphycin A shows weak cytotoxicity against these cell lines IC_{50} over 20 μ M. This difference may hint studies of structure activity correlations or mode of action.

V. CONCLUSIONS

In the area of drug discovery, marine natural products comprise a rich source of cytotoxic compounds with a potential to target cancers. As shown in development of eribulin, there must be emphasized the need for the researchers in isolation chemistry, synthetic organic chemistry, biochemistry, and pharmacy to work closely together for the development of promising new molecules. Natural cytotoxic compounds of new chemotypes with novel mode of action remain of much interest as compounds that might lead to the battle against the cancer tragedy.

REFERENCES

Abe, T., Naito, T., and Uemura, D. Unpublished work.

Aicher, T. D., Buszek, K. R., Fang, F. G., Forsyth, C. J., Jung, S. H., Kishi, Y., Materlich, M. C., Scola, P. M., Spero, D. M., and Yoon, S. K. (1992). Total synthesis of halichondrin B and norhalichondrin B. *J. Am. Chem. Soc.* **114**, 3162–3164.

Armstrong, R. W., Beau, J. M., Cheon, S. H., Christ, W. J., Fujioka, H., Ham, W. H., Hawkins, L. D., Jin, H., Kang, S. H., Kishi, Y., Martinelli, M. J., McWhorter, W. W., Jr., *et al.* (1989). Total synthesis of palytoxin carboxylic acid and palytoxin amide. *J. Am. Chem. Soc.* **111**, 7530–7533.

Bonnard, I., Rolland, M., Francisco, C., and Banaigs, B. (1997). Total structure and biological properties of laxaphycins A and B, cyclic lipopeptides from the marine cyanobacterium *Lyngbya majuscula*. *Lett. Pept. Sci.* **4**, 289–292.

Bonnard, I., Rolland, M., Salmon, J. M., Debiton, E., Barthomeuf, C., and Banaigs, B. (2007). Total structure and inhibition of tumor cell proliferation of laxaphycins. *J. Med. Chem.* **50**, 1266–1279.

Cimino, G., De Rosa, S., De Stefano, S., Self, R., and Sodano, G. (1983). The bromo-compounds of the true sponge Verongia aerophoba. *Tetrahedron Lett.* **24**, 3029–3032.

Fattorusso, E., Minale, L., Sodano, G., Moody, K., and Thomson, R. H. (1970). Aeroplysinin-I, a new bromo-compound from Aplysina aerophoba. *J. Chem. Soc. Chem. Commun.* 751–752.

Ferlay, J., Shin, H., Bray, F., Forman, D., Mathers, C., and Parkin, D. M. (2010). Estimates of worldwide burden of cancer in 2008: GLOBOCAN 2008. *Int. J. Cancer* **127**, 2893–2917.

Fu, X. and Schmitz, F. J. (1996). New brominated diphenyl ether from an unidentified species of Dysidea sponge. 13C NMR data for some brominated diphenyl ethers. *J. Nat. Prod.* **59**, 1102–1103.

Fu, X., Schmitz, F. J., Govindan, M., Abbas, S. A., Hanson, K. M., Horton, P. A., Crews, P., Laney, M., and Schatzman, R. C. (1995). Enzyme inhibitors: New and known polybrominated phenols and diphenyl ethers from four Indo-Pacific Dysidea sponges. *J. Nat. Prod.* **58**, 1384–1391.

Gao, S., Wang, Q., and Chen, C. (2009). Synthesis and structure revision of nakiterpiosin. *J. Am. Chem. Soc.* **131**, 1410–1412.

Gao, S., Wang, Q., Huang, L. J., Lum, L., and Chen, C. (2010a). Chemical and biological studies of nakiterpiosin and nakiterpiosinone. *J. Am. Chem. Soc.* **132**, 371–383.

Gao, X., Liu, Y., Kwong, S., Xu, Z., and Ye, T. (2010b). Total synthesis and stereochemical reassignment of bisebromoamide. *Org. Lett.* **12**, 3018–3021.

Hernández-Guerrero, C. J., Zubía, E., Ortega, M. J., and Carballob, J. L. (2007). Cytotoxic dibromotyrosine-derived metabolites from the sponge Aplysina gerardogreeni. *Bioorg. Med. Chem.* **15**, 5275–5282.

Hirata, Y. and Uemura, D. (1986). Halichondrins—Antitumor polyether macrolides from a marine sponge. *Pure Appl. Chem.* **58**, 701–710.

Ishibashi, M., Tsuda, M., Ohizumi, Y., Sasaki, T., and Kobayashi, J. (1991). Purealidin A, a new cytotoxic bromotyrosine-derived alkaloid from the Okinawan marine sponge Psammaplysilla purea. *Experientia* **47**, 299–300.

Jarvis, L. M. (2007). Success stories: From the ocean to patients. *C&E News*. Oct. 8, p. 24.

Luesch, H., Yoshida, W. Y., Moore, R. E., Paul, V. J., and Corbett, T. H. (2001). Total structure determination of apratoxin A, a potent novel cytotoxin from the marine cyanobacterium Lyngbya majuscule. *J. Am. Chem. Soc.* **123**, 5418–5423.

Luesch, H., Yoshida, W. Y., Moore, R. E., and Paul, V. J. (2002). New apratoxins of marine cyanobacterial origin from guam and palau. *Bioorg. Med. Chem.* **10**, 1973–1978.

MacMillan, J. B., Ernst-Russell, M. A., de Ropp, J. S., and Molinski, T. F. (2002). Lobocyclamides A-C, lipopeptides from a cryptic cyanobacterial mat containing Lyngbya confervoides. *J. Org. Chem.* **67**, 8210–8215.

Maru, N., Koyama, T., Ohno, O., Yamada, K., and Uemura, D. (2010a). Sunabedine, a novel toxic bromotyrosine-derivative alkaloid from Okinawan sponge, order Verongida. *Heterocycles* **82**, 371–375.

Maru, N., Ohno, O., Koyama, T., Yamada, K., and Uemura, D. (2010b). Papillamide, a novel fatty acid amide from the red alga Laurencia papillosa. *Chem. Lett.* **39**, 366–367.

Maru, N., Ohno, O., and Uemura, D. (2010c). Lyngbyacyclamides A and B, novel cytotoxic peptides from marine cyanobacteria Lyngbya sp. *Tetrahedron Lett.* **51**, 6384–6387.
Maru, N., Kitamura, M., Yamada, K., and Uemura, D. Unpublished work.
Masuda, M., Kawaguchi, S., and Phang, S. M. (1997). Taxonomic notes on Laurencia similis and L. papillosa (Ceramiales, Rhodophyta) from the Western Pacific. *Bot. Mar.* **40**, 229–240.
Newman, S. (2007). Eribulin, a simplified ketone analog of the tubulin inhibitor halichondrin B, for the potential treatment of cancer. *Curr. Opin. Investig. Drugs* **8**, 1057–1066.
Rouhi, A. M. (1999). Science & technology: Chemistry from unknown microbes. *C&E News.* Jan. 4, pp. 21–23.
Sumiya, E., Shimogawa, H., Sasaki, H., Tsutsumi, M., Yoshita, K., Ojika, M., Suenaga, K., and Uesugi, M. (2011). Cell-morphology profiling of a natural product library identifies. *ACS Chem. Biol.* **6**, 425–431.
Teruya, T., Nakagawa, S., Koyama, T., Suenaga, K., and Uemura, D. (2002). Terpiodiene: A novel tricyclic alcohol from the Okinawan sponge Terpios hoshinota. *Chem. Lett.* **31**, 38–39.
Teruya, T., Nakagawa, S., Koyama, T., Suenaga, K., Kita, M., and Uemura, D. (2003). Nakiterpiosin, a novel cytotoxic C-nor-D-homosteroid from the Okinawan sponge Terpios hoshinota. *Tetrahedron Lett.* **44**, 5171–5173.
Teruya, T., Nakagawa, S., Koyama, T., Arimoto, H., Kita, M., and Uemura, D. (2004). Nakiterpiosin and nakiterpiosinone, novel cytotoxic C-nor-D-homosteroids from the Okinawan sponge Terpios hoshinota. *Tetrahedron* **60**, 6989–6993.
Teruya, T., Sasaki, H., Fukazawa, H., and Suenaga, K. (2009). Bisebromoamide, a potent cytotoxic peptide from the marine Cyanobacterium Lyngbya sp.: Isolation, stereostructure, and biological activity. *Org. Lett.* **11**, 5062–5065.
Towle, M. J., Salvato, K. A., Budrow, J., Wels, B. F., Kuznetsov, G., Aalfs, K. K., Welsh, S., Zheng, W., Seletsky, B. M., Palme, M. H., Habgood, G. H., Singer, L. A., *et al.* (2001). In vitro and in vivo anticancer activities of synthetic macrocyclic ketone analogues of halichondrin B. *Cancer Res.* **61**, 1013–1021.
Uemura, D. (2010). Exploratory research on bioactive natural products with a focus on biological phenomena. *Proc. Jpn. Acad. Ser. B Phys. Biol. Sci.* **86**, 190–201.
Uemura, D., Takahashi, K., Yamamoto, T., Katayama, C., Tanaka, J., Okumura, Y., and Hirata, Y. (1985). Norhalichondrin A: An antitumor polyether macrolide from a marine sponge. *J. Am. Chem. Soc.* **107**, 4796–4798.
Uemura, D., Kita, M., Arimoto, H., and Kitamura, M. (2009). Recent aspects of chemical ecology: Natural toxins, coral communities, and symbiotic relationships. *Pure Appl. Chem.* **81**, 1093–1111.
Vogel, G. (2008). The inner lives of sponges. *Science* **320**, 1028–1030.
Wei, J. and Seemann, J. (2010). Nakiterpiosin targets tubulin and triggers mitotic catastrophe in human cancer cells. *Mol. Cancer Ther.* **9**, 3375–3385.
Wright, A. D., König, G. M., Angerhofer, C. K., Greenidge, P., Linden, A., and Desqueyroux-Faúndez, R. (1996). Antimalarial activity: The search for marine-derived natural products with selective antimalarial activity. *J. Nat. Prod.* **59**, 710–716.
Xynas, R. and Capon, R. J. (1989). Two new bromotyrosine-derived metabolites from an Australian marine sponge, Aplysina sp. *Aust. J. Chem.* **42**, 1427–1433.
Yagi, H., Matsunaga, S., and Fusetani, N. (1993). Purpuramines A-I, new bromotyrosine-derived metabolites from the marine sponge Psammaplysilla purpurea. *Tetrahedron* **49**, 3749–3754.

CHAPTER 11

Secondary Metabolites with New Medicinal Functions from Marine Organisms

Osamu Ohno,[*,1] **Kiyotake Suenaga,**[*,1] and **Daisuke Uemura**[†]

Contents

Abstract

We have focused on the identification of natural key compounds that possess biologically and medicinally intriguing functions. Some of bioactive naturally occurring compounds isolated from marine organisms were found to possess unique biological activities. Halichlorine from the marine sponge *Halichondria okadai* was shown to inhibit the activity of nuclear factor-κB in endothelial cells and block L-type Ca^{2+} channels. Thus, it may have therapeutic potentials for diseases such as atherosclerosis and hypertension.

* Department of Chemistry, Faculty of Science and Technology, Keio University, Yokohama, Japan
† Department of Chemistry, Faculty of Science, Kanagawa University, Hiratsuka, Japan
[1] Corresponding authors: Osamu Ohno and Kiyotake Suenaga, E-mail address: ohno@chem.keio.ac.jp; Suenaga@chem.keio.ac.jp

Advances in Food and Nutrition Research, Volume 65
ISSN 1043-4526, DOI: 10.1016/B978-0-12-416003-3.00011-1

I. INTRODUCTION

Various natural products with extraordinary chemical structures and significant biological activities have been isolated and characterized from both marine and terrestrial creatures. In our previous research, isolation and characterization of many biologically and medicinally intriguing compounds from both the marine and terrestrial creatures have been achieved. Some of those isolated compounds have been shown to exhibit various biological activities and have good potential for clinical or biological use. For example, halichondrins are the remarkable antitumor compounds which had been originally isolated from the marine sponge *Halichondria okadai* Kadota (Hirata and Uemura, 1986; Uemura *et al.*, 1985). A synthetic analogue of halichondrin B, eribulin mesylate (Halaven™), was recently approved for the treatment of patients with metastatic breast cancer by the US Food and Drug Administration (Huyck *et al.*, 2011; Towle, *et al.*, 2001). In our continuing research, we have constructed unique screening systems and have searched among secondary metabolites of marine organisms for biologically active compounds (Uemura, 2006, 2010; Uemura *et al.*, 2009). We have isolated both novel and known compounds and their new biological and physiological activities were examined by collaborating researchers.

II. HALICHLORINE, A BIOACTIVE MARINE NATURAL COMPOUND ISOLATED FROM A MARINE SPONGE *H. OKADAI* KADOTA

Halichlorine (**1**) (Fig. 11.1) is an alkaloid which was isolated from the marine sponge *H. okadai* Kadota (Kuramoto *et al.*, 1996). This compound was revealed to be a novel alkaloid containing an azaspiro[4.5]decane skeleton clarified by detailed spectroscopic analyses. The absolute stereostructure was confirmed by many synthetic studies (Arimoto *et al.*, 1998; Clive *et al.*, 2005; Liu *et al.*, 2009; Trauner *et al.*, 1999). Halichlorine was shown to inhibit the induction of vascular cell adhesion molecule-1 (VCAM-1) in cultured human umbilical vein endothelial cells. Drugs that block the inflammatory stimuli-induced expression of VCAM-1 may be useful for treating atherosclerosis, coronary artery diseases, angina, and noncardiovascular inflammatory diseases (Kock *et al.*, 1995). We introduce here the recent aspects of the biological and physiological activities of halichlorine.

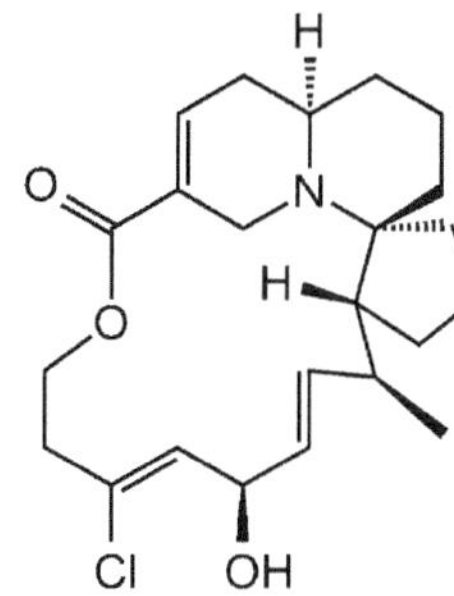

FIGURE 11.1 Structures of halichlorine (**1**).

A. Halichlorine inhibits LPS-induced NF-κb activation in endothelial cells

Various adhesion molecules are reported to be expressed on vascular endothelial cells at sites of inflammation (Libby, 2002). In fact, VCAM-1, expressed on the surface of vascular endothelial cells in response to inflammatory stimuli, is suggested to play an important role in leukocyte recruitment (Gerrity *et al.*, 1979). The adhesion of monocytes to vascular endothelial cells is the first critical step in the induction of atherosclerosis (Glass and Witztum, 2001). Supportively, the expression of VCAM-1 is reported to increase in atherosclerosis lesions (Cybulsky and Gimbrone, 1991). Intercellular adhesion molecule-1 (ICAM-1) and E-selectin are also expressed on the surface of vascular endothelial cells in response to inflammatory stimuli (Smith, 2008). VCAM-1 and ICAM-1 are adhesion molecules which are classified in the immunoglobulin superfamily, and E-selectin is a member of the selectin family. VCAM-1 is the receptor for VLA-4 (Elices *et al.*, 1990), whereas ICAM-1 is for LFA-1 and Mac-1 (Diamond *et al.*, 1990; Marlin and Springer, 1987; Rothlein *et al.*, 1986). E-selectin is the receptor for sialyl Lewis X (sLe^x) and sialyl Lewis A (sLe^a; Wittig *et al.*, 1996). The induction of ICAM-1, VCAM-1, and E-selectin by inflammatory stimuli such as tumor necrosis factor-α, interleukin-1β, and lipopolysaccharide (LPS) is associated with the functional activation of nuclear factor-κB (NF-κB) (Dustin *et al.*, 1986; Haraldsen *et al.*, 1996). NF-κB, mainly consisting of p65 and p50 proteins, is the transcription factor that binds to the κB sequence, which is found in the promoter regions of ICAM-1, VCAM-1, and E-selectin genes (Iademarco *et al.*, 1992; Karin and Greten, 2005; Sen and Baltimore, 1986; van de Stolpe *et al.*, 1994; Whelan *et al.*, 1991). These NF-κB-mediated gene expressions are significant for inflammation-related diseases including atherosclerosis, and thus, NF-κB is thought to be a good clinical target molecule. In fact, DHMEQ, a specific inhibitor of NF-κB, was reported to suppress induction of adhesion molecules and to reduce atherosclerosis in mice (Ariga *et al.*, 2002; Chiba *et al.*, 2006; Ohno *et al.*, 2005). Halichlorine had been found to inhibit the induction of VCAM-1, downstream of NF-κB signaling (Kuramoto *et al.*, 1996). Recently, Tsubosaka *et al.* (2010b) investigated the effect of halichlorine on LPS-induced inflammatory responses in bovine aortic endothelial cells (BAECs) and reported that halichlorine inhibited NF-κB activation as follows.

In BAECs, halichlorine was revealed to inhibit LPS-induced mRNA expressions of VCAM-1, ICAM-1, and E-selectin (Fig. 11.2A). Further,

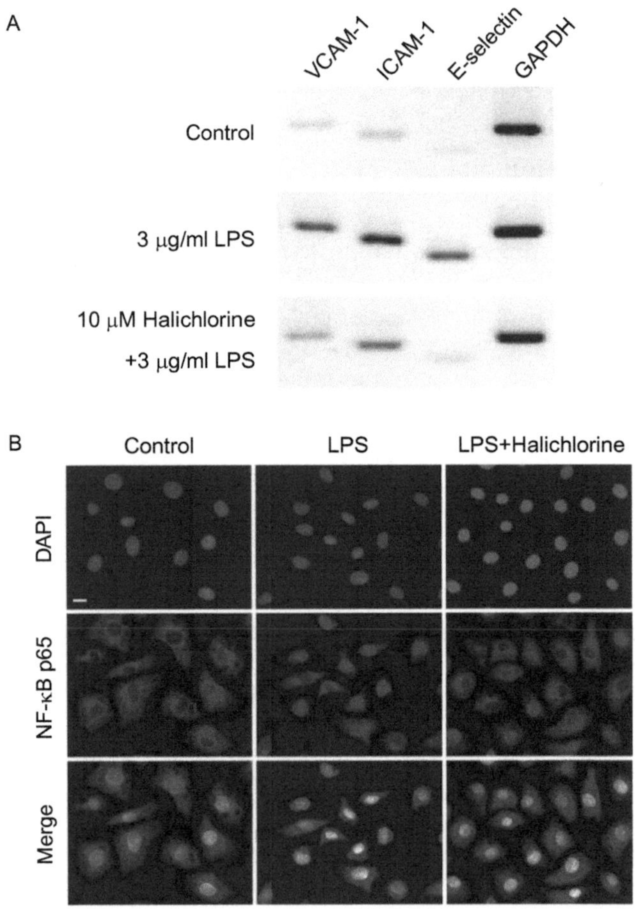

FIGURE 11.2 Inhibitory effect of halichlorine on NF-κB activation in BAECs. (A) BAECs were preincubated with or without halichlorine (10 μM) for 2 h and then exposed to LPS (3 μg/ml) for 3 h. Next, cells were collected and then analyzed by RT-PCR. Agarose gel electrophoresis demonstrates mRNA encoding GAPDH, VCAM-1, ICAM-1, and E-selectin. (B) BAECs were preincubated with or without halichlorine (10 μM) for 2 h and then exposed to LPS (3 μg/ml) for 1 h. The p65 subunit of NF-κB was immunostained with Alexa Flour 568-labeled antibody (red) and nuclei was stained with DAPI (blue). This figure is sited from Tsubosaka *et al.*, 2010b, and is permitted for use by the publisher.

halichlorine reduced U937 monocytes adhesion to LPS-stimulated BAECs. Then, phenomena upstream of adhesion molecules expressions were investigated, and halichlorine was demonstrated to inhibit LPS-induced nuclear accumulation of p65, a component of NF-κB (Fig. 11.2B). Because halichlorine did not inhibit the proliferation and viability of BAECs, its inhibitory effects against NF-κB were revealed to be not related to cytotoxicity. These data indicate that halichlorine is a unique inhibitor of NF-κB acting at the level of nuclear translocation. LPS is one of the major constituents of the outer membrane of Gram-negative bacteria and is recognized as a key molecule in the pathogenesis of inflammatory syndromes (Raetz and Whitfield, 2002). LPS induces the upregulation of adhesion molecules via the activation of NF-κB in endotherial cells in the same way as inflammatory cytokines, and inactivation of LPS leads to the suppression of the endothelial inflammation (Ohno *et al.*, 2004; Takahashi *et al.*, 2006). Various inflammatory cytokines are reported to be produced by activated macrophages and T cells in atherosclerosis (Glass and Witztum, 2001). Thus, LPS-induced adhesion between monocytes and BAECs can be supposed to reflect the situation that occurs in atherosclerosis. There may be other factors aside from inflammatory cytokines that induce cell adhesion to the endothelial cells in atherosclerosis. It was reported that native low-density lipoprotein (LDL) induced the expression of VCAM-1 and E-selectin in human aortic endothelial cells (Allen *et al.*, 1998). There is a possibility that halichlorine also inhibits the cell adhesion induced by LDL, if this adhesion is mediated by NF-κB.

As described above, halichlorine was shown to inhibit LPS-induced events in BAECs, including adhesion-molecule expression, adherence of monocytes, and NF-κB activation. Thus, halichlorine is thought to be an attractive candidate for the adjunctive therapy of diseases such as atherosclerosis.

B. Halichlorine inhibits L-type Ca^{2+} channels in vascular smooth muscle cells

Recently, Tsubosaka *et al.* (2010a) reported that halichlorine was also revealed to inhibit L-type Ca^{2+} channels, which leads to inhibit smooth muscle contraction. In their report, the direct effect of halichlorine on vascular contractility was investigated. Then, halichlorine was found to inhibit both high concentration of K^+- and phenylephrine-induced contractions in rat aorta dose dependently. The effect of halichlorine on high K^+-induced contraction was shown to be stronger than that on phenylephrine-induced contraction. Because known L-type Ca^{2+} channel blockers, verapamil and nifedipine, were observed to show the similar effect by them, it was suggested that halichlorine selectively inhibits L-type Ca^{2+} channels. Then, the effect of halichlorine on intracellular Ca^{2+} concentration in vascular smooth muscle tissue was examined using a fluorescent Ca^{2+} indicator, Fura-2.

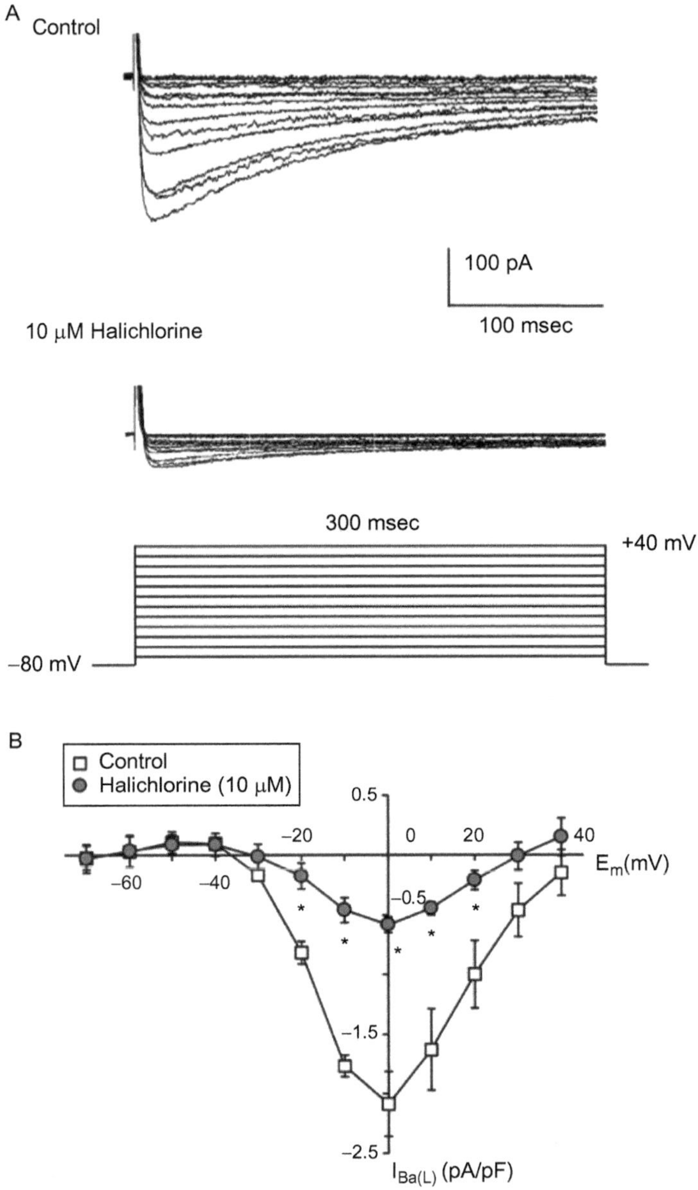

FIGURE 11.3 Effect of halichlorine on the voltage-dependent Ca^{2+} currents in rat aortic smooth muscle cells (A7r5 cells). (A) Typical traces of Ba^{2+} current for control (top trace) and after treatment of halichlorine (middle trace; 10 μM). A7r5 cells elicited by incremental 10 mV depolarizing steps from −70 mV to +40 mV. Holding potential was −80 mV. (B) Current–voltage relationships for the peak inward Ba^{2+} current density for control and after the addition of halichlorine (10 μM). $P<0.01$. This figure is sited from Tsubosaka *et al.*, 2010a, and is permitted for use by the publisher.

In this analysis, halichlorine was demonstrated to inhibit the Ca^{2+} influx induced by high K^+ significantly. High K^+-induced contraction in vascular smooth muscle is reported to be mediated mainly by Ca^{2+} influx through voltage-dependent L-type Ca^{2+} channels (Karaki *et al.*, 1997). Halichlorine was likely to inhibit L-type Ca^{2+} channel directly, and thus, this was examined using whole-cell patch-clamp analysis. As shown in Fig. 11.3A and B, halichlorine was revealed to reduce the Ca^{2+} channel current densities. These results suggested that halichlorine suppressed Ca^{2+} influx through the inhibition of voltage-dependent L-type Ca^{2+} channels. Thus, halichlorine was revealed to inhibit L-type Ca^{2+} channels selectively in vascular smooth muscle cells, which leads to the inhibition of intracellular Ca^{2+} influx and the reduction of vascular contractions. Although halichlorine-binding site in L-type Ca^{2+} channel has not been clarified, these findings may contribute to the development of halichlorine as a new Ca^{2+} channel blocker. Because many of L-type Ca^{2+} channel blockers have been applied for the treatment of patients with hypertension, there is a possibility that halichlorine has a potential as an antihypertensive agent.

III. CONCLUSION

As described above, several bioactive naturally occurring compounds from marine organisms and plants show unique biological activities. Halichlorine was shown to inhibit the activity of NF-κB in endothelial cells and to inhibit L-type Ca^{2+} channels. Although it needs to be examined more to clarify the mechanisms underlying its biological actions, halichlorine may have therapeutic potentials for diseases such as atherosclerosis and hypertension. We believe that these results and attempts will contribute to the development of novel key molecules with intriguing biological activities which are effective in preventing and curing diseases.

ACKNOWLEDGMENTS

The authors wish to thank Drs. Y. Tsubosaka, T. Murata, M. Hori, and H. Ozaki (University of Tokyo) for their contributions to these studies.

REFERENCES

Allen, S., Khan, S., Al-Mohanna, F., Batten, P., and Yacoub, M. (1998). Native low density lipoprotein-induced calcium transients trigger VCAM-1 and E-selectin expression in cultured human vascular endothelial cells. *J. Clin. Invest.* **101**, 1064–1075.

Ariga, A., Namekawa, J., Matsumoto, N., Inoue, J., and Umezawa, K. (2002). Inhibition of tumor necrosis factor-α-induced nuclear translocation and activation of NF-κB by dehydroxymethylepoxyquinomicin. *J. Biol. Chem.* **277**, 24625–24630.

Arimoto, H., Hayakawa, I., Kuramoto, M., and Uemura, D. (1998). Absolute stereochemistry of halichlorine; a potent inhibitor of VCAM-1 induction. *Tetrahedron Lett.* **39**, 861–862.

Chiba, T., Kondo, Y., Shinozaki, S., Kaneko, E., Ishigami, A., Maruyama, N., Umezawa, K., and Shimokado, K. (2006). A selective NFκB inhibitor, DHMEQ, reduced atherosclerosis in ApoE-deficient mice. *J. Atheroscler. Thromb.* **13**, 308–313.

Clive, D. L. J., Yu, M., Wang, J., Yeh, V. S. C., and Kang, S. (2005). Synthetic chemistry of halichlorine and the pinnaic acids. *Chem. Rev.* **105**, 4483–4514.

Cybulsky, M. I. and Gimbrone, M. A., Jr. (1991). Endothelial expression of mononuclear leukocyte adhesion molecule during atherogenesis. *Science* **251**, 788–791.

Diamond, M. S., Staunton, D. E., de Fougerolles, A. R., Stacker, S. A., Garcia-Aguilar, J., Hibbs, M. L., and Springer, T. A. (1990). ICAM-1 (CD54): A counter-receptor for Mac-1 (CD11b/CD18). *J. Cell Biol.* **111**, 3129–3139.

Dustin, M. L., Rothlein, R., Bhan, A. K., Dinarello, C. A., and Springer, T. A. (1986). Induction by IL-1 and interferon-γ: Tissue distribution, biochemistry, and function of a natural adherence molecule (ICAM-1). *J. Immunol.* **137**, 245–254.

Elices, M. J., Osborn, L., Takada, Y., Crouse, C., Luhowskyj, S., Hemler, M. E., and Lobb, R. R. (1990). VCAM-1 on activated endothelium interacts with the leukocyte integrin VLA-4 at a site distinct from the VLA-4/fibronectin binding site. *Cell* **60**, 577–584.

Gerrity, R. G., Naito, H. K., Richardson, M., and Schwartz, C. J. (1979). Dietary induced atherogenesis in swine. *Am. J. Pathol.* **95**, 775–785.

Glass, C. K. and Witztum, J. L. (2001). Atherosclerosis: The road ahead. *Cell* **104**, 503–516.

Haraldsen, G., Kvale, D., Lien, B., Farstad, I. N., and Brandtzaeg, P. (1996). Cytokine-regulated expression of E-selectin, intercellular adhesion molecule-1 (ICAM-1), and vascular cell adhesion molecule-1 (VCAM-1) in human microvascular endothelial cells. *J. Immunol.* **56**, 2558–2565.

Hirata, Y. and Uemura, D. (1986). Halichondrins—Antitumor polyether macrolides from a marine sponge. *Pure Appl. Chem.* **58**, 701–707.

Huyck, T. K., Gradishar, W., Manuguid, F., and Kirkpatrick, P. (2011). Eribulin mesylate. *Nat. Rev. Drug Discov.* **10**, 173–174.

Iademarco, M. F., McQuillan, J. J., Rosen, G. D., and Dean, D. C. (1992). Characterization of the promoter for vascular cell adhesion molecule-1 (VCAM-1). *J. Biol. Chem.* **267**, 16323–16329.

Karaki, H., Ozaki, H., Hori, M., Mitsui-Saito, M., Amano, K., Harada, K., Miyamoto, S., Nakazawa, H., Won, K. J., and Sato, K. (1997). Calcium movements, distribution, and functions in smooth muscle. *Pharmacol. Rev.* **49**, 157–230.

Karin, M. and Greten, F. R. (2005). NF-κB: Linking inflammation and immunity to cancer development and progression. *Nat. Rev. Immunol.* **5**, 749–759.

Kock, A. E., Halloran, M. M., Haskell, C. J., Sah, M. R., and Polverini, P. J. (1995). Angiogenesis mediated by soluble forms of E-selectin and vascular cell adhesion molecule-1. *Nature* **376**, 517–519.

Kuramoto, M., Chou, T., Yamada, K., Chiba, T., Hayashi, Y., and Uemura, D. (1996). Halichlorine, an inhibitor of VCAM-1 induction from the marine sponge *Halichondria okadai* Kadota. *Tetrahedron Lett.* **37**, 3867–3870.

Libby, P. (2002). Inflammation in atherosclerosis. *Nature* **420**, 868–874.

Liu, D., Acharya, H. P., Yu, M., Wang, J., Yeh, V. S., Kang, S., Chiruta, C., Jachak, S. M., and Clive, D. L. (2009). Total synthesis of the marine alkaloid halichlorine: Development and use of a general route to chiral piperidines. *J. Org. Chem.* **74**, 7417–7428.

Marlin, S. D. and Springer, T. A. (1987). Purified intercellular adhesion molecule-1 (ICAM-1) is a ligand for lymphocyte function-associated antigen 1 (LFA-1). *Cell* **51**, 813–819.

Ohno, O., Ikeda, Y., Sawa, R., Igarashi, M., Kinoshita, N., Suzuki, Y., Miyake, K., and Umezawa, K. (2004). Isolation of heptadepsin, a novel bacterial cyclic depsipeptide that inhibits lipopolysaccharide activity. *Chem. Biol.* **11**, 1059–1070.

Ohno, O., Shima, Y., Ikeda, Y., Sakurai, K., Watanabe, K., Kawai, Y., and Umezawa, K. (2005). Inhibition of cellular adhesion in human umbilical vein endothelial cells by NF-κB inhibitor DHMEQ under flow. *Oncol. Res.* **15**, 189–197.

Raetz, C. R. and Whitfield, C. (2002). Lipopolysaccharide endotoxins. *Annu. Rev. Biochem.* **71**, 635–700.

Rothlein, R., Dustin, M. L., Marlin, S. D., and Springer, T. A. (1986). A human intercellular adhesion molecule (ICAM-1) distinct from LFA-1. *J. Immunol.* **137**, 1270–1274.

Sen, R. and Baltimore, D. (1986). Multiple nuclear factors interact with the immunoglobulin enhancer sequences. *Cell* **46**, 705–715.

Smith, C. W. (2008). 3. Adhesion molecules and receptors. *J. Allergy Clin. Immunol.* **121**, S375–S379.

van de Stolpe, A., Caldenhoven, E., Stade, B. G., Koenderman, L., Raaijmakers, J. A., Johnson, J. P., and van der Saag, P. T. (1994). 12-O-tetradecanoylphorbol-13-acetate- and tumor necrosis factor α-mediated induction of intercellular adhesion molecule-1 is inhibited by dexamethasone. Functional analysis of the human intercellular adhesion molecular-1 promoter. *J. Biol. Chem.* **269**, 6185–6192.

Takahashi, T., Ohno, O., Ikeda, Y., Sawa, R., Homma, Y., Igarashi, M., and Umezawa, K. (2006). Inhibition of lipopolysaccharide activity by a bacterial cyclic lipopeptide surfactin. *J. Antibiot.* **59**, 35–43.

Towle, M. J., Salvato, K. A., Budrow, J., Wels, B. F., Kuznetsov, G., Aalfs, K. K., Welsh, S., Zheng, W., Seletsky, B. M., Palme, M. H., Habgood, G. J., Singer, L. A., *et al.* (2001). In vitro and in vivo anticancer activities of synthetic macrocyclic ketone analogues of halichondrin B. *Cancer Res.* **61**, 1013–1021.

Trauner, D., Schwarz, J. B., and Danishefsky, S. J. (1999). Total synthesis of (+)-halichlorine: An inhibitor of VCAM-1 expression. *Angew. Chem. Int. Ed. Engl.* **38**, 3542–3545.

Tsubosaka, Y., Murata, T., Kinoshita, K., Yamada, K., Uemura, D., Hori, M., and Ozaki, H. (2010a). Halichlorine is a novel L-type Ca^{2+} channel inhibitor isolated from the marine sponge *Halichondria okadai* Kadota. *Eur. J. Pharmacol.* **628**, 128–131.

Tsubosaka, Y., Murata, T., Yamada, K., Uemura, D., Hori, M., and Ozaki, H. (2010b). Halichlorine reduces monocyte adhesion to endothelium through the suppression of nuclear factor-κB activation. *J. Pharmacol. Sci.* **113**, 208–213.

Uemura, D. (2006). Bioorganic studies on marine natural products—Diverse chemical structures and bioactivities. *Chem. Rec.* **6**, 235–248.

Uemura, D. (2010). Exploratory research on bioactive natural products with a focus on biological phenomena. *Proc. Jpn. Acad. Ser. B Phys. Biol. Sci.* **86**, 190–201.

Uemura, D., Kita, M., Arimoto, H., and Kitamura, M. (2009). Recent aspects of chemical ecology: Natural toxins, coral communities, and symbiotic relationships. *Pure Appl. Chem.* **81**, 1093–1111.

Uemura, D., Takahashi, K., Yamamoto, T., Katayama, C., Tanaka, J., Okumura, Y., and Hirata, Y. (1985). Norhalichondrin A, an antitumor polyether macrolide from a marine sponge. *J. Am. Chem. Soc.* **107**, 4796–4798.

Wittig, B. M., Kaulen, H., Thees, R., Schmitt, C., Knolle, P., Stock, J., Meyer zum Buschenfelde, K. H., and Dippold, W. (1996). Elevated serum E-selectin in patients with liver metastases of colorectal cancer. *Eur. J. Cancer* **32A**, 1215–1218.

Whelan, J., Ghersa, P., Hooft van Huijsduijnen, R., Gray, J., Chandra, G., Talabot, F., and DeLamarter, J. F. (1991). An NFκB-like factor is essential but not sufficient for cytokine induction of endothelial leukocyte adhesion molecule 1 (ELAM-1) gene transcription. *Nucleic Acids Res.* **19**, 2645–2653.

CHAPTER **12**

Structure–Function Relationship of Anticoagulant and Antithrombotic Well-Defined Sulfated Polysaccharides from Marine Invertebrates

Vitor H. Pomin[*,†,1]

Contents

Abstract

Marine sulfated polysaccharides (MSPs), such as sulfated fucans (SFs), sulfated galactans (SGs), and glycosaminoglycans (GAGs) isolated from invertebrate animals, are highly anionic polysaccharides capable of interacting with certain cationic proteins, such as

[*] Complex Carbohydrate Research Center, University of Georgia, Athens, Georgia, USA
[†] Program of Glycobiology, Institute of Medical Biochemistry, Federal University of Rio de Janeiro, Rio de Janeiro, Brazil
[1] Corresponding author: Vitor H. Pomin, *E-mail address:* vhpomin@gmail.com

Advances in Food and Nutrition Research, Volume 65
ISSN 1043-4526, DOI: 10.1016/B978-0-12-416003-3.00012-3

(co)-factors of the coagulation cascade during clotting-inhibition process. Primarily, these molecular complexes between MSPs and coagulation-related proteins seem to be driven mostly by electrostatic interactions. However, through a systematic comparison using several novel well-defined sulfated polysaccharides composed of repetitive oligosaccharides with clear sulfation patterns, it was proved that those molecular interactions are essentially regulated by the stereochemistry of the glycans (which depends on a conjunction of anomeric configurations, sugar types, conformational preferences, glycosylation, and sulfation sites), rather than just a mere consequence of the electronegative density charges (mainly from number of sulfate groups). Here, we present an overview about the structure–function relationship of the invertebrate MSPs with regular structures as potential anticoagulant and antithrombotic agents, as pathologies related to the cardiovascular system are one of the major causes of mortality in the world.

I. INTRODUCTION TO MARINE SULFATED POLYSACCHARIDES: A HIGH TENDENCY FOR REGULAR CHEMICAL STRUCTURES IN INVERTEBRATES

The efforts in both structural and functional studies of marine sulfated polysaccharides (MSPs) have been increasing significantly over the past 10 years (Pomin, 2008; Pomin and Mourão, 2008). This attention is a consequence of the large interest in novel sulfated polysaccharides (SPs) such as glycosaminoglycans (GAGs). GAGs are of particular interest to researchers looking for alternative and/or safer sources of heparin (Mourão and Pereira, 1999). Heparin is the most clinically exploited anticoagulant over the past 50 years (Fareed *et al.*, 2000). Until the early 1990s, sulfated fucans (SFs), homopolymers of fucopyranosyl units exclusively in α-L-form (Pomin, 2008; Pomin and Mourão, 2008), and sulfated galactans (SGs), homopolymers of α-L- or α– and/or β-D-galactopyranosyl units (Pomin, 2009) were usually extracted and characterized from the cell walls of the three major groups of macroalgae. Phaeophyta (brown algae) expresses SFs, while Rhodophyta (red algae) and Chlorophyta (green algae) both express SGs. Historically, algal SFs and SGs are used mainly in clinical tests and large-scale extractions for industrial purposes. Recently however, new and interesting sources of these compounds have been found in the extracellular matrices of certain marine invertebrates (Mourão, 2004, 2007; Mourão and Pereira, 1999; Pomin, 2008, 2009; Pomin and Mourão, 2008). In contrast with most algal sulfated polysaccharides, the invertebrate polymers exhibit highly regular chemical structures (Figs. 12.1 and 12.2A–D), which make it easier to correlate their biological functions to their respective structural features (Pomin, 2008;

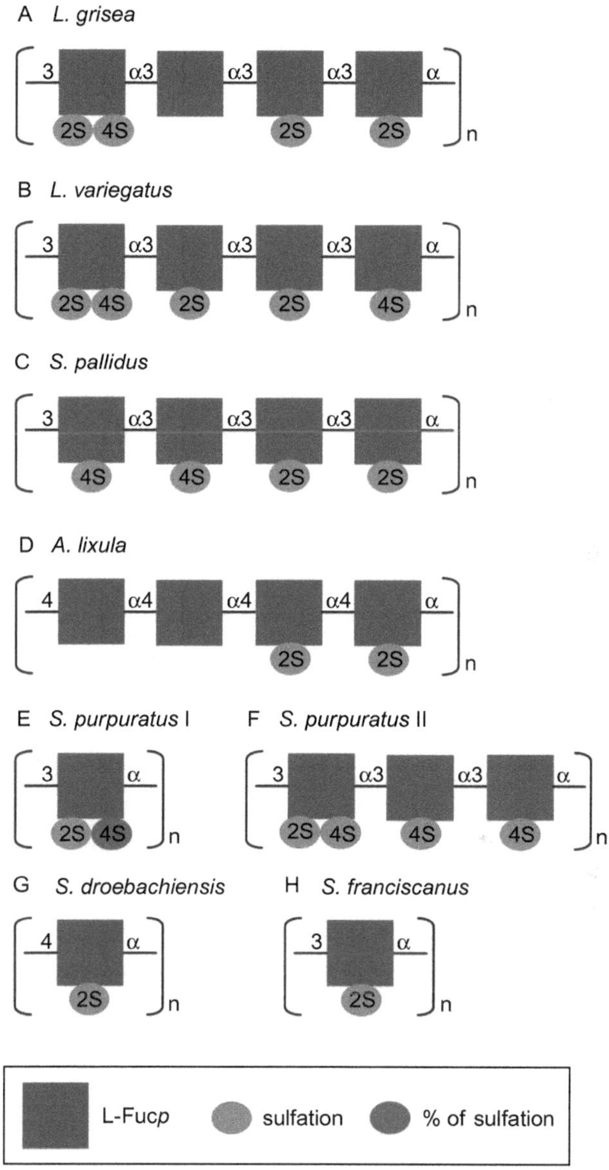

FIGURE 12.1 Chemical structures of repeating units of the SFs from the cell wall of the sea cucumber (A) and from the egg jelly coat of sea urchins (B–H). These polysaccharides are composed of α-L-fucopyranosyl units (green squares). The species-specific structures vary in sulfation patterns (exclusively 2- and/or 4-positions, in glycosidic linkages: α(1→3) (A–C, E, F, and H) and α(1→4) (D and G), and in number of residues of the repetitive units: tetrasaccharides (A–D), trisaccharides (F), and monosaccharides (E, G, and H), but they are all linear. The numbers over the black thin lines represent the glycosidic linkage. The gray and violet ellipses represent the right and % of sulfation, respectively. The numbers

Pomin and Mourão, 2008; Vilela-Silva *et al.*, 2008). This is of enormous benefit to the entire field of glycobiology. At one time, it was relatively hard to accurately correlate the structure–activity relationship for the majority of SPs. This same advantage is also observed for some marine invertebrate GAGs (Fig. 12.2F). In contrast, many mammalian GAGs exhibit a large variety of sulfation patterns (Gandhi and Mancera, 2008) and consequentially demand much more effort to assay their biological activities, determine their specific structural features, and subsequently propose a dependable structure–function relationship.

The MSPs with well-defined chemical structures (SFs, SGs, and GAGs in Figs. 12.1 and 12.2) are extracted from echinoderms (Echinodermata) like sea cucumbers (Holothuroidea) (Mulloy *et al.*, 2004) and sea urchins (Echinoidea) (Pomin and Mourão, 2008; Vilela-Silva *et al.*, 2008) or ascidians (Urochordata, Ascidiacea) (Mourão and Perlin, 1987; Pavão, 2002; Pavão *et al.*, 1995, 1998; Santos *et al.*, 1992; Vicente *et al.*, 2001). The SFs (Fig. 12.1B–H) and SGs (Fig. 12.2A and B) are widely spread throughout the jelly coat that surrounds sea urchin eggs. As discussed below, these particular MSPs are physiologically responsible for controlling the external fertilization in sea urchins through a very restrict species-specific manner (Mourão, 2007; Vilela-Silva *et al.*, 2008). Among all the MSPs with well-defined chemical structures described so far (Figs. 12.1 and 12.2), there is just a single description from the body wall of the sea cucumber (Fig. 12.1A) (Mulloy *et al.*, 2004), a few examples of SGs isolated from the tunic of ascidians (Fig. 12.2C and D) (Mourão and Perlin, 1987; Pavão *et al.*, 1998; Santos *et al.*, 1992), and some reports from specific red algal cell walls (Fig. 12.2E) (Fonseca *et al.*, 2008; Pereira *et al.*, 2005). All of the remaining structures (Figs. 12.1B–H and 12.2A and B) belong to the egg jelly coat of sea urchins (Alves *et al.*, 1997, 1998; Mulloy *et al.*, 2004;

---

before *S* inside the ellipses represent the sulfation position. The structures are the following: (A) *Ludwgothurea grisea* [→3)-α-L-Fuc*p*-2,4(OSO_3^-)-(1→3)-α-L-Fuc*p*-(1→3)-α-L-Fuc*p*-2(OSO_3^-)-(1→3)-α-L-Fuc*p*-2(OSO_3^-)-(1→]$_n$ (Mulloy *et al.*, 2004); (B) *Lytechinus variegatus* [→3)-α-L-Fuc*p*-2,4(OSO_3^-)-(1→3)-α-L-Fuc*p*-2(OSO_3^-)-(1→3)-α-L-Fuc*p*-2(OSO_3^-)-(1→3)-α-L-Fuc*p*-4(OSO_3^-)-(1→]$_n$ (Pomin and Mourão, 2008); (C) *Strongylocentrotus pallidus* [→3)-α-L-Fuc*p*-4(OSO_3^-)-(1→3)-α-L-Fuc*p*-4(OSO_3^-)-(1→3)-α-L-Fuc*p*-2(OSO_3^-)-(1→3)-α-L-Fuc*p*-2(OSO_3^-)-(1→]$_n$ (Vilela-Silva *et al.*, 2002); (D) *Arbacia lixula* [→4)-α-L-Fuc*p*-2(OSO_3^-)-(1→4)-α-L-Fuc*p*-2(OSO_3^-)-(1→4)-α-L-Fuc*p*-(1→4)-α-L-Fuc*p*-(1→]$_n$ (Alves *et al.*, 1997); (E) *Strongylocentrotus purpuratus*-I ~80% [→3)-α-L-Fuc*p*-2,4(OSO_3^-)-(1→]$_n$ and ~20% [→3)-α-L-Fuc*p*-2(OSO_3^-)-(1→]$_n$ and (F) *S. purpuratus*-II [→3)-α-L-Fuc*p*-2,4(OSO_3^-)-(1→3)-α-L-Fuc*p*-4(OSO_3^-)-(1→3)-α-L-Fuc*p*-4(OSO_3^-)-(1→]$_n$ (Alves *et al.*, 1998); (G) *Strongylocentrotus droebachiensis* [→4)-α-L-Fuc*p*-2(OSO_3^-)-(1→]$_n$ (Vilela-Silva *et al.*, 2002); and (H) *Strongylocentrotus franciscanus* [3)-α-L-Fuc*p*-2(OSO_3^-)-(1→]$_n$ (Vilela-Silva *et al.*, 1999). (For interpretation of the references to color in this figure legend, the reader is referred to the Web version of this chapter.)

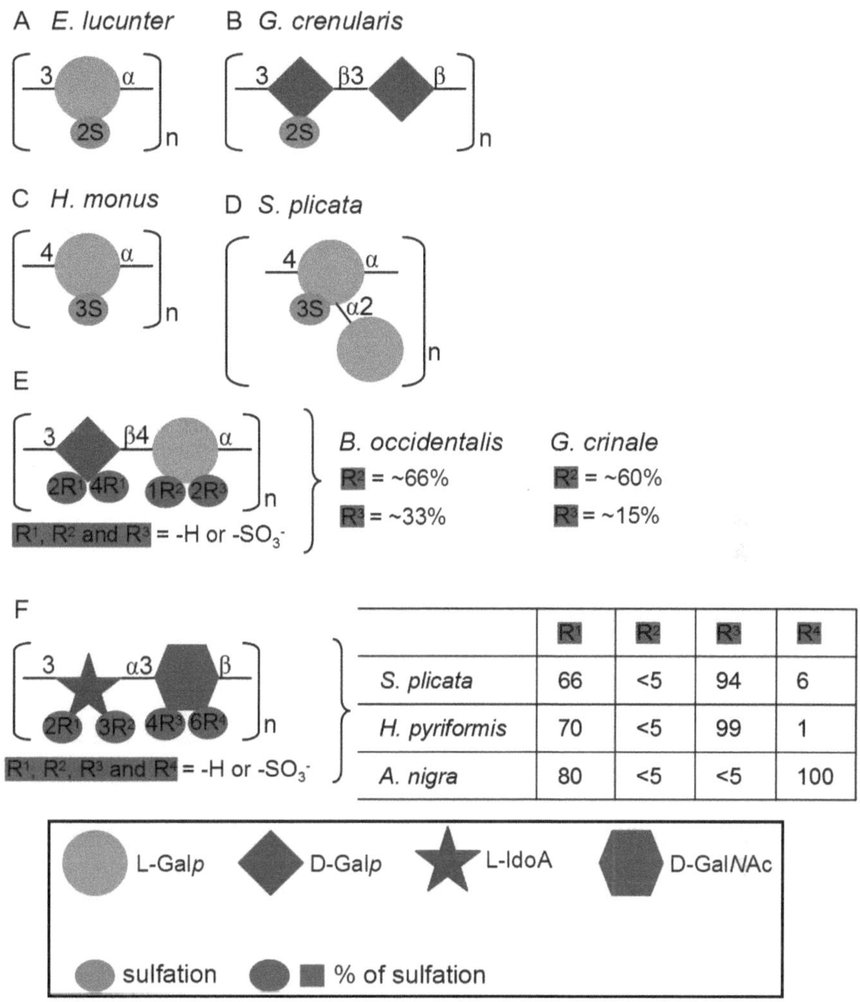

FIGURE 12.2 Chemical structures of the repeating units of SGs from the egg jelly coat of sea urchins (A and B), from the tunic of ascidians (C and D), and from the red algae (E); and of the repeating units of DSs from ascidians (F). These polysaccharides are composed of L-galactopyranoses (orange circles), D-galactopyranoses (blue diamond), L-iduronic acids (red star), and D-*N*-acetylgalactosamines (green hexagons). The numbers over the black thin lines represent the glycosidic linkage. The gray and violet boxes and ellipses represent the right and % of sulfation, respectively. The numbers before *S* and *R* inside the ellipses represent the positions of sulfation or radicals. The superscripted numbers after *R* represent the respective radical. The structures are the following: (A) *Echinometra lucunter* (Alves *et al*., 1997); (B) *Glyptocidaris crenularis* [→3-α-L-Gal*p*-2(OSO_3^-)-1→3-α-L-Gal*p*-1→]$_n$ (Castro *et al*., 2009); (C) *Herdmania monus* [→4)-α-L-Gal*p*-3(SO_3^-)-(1→]$_n$ (Santos *et al*., 1992); (D) *Styela plicata* {→4)-α-L-Gal*p*-2[→1)-α-L-Gal*p*-3(OSO_3^-)]-3(OSO_3^-)-(1→}$_n$ (Mourão and Perlin, 1987); (E) both *Botriocladia occidentalis* and *Gelidium crinali* express [3-β-D-Gal*p*-1→4-α-Gal-1→]$_n$ with different sulfation contents (Fonseca *et al*.,

Vilela-Silva *et al.*, 1999, 2002). The body of some species of ascidians also contains peculiar GAGs that show distinct sulfation patterns from mammalian GAGs (Pavão *et al.*, 1995, 1998) as exemplified with the tunicate dermatan sulfates (DSs) in Fig. 12.2F.

II. THE INTERACTION OF MSPS WITH COAGULATION (CO)-FACTORS: PREVENTION OF BLOOD COAGULATION

Unrelated to the natural biological role as inducers of fertilization, the SFs and SGs also exhibit potential pharmacological actions in mammalian systems. Among several clinical activities: antiviral, antimetastatic, antiangiogenic, anti-inflammatory, antiadhesive (Coombe *et al.*, 1987; Cumashi *et al.*, 2007; Harrop *et al.*, 1992), the anticoagulant and antithrombotic actions are the most explored so far (Farias *et al.*, 2000; Pomin, 2008). This is due to a pressing need for new antithrombotic drugs as a consequence of the increasing incidence of thromboembolic diseases. In fact, cardiovascular diseases are the leading cause of death worldwide (30% of total causes) (Becker *et al.*, 2007; Berteau and Mulloy, 2003; Cumashi *et al.*, 2007; Farias *et al.*, 2000; Mourão, 2004; Mourão and Pereira, 1999, Pereira *et al.*, 1999, 2002a,b, 2005; Pomin, 2008, 2009; Pomin and Mourão, 2008).

The SFs and SGs show great advantages as alternative sources for anticoagulant therapies, especially because of the massive use of heparin (Fareed *et al.*, 2000). This mammalian GAG has the highest negative charge density ever seen for a natural molecule. It is also well known to present several limitations due to collateral effects and limited sources of material (Mourão, 2004). The situation has been complicated due to the alarming discovery of heparin preparations that have been contaminated with oversulfated chondroitin sulfate (Guerrini *et al.*, 2008). This contaminant induces hypotension associated with kallikrein release when administered by intravenous injection (Kishimoto *et al.*, 2008).

The new MSPs (Figs. 12.1 and 12.2) offer some advantages over heparin. They show considerably low contamination levels of virus and/or prions because they are exclusively extracted from marine sources. Contamination in clinical solutions of heparin can occur more easily, as this compound is extracted from mammalian sources, like porcine and bovine intestinal mucosa and bovine lung (Mourão, 2004). Moreover, the invertebrate MSPs are promising to be more useful clinical reagents than algal

---

2008; Pereira *et al.*, 2005); and (F) the DS from *Styela plicata*, *Halocynthia pyriformis*, and *Ascidian nigra* are composed of [4-α-L-IdoA-1→3-β-D-GalNAc-1]$_n$ with also different sulfation patterns (Pavão *et al.*, 1995, 1998). (For interpretation of the references to color in this figure legend, the reader is referred to the Web version of this chapter.)

SPs due to their regular and well-defined structures (Mourão, 2004; Pomin, 2008; Pomin and Mourão, 2008), even though algal molecules have been utilized more until now.

In addition to all the benefits described above, the very clear and regular structures of invertebrate SFs (Fig. 12.1) and SGs (Fig. 12.2A–D), the algal SGs (Fig. 12.2E), and the ascidians GAGs (Fig. 12.2F) enable determination of well-defined structure–function relationships (Pomin, 2008; Pomin and Mourão, 2008). This knowledge allows prediction of the major molecular features involved in molecular interactions such as those between the MSPs and blood (co)-factors, similarly used for the fertilization assays mentioned in the previous section. The anticoagulant action of the SPs resides mainly in the potentiation of natural inhibitors of plasma proteases. The plasma proteases include activated factor II (IIa, commonly known as thrombin) and activated factor X (Xa). Inhibitors include the serpins (antithrombin, AT; and heparin cofactor II, HCII). The catalysis of SPs can be classified into two distinct mechanisms: (i) the allosteric change of the serpins induced by the SPs and (ii) the template mechanisms, where the glycosidic chain of the SPs can act as a "bridge" that brings together both protease and serpin. The mechanisms for AT and HCII are illustrated in Fig. 12.4A and B, respectively.

Next, we will describe particular cases of how each specific structural feature of the MSPs (Figs. 12.1 and 12.2) can individually account for different levels of activities and interactions with serpins and/or blood proteases (Fig. 12.3). This results in differing kinetics to prevent clot formation. The particular structural contributions of the MSPs can be properly recognized when the anticoagulant properties of these biopolymers are systematically compared. In addition, the results might offer qualitatively accurate data. This systematic comparison was carried out by two *in vitro* anticoagulant assays: (i) the APTT (activated partial thromboplastin time) which measures the general anticoagulant effect where all the plasma proteins are included and (ii) the direct measurement of the inhibition level of factors IIa or Xa by AT or HCII in the presence of the MSPs. Results from these assays are compiled in Table 12.1.

III. DECODING THE REGULATING STRUCTURAL FEATURE OF MSPS IN INTERACTIONS WITH COAGULATION (CO)-FACTORS

A. An example of influence by sugar type

When we compare all the structures in Figs. 12.1 and 12.2, we can trace structural similarities and differences for the SFs and SGs. For example, both SF from *Strongylocentrotus franciscanus* (Fig. 12.1H) and SG from

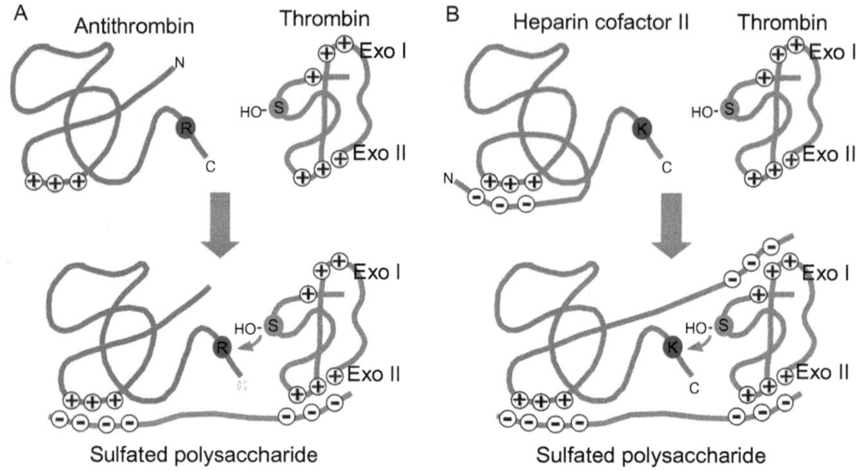

FIGURE 12.3 Molecular schematic representation of the anticoagulant mechanisms of SPs. The thrombin (IIa, in blue) can be inhibited by serpins such as antithrombin (AT, in green) or heparin cofactor-II (HCII, in orange). In both cases, the SP (gray line) brings together the serpins and the protease-IIa mainly through electrostatic interactions of their opposite charges. In the thrombin, this charged cluster is the EXO II. Next, the hydroxyl groups of a serine (S) residue from thrombin will bind to the C-terminus of the serpins, actually to an arginine (R) residue of the AT (A), or to a lysine (K) residue in the case of the HCII (B). In the bind states, a conformational change will occur in both serpins, although this change is more predominant and necessary at the HCII case. (B) Note that the N-terminus of the HCII will interact also with the EXO I of IIa through also electrostatic contacts. With the examples described throughout the text, it is clear that the template mechanism between SPs (SFs, SGs, GAGs), serpins (AT, HCII), and protease (IIa) has differential stabilities or formation kinetics directly related to the structural features of the SPs. (For interpretation of the references to color in this figure legend, the reader is referred to the Web version of this chapter.)

Echinometra lucunter (Fig. 12.2A) present the same sulfation pattern (exclusive and entirely 2-sulfated), the same anomeric configuration (α-form), the same glycosidic linkage (1→3), and the same molecular mass (~100 kDa). Their single difference is their sugar type (fucopyranose or galactopyranose, respectively) (Pomin, 2008). Interestingly, this single structural difference is itself enough to promote great changes in the anticoagulant properties of these homopolysaccharides. The 2-sulfated α-galactan from *E. lucunter* exhibits a significant anticoagulant activity (APTT of 20IUmg, although almost 10-fold less than UFH, Table 12.1). The specific anticoagulant assay with the purified proteases revealed that this SG enhances both IIa and factor Xa inhibition by either AT or HCII (Table 12.1). However, the anticoagulant effect of 2-sulfated α-fucan from *S. franciscanus* is exclusively based on catalysis of AT inhibition over factor Xa, although it

TABLE 12.1 Anticoagulant activities of MSPs measured by APTT[a] and by IC_{50} for thrombin (IIa) and factor Xa inhibition in the presence of antithrombin (AT) or heparin cofactor II (HCII) (Fonseca *et al.*, 2008; Mourão, 2004; Mourão and Pereira, 1999; Pavão *et al.*, 1998; Pereira *et al.*, 2005)

				IC_{50} (μg/mL)		
Polysaccharide	Source	Structure (figure)	APTT (IU/mg)	IIa/AT	IIa/HCII	Xa/AT
3-linked sulfated α-L-fucans	*S. purpuratus* (I)	1E	76	0.3	0.3	2
	S. purpuratus (II)	1F	10	0.9	2	ND
	S. pallidus	1C	18	>500	>500	>500
	L. variegatus	1B	3	>500	>500	>500
	S. franciscanus	1H	~2	>500	>500	250
	L. grisea	1A	<1	>500	>500	>500
4-linked sulfated α-L-fucans	*S. droebachiensis*	1G	<1	ND	ND	ND
	A. lixula	1D	~2	150	150	>500
Sulfated α-L-galactans	*E. lucunter*	2A	20	3	6	20
	H. monus	2C	~2	>500	>500	>500
	S. plicata	2D	<1	>500	>500	>500
Algal SGs (Fonseca *et al.*, 2008; Pereira *et al.*, 2005)	*B. occidentalis*	2E	93	0.02	1.1	2.5
	G. crinali		65	0.02	25	1.5
Ascidian DSs (Pavão *et al.*, 1998)	*S. plicata*	2F	8	Inactive	0.31	Inactive
	H. pyriformis		11	Inactive	0.35	Inactive
	A. nigra		<5	Inactive	320	Inactive
Mammalian standard DSs (Pavão *et al.*, 1998)	Native		2	nr[b]	3	nr
	Oversulfated		13	nr	2	nr

[a] The activity is expressed as international units/mg using a parallel standard curve based on the International Heparin Standard (193units/mg).
[b] Nonreported.

is 12.5-fold less active than the α-SG. This single effect over the Xa/AT system explains the much lower activity of the compound from *S. franciscanus* (APTT of ~2IUmg^{-1}, 100-fold less active than UFH), as the anti-Xa activity has a relatively minor influence on the APTT. This is an illustrative and typical example of a sugar-type-dependent biological effect of polysaccharides.

B. An example of preferential conformation binding

Interaction energies obtained from molecular dynamic simulations are statistically equivalent for both 2-sulfated α-L-galactan from *E. lucunter* (Fig. 12.2A) and 2-sulfated α-L-fucan from *S. franciscanus* (Fig. 12.1H). This would be expected based on the structural similarities of these two compounds, especially conformations in solution. However, they revealed markedly different interactions for binding with the complex AT/IIa. The explanation for this result is the extreme difference in orientations on the AT-binding site (~90°, Fig. 12.4C and D). Such difference in orientation upon binding to AT explains the lack of activity of SF under the same experimental conditions and fully supports the expected bridging mechanism in the activity of SG, as previously described (Melo *et al.*, 2004). Thus, both compounds are capable of interacting with AT as indicated by theoretical results and experimental data showing that

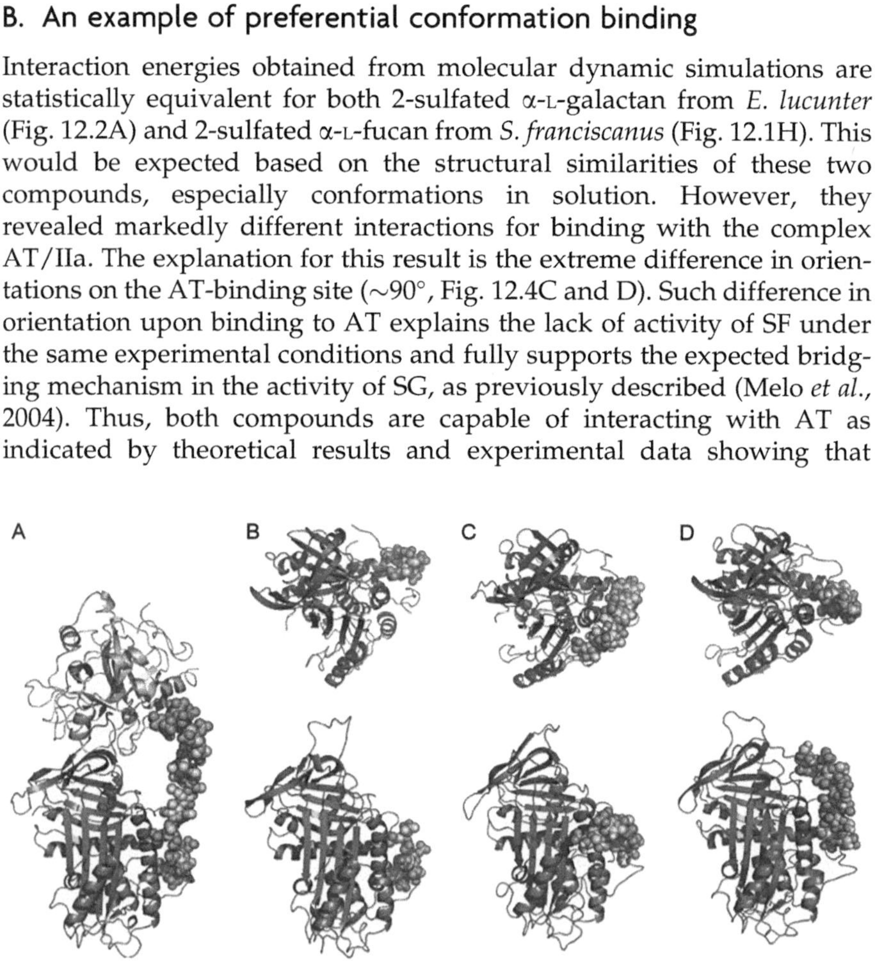

FIGURE 12.4 Structures of the complexes between different SPs and AT. (A) Ternary complex between AT, thrombin, and an heparin derivative (PDB ID 1TB6); (B) AT bounded to the synthetic pentasaccharide (PDB ID 1E03); (C) final structure from a 5-ns MD of AT complexed to a SF decasaccharide with pyranose rings; (D) final structure from a 5-ns MD of AT complexed to a SG decasaccharide with pyranose rings. For (B)–(D), two orientations of the complexes are presented. Data from Becker *et al.* (2007).

both polysaccharides are retained on an AT-affinity column. However, only the complex formed by the SG and the AT (Fig. 12.4D) allows further interaction with IIa. An analogous complex is the ternary complex between heparin-AT-IIa (Fig. 12.4A) which supports the potentiation of this protease inactivation by AT in the presence of the SG. This is an example of how the orientational constraints in solution can influence spatial preferences of the SPs for their different bindings with target proteins (Melo *et al.*, 2004). It is obvious that the swap of the sugar type will drastically change the conformation of the polymer as well, implying a different conformational binding as explained in this section.

C. An example of influence by sulfation pattern

Based on this same systematic comparison, the SGs from the red algal species, *Botryocladia occidentalis* and *Gelidium crinale*, exhibited identical backbones and the same chain size. However, there are slight differences in their sulfation patterns (Fig. 12.2E). As a consequence of this difference, the two algal SGs differ in their anticoagulant and venous antithrombotic activities as previously described (Fonseca *et al.*, 2008). SGs from *G. crinale* exhibit procoagulant and prothrombotic effects in low doses (up to 1.0 mg/kg body weight). In high doses (>1.0mg/kg), this polysaccharide inhibits both venous and arterial thrombosis in rats. In contrast, SG from *B. occidentalis* is a very potent anticoagulant and antithrombotic compound in low doses (up to 0.5mg/kg body weight), inhibiting venous experimental thrombosis, but these effects are reverted in high doses. Conversely, arterial thrombosis is only inhibited at high doses (>1.0 mg/kg) of the polysaccharide from *B. occidentalis*. These results indicate that slight differences in the proportions and/or distribution of sulfated residues along the galactosyl chain may be critical for the interaction between proteases, inhibitors, and activators of the coagulation system, resulting in a distinct pattern in anti- and procoagulant activities and in the antithrombotic action. As summarized in Table 12.1, these structural differences account for the 30% difference in anticoagulant activity (APTT) of these algal macromolecules and even greater difference in catalytic effect of the sulfated polysaccharide on HCII-mediated anti-IIA activity.

Indeed, the structural requirement for the interaction of these SPs with the coagulation cofactors and their target proteases and inhibitors are stereospecific (Mourão, 2004; Mourão and Pereira, 1999; Pomin, 2008; Pomin and Mourão, 2008). The site of sulfation has a major impact on activity. This can be illustrated by the fact that 2,4-di-sulfated units have an amplifying effect on the AT-mediated anticoagulant activity in the series of 3-linked α-L-fucans (Fig. 12.1; Table 12.1). Specific sulfation sites are required for the interaction with plasma serine-protease inhibitors. Note the occurrence of the 4-sulfated unit content in the

3-linked α-L-fucans: *Lytechinus variegatus* (a single 4-sulfated unit/tetrasaccharide—Fig. 12.1B), *Strongylocentrotus pallidus* (a double 4-sulfated unit/tetrasaccharide—Fig. 12.1C), and *Strongylocentrotus purpuratus,* isotype II (a double 4-sulfated unit/trisaccharide—Fig. 12.1F). This 4-sulfation is the structural motif required to enhance the inhibition of IIa by HCII. In contrast, the presence of 2-sulfated residues seems to have a deleterious effect on HCII-mediated anti-IIA activity of the polysaccharides (Mourão, 2004).

Ascidian DSs with the backbone structure [4-α-L-IdoA-1→3-β-D-GalNAc-1]$_n$ but with different patterns of sulfation (Fig. 12.2F) also reveal different anticoagulant activities (Table 12.1) (Pavão *et al.*, 1995,1998; Vicente *et al.*, 2001). They are also useful and reliable cases to describe the influence of sulfation sites in biological actions. All three ascidian DSs have a high content of 2-*O*-sulfated α-L-iduronic acid residues but differ in the pattern of sulfation of the *N*-acetyl-β-D-galactosaminyl units. *Styela plicata* and *Halocynthia pyriformis* have 4-*O*-sulfated units, but in *Ascidian nigra,* they are 6-*O*-sulfated. This collection of ascidian DSs, where the extent and position of sulfate substitution have been widely characterized, was examined in anticoagulant assays. DS from *A. nigra* has no discernible anticoagulant activity, which indicates that 4-*O*-sulfation of the *N*-acetyl-β-D-galactosamine is essential for that activity (Table 12.1). In contrast, DSs from *S. plicata* and *H. pyriformis* are potent anticoagulants due to potentiation of thrombin inhibition by HCII. These ascidian DSs have ~10- and ~6-fold higher activity with HCII than native and oversulfated mammalian DSs, respectively (Table 12.1). They have no effect on factors IIa- or Xa inhibition by AT. These highly sulfated ascidian DSs show that they require very selective sulfation sites for promoting the HCII interactions, and consequently, their highly specific anticoagulant properties are intimately related to their sulfation levels (Table 12.1 vs. Fig. 12.2F).

IV. REMARK CONCLUSIONS

The example cases in this chapter have clearly shown how the structural features of marine homopolysaccharides (SFs and galactans) can be evaluated for interactions with coagulation proteins (acting as promising anticoagulant or antithrombotic reagents). It was observed that the carbohydrate–protein complexes are indeed stereospecific and not a mere consequence of charge density. It is important to note that the glycosidic structural characteristics examined individually in this chapter do not really account independently and/or exclusively for the biological actions or molecular interactions. Obviously, the active or inactive conformations of these homopolysaccharides arise from the presence of all structural features together, which comprise the "final code" required for high-affinity interactions with the target proteins. However, specific or major structural features of the carbohydrates certainly act as preponderant regulators in the biological functions as succinctly described in this work.

ACKNOWLEDGMENTS

I am grateful to Prof. Paulo A.S. Mourão and his glycobiology group from the University of Brazil, where I got my Ph.D. focused mostly on structures and function of marine sulfated polysaccharides. I acknowledge Prof. James H. Prestegard for the postdoctoral position in his laboratory at Complex Carbohydrate Research Center, University of Georgia, USA, where I studied biomolecular NMR and glycosaminoglycans. I also want to kindly thank Laura Morris from Prestegard's lab for the great editing of this work.

REFERENCES

Alves, A. P., Mulloy, B., Diniz, J. A., and Mourão, P. A. (1997). Sulfated polysaccharides from the egg jelly layer are species-specific inducers of acrosomal reaction in sperms of sea urchins. *J. Biol. Chem.* **272**, 6965–6971.

Alves, A. P., Mulloy, B., Moy, G. W., Vacquier, V. D., and Mourão, P. A. (1998). Females of the sea urchin *Strongylocentrotus purpuratus* differ in the structures of their egg jelly sulfated fucans. *Glycobiology* **8**, 939–946.

Becker, C. F., Guimaraes, J. A., Mourão, P. A., and Verli, H. (2007). Conformation of sulfated galactan and sulfated fucan in aqueous solutions: Implications to their anticoagulant activities. *J. Mol. Graph. Model.* **26**, 391–399.

Berteau, O. and Mulloy, B. (2003). Sulfated fucans, fresh perspectives: Structures, functions, and biological properties of sulfated fucans and an overview of enzymes active toward this class of polysaccharide. *Glycobiology* **13**, 29R–40R.

Castro, M. O., Pomin, V. H., Santos, L. L., Vilela-Silva, A. C., Hirohashi, N., Pol-Fachin, L., Verli, H., and Mourão, P. A. (2009). A unique 2-sulfated {beta}-galactan from the egg jelly of the sea urchin *Glyptocidaris crenularis*: Conformation flexibility versus induction of the sperm acrosome reaction. *J. Biol. Chem.* **284**, 18790–18800.

Coombe, D. R., Parish, C. R., Ramshaw, I. A., and Snowden, J. M. (1987). Analysis of the inhibition of tumour metastasis by sulphated polysaccharides. *Int. J. Cancer* **39**, 82–88.

Cumashi, A., Ushakova, N. A., Preobrazhenskaya, M. E., D'Incecco, A., Piccoli, A., Totani, L., Tinari, N., Morozevich, G. E., Berman, A. E., Bilan, M. I., Usov, A. I., Ustyuzhanina, N. E., *et al.* (2007). A comparative study of the anti-inflammatory, anticoagulant, antiangiogenic, and antiadhesive activities of nine different fucoidans from brown seaweeds. *Glycobiology* **17**, 541–552.

Fareed, J. W., Hoppensteadt, D., and Bick, R. L. (2000). An update on heparins at the beginning of the new millennium. *Sem. Thromb. Haemost.* **26**, 5–21.

Farias, W. R., Valente, A. P., Pereira, M. S., and Mourão, P. A. (2000). Structure and anticoagulant activity of sulfated galactans. Isolation of a unique sulfated galactan from the red algae *Botryocladia occidentalis* and comparison of its anticoagulant action with that of sulfated galactans from invertebrates. *J. Biol. Chem.* **275**, 29299–29307.

Fonseca, R. J., Oliveira, S. N., Melo, F. R., Pereira, M. G., Benevides, N. M., and Mourão, P. A. (2008). Slight differences in sulfation of algal galactans account for differences in their anticoagulant and venous antithrombotic activities. *Thromb. Haemost.* **99**, 539–545.

Gandhi, N. S. and Mancera, R. L. (2008). The structure of glycosaminoglycans and their interactions with proteins. *Chem. Biol. Drug Des.* **72**, 455–482.

Guerrini, M., Beccati, D., Shriver, Z., Naggi, A., Viswanathan, K., Bisio, A., Capila, I., Lansing, J. C., Guglieri, S., Fraser, B., Al-Hakim, A., Gunay, N. S., *et al.* (2008). Oversulfated chondroitin sulfate is a contaminant in heparin associated with adverse clinical events. *Nat. Biotechnol.* **10**, 1–7.

Harrop, H. A., Rider, C. C., and Coombe, D. R. (1992). Sulphated polysaccharides exert anti-HIV activity at differing sites. *Biochem. Soc. Trans.* **20**, 163S.

Kishimoto, T. K., Viswanathan, K., Ganguly, T., Elankumaran, S., Smith, S., Pelzer, K., Lansing, J. C., Sriranganathan, N., Zhao, G., Galcheva-Gargova, Z., Al-Hakim, A., Bailey, G. S., *et al.* (2008). Contaminated heparin associated with adverse clinical events and activation of the contact system. *N. Engl. J. Med.* **358**, 1–11.

Melo, F. R., Pereira, M. S., Foguel, D., and Mourão, P. A. (2004). Antithrombin-mediated anticoagulant activity of sulfated polysaccharides: Different mechanisms for heparin and sulfated galactans. *J. Biol. Chem.* **279**, 20824–20835.

Mourão, P. A. (2004). Use of sulfated fucans as anticoagulant and antithrombotic agents: Future perspectives. *Curr. Pharm. Design* **10**, 967–981.

Mourão, P. A. (2007). A carbohydrate-based mechanism of species recognition in sea urchin fertilization. *Braz. J. Med. Biol. Res.* **40**, 5–17.

Mourão, P. A. and Pereira, M. S. (1999). Searching for alternatives to heparin: Sulfated fucans from marine invertebrates. *Trends Cardiovasc. Med.* **9**, 225–232.

Mourão, P. A. and Perlin, A. S. (1987). Structural features of sulfated glycans from the tunic of *Styela plicata* (Chordata-Tunicata). A unique occurrence of L-galactose in sulfated polysaccharides. *Eur. J. Biochem.* **166**, 431–436.

Mulloy, B., Ribeiro, A. C., Alves, A. P., Vieira, R. P., and Mourão, P. A. (2004). Sulfated fucans from echinoderms have a regular tetrasaccharide repeating unit defined by specific patterns of sulfation at the 0–2 and 0–4 positions. *J. Biol. Chem.* **269**, 22113–22123.

Pavão, M. S. (2002). Structure and anticoagulant properties of sulfated glycosaminoglycans from primitive Chordates. *An. Acad. Bras. Cien.* **74**, 105–112.

Pavão, M. S., Mourão, P. A., Mulloy, B., and Tollefsen, D. M. (1995). A unique dermatan sulfate-like glycosaminoglycan from ascidian. Its structure and the effect of its unusual sulfation pattern on anticoagulant activity. *J. Biol. Chem.* **270**, 31027–31036.

Pavão, M. S., Aiello, K. R., Werneck, C. C., Silva, L. C., Valente, A. P., Mulloy, B., Colwell, N. S., Tollefsen, D. M., and Mourão, P. A. (1998). Highly sulfated dermatan sulfates from Ascidians. Structure versus anticoagulant activity of these glycosaminoglycans. *J. Biol. Chem.* **273**, 27848–27857.

Pereira, M. S., Mulloy, B., and Mourão, P. A. (1999). Structure and anticoagulant activity of sulfated fucans. Comparison between the regular, repetitive, and linear fucans from echinoderms with the more heterogeneous and branched polymers from brown algae. *J. Biol. Chem.* **274**, 7656–7667.

Pereira, M. S., Melo, F. R., and Mourão, P. A. (2002a). Is there a correlation between structure and anticoagulant action of sulfated galactans and sulfated fucans? *Glycobiology* **12**, 573–580.

Pereira, M. S., Vilela-Silva, A. C., Valente, A. P., and Mourão, P. A. (2002b). A 2-sulfated, 3-linked alpha-L-galactan is an anticoagulant polysaccharide. *Carbohydr. Res.* **337**, 2231–2238.

Pereira, M. G., Benevides, N. M., Melo, M. R., Valente, A. P., Melo, F. R., and Mourão, P. A. (2005). Structure and anticoagulant activity of a sulfated galactan from the red alga, *Gelidium crinale.* Is there a specific structural requirement for the anticoagulant action? *Carbohydr. Res.* **340**, 2015–2023.

Pomin, V. H. (2008). Review: An overview about the structure-function relationship of marine sulfated homopolysaccharides with regular chemical structures. *Biopolymers* **91**, 601–609.

Pomin, V. H. (2009). Structural and functional insights into sulfated galactans: A systematic review. *Glycoconj. J.* **27**, 1–12.

Pomin, V. H. and Mourão, P. A. (2008). Structure, biology, evolution, and medical importance of sulfated fucans and galactans. *Glycobiology* **18**, 1016–1027.

Santos, J. A., Mulloy, B., and Mourão, P. A. (1992). Structural diversity among sulfated alpha-L-galactans from ascidians (tunicates). Studies on the species *Ciona intestinalis* and *Herdmania monus*. *Eur. J. Biochem.* **204**, 669–677.

Vicente, C. P., Zancan, P., Peixoto, L. L., Alves-Sá, R., Araújo, F. S., Mourão, P. A., and Pavão, M. S. (2001). Unbalanced effects of dermatan sulfates with different sulfation patterns on coagulation, thrombosis and bleeding. *Thromb. Haemost.* **86**, 1215–1220.

Vilela-Silva, A. C., Alves, A. P., Valente, A. P., Vacquier, V. D., and Mourão, P. A. (1999). Structure of the sulfated alpha-L-fucan from the egg jelly coat of the sea urchin *Strongylocentrotus franciscanus*: Patterns of preferential 2-O- and 4-O-sulfation determine sperm cell recognition. *Glycobiology* **9**, 927–933.

Vilela-Silva, A. C., Castro, M. O., Valente, A. P., Biermann, C. H., and Mourão, P. A. (2002). Sulfated fucans from the egg jellies of the closely related sea urchins *Strongylocentrotus droebachiensis* and *Strongylocentrotus pallidus* ensure species-specific fertilization. *J. Biol. Chem.* **277**, 379–387.

Vilela-Silva, A. C., Hirohashi, N., and Mourão, P. A. (2008). The structure of sulfated polysaccharides ensures a carbohydrate-based mechanism for species recognition during sea urchin fertilization. *Int. J. Dev. Biol.* **52**, 551–559.

CHAPTER 13

Health Benefits of n-3 Polyunsaturated Fatty Acids: Eicosapentaenoic Acid and Docosahexaenoic Acid

Nalin Siriwardhana,[*,†,1] **Nishan S. Kalupahana,**[*,†,‡] and **Naima Moustaid-Moussa**[*,†]

Contents

Abstract

Marine-based fish and fish oil are the most popular and well-known sources of n-3 polyunsaturated fatty acids (PUFAs), namely, eicosapentaenoic acid (EPA) and docosahexaenoic acid (DHA). These n-3 PUFAs are known to have variety of health benefits against cardiovascular diseases (CVDs) including well-established hypotriglyceridemic and anti-inflammatory effects. Also, various studies indicate promising antihypertensive, anticancer, antioxidant, antidepression, antiaging, and antiarthritis effects. Moreover, recent studies also indicate anti-inflammatory and insulin-sensitizing effects of these

* Department of Animal Science, University of Tennessee, Knoxville, Tennessee, USA
† UT Obesity Research Center, University of Tennessee, Knoxville, Tennessee, USA
‡ Department of Physiology, Faculty of Medicine, University of Peradeniya, Peradeniya, Sri Lanka
[1] Corresponding author: Nalin Siriwardhana, *E-mail address:* rsiriwar@utk.edu

Advances in Food and Nutrition Research, Volume 65
ISSN 1043-4526, DOI: 10.1016/B978-0-12-416003-3.00013-5

fatty acids in metabolic disorders. Classically, n-3 PUFAs mediate some of these effects by antagonizing n-6 PUFA (arachidonic acid)-induced proinflammatory prostaglandin E_2 (PGE_2) formation. Another well-known mechanism by which n-3 PUFAs impart their anti-inflammatory effects is via reduction of nuclear factor-κB activation. This transcription factor is a potent inducer of proinflammatory cytokine production, including interleukin 6 and tumor necrosis factor-α, both of which are decreased by EPA and DHA. Other evidence also demonstrates that n-3 PUFAs repress lipogenesis and increase resolvins and protectin generation, ultimately leading to reduced inflammation. Finally, beneficial effects of EPA and DHA in insulin resistance include their ability to increase secretion of adiponectin, an anti-inflammatory adipokine. In summary, n-3 PUFAs have multiple health benefits mediated at least in part by their anti-inflammatory actions; thus their consumption, especially from dietary sources, should be encouraged.

I. SOURCES AND INTAKES OF EICOSAPENTAENOIC ACID AND DOCOSAHEXAENOIC ACID

Marine fish is the well-known source of eicosapentaenoic acid (EPA) and docosahexaenoic acid (DHA). EPA ($C_{20}H_{30}O_2$; 20:5n-3) and DHA ($C_{22}H_{34}O_2$; 22:6n-3) are long-chain ω-3/n-3 polyunsaturated fatty acids (PUFAs). Technically, EPA is a nonessential n-3 fatty acid as the human body can convert essential n-3 alpha-linolenic acid (ALA) into EPA and DHA. However, in humans, this conversion is not efficient enough to meet the EPA and DHA demand to impart beneficial health effects; thus, it is expected to obtain these fatty acids from dietary sources. The n-3 PUFA levels consumed in the United States and many other countries in the world are very low compared to the recommended intake of n-6/n-3 PUFAs (11:1 (n6:n3) vs. 2.3:1; Davis and Kris-Etherton, 2003; Whelan and Rust, 2006). EPA and DHA are among the most studied bioactive compounds of marine origin. Fish oil being the major and well-known source of EPA and DHA, dietary fish such as anchovy, bluefish, herring, mackerel, mullet, sardines, salmon, sturgeon, tuna, and trout are popular sources of fish oil. Further, due to its recognized health benefits and recommendation by several health agencies, fish oil (primarily DHA and EPA) has become a very popular dietary supplement. Fish oil is commercially available as soft gel capsules. These preparations are often available with added vitamins, antioxidants, and various flavors. EPA and DHA are not only found in fish oil but also in marine algal species. However, they are considerably less concentrated in most of the algae, compared to some fish sources. Most of the identified n-3 PUFA-rich algae contain higher amount of DHA compared to EPA.

Fat is a key component in the human diet. Research shows that excessive consumption of saturated fat negatively impacts several biomarkers of health while monounsaturated and n-3 PUFAs are beneficial to human health. Moreover, research shows that unbalanced dietary ratios of n-6: n-3 may lead to various health complications as well as disease progression while increased n-3 levels impart prevention and health promoting effects (Burghardt *et al.*, 2010; Goodstine *et al.*, 2003; Simopoulos, 2002; Wan *et al.*, 2010). The American Heart Association recommends eating fatty fish meals at least twice a week due to their promising health and especially cardiovascular benefits. Here, we review some of the health benefits of n-3 PUFAs, due in part to their anti-inflammatory effects in cancer, cardiovascular diseases (CVDs), obesity, and other metabolic disorders.

II. HEALTH BENEFITS OF N-3 PUFAS

Fish oil-derived n-3 PUFAs have been a healthy nutritional supplement for long time due to numerous health benefits including promising effects to improve cardiovascular health (Harris, 2007; Lu *et al.*, 2011), brain function, and overall health during pregnancy (Koletzko *et al.*, 2008). Also, fish oil-derived n-3 PUFAs are well known to prevent inflammation (Calder, 2008; Fetterman and Zdanowicz, 2009; Figueras *et al.*, 2011), aging (Dyall *et al.*, 2010), arthritis (Wall *et al.*, 2010), insulin resistance (Kalupahana *et al.*, 2010a), and depression and slows the progression of certain cancers (Astorg, 2004; Cabanes *et al.*, 2003; Leitzmann *et al.*, 2004; Fig. 13.1). Further, it has been shown that fish oil-derived n-3 PUFAs can prevent cardiac remodeling and dysfunction under pressure overload conditions in rats (Duda *et al.*, 2009) and atrial fibrillation associated with heart failure in a rabbit (Kitamura *et al.*, 2011). Also, EPA and DHA are known to induce breast cancer cell apoptosis and impart anticancer effects. Wu *et al.* (2005) showed that *in vivo* and *in vitro* treatment of n-3 PUFAs can inhibit the breast cancer growth through activation of a neutral sphingomyelinase-mediated pathway (Wu *et al.*, 2005). Moreover, it has been reported that an approach with fish oil nutritional intervention can improve the health in lung cancer patients (Murphy *et al.*, 2011). In a recent study, we reported that EPA significantly prevents and reverses insulin resistance in high-fat diet-induced obese mice via modulation of adipose tissue inflammation (Kalupahana *et al.*, 2010a). Moreover, it has been reported that supplementation of long-chain ω-3 PUFAs in elderly female patients reduces the occurrence of depressive symptoms, improves phospholipid fatty acid profile and health-related quality of life (Rondanelli *et al.*, 2011).

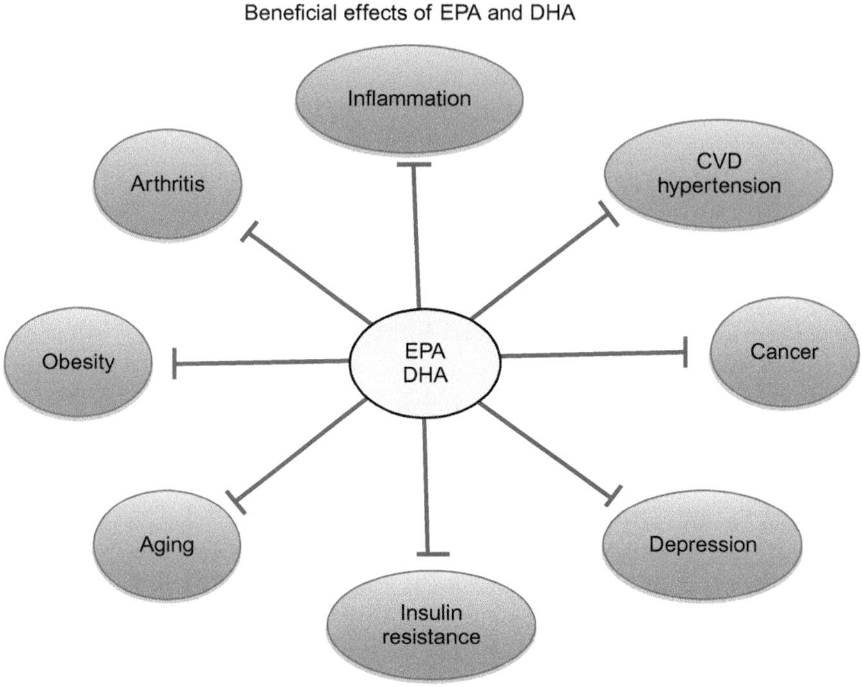

FIGURE 13.1 Beneficial effects of EPA and DHA. Scientifically validated beneficial effects of EPA and DHA as evident by epidemiological evidence as well as *in vivo* and *in vitro* studies. Blocked lines are to indicate the suppression effects. EPA, eicosapentaenoic acid; DHA, docosahexaenoic acid; CVD, cardiovascular disease. (For color version of this figure, the reader is referred to the Web version of this chapter.)

n-3 PUFAs are mostly known for their consistent hypotriglyceridemic effects (Harris, 1996, 1997). Moreover, diets rich in PUFAs are able to alter adipose tissue gene expression and metabolism. Hypotriglyceridemic effects of n-3 PUFA are believed to be primarily mediated via peroxisome proliferator activated receptor (PPARα)-dependent mechanisms (Buettner *et al.*, 2006), which may also account for the improved insulin resistance and reduction of hepatic steatosis in people consuming diets high in n-3 PUFA.

A. n-3 PUFAs and cardiovascular diseases

n-3 PUFAs reduce the risk of sudden cardiac death. For example, in the Physicians Health Study, individuals with the highest n-3 PUFA level in blood had a 90% reduction in sudden cardiac death compared to ones with the lowest n-3 PUFA levels (Albert *et al.*, 2002). These fatty acids also reduce the total mortality and sudden death in patients following myocardial infarction (Marchioli *et al.*, 2002). These beneficial effects of n-3 PUFA in reducing sudden cardiac death are attributed to two main factors. First, n-3 PUFAs reduce cardiac arrhythmias (Raitt *et al.*, 2005), which are responsible for the majority of sudden cardiac deaths. Second, these fatty acids increase atherosclerotic plaque stability (Thies *et al.*, 2003), thereby preventing plaque rupture and subsequent coronary events.

n-3 PUFAs also reduce several risk factors for CVDs. These include reductions in plasma triglyceride levels as well as several inflammatory cytokine levels. Detailed anti-inflammatory mechanisms of n-3 PUFA are discussed below.

III. ANTI-INFLAMMATORY EFFECTS OF EPA AND DHA

One of the main anti-inflammatory mechanisms is that EPA acts as a competitive inhibitor for proinflammatory arachidonic acid (AA) on cyclooxygenase (COX), which produces proinflammatory eicosanoids (Fig. 13.2). When EPA is subjected to oxygenation by COX, the less-inflammatory prostaglandin E_3 (PGE_3) is generated while AA generates highly proinflammatory prostaglandin E_2 (PGE_2). As EPA is structurally similar to AA, EPA can also compete with AA to incorporate into the cell membrane phospholipids and become the preferential substrate for COX

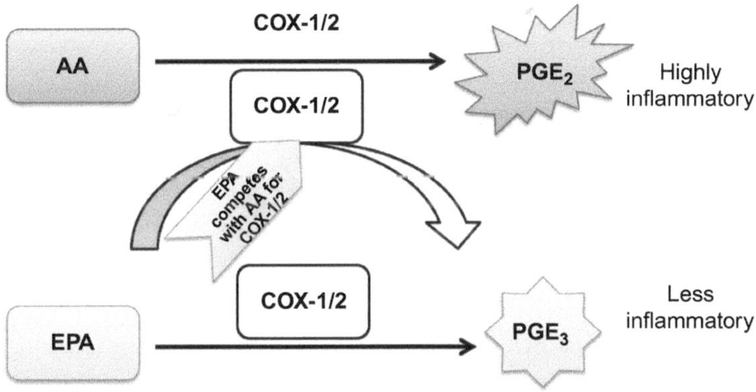

FIGURE 13.2 EPA competitively inhibits PGE_2 formation by COX-1 and COX-2. EPA competitively binds with COX-1 and COX-2 and produces less-inflammatory PGE_3 while suppressing AA binding which produces highly inflammatory PGE_2. AA, arachidonic acid; COX, cyclooxygenase; EPA, eicosapentaenoic acid; $PGE_{2/3}$, prostaglandin $E_{2/3}$. Calder (2009) was used as reference for the design. (For color version of this figure, the reader is referred to the Web version of this chapter.)

activity. Oxygenation of AA by COX can be inhibited by n-3 PUFAs and PGE_3 is generated instead of PGE_2. Thus, n-3 PUFAs act like nonsteroidal anti-inflammatory drugs to decrease COX enzyme activities and subsequently decrease inflammation including in adipose tissue as indicated by our recent studies (Ferrucci *et al.*, 2006; Kalupahana *et al.*, 2010a; Wortman *et al.*, 2009). Indeed, treatment of EPA in the diet can decrease PGE_2 levels through competitive inhibition of AA (Lands *et al.*, 1973). Also, it has been shown that EPA and DHA can suppress the COX-2 activity in human umbilical vein endothelial cells (Lee *et al.*, 2009). Thus, PGE_2 levels can be manipulated by increasing consumption of EPA. COX, also known as prostaglandin H_2 synthase (PGH_2 synthase), has both COX and peroxidase activity. The COX complex is made up of COX-1 and COX-2 enzymes. COX-1 is constitutively expressed while the expression of COX-2 is inducible and highly regulated. Pharmacological inhibition of COX-2 with celecoxib decreases PGE_2 secretion in adipocytes, and subsequently, lipolysis can be also decreased (Wortman *et al.*, 2009). Therefore, when COX-associated biochemical reactions are intensively associated with EPA due to increased EPA levels and decreased proinflammatory AA levels, less harmful and significantly reduced inflammatory status can be expected. Thus, increased n-3 PUFA levels compared to n-6 PUFA can impart significant beneficial health effects against inflammation. Also, it has been recently suggested that increased n-3 PUFA levels in patients can improve the outcome of critically ill patients (Ott *et al.*, 2011).

IV. N-3 PUFAS AND METABOLIC DISORDERS

In both animals and humans, n-3 PUFA was shown to reduce inflammation and insulin resistance (Bierhaus *et al.*, 2004; Jones *et al.*, 2003, 2005). Both EPA and DHA are known to have anti-inflammatory activities that reduce the proinflammatory mediators such as tumor necrosis factor-α (TNF-α), interleukin 1 (IL-1), IL-6, IL-8, resistin, plasminogen activator inhibitor 1, and monocyte chemotactic protein-1 (MCP-1) levels (Kalupahana *et al.*, 2010b). Obesity is associated with a chronic low-grade inflammation characterized by an increase in proinflammatory cytokines and adipokines. EPA and DHA are well known to beneficially alter several metabolic processes including carbohydrate and lipid metabolism. n-3 PUFAs increase β-oxidation of fatty acids and inhibit hepatic and adipose lipogenic gene transcription (Al-Hasani and Joost, 2005; Jones *et al.*, 1996; Sampath and Ntambi, 2004), thus shifting lipids from synthetic pathways to oxidative pathways. Further, dietary n-3 PUFAs act as ligand activators of PPARα, which induces the transcription of genes involved in β-oxidation (Neschen *et al.*, 2002). Lipid-metabolizing genes downregulated by PUFAs include the fatty acid synthase, lipoprotein lipase, hormone-sensitive lipase, adipocyte lipid-binding protein (aP2), acetyl-CoA carboxylase, and the enzyme stearoyl-CoA desaturase-1 (Jones *et al.*, 1996; Sessler *et al.*, 1996).

Also, it was recently demonstrated that saturated fatty acids can activate Toll-like receptor 4 (TLR4) and TLR2 signaling in adipocytes leading to increased transcription of inflammatory markers via with nuclear factor-κB (NF-κB). NF-κB activation has been implicated in many inflammatory processes (Lira *et al.*, 2010). Some proinflammatory cytokines such as TNF-α and IL1 directly activate NF-κB. NF-κB activation has been reported to play a pivotal role in obesity, cancer, atherosclerosis, diabetes, arthritis, and other CVD, inflammatory, and immunological disorders (Bierhaus *et al.*, 2004; Jones *et al.*, 2003, 2005). It has been suggested that NF-κB may be an important therapeutic target for specific CVDs (Bierhaus *et al.*, 2004; Jones *et al.*, 2005; Ruan and Lodish, 2003). MCP-1, which plays a critical role in the development of CVD, is increased by activated NF-κB (Niu and Kolattukudy, 2009; Zhu *et al.*, 2008). Moreover, MCP-1 can induce adipogenesis, macrophage infiltration into adipose tissue, and insulin resistance (Kanda *et al.*, 2006; Younce *et al.*, 2009; Zhu *et al.*, 2008). Several studies have reported that NF-κB activation can be suppressed by fish oil and also identified that suppression of NF-κB activation has been strongly associated with mechanisms responsible for suppression of several disease conditions by fish oil (Encarnacion *et al.*, 2008, 2011; Lee *et al.*, 2009; Suzuki *et al.*, 2010; Zhao *et al.*, 2004). Further, Bellenger *et al.* (2011) showed that pancreatic n-3 fatty acid enrichment inhibits proinflammatory cytokines and NF-κB protein expression and increases IκBα protein expression in MLD-STZ-induced fat-1 transgenic mice.

It has been shown that fish oil can not only suppress proinflammatory mediators but also can increase the anti-inflammatory ones such as adiponectin (Duda *et al.*, 2009; Kalupahana *et al.*, 2010b). Increased adiponectin levels can reduce inflammation and beneficially improve the metabolism. Specifically, the increased levels of adiponectin can significantly reduce the insulin resistance. Oster *et al.* (2010) showed that DHA increases cellular adiponectin mRNA and secreted adiponectin protein in 3T3-L1 adipocytes, possibly by a mechanism involving PPARγ. A recent dietary intervention study conducted on healthy Japanese female subjects by Kondo *et al.* (2010) showed that a fish-based diet intervention increased the serum adiponectin concentration in young, nonobese, healthy Japanese female subjects. Also, the same study indicated that the increment in serum ω-3 PUFA may regulate the serum adiponectin concentration (Kondo *et al.*, 2010).

Other reported mechanisms mediating PUFA effects in metabolic disorders include binding to GPCR-120 and generation of newly discovered resolvins and protectins (Ariel and Serhan, 2007; Schwab *et al.*, 2007). Stimulation of GPR120 with n-3 FAs causes broad anti-inflammatory effects in monocytic RAW 264.7 cells and in primary intraperitoneal macrophages (Oh *et al.*, 2010). Further, it has been reported that inefficient biosynthesis of n-3-derived mediators of resolution of inflammation in muscle and adipose tissue contributes to the maintenance of chronic inflammation in obesity (White *et al.*, 2010). Hellmann *et al.* (2011) showed that resolving D1 can decrease adipose tissue macrophage accumulation and improves insulin sensitivity in obese-diabetic mice.

V. HEALTH CONCERNS

It is always a concern that heat, light, and oxygen can promote fish oil degradation which can produce harmful compounds such as peroxides. With the advances in food and pharmaceutical science, antioxidants are now incorporated into fish oil to prevent fish oil oxidation. Also, the effective storage and packaging methods have been developed to assure the quality. However, it is important to follow the storage guides provided with the products and simply avoid the exposure to heat, light, and oxygen. Also, extended cooking and deep frying can make fish oil unhealthy. Moreover, the quality and source of the fish oil is highly important. Some fish and fish oil may have unhealthy levels of heavy metals such as mercury and selenium. Some studies have reported that fish oil contaminated with mercury can increase CVD risk instead of expected CVD protection benefits while some studies indicate that there should be intensive investigations to confirm the harmful effects of mercury on CVD (Mozaffarian *et al.*, 2011; Virtanen *et al.*, 2005). Specially, when eating fish or taking fish oil supplements during pregnancy, it is always good to carefully select fish oil source for the safety of both the mother and the baby.

Taken together, n-3 PUFAs such as EPA and DHA are promising health promoting compounds and intake of dietary EPA and DHA can suppress the inflammation and inflammation-associated health complications. While there are several mechanisms suggested and detailed for the anti-inflammatory activities of EPA and DHA, here we have summarized a few important mechanisms such as competitive inhibition of AA-mediated PGE_2 formation, suppression of NF-κB activation, generation/production/secretion of anti-inflammatory mediators (adiponectin, resolving, and protectin), and anti-inflammatory activities mediated via GRP120. Also, EPA and DHA are known to beneficially alter the metabolic processes by reducing adiposity and increasing lipid oxidation, thus contributing antiobesity effects in general. Moreover, the reduction of excessive adiposity can directly reduce the low-grade chronic inflammation associated with obesity. Therefore, EPA and DHA can also be used in dietary interventions effectively to overcome obesity, inflammation, and many other disease conditions.

REFERENCES

Albert, C. M., Campos, H., Stampfer, M. J., Ridker, P. M., Manson, J. E., Willett, W. C., and Ma, J. (2002). Blood levels of long-chain n-3 fatty acids and the risk of sudden death. *N. Engl. J. Med.* **346**, 1113–1118.

Al-Hasani, H. and Joost, H. G. (2005). Nutrition-/diet-induced changes in gene expression in white adipose tissue. *Best Pract. Res. Clin. Endocrinol. Metab.* **19**, 589–603.

Ariel, A. and Serhan, C. N. (2007). Resolvins and protectins in the termination program of acute inflammation. *Trends Immunol.* **28**, 176–183.

Astorg, P. (2004). Dietary n-6 and n-3 polyunsaturated fatty acids and prostate cancer risk: A review of epidemiological and experimental evidence. *Cancer Causes Control* **15**, 367–386.

Bellenger, J., Bellenger, S., Bataillem, A., Massey, K. A., Nicolaou, A., Rialland, M., Tessier, C., Kang, J. X., and Narce, M. (2011). High pancreatic n-3 fatty acids prevent STZ-induced diabetes in fat-1 mice: Inflammatory pathway inhibition. *Diabetes* **60**, 1090–1099.

Bierhaus, A., Humpert, P. M., and Nawroth, P. P. (2004). NF-kappa B as a molecular link between psychosocial stress and organ dysfunction. *Pediatr. Nephrol.* **19**, 1189–1191.

Buettner, R., Parhofer, K. G., Woenckhaus, M., Wrede, C. E., Kunz-Schughart, L. A., Schölmerich, J., and Bollheimer, L. C. (2006). Defining high-fat-diet rat models: Metabolic and molecular effects of different fat types. *J. Mol. Endocrinol.* **36**, 485–501.

Burghardt, P. R., Kemmerer, E. S., Buck, B. J., Osetek, A. J., Yan, C., Koch, L. G., Britton, S. L., and Evans, S. J. (2010). Dietary n-3:n-6 fatty acid ratios differentially influence hormonal signature in a rodent model of metabolic syndrome relative to healthy controls. *Nutr. Metab. (Lond)* **7**, 53.

Cabanes, A., Wang, M., Olivo, S., Gustafsson, J., and Hilakivi-Clarke, L. (2003). Effect of n-3 polyunsaturated fatty acids (PUFAs) on breast cancer progression. *Cancer Epidemiol. Biomarkers Prev.* **12** (1305S).

Calder, P. (2008). The relationship between the fatty acid composition of immune cells and their function. *Prostaglandins Leukot. Essent. Fatty Acids* **79**, 101–108.

Calder, P. C. (2009). Polyunsaturated fatty acids and inflammatory processes: New twists in an old tale. *Biochimie.* **91**, 791–795.

Davis, B. and Kris-Etherton, P. (2003). Achieving optimal essential fatty acid status in vegetarians: Current knowledge and practical implications. *Am. J. Clin. Nutr.* **78**, 640S–646S.

Duda, M., O'Shea, K., Tintinu, A., Xu, W., Khairallah, R., Barrows, B., Chess, D., Azimzadeh, A., and Harris, W. (2009). Fish oil, but not flaxseed oil, decreases inflammation and prevents pressure overload-induced cardiac dysfunction. *Cardiovasc. Res.* **81**, 319–327.

Dyall, S., Michael, G., and Michael-Titus, A. (2010). Omega-3 fatty acids reverse age-related decreases in nuclear receptors and increase neurogenesis in old rats. *J. Neurosci. Res.* **88**, 2091–2102.

Encarnacion, M., Warner, G., Gray, C., Cheng, J., Keryakos, H., Nath, K., and Grande, J. (2008). Signaling pathways modulated by fish oil in salt-sensitive hypertension. *Am. J. Physiol. Renal Physiol.* **294**, F1323–F1335.

Encarnacion, M., Warner, G., Cheng, J., Gray, C., Nath, K., and Grande, J. (2011). n-3 Fatty acids block TNF-alpha-stimulated MCP-1 expression in rat mesangial cells. *Am. J. Physiol. Renal Physiol.* **300**, F1142–F1151.

Ferrucci, L., Cherubini, A., Bandinelli, S., Bartali, B., Corsi, A., Lauretani, F., Martin, A., Andres-Lacueva, C., Senin, U., and Guralnik, J. M. (2006). Relationship of plasma polyunsaturated fatty acids to circulating inflammatory markers. *J. Clin. Endocrinol. Metab.* **91**, 439–446.

Fetterman, J. and Zdanowicz, M. (2009). Therapeutic potential of n-3 polyunsaturated fatty acids in disease. *Am. J. Health Syst. Pharm.* **66**, 1169–1179.

Figueras, M., Olivan, M., Busquets, S., Lopez-Soriano, F., and Argiles, J. (2011). Effects of eicosapentaenoic acid (EPA) treatment on insulin sensitivity in an animal model of diabetes: Improvement of the inflammatory status. *Obesity* **19**, 362–369.

Goodstine, S. L., Zheng, T., Holford, T. R., Ward, B. A., Carter, D., Owens, P. H., and Mayne, S. T. (2003). Dietary (n-3)/(n-6) fatty acid ratio: Possible relationship to premenopausal but not postmenopausal breast cancer risk in U.S. women. *J. Nutr.* **133**, 1409–1414.

Harris, W. S. (1996). n-3 fatty acids and lipoproteins: Comparison of results from human and animal studies. *Lipids* **31**, 243–252.

Harris, W. S. (1997). n-3 fatty acids and serum lipoproteins: Animal studies. *Am. J. Clin. Nutr.* **65**, 1611S–1616S.

Harris, W. S. (2007). Omega-3 fatty acids and cardiovascular disease: A case for omega-3 index as a new risk factor. *Pharmacol. Res.* **55**, 217–223.

Hellmann, J., Tang, Y., Kosuri, M., Bhatnagar, A., and Spitem, M. (2011). Resolvin D1 decreases adipose tissue macrophage accumulation and improves insulin sensitivity in obese-diabetic mice. *FASEB J.* **25**, 2399–2407.

Jones, B. H., Maher, M. A., Banz, W. J., Zemel, M. B., Whelan, J., Smith, P. J., and Moustaid, N. (1996). Adipose tissue stearoyl-CoA desaturase mRNA is increased by obesity and decreased by polyunsaturated fatty acids. *Am. J. Physiol.* **271**, E44–E49.

Jones, W. K., Brown, M., Ren, X., He, S., and McGuinness, M. (2003). NF-kappaB as an integrator of diverse signaling pathways: The heart of myocardial signaling? *Cardiovasc. Toxicol.* **3**, 229–254.

Jones, W. K., Brown, M., Wilhide, M., He, S. W., and Ren, X. P. (2005). NF-kappa B in cardiovascular disease—Diverse and specific effects of a "general" transcription factor? *Cardiovasc. Toxicol.* **5**, 183–201.

Kalupahana, N. S., Claycombe, K., Fletcher, S., Wortman, P., and Moustaid-Moussa, N. (2010a). Eicosapentaenoic acid improves adipose tissue inflammation in part via downregulation of adipose angiotensinogen secretion. *Obesity* **18** (S71-S).

Kalupahana, N. S., Claycombe, K., Newman, S. J., Stewart, T., Siriwardhana, N., Matthan, N., Lichtenstein, A. H., and Moustaid-Moussa, N. (2010b). Eicosapentaenoic acid prevents and reverses insulin resistance in high-fat diet-induced obese mice via modulation of adipose tissue inflammation. *J. Nutr.* **140**, 1915–1922.

Kanda, H., Tateya, S., Tamori, Y., Kotani, K., Hiasa, K. I., Kitazawa, R., Kitazawa, S., Miyachi, H., and Maeda, S. (2006). MCP-1 contributes to macrophage infiltration into adipose tissue, insulin resistance, and hepatic steatosis in obesity. *J. Clin. Invest.* **116**, 1494–1505.

Kitamura, K., Shibata, R., Tsuji, Y., Shimano, M., Inden, Y., and Murohara, T. (2011). Eicosapentaenoic acid prevents atrial fibrillation associated with heart failure in a rabbit model. *Am. J. Physiol. Heart Circ. Physiol.* **300**, H1814–H1821.

Koletzko, B., Lien, E., Agostoni, C., Böhles, H., Campoy, C., Cetin, I., Decsi, T., Dudenhausen, J. W., and Dupont, C. (2008). The roles of long-chain polyunsaturated fatty acids in pregnancy, lactation and infancy: Review of current knowledge and consensus recommendations. *J. Perinat. Med.* **36**, 5–14.

Kondo, K., Morino, K., Nishio, Y., Kondo, M., Fuke, T., Ugi, S., Iwakawa, H., Kashiwagi, A., and Maegawa, H. (2010). Effects of a fish-based diet on the serum adiponectin concentration in young, non-obese, healthy Japanese subjects. *J. Atheroscler. Thromb.* **17**, 628–637.

Lands, W. E., Letellierm, P. R., Romem, L. H., and Vanderhoekm, J. Y. (1973). Inhibition of prostaglandin biosynthesis. *Adv. Biosci.* **9**, 15–28.

Lee, S., Kim, H., Chang, K., Baek, J., Park, J., Shin, J., Choi, W., Lee, J., and Paik, W. (2009). DHA and EPA down-regulate COX-2 expression through suppression of NF-kappa B activity in LPS-treated human umbilical vein endothelial cells. *Korean J. Physiol. Pharmacol.* **13**, 301–307.

Leitzmann, M., Stampfer, M., Michaud, D., Augustsson, K., Colditz, G., Willett, W., and Giovannucci, E. (2004). Dietary intake of n-3 and n-6 fatty acids and the risk of prostate cancer. *Am. J. Clin. Nutr.* **80**, 204–216.

Lira, F. S., Rosa, J. C., Pimentel, G. D., Tarini, V. A., Arida, R. M., Faloppa, F., Alves, E. S., do Nascimento, C. O., and Oyama, L. M. (2010). Inflammation and adipose tissue: Effects of progressive load training in rats. *Lipids Health Dis.* **9**, 109.

Lu, J., Borthwick, F., Hassanali, Z., Wang, Y., Mangat, R., Ruth, M., Shi, D., Jaeschke, A., and Russell, J. (2011). Chronic dietary n-3 PUFA intervention improves dyslipidaemia and subsequent cardiovascular complications in the JCR:LA-cp rat model of the metabolic syndrome. *Br. J. Nutr.* **105**, 1572–1582.

Marchioli, R., Barzi, F., Bomba, E., Chieffo, C., Di Gregorio, D., Di Mascio, R., Franzosi, M. G., Geraci, E., and Levantesi, G. (2002). Early protection against sudden death by n-3 polyunsaturated fatty acids after myocardial infarction: Time-course analysis of the results of the Gruppo Italiano per lo Studio della Sopravvivenza nell'Infarto Miocardico (GISSI)-Prevenzione. *Circulation* **105**, 1897–1903.

Mozaffarian, D., Shi, P., Morris, J. S., Spiegelman, D., Grandjean, P., Siscovick, D. S., Willett, W. C., and Rimm, E. B. (2011). Mercury exposure and risk of cardiovascular disease in two U.S. cohorts. *N. Engl. J. Med.* **364**, 1116–1125.

Murphy, R., Mourtzakis, M., Chu, Q., Baracos, V., Reiman, T., and Mazurak, V. (2011). Nutritional intervention with fish oil provides a benefit over standard of care for weight and skeletal muscle mass in patients with nonsmall cell lung cancer receiving chemotherapy. *Cancer* **117**, 1775–1782.

Neschen, S., Moore, I., Regittnig, W., Yu, C. L., Wang, Y., Pypaert, M., Petersen, K. F., and Shulman, G. I. (2002). Contrasting effects of fish oil and safflower oil on hepatic peroxisomal and tissue lipid content. *Am. J. Physiol. Endocrinol. Metab.* **282**, E395–E401.

Niu, J. L. and Kolattukudy, P. E. (2009). Role of MCP-1 in cardiovascular disease: Molecular mechanisms and clinical implications. *Clin. Sci.* **117**, 95–109.

Oh, D., Talukdar, S., Bae, E., Imamura, T., Morinaga, H., Fan, W., Li, P., Lu, W., Watkins, S., and Olefsky, J. (2010). GPR120 is an omega-3 fatty acid receptor mediating potent anti-inflammatory and insulin-sensitizing effects. *Cell* **142**, 687–698.

Oster, R. T., Tishinsky, J. M., Yuan, Z., and Robinson, L. E. (2010). Docosahexaenoic acid increases cellular adiponectin mRNA and secreted adiponectin protein, as well as PPARγ mRNA, in 3T3-L1 adipocytes. *Appl. Physiol. Nutr. Metab.* **35**, 783–789.

Ott, J., Hiesgen, C., and Mayer, K. (2011). Lipids in critical care medicine. *Prostaglandins Leukot. Essent. Fatty Acids* **85**, 267–273.

Raitt, M. H., Connor, W. E., Morris, C., Kron, J., Halperin, B., Chugh, S. S., McClelland, J., Cook, J., and MacMurdy, K. (2005). Fish oil supplementation and risk of ventricular tachycardia and ventricular fibrillation in patients with implantable defibrillators: A randomized controlled trial. *JAMA* **293**, 2884–2891.

Rondanelli, M., Giacosa, A., Opizzi, A., Pelucchi, C., La Vecchia, C., Montorfano, G., Negroni, M., Berra, B., Politi, P., and Rizzo, A. (2011). Long chain omega 3 polyunsaturated fatty acids supplementation in the treatment of elderly depression: Effects on depressive symptoms, on phospholipids fatty acids profile and on health-related quality of life. *J. Nutr. Health Aging* **15**, 37–44.

Ruan, H. and Lodish, H. F. (2003). Insulin resistance in adipose tissue: Direct and indirect effects of tumor necrosis factor-alpha. *Cytokine Growth Factor Rev.* **14**, 447–455.

Sampath, H. and Ntambi, J. M. (2004). Polyunsaturated fatty acid regulation of gene expression. *Nutr. Rev.* **62**, 333–339.

Schwab, J. M., Chiang, N., Arita, M., and Serhan, C. N. (2007). Resolvin E1 and protectin D1 activate inflammation-resolution programmes. *Nature* **447**, 869–874.

Sessler, A. M., Kaur, N., Palta, J. P., and Ntambi, J. M. (1996). Regulation of stearoyl-CoA desaturase 1 mRNA stability by polyunsaturated fatty acids in 3T3-L1 adipocytes. *J. Biol. Chem.* **271**, 29854–29858.

Simopoulos, A. P. (2002). The importance of the ratio of omega-6/omega-3 essential fatty acids. *Biomed. Pharmacother.* **56**, 365–379.

Suzuki, M., Noda, K., Kubota, S., Hirasawa, M., Ozawa, Y., Tsubota, K., Mizuki, N., and Ishida, S. (2010). Eicosapentaenoic acid suppresses ocular inflammation in endotoxin-induced uveitis. *Mol. Vis.* **16**, 1382–1388.

Thies, F., Garry, J. M., Yaqoob, P., Rerkasem, K., Williams, J., Shearman, C. P., Gallagher, P. J., Calder, P. C., and Grimble, R. F. (2003). Association of n-3 polyunsaturated fatty acids with stability of atherosclerotic plaques: A randomised controlled trial. *Lancet* **361**, 477–485.

Virtanen, J. K., Voutilainen, S., Rissanen, T. H., Mursu, J., Tuomainen, T. P., Korhonen, M. J., Valkonen, V. P., Seppänen, K., Laukkanen, J. A., and Salonen, J. T. (2005). Mercury, fish oils, and risk of acute coronary events and cardiovascular disease, coronary heart disease, and all-cause mortality in men in eastern Finland. *Arterioscler. Thromb. Vasc. Biol.* **25**, 228–233.

Wall, R., Ross, R., Fitzgerald, G., and Stanton, C. (2010). Fatty acids from fish: The anti-inflammatory potential of long-chain omega-3 fatty acids. *Nutr. Res.* **68**, 280–289.

Wan, J. B., Huang, L. L., Rong, R., Tan, R., Wang, J., and Kang, J. X. (2010). Endogenously decreasing tissue n-6/n-3 fatty acid ratio reduces atherosclerotic lesions in apolipoprotein E-deficient mice by inhibiting systemic and vascular inflammation. *Arterioscler. Thromb. Vasc. Biol.* **30**, 2487–2494.

Whelan, J. and Rust, C. (2006). Innovative dietary sources of n-3 fatty acids. *Annu. Rev. Nutr.* **26**, 75–103.

White, P. J., Arita, M., Taguchi, R., Kang, J. X., and Marette, A. (2010). Transgenic restoration of long-chain n-3 fatty acids in insulin target tissues improves resolution capacity and alleviates obesity-linked inflammation and insulin resistance in high-fat-fed mice. *Diabetes* **59**, 3066–3073.

Wortman, P., Miyazaki, Y., Kalupahana, N. S., Kim, S., Hansen-Petrik, M., Saxton, A. M., Claycombe, K. J., Voy, B. H., Whelan, J., and Moustaid-Moussa, N. (2009). n3 and n6 polyunsaturated fatty acids differentially modulate prostaglandin E secretion but not markers of lipogenesis in adipocytes. *Nutr. Metab. (Lond)* **6**, 5.

Wu, M., Harvey, K., Ruzmeto, N., Welch, Z., Sech, L., Jackson, K., Stillwell, W., Zaloga, G., and Siddiqui, R. (2005). Omega-3 polyunsaturated fatty acids attenuate breast cancer growth through activation of a neutral sphingomyelinase-mediated pathway. *Int. J. Cancer* **117**, 340–348.

Younce, C. W., Azfer, A., and Kolattukudy, P. E. (2009). MCP-1 (monocyte chemotactic protein-1)-induced protein, a recently identified zinc finger protein, induces adipogenesis in 3T3-L1 pre-adipocytes without peroxisome proliferator-activated receptor gamma. *J. Biol. Chem.* **284**, 27620–27628.

Zhao, Y., Joshi-Barve, S., Barve, S., and Chen, L. (2004). Eicosapentaenoic acid prevents LPS-induced TNF-alpha expression by preventing NF-kappa B activation. *J. Am. Coll. Nutr.* **23**, 71–78.

Zhu, J., Yong, W., Wu, X., Yu, Y., Lv, J., Liu, C., Mao, X., Zhu, Y., Xu, K., and Han, X. (2008). Anti-inflammatory effect of resveratrol on TNF-alpha-induced MCP-1 expression in adipocytes. *Biochem. Biophys. Res. Commun.* **369**, 471–477.

CHAPTER 14

Biological Importance and Applications of Squalene and Squalane

Se-Kwon Kim[*,†,1] and **Fatih Karadeniz**[*]

Contents

Abstract

Squalene is a polyunsaturated hydrocarbon with a formula of $C_{30}H_{50}$. Squalene can be found in certain fish oils, especially shark liver oil, in high amounts and some vegetable oils in relatively smaller amounts. Human sebum also contains 13% squalene as one of its major constituents. Squalane is a saturated derivative of squalene and also found in these sources. Interest in squalene has been raised after its characterization in shark liver oil which is used as a traditional medicine for decades. Several studies exhibited results that prove certain bioactivities for squalene and squalane. Up to date, anticancer, antioxidant, drug carrier, detoxifier, skin hydrating, and emollient activities of these substances have been

[*] Department of Chemistry, Pukyong National University, Busan, Republic of Korea
[†] Marine Bioprocess Research Center, Pukyong National University, Busan, Republic of Korea
[1] Corresponding author: Se-Kwon Kim, *E-mail address:* sknkim@pknu.ac.kr

Advances in Food and Nutrition Research, Volume 65
ISSN 1043-4526, DOI: 10.1016/B978-0-12-416003-3.00014-7

reported both in animal models and *in vitro* environments. According to promising results from recent studies, squalene and squalane are considered important substances in practical and clinical uses with a huge potential in nutraceutical and pharmaceutical industries.

I. BACKGROUND

Squalene is a triterpene which can be found widely in organism varying from humans to plants (Fig. 14.1). It is a polyunsaturated hydrocarbon with a formula of $C_{30}H_{50}$ and formed by six isoprene units. Squalene has gained massive amount of attraction after being identified in the shark liver extract discovered by Dr. Tsujimoto in 1903, hence it received its name (*Squaluss* spp.). Following this discovery, the shark liver oil has been considered as the main source of squalene along olive oil which also includes squalene as a major hydrocarbon (Gershbein and Singh, 1969). Up to date, squalene has been investigated extensively and reported to play crucial roles in steroid (especially dietary cholesterol) synthesis in human. Under the light of its importance in marine animal, deep-sea sharks, squalene also has been tested for its beneficial bioactivities including antioxidant, antitumor, and cytoprotective effects (Auffray, 2007; Passi *et al.*, 2002; Rao *et al.*, 1998). It has been promoted as a singlet oxygen receiver and an important lead for future cosmeceutical researches. The cosmeceutical potential of squalene has been extensively raised after the studies which showed that squalene existed widely in human sebum, hair fat, and other surface lipids (Passi *et al.*, 2002). In addition to cosmeceutical studies, squalene is regarded as an important compound for chemoprotective activities as well as nutraceutical for maintaining health under toxic exposure (Das *et al.*, 2003). Fully natural occurrence and studied

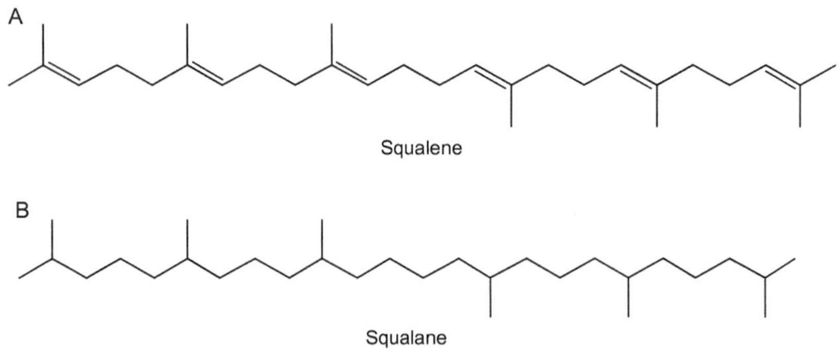

FIGURE 14.1 Chemical structures of squalene (A) and squalane (B).

beneficial properties of squalene promote this aliphatic hydrocarbon for leading significant improvements on the way to develop nutraceuticals which can help people to maintain a quality and healthy life in today's nonnatural trended life.

On the other hand, squalane is a saturated derivative of squalene which also was identified in human sebum. Its inert properties and suggested low toxicity compared to its analog squalene gained attraction from researchers, especially in the area of cosmetics (Allison, 1999). Proved emollient and moisturizing properties of squalane paved its way into protective nutraceutical and cosmeceutical area along its unsaturated relative, squalene.

These two components in shark liver oil have been used as traditional medicine for ages in order to keep people healthy. After their characterization, and highly intensified researches for their bioactivities, they are still important factors of supplementary diet for a healthier living. Up to date, several researches brought novel insights to their properties and proved various beneficial uses of these compounds.

II. BIOLOGICAL ROLES OF SQUALENE AND SQUALANE

Human skin is continuously exposed to environmental factors including ultraviolet (UV) light radiation through sunlight. Skin includes small glands named sebaceous glands which secrete sebum, a matter rich in lipids through hair follicles in order to lubricate and protect the skin and hair. It might be considered that due to high lipid content of human skin, it should be quite susceptible to UV radiation-induced oxidative stress. However, according to several studies, natural skin protection against this high exposure is suggested to be carried out by squalene which is one of the major components of human skin (Kohno *et al.*, 1995).

Squalene takes part in metabolism as precursor for synthesis of steroids and structurally quite similar to β-carotene, coenzyme q10, vitamins K_1, E, and D. The squalene in skin and fat tissue comes from endogenous cholesterol synthesis as well as dietary resources in people who consume high amounts of olive and fish oil especially shark liver (Gershbein and Singh, 1969). Squalene is synthesized by squalene synthase which converts two units of farnesyl pyrophosphate, direct precursor for terpenes and steroids, into squalene. As a secosteroid, vitamin D biosynthesis is also regulated by squalene. Moreover, being precursor for each steroid family makes squalene a crucial component of the body.

Further, squalane also exists in sebum composition. Additionally, due to its unique properties and more stable nature as industrial product than squalene, squalane is highly preferable in cosmeceutical and nutraceutical approaches. Squalane is mainly synthesized by hydrogenation of squalene derived from variety of natural sources with the increasing demand for squalane in industrial area.

A. Antioxidant

One of the main causes of cell damage, especially on skin is oxidation. In case of human skin, exposure to sunlight induces a lipid peroxidation process in human skin layer which leads to deterioration. In this matter, squalene has been reported to protect human skin from lipid peroxidation caused by either UV exposure or any other oxidative stress.

Squalene has also been reported to possess a resistant to peroxidation and acts as singlet oxygen scavenger which explains its antioxidant properties (Kohno *et al.*, 1995). Moreover, several *in vivo* and *in vitro* studies suggested the similar antioxidant efficiency results. Quenching singlet oxygen rate of squalene is found to be reasonably higher than that of all other human skin lipids and even in a similar manner with 3,5-di-*t*-butyl-4-hydroxytoluene. Same study suggests that stability of squalene against free radicals makes human skin surface able to resist lipid peroxidation propagation with an adequate amount of squalene in composition. In another study, Warleta *et al.* (2010) showed that squalene treatment successfully protected human mammary epithelial cells against oxidative DNA damage. Antiradical activities of squalene have been measured by 2,2-diphenyl-1-picrylhydrazyl, 2,2′-azino-bis(3-ethylbenzothiazoline-6-sulphonic acid), and oxygen radical absorbance capacity assays. According to assay results, radical scavenging capacity of squalene has been reported to be significantly important to be an efficient free radical scavenger. Following radical scavenging capacity assays, *in vitro* tests have been carried out in human mammary epithelial cells and human breast cancer cells which have been damaged by H_2O_2. As expected, squalene treatment relieved the oxidative stress in epithelial cells but not in tumor cells, suggesting a possible role of antitumor activity of squalene through oxidative stress protection. Lower cases of cancer in populations consuming high amount of squalene including nutrients such as olive oil and shark liver oil might be related to these results with being in awe of promising results that shows squalene protects human cells with tumor susceptibility against high oxidative stress.

In vivo experimental evidences support the *in vitro* antioxidant effect of squalene. Senthilkumar *et al.* (2006) suggested that dietary-supplied squalene to experimental rats normalized the enzymic and nonenzymic antioxidants after cyclophosphamide-induced toxicity. The rats were induced to have normal cell toxicity in their heart by being fed by 150mg cyclophosphamide/kg of rat. In experimental design, squalene treatment at 0.4 ml/day/rat among various concentrations (0.2–1.0ml/day/rat) has been found to attenuate the oxidative damage at the maximum efficiency with minimum dose. In addition to protection of *in vivo* models' heart and red blood cells by squalene treatment, underlying *in vivo* mechanism of squalene efficiency is yet to be revealed.

Alcohol has been known to cause lipid peroxidation. Moreover, squalene has been suggested to protect cells from lipid protection. In this context, Aguilera *et al.* (2005) tested squalene efficiency against alcohol-induced toxicity in the chick embryo in order to have ideas about protection of embryos from high alcohol consumption during pregnancy. Chick embryos were injected with 10 μl of absolute ethanol and 10 μl of squalene in test groups. Later, retina was assayed for lipid composition and structure by lipid and optical microscope analysis, respectively. Results showed that squalene could reduce the damaging effects of alcohol in lipid composition and structure of retina and hence act as a preventive agent for alcohol ingestion during pregnancy.

B. Anticancer

As shark liver oil contains more than 40% squalene while the liver of shark accounts for 20–25% of its total body weight, sharks are thought to be the richest source of squalene (Mathews, 1992). Accordingly, the correlation between high amounts of squalene and absence of cancer in shark species ought to be suggested. There are reports which suggest partial and possible relationship between low incidence of cancer and consuming high olive oil product which also includes high amounts of squalene (Owen *et al.*, 2000, 2004).

During recent years, squalene was found to be active for protection against various carcinogens. Therefore, an adjunctive squalene therapy in some kind of cancers happened to be the most primary use of squalene as a nutraceutical. Although there are no human trials to prove squalene role in cancer treatments, *in vitro* and animal experimental evidences suggest an anticancer role for this nutrient.

When squalene treated in conjunction with antitumor drugs, tumor growth has been observed to be arrested or prevented in various experimental models, although squalene itself has been known to be less effective on tumor inhibition. It has been reported that squalene-affiliated oleic acid treatment inhibits benzo(a)pyrene-induced skin carcinogenicity (Van Duuren and Goldschmidt, 1976). Similarly, in CD-1 mice with 7,12-dimethylbenz[a]anthracene-induced and 12-*O*-tetradecanoylphorbol-13-acetate-promoted skin tumors, 26.67% reduction in the incidence of tumors were observed after 5% squalene treatment during the prevention study (Murakoshi *et al.*, 1992). Yamaguchi *et al.* (1985) suggested that squalene increased the effect of 3-[(4-amino-2-methyl-5-pyrimidinyl)-methyl]-1-(2-chloroethyl)-1-nitrosourea (ACNU), an antitumor agent.

Squalene showed potential at a concentration of 4.2g/kg along 10mg/kg ACNU in animal models. The adjunctive treatment resulted in prolonged survival for lymphocyctic leukemia P388 with an absence of notable toxicity to host. In a study by Nakagawa *et al.* (1985), interaction between squalene and anticancer agents was assayed. A potentiating effect of squalene was observed for Adriamycin, 5-fluorouracil, bleomycin, and *cis*-diamminedichloroplatinum. Squalene addition enhanced the cytotoxicity and antitumor activity. In similar studies, it has been shown that squalene treatment before and/or during anticancer treatment effectively enhanced the inhibition of chemically induced skin tumorigenesis and regression of some already existing tumors in animal models.

Squalene is suggested to show not only antitumor but also chemoprotective effects. Dietary-supplied 1% squalene is reported to be considerably effective against aberrant cryptic foci (early preneoplastic lesions in colon carcinogenesis; Rao *et al.*, 1998). Mice with azoxymethane-induced colonic aberrant crypt foci were fed by 1% dietary squalene for 10 weeks. At the end of experimental procedure, it was observed that squalene treatment successfully resulted in more than 45% inhibition of the cryptic foci in colon. In a similar manner, nontoxic levels of squalene inhibited the mammary preneoplastic transformation which also shows its efficiency in chemoprotection. For instance, when squalene was initiated orally to mice along injection of 4-(methylnitrosamino)-1-(3-pyridyl)-1-butanone, mammary tumorigenesis causing agent, lung hyperplasia, and tumor multiplicity were decreased by 70% and 58%, respectively, and mammary carcinogenesis was significantly inhibited (Smith *et al.*, 1998).

In a related research, Storm *et al.* (1993) have reported a radioprotective effect of squalene. Mice were administrated by 2% squalene for 14 days before and 30 days after a lethal whole-body γ-irradiation. Results showed that white cell counts in squalene-treated groups were considerably high when compared to nontreated control groups. Also, white cell protection was observed along a prolonged survival. Squalene treatment was suggested to exhibit cellular and systemic radioprotection as a result of this research.

All results indicate that squalene makes a big contribution to enhance tumor growth inhibition and tumor suppression by anticancer treatments as well as by acting as a successful chemoprotective agent.

C. Skin care

Squalene is preferred for skin care due to its notably high emollient properties. It can be easily absorbed deep by the skin acting as a normalizer for flexibility and suppleness of the skin without presenting an oily residue. Blasco *et al.* (2006) studied emulsions for skin constituents in respect to squalene amount. The stability, centrifugation, viscosity, and pH of squalene were measured along a microscopic analysis in order to determine the optimal conditions for oil–water skin emulsions. Results suggested that the stability and viscosity of the emulsions are greatly influenced by the percentage of squalene.

As an emollient, squalene is expected to increase skin hydration due to skin surface occlusions. In addition, squalene is a substance believed to maintain moisture in the stratum corneum. Novel substitutes were researched for vernix caseosa which is a reported highly efficient barrier cream for facilitating stratum corneum hydration for barrier-deficient skins. For this purpose, various lipid fractions were mixed with squalene, triglycerides, cholesterol, ceramides, and fatty acids to produce a mixture that can generate similar compositions of vernix caseosa (Rissmann *et al.*, 2008). According to results, squalene-including mixtures were able to increase the barrier to maintain hydration in a comparable manner to vernix caseosa. Five percent sodium lauryl sulfate-treated rat and human skin showed increased transepidermal water loss and riboflavin penetration. However, squalene treatment reverts the effects of sodium lauryl sulfate.

Okada and Matsumoto (2004) tested an aqueous gel for hemodialysis patients with mild uremic pruritus to replace their usual treatment with antihistamine and urea-containing ointments. A high water containing gel which includes squalane among other naturally derived ingredients has been applied to patients with mild uremic pruritus for 2 weeks. Results showed that skin dryness and itching were significantly improved after 2 weeks of gel treatment. Following 2 weeks after stopping the gel treatment, itching scale raised by 1.2 ± 0.5 and skin dryness reappeared in 40% which suggests the positive effect of gel treatment.

D. Drug delivery agent

Drug delivery compounds have been of much interest recently in order to achieve fully effective treatment and therapies. Drug carriers are needed to transport the effective agent to a targeted location or to keep the effective agent unharmed in different conditions of the body. In this manner, lipids are mostly preferable due to their nontoxic nature and ability to form a protective vesicle and fuse the cell membrane easily. In the light of this approach, squalene has been studied widely for drug delivery either by emulsions or by conjugates.

There are several reports for successfully prolonged release of short half-life drugs with squalene emulsions. Wang *et al.* (2008) reported that a squalene emulsion stabilized by phosphatidylethanolamine or pluronic F68 prolonged *in vitro* release of a morphine prodrug. Intravenous administration of these emulsions exhibited longer analgesic activity in animal models. In similar studies, squalene-based emulsions were successfully applied to deliver lipophilic ester prodrugs of nalbuphine and increase skin permeability and delivery of the encapsulated psoralen (Fang *et al.*, 2008; Wang *et al.*, 2006).

Besides emulsions, lipid–drug conjugates gained attention in recent years. The conjugates usually are obtained by covalent coupling of the drug to biocompatible lipid moieties. Lipid–drug conjugates are shown to possess ability to improve pharmacological effect and decrease toxicity besides carrying the applied drug safely. Squalenoyl nanomedicine is one fine example for squalene-based lipid–drug conjugates. In this technology, squalene's natural folding conformation was used to attach hydrocarbon to analogues of anticancer and antiviral drugs which resulted in formation of nanoassemblies in water with a diameter of 100–300nm in the absence of any surfactants. Nucleoside anticancer agent gemcitabine (2,2′-difluorodeoxycytidine) was the first example of this attachment. Gemcitabine's several holdbacks as a drug was eliminated by a successful conjugation with squalene (Reddy *et al.*, 2008). Gemcitabine has a rapid metabolism which results in a short biological half-life after administration. Therefore, higher doses are needed to express the required activity. However, squalene–gemcitabine nanoassemblies showed significantly improved cytotoxicity in KB-3 (human nasopharyngeal carcinoma) and MCF7 (human breast carcinoma) cells (Couvreur *et al.*, 2006). When squalene is administrated as a dietary supplement, it has been reported that it is absorbed efficiently, reaching the liver rapidly and converted to sterol and bile acids and distributed with diverse amounts along various lipid fractions such as in very low-density lipoprotein, low-density lipoprotein, and high-density lipoprotein which contain 50.8%, 25.6%, and 23.6% squalene, respectively (Saudek *et al.*, 1978). Therefore, high absorption rate of squalene enhances its potential use as delivery agent for orally administrated therapeutic compounds. Related studies exhibit highly promising results for oral delivery of therapeutic agents by squalene conjugation.

E. Detoxifier

Squalene is suggested to enhance elimination of lipophilic xenobiotics by several experimental evidences. Its nonpolar structure promotes a promising affinity for unionized drugs. Richter and Schäfer (1982) studied squalene for elimination of [14C]hexachlorobenzene (HCB) as an alternative method to paraffin treatment. Animal models have been fed by squalene and paraffin as 8% of the diet. Results indicated that squalene supplementation was as effective as paraffin on fecal excretion of HCB.

After 3 weeks of dietary squalene supplementation, three times higher HCB excretion was observed through feces as well as notably decreased HCB elimination half-life. In addition, it has been reported that this detoxifying effect of squalene was dose dependent and, as the percentage lowered to 5.0 of the diet, no significant effect was observed in liver on blood HCB concentrations. In a similar research, Kamimura *et al.* (1992) suggested squalene as a good agent for drug elimination from the body. Rats fed by squalene exerted increased amounts of fecal excretion of theophylline and strychnine.

F. Anti-infectant

Nowicki and Bara ska-Rybak (2007) studied the protective effect of shark liver oil. They observed a significant protection against bacterial and fungal infections by shark liver oil treatment which contains mostly squalene and alkylglycerol. Further, this treatment showed improved effects on xerosis and skin lesion-induced atopic dermatitis. This antibacterial and antifungal effect could be accounted for the high-squalene-including composition of the shark liver oil; however, detailed studies are needed to be carried out for reputed activity of squalene as an anti-infectant.

III. CONCLUSION

On the basis of the presented reports, squalene and squalane are two naturally derived compounds with bright future in nutraceutical and pharmaceutical area. Squalene, which can be obtained by natural sources, exists readily in human body and yet to be examined in detail for its biological role.

Squalene supplementation is suggested to be accounted for tumor growth inhibition and prevention of normal cells to turn into tumor cells under oxidative stress. Although there is lack of evidence for human trials to show anticancer and antioxidant effects of squalene, animal models and *in vitro* experiments highlight a significant activity which urges for further exploration.

Detoxifying and anti-infectant effects of squalene itself have not been reported yet. However, substances including squalene or squalene supplementation showed notable improvements. Further researches to elucidate the underlying mechanisms of shark liver oil and squalene application are expected to report more results that promote squalene and squalane as effective candidates for practical and clinical use.

REFERENCES

Aguilera, Y., Dorado, M. E., Prada, F. A., Martínez, J. J., Quesada, A., and Ruiz-Gutiérrez, V. (2005). The protective role of squalene in alcohol damage in the chick embryo retina. *Exp. Eye Res.* **80**, 535–543.

Allison, A. C. (1999). Squalene and squalane emulsions as adjuvants. *Methods* **19**, 87–93.

Auffray, B. (2007). Protection against singlet oxygen, the main actor of sebum squalene peroxidation during sun exposure, using commiphora myrrha essential oil. *Int. J. Cosmet. Sci.* **29**, 23–29.

Blasco, L., Duracher, L., Forestier, J. P., Vian, L., and Marti-Mestres, G. (2006). Skin constituents as cosmetic ingredients. Part I: A study of bio-mimetic monoglycerides behavior at the squalene–water interface by the pendant drop method in a static mode. *J. Dispers. Sci. Technol.* **27**, 799–810.

Couvreur, P., Stella, B., Reddy, L. H., Hillaireau, H., Dubernet, C., Desmaële, D., Lepêtre-Mouelhi, S., Rocco, F., Dereuddre-Bosquet, N., and Clayette, P. (2006). Squalenoyl nanomedicines as potential therapeutics. *Nano Lett.* **6**, 2544–2548.

Das, B., Yeger, H., Baruchel, H., Freedman, M., Koren, G., and Baruchel, S. (2003). In vitro cytoprotective activity of squalene on a bone marrow versus neuroblastoma model of cisplatin-induced toxicity: Implications in cancer chemotherapy. *Eur. J. Cancer* **39**, 2556–2565.

Fang, J. Y., Fang, C. L., Liu, C. H., and Su, Y. H. (2008). Lipid nanoparticles as vehicles for topical psoralen delivery: Solid lipid nanoparticles (sln) versus nanostructured lipid carriers (NLC). *Eur. J. Pharm. Biopharm.* **70**, 633–640.

Gershbein, L. L. and Singh, E. J. (1969). Hydrocarbons of dogfish and cod livers and herring oil. *J. Am. Oil Chem. Soc.* **46**, 554–557.

Kamimura, H., Koga, N., Oguri, K., and Yoshimura, H. (1992). Enhanced elimination of theophylline, phenobarbital and strychnine from the bodies of rats and mice by squalane treatment. *J. Pharmacobiodyn.* **15**, 215–221.

Kohno, Y., Egawa, Y., Itoh, S., Nagaoka, S., Takahashi, M., and Mukai, K. (1995). Kinetic study of quenching reaction of singlet oxygen and scavenging reaction of free radical by squalene in n-butanol. *Biochim. Biophys. Acta* **1256**, 52–56.

Mathews, J. (1992). Sharks still intrigue cancer researchers. *J. Natl. Cancer Inst.* **84**, 1000.

Murakoshi, M., Nishino, H., Tokuda, H., Iwashima, A., Okuzumi, J., Kitano, H., and Iwasaki, R. (1992). Inhibition by squalene of the tumor promoting activity of 12 o-tetradecanoylphorbol-13 acetate in mouse skin carcinogenesis. *Int. J. Cancer* **52**, 950–952.

Nakagawa, M., Yamaguchi, T., Fukawa, H., Ogata, J., Komiyama, S., Akiyama, S., and Kuwano, M. (1985). Potentiation by squalene of the cytotoxicity of anticancer agents against cultured mammalian cells and murine tumor. *Jpn. J. Cancer Res.* **76**, 315–320.

Nowicki, R. and Bara ska-Rybak, W. (2007). Shark liver oil as a supporting therapy in atopic dermatitis. *Pol. Merkur. Lekarski.* **22**, 312–313.

Okada, K. and Matsumoto, K. (2004). Effect of skin care with an emollient containing a high water content on mild uremic pruritus. *Ther. Apher. Dial.* **8**, 419–422.

Owen, R. W., Giacosa, A., Hull, W. E., Haubner, R., Wurtele, G., Spiegelhalder, B., and Bartsch, H. (2000). Olive-oil consumption and health: The possible role of antioxidants. *Lancet Oncol.* **1**, 107–112.

Owen, R., Haubner, R., Würtele, G., Hull, W., Spiegelhalder, B., and Bartsch, H. (2004). Olives and olive oil in cancer prevention. *Eur. J. Cancer Prev.* **13**, 319–326.

Passi, S., De Pità, O., Puddu, P., and Littarru, G. P. (2002). Lipophilic antioxidants in human sebum and aging. *Free Radic. Res.* **36**, 471–477.

Rao, C. V., Newmark, H. L., and Reddy, B. S. (1998). Chemopreventive effect of squalene on colon cancer. *Carcinogenesis* **19**, 287–290.

Reddy, L. H., Marque, P. E., Dubernet, C., Mouelhi, S. L., Desmaele, D., and Couvreur, P. (2008). Preclinical toxicology (subacute and acute) and efficacy of a new squalenoyl gemcitabine anticancer nanomedicine. *J. Pharmacol. Exp. Ther.* **325**, 484–490.

Richter, E. and Schäfer, S. (1982). Effect of squalane on hexachlorobenzene (hcb) concentrations in tissues of mice. *J. Environ. Sci. Health B* **17**, 195–203.

Rissmann, R., Oudshoorn, M. H. M., Kocks, E., Hennink, W. E., Ponec, M., and Bouwstra, J. A. (2008). Lanolin-derived lipid mixtures mimic closely the lipid composition and organization of vernix caseosa lipids. *Biochim. Biophys. Acta* **1778**, 2350–2360.

Saudek, C. D., Frier, B. M., and Liu, G. (1978). Plasma squalene: Lipoprotein distribution and kinetic analysis. *J. Lipid Res.* **19**, 827–835.

Senthilkumar, S., Yogeeta, S. K., Subashini, R., and Devaki, T. (2006). Attenuation of cyclophosphamide induced toxicity by squalene in experimental rats. *Chem. Biol. Interact.* **160**, 252–260.

Smith, T. J., Yang, G., Seril, D. N., Liao, J., and Kim, S. (1998). Inhibition of 4-(methylnitrosamino)-1-(3-pyridyl)-1-butanone-induced lung tumorigenesis by dietary olive oil and squalene. *Carcinogenesis* **19**, 703–706.

Storm, H. M., Oh, S. Y., Kimler, B. F., and Norton, S. (1993). Radioprotection of mice by dietary squalene. *Lipids* **28**, 555–559.

Van Duuren, B. and Goldschmidt, B. (1976). Cocarcinogenic and tumor-promoting agents in tobacco carcinogenesis. *J. Natl. Cancer Inst.* **56**, 1237–1242.

Wang, J. J., Sung, K., Hu, O. Y. P., Yeh, C. H., and Fang, J. Y. (2006). Submicron lipid emulsion as a drug delivery system for nalbuphine and its prodrugs. *J. Control. Release* **115**, 140–149.

Wang, J. J., Sung, K., Yeh, C. H., and Fang, J. Y. (2008). The delivery and antinociceptive effects of morphine and its ester prodrugs from lipid emulsions. *Int. J. Pharm.* **353**, 95–104.

Warleta, F., Campos, M., Allouche, Y., Sánchez-Quesada, C., Ruiz-Mora, J., Beltrán, G., and Gaforio, J. J. (2010). Squalene protects against oxidative DNA damage in mcf10a human mammary epithelial cells but not in mcf7 and mda-mb-231 human breast cancer cells. *Food Chem. Toxicol.* **48**, 1092–1100.

Yamaguchi, T., Nakagawa, M., Hidaka, K., Yoshida, T., Sasaki, T., Akiyama, S. I., and Kuwano, M. (1985). Potentiation by squalene of anti-tumor effect of 3-[(4-amino-2-methyl-5-pyrimidinyl)methyl]-1-(2-chloroethyl)-nitrosourea in a murine tumor system. *Jpn. J. Cancer Res.* **76**, 1021–1026.

CHAPTER 15

Development of Bioactive Peptides from Fish Proteins and Their Health Promoting Ability

Mahinda Senevirathne* and **Se-Kwon Kim***,†,1

Contents

Abstract

Great amount of marine fish species have been identified with potential nutraceutical and medicinal values. Consequently, a number of bioactive compounds have been identified including fish muscle proteins, peptides, collagen and gelatin, fish oil, fish bone. Bioactive peptides derived from various fish muscle proteins have shown various biological activities including antihypertensive, antibacterial, anticoagulant, anti-inflammatory, and antioxidant activities, and hence they may be a potential material for biomedical and

* Marine Bioprocess Research Center, Pukyoung National University, Busan, Republic of Korea
† Department of Chemistry, Marine Biochemistry Laboratory, Pukyoung National University, Busan, Republic of Korea
1 Corresponding author: Se-Kwon Kim, *E-mail address:* sknkim@pknu.ac.kr

Advances in Food and Nutrition Research, Volume 65
ISSN 1043-4526, DOI: 10.1016/B978-0-12-416003-3.00015-9

food industries. Further, they are commonly used in medical and pharmaceutical industries as carrier molecules for drugs, proteins, and genes. Hence, fish muscle protein-derived peptides are valuable natural resources that can be potential material for biomedical, nutraceutical, and food industries.

I. INTRODUCTION

The marine environment has a tremendous biodiversity, and it is a source of a plethora of bioactive compounds with great potential for food and pharmaceutical areas. Ocean covers 70% of the earth and around 70% of creatures are living in it. Recent studies have confirmed that most of these species possess active compounds but still underutilized sources for various applications. Fish used as human food accounts for 78% of the total fish catch in developing and developed countries (Vannuccini, 2004). Fish catch is mainly used for human consumptions and other minor uses such as meal production and bait. Moreover, many studies are conducting with the fish wastes discarded from fisheries industries. Great amount of researches have been initiated to investigate an increased utilization of fish muscle protein and found that they possess potential bioactive materials such as gelatin, collagen, peptides, and various fatty acids including EPA and DHA in various industries (Wassawa *et al.*, 2007).

Recent studies have reported a number of bioactive peptides from various fish sources. They possess short sequences with approximately 2–20 amino acids in length that exert physiological benefits when consumed *in vivo*. They can be produced by proteolytic hydrolysis using commercially available enzymes including trypsin, chymotrypsin, alcalase, or proteolytic microorganisms and fermentation methods (Vercruysse *et al.*, 2005). After digestion, bioactive peptides can be absorbed in the intestine and enter the blood stream directly, which ensures their bioavailability *in vivo* and a physiological effect at the target site (Erdmann *et al.*, 2008). A growing number of literatures demonstrated that peptides from marine organisms had a wide spectrum of effects, such as antihypertensive, immunomodulating, antithrombotic, antioxidative, anticancer, and antimicrobial (Je *et al.*, 2007; Vercruysse *et al.*, 2005; Yang *et al.*, 2010). A number of bioactive compounds have been isolated from remaining fish muscle proteins, collagen and gelatin, fish oil, fish bone, and internal organs (Je *et al.*, 2005; Jeon and Kim, 2002; Kim *et al.*, 2001). Generally, a far better profitability is obtained by producing human consumables, and the highest profitability is currently expected from bioactive compounds. These bioactive compounds can be extracted and purified with technologies varying from simple to complex, and such compounds may include preparation and isolation of bioactive peptides for biotechnological and pharmaceutical applications. Further, some of these bioactive compounds have been identified to possess nutraceutical potentials that are beneficial in human health promotion. Therefore, much attention must be paid for the development of new technologies in search of novel bioactive compounds from marine fish protein. This chapter discusses trends in developing bioactive peptides from marine fish protein and their biological activities relating to human health.

II. STRUCTURE–ACTIVITY RELATIONSHIP OF FISH PEPTIDES

Peptide quantitative structure–activity relationship modeling has been used for predicting peptide structures with high ACE inhibitory activity. Using physicochemical descriptors, Pripp *et al.* (2004) emphasized that increased side chain hydrophobicity at the carboxy-terminus and decreased side chain length of the penultimate amino acid exemplified ACE inhibitory potential of peptides up to six amino acids in length. Moreover, ACE appears to be strongly influenced by the C-terminal sequence of the peptides. Hence, it has been suggested that proline, lysine, or arginine are the preferred amino acids at the C-terminal residue and thus contribute to ACE-I-inhibitory potency (Meisel, 1997). Further, studies with SHR revealed that dipeptides with a C-terminal tyrosine residue produced a slow but prolonged decrease in systolic blood pressure compared to dipeptides with phenylalanine at the C-terminal. In contrast, dipeptides with a C-terminal phenylalanine caused a more rapid reduction in systolic blood pressure and a shorter duration of action (Suetsuna, 1998).

Antimicrobial peptides are biomolecules employed for the defense against bacteria by animals as well as plants (Brogden, 2005). These short-chain positively charged peptides secreted by organisms are composed of 12–45 amino acid residues. When the antimicrobial peptides binding onto the outermost leaflet of negatively charged bacterial membrane by the aid of hydrophobic and electrostatic interactions, they fold to amphipathic secondary structures, typically α-helices and β-sheets. Those peptides can kill the bacteria via physical, chemical, or biological processes. Many models are suggested to explain the antibacterial mechanisms of peptides including Shai–Matsuzaki–Huang model, barrel–stave model, carpet model, and toroidal-pore model. Moreover, many theoretical and experimental models have been used to explain the antimicrobial mechanisms of cell lysis by antimicrobial peptides including, molecular mean-field theory (La Rocca *et al.*, 1999), dynamics simulations (Leontiadou *et al.*, 2006), and coarse-grained simulations (Lopez *et al.*, 2005). Most of these approaches support the SMH model that a physical hole in the membrane is stable and is an effective mechanism of antimicrobial activity.

Several mechanisms have been presented to explain the antioxidative activity of peptides including metal ion chelating ability, radical scavenging activity, and aldehyde adduction (Zhou and Decker, 1999). Most of the food-derived peptides possess low molecular weight with hydrophobic residues including Lue and Val in their N-terminus and Pro, His, Tyr, Trp, Met, and Sys in their sequences (Chen *et al.*, 1995). Those amino acid residues can increase the presence of peptides at the water–lipid interface and promote the radical scavenging activity at the lipid phase (Ranathunga *et al.*, 2006). The mechanism of His containing peptides could be attributed to the hydrogen donating ability, lipid peroxyl radical-trapping ability, and the metal chelating ability of the imidazole group (Chan and Decker, 1994).

III. SOURCES OF FISH PEPTIDES

Peptides can be isolated from various fish parts, or various parts of marine fishes possess peptides.

A. Muscle protein peptides

Discarded fish bones and cutoffs may contain considerable amounts of muscle proteins. These muscle proteins are nutritionally valuable and easily digestible with well-balanced amino acid composition (Venugopal *et al.*, 1996). Therefore, fish proteins derived from seafood processing by-products can be hydrolyzed enzymatically to recover protein. Protein hydrolysates from several marine species have been analyzed for their nutritional and functional properties, and researches have mainly explored the possibility of obtaining biologically active peptides (Benkajul and Morrissey, 1997). Moreover, skipjack tuna muscle (Kohama *et al.*, 1988), sardine muscle (Bougatef *et al.*, 2008), and shark meat (Wu *et al.*, 2008) have been used to separate potential peptides.

B. Peptides from fish skin collagen and gelatin

Gelatin is derived from the fibrous protein collagen, which is the principal constituent of animal skin, bone, and connective tissue. Fish skin waste could be used as a potential source to isolate collagen and gelatin. Zhu *et al.* (2010) evaluated the effect of collagen peptides on markers of metabolic nuclear receptors.

C. Fish bone as potential peptide source

The organic component of fish bone represents 30% of collagen, while inorganic component mainly consists of calcium phosphate and hydroxylapatite and it was around 60–70% (Aronove *et al.*, 2007; Barakat *et al.*, 2008).

D. Peptides from other body parts

Antimicrobial peptides mainly present in the mucous layer and eliminate the pathogenic bacteria before they enter the skin barrier. Several antimicrobial peptides have been isolated from marine fishes, such as Pleurocidins from winter flounder, *Pleuronectes americanus* (Walbaum), *American plaice*, and *Hippoglossoides platessoides* (Fabricius), and Atlantic halibut, *Hippoglossus hippoglossus* (Cole *et al.*, 1997; Douglas *et al.*, 2003). Intestinal antimicrobial peptides were derived from hagfish, *Eptatretus burgeri* (Girard) (Hwang *et al.*, 1999), chrysophsins from gills of red sea bream (Iijima *et al.*, 2003), and piscidin or moronecidin from gills and skin of striped bass (Lauth *et al.*, 2002; Silphaduang and Noga, 2001).

IV. PRODUCTION OF PEPTIDES FROM VARIOUS FISH BODY PARTS

It is a key step to develop methods to separate peptides with different molecular weights. An ultrafiltration membrane system equipped with the appropriate molecular weight cutoff has been effectively used in separating peptides having desired molecular weights (Jeon *et al.*, 2000). In order to obtain functionally active peptides, it is a common method to use the type of enzymes letting sequential enzymatic digestions.

Antimicrobial peptides have been purified from the gill filaments of sea beams by acid extraction method (Iijima *et al.*, 2003). Gill filaments were crushed in liquid nitrogen and resulting powder was boiled in water. Then the extraction was performed with hydrochloric acid, trifluoroacetic acid, formic acid, and NaCl. Finally, this extract was purified in a Sep-Pack C18 cartridge, SP-Sephadex C-25 resin, gel filtration, reversed-phase HPLC, ESI–MS/protein sequencer, cation exchange HPLC, and MLC, respectively. Moreover, several studies have used acid extraction method to isolate antimicrobial peptides from various body parts of fish (Lauth *et al.*, 2002; Zhang *et al.*, 2009). However, extraction medium containing sodium acetate, Triton X-100, and phenylmethylsulfonyl fluoride also have been used for the extraction of antimicrobial peptides from fish (Park *et al.*, 1998).

Collagen is generally extracted with acid treatment and solubilized without altering its triple helix. However, thermal treatment cleaves hydrogen and covalent bonds that stabilizes the triple helix configuration of collagen and converts its helical conformation into coiled conformation resulting in gelatin state (Djabourov *et al.*, 1993). Fish skin gelatin is processed into peptides by enzymatic hydrolysis. In many studies, many commercial proteases, such as alcalase (Dong *et al.*, 2008) and flavourzyme (Thiansilakul *et al.*, 2007a,b), were used to obtain antioxidative peptides by hydrolyzing various fish protein sources.

A. Health promoting ability of fish peptides

Peptides isolated from various fish protein hydrolysates have shown a different biological activities such as antihypertensive, antioxidative, antithrombotic, and immunomodulatory activities.

1. Antioxidant activity

Peptides derived from fish proteins have shown the ability of exerting potent antioxidative activities (Table 15.1) in different oxidative systems (Rajapakse *et al.*, 2005). Currently, an increasing interest exists to explore natural antioxidative substances without side effects and the identified antioxidative activities have potential to develop safe and nonhazardous natural antioxidants for the complications arose from oxidation of biomolecules. Je *et al.* (2007) purified an antioxidative peptide from tuna backbone protein and identified as VKAGFAWTANQQLS (1519Da) which was very important regarding the functional foods. It was reported that the derived peptides were good radical scavengers and antioxidant.

In another study, an antioxidant peptide has been isolated from hoki frame protein by gastrointestinal digestion (Kim *et al.*, 2007). A number of enzymes have been used for the hydrolysis process, and the isolated peptides showed strong antioxidant activities in various model systems.

Gelatin peptides have repeated unique Gly–Pro–Ala sequence in their structure, and it is presumed that the observed antioxidative properties of gelatin peptides can be associated with their unique amino acid compositions.

A dipeptide Met-Tyr, derived from sardine muscle (Matsufuji *et al.*, 1994), stimulates expression of the antioxidant defense protein HO-1 in a concentration-dependent manner. Previous findings revealed that HO-1 protein expression is accompanied by the induction of a secondary antioxidant protein, ferritin. In a present study, the effect of Met-Tyr on the expression of the antioxidant stress proteins, heme oxygenase-1 (HO-1), and ferritin in endothelial cells derived from the human umbilical vein and their contribution to the decrease in radical formation that occurs under the influence of this dipeptide were studied and reported potential activity (Erdmann *et al.*, 2006).

TABLE 15.1 Potential bioactive antioxidative peptides derived from various fish proteins

Name of organism and substrate	References
Alaska pollak skin gelatin	Kim *et al.* (2001)
Hoki frame protein	Je *et al.* (2005)
Alaska pollak frame protein	Je *et al.* (2005)
Conger eel muscle protein	Ranathunga *et al.* (2006)
Hoki frame protein	Kim *et al.* (2007)
Jumbo squid skin gelatin	Mendis *et al.* (2005)

2. Antihypertensive activity

ACE inhibitory peptides have been separated from the skin of skate by Lee *et al.* (2011). The purified peptides showed IC_{50} values of 95 and 148 μ M, respectively, for both peptides isolated from the skin of skate. Further, Lineweaver–Burk plots indicated that the peptides act as noncompetitive inhibitor against ACE. Recently, many inhibitory peptides against ACE are reported (Table 15.2) as natural alternative biofunctional peptides that are safer than that of the existing artificial ACE inhibitory compounds in the market that show some side effects. Various peptides have been isolated from seafood by-products from the fisheries industry such as backbones from tuna (Lee *et al.*, 2010). Tuna backbone has hydrolyzed using various proteases such as alcalase, α-chymotrypsin, neutrase, papain, pepsin, and trypsin to obtain an antioxidant peptide (Je *et al.*, 2007).

Gelatin peptides have repeated unique Gly–Pro–Ala sequence in their structure, and it is presumed that the observed antihypertensive properties of gelatin peptides can be associated with their unique amino acid compositions.

3. Antimicrobial activity

Antimicrobial peptides are widely distributed throughout the animal kingdom and show potential antimicrobial activities (Table 15.3). Three isoforms of a novel C-terminally amidated peptides, such as chrysophsin-1, chrysophsin-2, and chrysophsin-3, consist of 25, 25, and 20 amino acids, respectively, and are highly cationic, containing an unusual C-terminal RRRH sequence was isolated from the gills of red sea bream, *Chrysophrys* (*Pagrus*) *major* (Iijima *et al.*, 2003). These peptides show potent antimicrobial activity against Gram-negative and Gram-positive bacteria and act as nonspecific defense substances in fish skin. The antimicrobial peptides fold to amphipathic secondary structures, typically α-helices and β-sheets. Such peptides can kill the bacteria via physical, chemical, or biological processes. A novel 25-residue linear antimicrobial peptide

TABLE 15.2 Potential bioactive antimicrobial peptides derived from various fish proteins

Fish name	Body part	Peptide name	Reference
Large yellow croaker	Head and kidney	Hepadin	Zhang *et al.* (2009)
Red sea bream	Gills	Chrysophsin-1, 2, and 3	Iijima *et al.* (2003)
Loach (mud fish)	Whole body	Misgurin	Park *et al.* (1997)
Catfish	Proteinaceous epithelial muscosal layer	Parasin 1	Park *et al.* (1998)
Winter flounder	Skin mucous secretion	Pleurocidin	Cole *et al.* (1997)
Striped bass	Skin and gill	Moronecidin	Lauth *et al.* (2002)
Bony fish (winter flounder) Atlantic salmon	Hepcidin-like		Douglas *et al.* (2003)

TABLE 15.3 Other potential biological activities from peptides derived from various fish proteins

Name of organism and substrate	Bioactivity	Reference
Fish protein hydrolysate	Anticoagulant	Rajapakse *et al.* (2005)
Alaska pollack back bone	Calcium bioavailability	Jung *et al.* (2006)
Fish protein hydrolysate	Breast cancer	Picot *et al.* (2006)
Tilapia muscle	Anticancer	Chang *et al.* (2011)
Anchovy sauce, whole fish	Human lymphoma cancer	Lee *et al.* (2003)
Tuna dark muscle	Breast cancer	Hsu *et al.* (2011)

named as pleurocidin found in the skin mucous secretions of the winter flounder was isolated (Cole *et al.*, 1997). Pleurocidin is predicted to assume an amphipathic α-helical conformation similar to many other linear antimicrobial peptides and exhibited antimicrobial activity against *Escherichia coli* in a bacterial cell lysis plate assay. Moreover, a study with skate skin included the production of collagen and the purification of an antimicrobial peptide (Cho *et al.*, 2004).

4. Anticoagulant effect

Enzymatically prepared fish muscle peptides also have shown anticoagulant and antiplatelet properties tested *in vitro*, and the results have suggested the capability of fish peptides to inhibit coagulation factors in the intrinsic pathway of coagulation (Rajapakse *et al.*, 2005). A peptide with 12.2kDa molecular weight was purified from yellowfin sole that inhibited activated coagulation factor XII by forming an inactive complex regardless of Zn^{2+} mediation in their study. In addition, this anticoagulant peptide acts to antagonize platelet membrane glycoprotein integrin to arrest platelet aggregation.

5. Anticancer effect

Proteins, peptides, and amino acids have been reported to show antitumor or antiproliferative activities. However, the antiproliferative activity of peptides derived from fish protein was rarely studied. Picot *et al.* (2006) reported that hydrolysates obtained from three blue whiting, three cod, three plaice, and one salmon showed significant inhibition against two human breast cancer cell lines, MCF-7/6 and MDA-MB-231. The hydrolysates contained a complex mixture of free amino acids, peptides with various sizes ranging up to 7kDa, and in a lower proportion, lipids and sodium chloride. Moreover, two antiproliferative peptides were purified from tuna dark muscle by enzymatic hydrolysis with papain and protease against human breast cancer cell line MCF-7 (Hsu *et al.*, 2011). The amino acid sequences of the two antiproliferative peptides were Leu-Pro-His-Val-Leu-Thr-Pro-Glu-Ala-Gly-Ala-Thr (1206Da) and Pro-Thr-Ala-Glu-Gly-Gly-Val-Tyr-Met-Val-Thr (1124Da), and they showed significant activities in a dose-dependent manner. Further, antiproliferative hydrophobic peptide (440.9Da) derived from anchovy fish source was able to induce apoptosis in human U937 lymphoma cells through the increase of caspase-3 and caspase-8 activity (Lee *et al.*, 2003, 2004). Tilapia (*Oreochromis mossambicus*) hepcidin TH2-3 was evaluated in several tumor cell lines. The results indicated that TH2-3 inhibited human fibrosarcoma cell (HT1080 cell line) proliferation and migration. Chang *et al.* (2011) also investigated an antimicrobial peptide, TH1-5 with the view of finding antitumor activity in cancel cells, including human cervix adenocarcinoma cell (HeLa), human hepatocellular carcinoma cell (HepG2), human fibrosarcoma cell (HT1080), *Cercopithecus aethiops* kidney cell (COS-7), and human kidney cell (WS-1).

6. Ca absorbing and bone mineralization ability

Kim *et al.* (1999) and Jung *et al.* (2005) reported that fish peptides are capable of accelerating calcium absorption. Further, researchers have identified that fish protein hydrolysates possess hormone-like peptides and growth factors to accelerate calcium absorption (Fouchereau-Peron *et al.*, 1999). These peptides are capable of binding to cell surface receptors on osteoclasts and involved mainly in calcium metabolism by decreasing the number of osteoclasts. Therefore, these peptides could be used in the treatment of osteoporosis and Paget's disease (Table 15.3).

In addition, gelatin peptides have shown to accelerate absorption of dietary calcium in animal models increasing calcium bioavailability (Kim *et al.*, 1998). Jung *et al.* (2006) reported that fish bone peptides (FBP) II could inhibit the formation of insoluble Ca salts in neutral pH. During the experimental period, Ca retention was increased and loss of bone mineral was decreased by FBP II supplementation in ovariectomized rats. The levels of femoral total Ca, bone mineral density, and strength were also significantly increased by the FBP II diet to levels similar to those of the casein phosphopeptide diet group.

7. Others activities

Further, acidic peptide fractions from Atlantic cod hydrolysate have shown strong immunostimulatory effects, and treatment of these peptides has stimulated the oxidative burst of Atlantic salmon leucocytes (Gildberg *et al.*, 1996). Basically, immunomodulators that enhance the production of oxygen metabolites in macrophages that are responsible for these oxygen metabolites determine the oxidative burst. Oxidative burst reactions are of major importance for the bactericidal power of phagocytes.

Some peptides isolated from various fishes and fermented fish have been reported to show prolyl endopeptidase inhibitory activity (Sørensen *et al.*, 2004). The peptides obtained from muscles from salmon, trout, and cod as well as Norwegian fermented trout muscle have been investigated for the potential inhibitory activity for prolyl endopeptidase and found that they are good sources.

V. FUTURE TRENDS OF PEPTIDES FROM FISH PROTEINS

Technological advances have created a cost- and time-efficient platform to screen a large number of compounds for their pharmacological and medicinal potency. These technologies give insightful indications on the feasibility of transforming a new chemical entity into a drug and also reduce the requirement for more extensive clinical trials. Due to their inherent bioactive properties, peptides could be useful as therapeutic or prophylactic agents. The production of peptide therapeutics from marine food proteins is therefore a promising approach exploitable by the pharmaceutical industry.

VI. CONCLUSIONS

Marine fish protein and peptides have recently attracted great attention due to their potential biological activities and natural abundance. Also they are underutilized natural resources that have potential to be used in various industries. Moreover, discarded wastes from seafood processing have been recognized as potential materials to derive bioactive peptides for various fields. They have been reported to have range of biological activities such as antioxidant, antihypertensive, anticoagulant, antibacterial, and calcium-absorbing abilities.

REFERENCES

Aronove, D., Karlov, A., and Rosenman, G. (2007). Hydroxyapatite nanoceramics: Basic physical properties and biointerface modification. *J. Eur. Ceram. Soc.* **27**, 4181–4186.

Barakat, A. M. N., Khalil, K. A., Sheikh, F. A., Omran, A. M., Gaihre, B., Khil, S. M., and Kim, H. Y. (2008). Physicochemical characterization of hydroxyapatite extracted from bones by three different methods: Extraction of biological desirable HAp. *Mater. Sci. Eng. C* **28**, 1381–1387.

Benkajul, S. and Morrissey, M. T. (1997). Protein hydrolysates from Pacific whiting solid wastes. *J. Agric. Food Chem.* **45**, 3423–3430.

Bougatef, A., Nedjar-Arroume, N., Ravallec-Ple, R., Leroy, Y., Guillochon, D., Barkia, A., and Nasri, M. (2008). Angiotensin I-converting enzyme (ACE) inhibitory activities of sardinelle (*Sardinella aurita*) by-products protein hydrolysates obtained by treatment with microbial and visceral fish serine proteases. *Food Chem.* **111**, 350–356.

Brogden, K. A. (2005). Antimicrobial peptides: Pore formers or metabolic inhibitors in bacteria? *Nat. Rev. Microbiol.* **3**, 238–250.

Chan, K. M. and Decker, E. A. (1994). Endogenous muscle antioxidants. *Crit. Rev. Food Sci. Nutr.* **34**, 403–426.

Chang, W. T., Panb, C. Y., Rajanbabu, V., Cheng, C. W., and Chen, J. Y. (2011). Tilapia (*Oreochromis mossambicus*) antimicrobial peptide, hepcidin 1–5, shows antitumor activity in cancer cells. *Peptides* **32**, 342–352.

Chen, H. M., Muramoto, K., and Yamauchi, F. (1995). Structural analysis of antioxidative peptides from soybean β-conglycinin. *J. Agric. Food Chem.* **43**, 574–578.

Cho, S. H., Michael, L. J., and Eun, J. B. (2004). Nutritional composition and microflora of the fresh and fermented skate (*Raja Kenojei*) skins. *Int. J. Food Sci. Nutr.* **55**, 45–51.

Cole, A. M., Weis, P., and Diamond, G. (1997). Isolation and characterization of pleurocidin, an antimicrobial peptide in the skin secretions of winter flounder. *J. Biol. Chem.* **272**, 12008–12013.

Djabourov, M., Lechaire, J. P., and Gaill, F. (1993). Structure and rheology of gelatin and collagen gels. *Biorheology* **30**, 191–205.

Dong, S., Zeng, M., Wang, D., Liu, Z., Zhao, Y., and Yang, H. (2008). Antioxidant and biological properties of protein hydrolysates prepared from silver carp (*Hypophthalmichthys molitrix*). *Food Chem.* **107**, 1485–1493.

Douglas, S. E., Gallant, J. W., Liebscher, R. S., Dacanay, A., and Tsoi, S. C. M. (2003). Identification and expression analysis of hepcidin-like antimicrobial peptides in bony fish. *Dev. Comp. Immunol.* **27**, 589–601.

Erdmann, K., Grosser, N., Schipporeit, K., and Schroder, H. (2006). The ACE inhibitory dipeptide Met-Tyr diminishes free radical formation in human endothelial cells via induction of heme oxygenase-1 and ferritin. *J. Nutr.* **136**, 2148–2152.

Erdmann, K., Cheung, W. Y., and Schröder, H. (2008). The possible roles of food derived bioactive peptides in reducing the risk of cardiovascular diseases. *J. Nutr. Biochem.* **19**, 643–654.

Fouchereau-Peron, M., Duvail, L., Michel, C., Gildberg, A., Batista, I., and Gal, Y. I. (1999). Isolation of an acid fraction from a fish protein hydrolysate with a calcitonin-gene-related-peptide-like biological activity. *Biotechnol. Appl. Biochem.* **29**, 87–92.

Gildberg, A., Bogwald, J., Johansen, A., and Stenberg, E. (1996). Isolation of acid peptide fractions from a fish protein hydrolysate with strong stimulatory effect on Atlantic salmon (*Salmo salar*) head kidney leucocytes. *Comp. Biochem. Physiol. B Biochem. Mol. Biol.* **11**, 97–101.

Hsu, K. C., Li-Chan, E. C. Y., and Jao, C. L. (2011). Antiproliferative activity of peptides prepared from enzymatic hydrolysates of tuna dark muscle on human breast cancer cell line MCF-7. *Food Chem.* **126**, 617–622.

Hwang, E. Y., Seo, J. K., Kim, C. H., Go, H. J., Kim, E. J., *et al.* (1999). Purification and characterization of a novel antimicrobial peptide from the skin of the hagfish Eptatretus burgeri. *Int. J. Food Sci. Nut.* **4**, 28–32.

Iijima, N., Tanimoto, N., Emoto, Y., Morita, Y., Uematsu, K., Murakami, T., and Nakai, T. (2003). Purification and characterization of three isoforms of chrysophsin, a novel antimicrobial peptide in the gills of the red sea bream, *Chrysophrys major*. *Eur. J. Biochem.* **270**, 675–686.

Je, J. Y., Park, P. J., Jung, W. K., and Kim, S. K. (2005). Isolation of angiotensin I converting enzyme (ACE) inhibitor from fermented oyster sauce, *Crassostrea gigas*. *Food Chem.* **90**, 809–814.

Je, J. Y., Qian, Z. J., Byun, H. G., and Kim, S. K. (2007). Purification and characterization of an antioxidant peptide obtained from tuna backbone protein by enzymatic hydrolysis. *Process Biochem.* **42**, 840–846.

Jeon, Y. J. and Kim, S. K. (2002). Antitumor activity of chitosan oligosaccharides produced in an ultra filtration membrane reactor system. *J. Microbiol. Biotechnol.* **12**, 503–507.

Jeon, Y. J., Byun, H. G., and Kim, S. K. (2000). Improvement of functional properties of cod frame protein hydrolysates using ultrafiltration membranes. *Process Biochem.* **35**, 471–478.

Jung, W. K., Park, P. J., Byun, H. G., Moon, S. H., and Kim, S. K. (2005). Preparation of hoki (Johnius belengerii) bone oligophosphopeptide with a high affinity to calcium by carnivorous intestine crude proteinase. *Food Chem.* **91**, 333–340.

Jung, W. K., Lee, B. J., and Kim, S. K. (2006). Fish-bone peptide increases calcium solubility and bioavailability in ovariectomised rats. *Br. J. Nutr.* **95**, 124–132.

Kim, G. H., Jeon, Y. J., Byun, H. G., Lee, Y. S., and Kim, S. K. (1998). Effect of calcium compounds from oyster shell bound fish skin gelatin peptide in calcium deficient rats. *J. Korean Fish. Soc.* **31**, 149–159.

Kim, S. K., Jeon, Y. J., Byun, H. G., Park, P. J., Kim, G. H., Choi, Y. R., and Yeon, S. K. (1999). Calcium absorption acceleration effect on phosphorylated and non-phosphorylated peptides from hoki (*Johnius belengerii*) frame. *J. Korean Fish. Soc.* **32**, 713–717.

Kim, S. K., Kim, Y. T., Byun, H. G., Nam, K. S., Joo, D. S., and Shahidi, F. (2001). Isolation and characterization of antioxidative peptides from gelatin hydrolysate of Allaska Pollack skin. *J. Agric. Food Chem.* **49**, 1984–1989.

Kim, S. Y., Je, J. Y., and Kim, S. K. (2007). Purification and characterization of antioxidant peptide from hoki (*Johnius belengerii*) frame protein by gastrointestinal digestion. *J. Nutr. Biochem.* **18**, 31–38.

Kohama, Y., Matsumoto, S., Oka, H., Teramoto, T., Okabe, M., and Mimura, T. (1988). Isolation of angiotensin-converting enzyme inhibitor from tuna muscle. *Biochem. Biophys. Res. Commun.* **155**(1), 332–337.

La Rocca, P., Biggin, P. C., and Tieleman, D. P. (1999). Simulation studies of the interaction of antimicrobial peptides and lipid bilayers. *Biochim. Biophys. Acta* **1462**, 185–200.

Lauth, X., Shike, H., Burns, J. C., Westerman, M. E., Ostland, V. E., Carlberg, J. M., Van Olst, J. C., Nizet, V., Taylor, S. W., Shimizu, C., and Bulet, P. (2002). Discovery and characterization of two isoforms of moronecidin, a novel antimicrobial peptide from hybrid striped bass. *J. Biol. Chem.* **277**, 5030–5039.

Lee, Y. G., Kim, J. Y., Lee, K. W., Kim, K. H., and Lee, H. J. (2003). Peptides from anchovy sauce induce apoptosis in a human lymphoma cell (U937) through the increase of caspase-3 and -8 activity. *Ann. NY Acad. Sci.* **1010**, 399–404.

Lee, Y. G., Lee, K. W., Kim, J. Y., Kim, K. H., and Lee, H. J. (2004). Induction of apoptosis in a human lymphoma cell line by hydrophobic peptide fraction separated from anchovy sauce. *Biofactors* **21**, 63–67.

Lee, S. H., Qian, Z. J., and Kim, S. K. (2010). A novel angiotensin I converting enzyme inhibitory peptide from tuna frame protein hydrolysate and its antihypertensive effect in spontaneously hypertensive rats. *J. Agric. Food Chem.* **118**, 96–102.

Lee, J. K., Jeon, J. K., and Byun, H. G. (2011). Effect of angiotensin I converting enzyme inhibitory peptide purified from skate skin hydrolysate. *Food Chem.* **125**, 495–499.

Leontiadou, H., Mark, A. E., and Marrink, S. J. (2006). Antimicrobial peptide in action. *J. Am. Chem. Soc.* **128**, 12156–12161.

Lopez, C. F., Nielsen, S. O., Moore, P. B., and Klein, M. L. (2005). Structure and dynamics of model pore insertion into a membrane. *Biophys. J.* **88**, 3083–3094.

Matsufuji, H., Matsui, T., Seki, E., Osajima, K., Nakashima, M., and Osajima, Y. (1994). Angiotensin I-converting enzyme inhibitory peptides in an alkaline protease hydrolyzate derived from sardine muscle. *Biosci. Biotechnol. Biochem.* **58**, 2244–2245.

Meisel, H. (1997). Biochemical properties of regulatory peptides derived from milk proteins. *Biopolymers* **43**, 119–128.

Mendis, E., Rajapakse, N., Byun, H. G., and Kim, S. K. (2005). Investigation of jumbo squid (Dosidicus gigas) skin gelatin peptides for their in vitro antioxidant effects. *Life Sci.* **77**, 2166–2178.

Park, C. B., Lee, J. H., Park, I. Y., Kim, M. S., and Kim, C. H. (1997). A noval antimicrobial peptide from the loach, *Misgurnus anguillicaudatus*. *FEBS Lett.* **411**, 173–178.

Park, I. Y., Park, C. B., Kim, M. S., and Kim, S. C. (1998). Parasin I, an antimicrobial peptide derived from histone H2A in the catfish, *Parasilurus asotus*. *FEBS Lett.* **437**, 258–262.

Picot, L., Bordenave, S., Didelot, S., Fruitier-Arnaudin, I., Sannier, F., Thorkelsson, G., Berge, J. P., Guerard, F., Chabeaud, A., and Piot, J. M. (2006). Antiproliferative activity of fish protein hydrolysates on human breast cancer cell lines. *Process Biochem.* **41**, 1217–1222.

Pripp, A. H., Isaksson, T., Stepaniak, L., and Sørhaug, T. (2004). Quantitative structure–activity relationship modeling of ACE-inhibitory peptides derived from milk proteins. *Eur. Food Res. Technol.* **219**, 579–583.

Rajapakse, N., Jung, W. K., Mendis, E., Moon, S. H., and Kim, S. K. (2005). A novel anticoagulant purified from fish protein hydrolysate inhibits factor XIIa and platelet aggregation. *Life Sci.* **76**, 2607–2619.

Ranathunga, S., Rajapakse, N., and Kim, S. K. (2006). Purification and characterization of antioxidative peptide derived from muscle of conger eel (*Conger myriaster*). *Eur. Food Res. Technol.* **222**, 310–315.

Silphaduang, U. and Noga, E. J. (2001). Peptide antibiotics in mast cells of fish. *Nature* **414**, 268–269.

Sørensen, R., Kildal, E., Stepaniak, L., Pripp, A. H., and Sørhaug, T. (2004). Screening for peptides from fish and cheese inhibitory to prolyl endopeptidase. *Nahrung/Food* **48**, 53–56.

Suetsuna, K. (1998). Isolation and characterization of angiotensin I converting enzyme inhibitor dipeptides derived from *Allium sativum* L. (garlic). *J. Nutr. Biochem.* **9**, 415–419.

Thiansilakul, Y., Benjakul, S., and Shahidi, F. (2007a). Antioxidative activity of protein hydrolysate from round scad muscle using Alcalase and Flavourzyme. *J. Food Biochem.* **31**, 266–287.

Thiansilakul, Y., Benjakul, S., and Shahidi, F. (2007b). Compositions, functional properties and antioxidative activity of protein hydrolysates prepared from round scad (*Decapterus maruadsi*). *Food Chem.* **103**, 1385–1394.

Vannuccini, S. (2004). Overview of fish production, utilization, consumption and trade. FAO fishery information, data and statistic unit report, ftp://ftp.fao.org/fi/stat/overview/2001/commodit/2001fisheryoverview.pdf.

Venugopal, V., Chawla, S. P., and Nair, P. M. (1996). Spray dried protein powder from threadfin beam: Preparation, properties and comparison with FPC type B. *J. Muscle Foods* **7**, 55–58.

Vercruysse, L., Van Camp, J., and Smagghie, G. (2005). ACE inhibitory peptides derived from enzymatic hydrolysates of animal muscle protein: A review. *J. Agric. Food Chem.* **53**, 8106–8115.

Wassawa, J., Tang, J., and Gu, X. (2007). Utilization of fish processing by-products in the gelatin industry. *Food Rev. Int.* **23**, 159–174.

Wu, H., He, H. L., Chen, X. L., Sun, C. Y., Zhang, Y. Z., and Zhou, B. C. (2008). Purification and identification of novel angiotensin-I-converting enzyme inhibitory peptides from shark meat hydrolysate. *Process Biochem.* **43**, 457–461.

Yang, L., Chen, L., Zeng, R., Li, C., Qiao, R., Hu, L., and Li, Z. (2010). Synthesis, nanosizing and in vitro drug release of a novel anti-HIV polymeric prodrug: Chitosan-O-isopropyl-50-O-d4T monophosphate conjugate. *Bioorg. Med. Chem.* **18**, 117–123.

Zhang, J., Yan, Q., Ji, R., Zou, W., and Guo, G. (2009). Isolation and characterization of a hepcidin peptide from the head kidney of large yellow croaker, *Pseudosciaena crocea*. *Fish Shellfish Immunol.* **26**, 864–870.

Zhou, S. and Decker, E. A. (1999). Ability of amino acids, dipeptides, polyamines, and sulfhydryls to quench hexanal, a saturated aldehydic lipid oxidation product. *J. Agric. Food Chem.* **47**, 1932–1936.

Zhu, C. F., Li, G. Z., Peng, H. B., Zhang, F., Chen, Y., and Li, Y. (2010). Effect of marine collagen peptides on markers of metabolic nuclear receptors in type 2 diabetic patients with/without hypertension1. *Biomed. Environ. Sci.* **23**, 113–120.

CHAPTER 16

Marine Fish-Derived Bioactive Peptides as Potential Antihypertensive Agents

Se-Kwon Kim,[*,†,1] **Dai-Hung Ngo,**[*] and **Thanh-Sang Vo**[*]

Contents

Abstract

Hypertension is the most widespread risk factor for many serious cardiovascular diseases. Angiotensin-converting enzyme (ACE) plays a crucial role in cardiovascular physiological regulation by converting angiotensin I to a potent vasoconstrictor, angiotensin II. Hence, the inhibition of ACE is a key target for antihypertensive activity. Recently, potent antihypertensive peptides have been purified widely by enzymatic hydrolysis of muscle protein, skin collagen, and gelatin of many different kinds of marine fishes. Marine fish-derived bioactive peptides can be developed as antihypertensive components in functional foods or nutraceuticals. This contribution presents an overview of the ACE inhibitory peptides derived from marine fishes and discusses their future prospects to be used as potential drug candidates for preventing and treating high blood pressure.

[*] Department of Chemistry, Pukyong National University, Busan, Republic of Korea
[†] Marine Bioprocess Research Center, Pukyong National University, Busan, Republic of Korea
[1] Corresponding author: Se-Kwon Kim, *E-mail address:* sknkim@pknu.ac.kr

Advances in Food and Nutrition Research, Volume 65
ISSN 1043-4526, DOI: 10.1016/B978-0-12-416003-3.00016-0

I. INTRODUCTION

Hypertension (high blood pressure) is increasingly prevalent in developed countries and is one of the major independent risk factors for myocardial infarction, congestive heart failure, arteriosclerosis, stroke, and end-stage renal disease. Angiotensin-converting enzyme (ACE, a dipeptidyl carboxypeptidase) plays an important role in the regulation of blood pressure as well as cardiovascular function by converting the decapeptide angiotensin I to the vasoconstricting octapeptide angiotensin II. Moreover, ACE inactivates bradykinin, a vasodilatory peptide, which leads to increase blood pressure (Shahidi and Zhong, 2008).

ACE inhibitors are considered to be useful therapeutic approaches to treat hypertension, heart failure, stroke, and myocardial infarction. Therefore, in the development of drugs to control high blood pressure, three kinds of synthetic ACE inhibitors were designed; they are grouped by their ligand for the active site on ACE. Captopril, the key representative of this group, has a sulfhydryl moiety; enalapril and lisinopril have a carboxyl moiety; and fosinopril has a phosphorus group. Many studies have been attempted in the synthesis of ACE inhibitors, which are currently used as clinical antihypertensive drugs (Wijesekara and Kim, 2010). However, these synthetic drugs are believed to have adverse side effects such as cough, taste disturbances, dizziness, headache, skin rashes, and angioneurotic edema. Therefore, it is necessary to search safer, more economical, more innovative, and no-side-effect ACE inhibitors in the treatment of essential hypertension and heart failure in humans (Kamath *et al.*, 2007).

Recently, marine fish-derived bioactive peptides have been shown to possess many physiological functions including antihypertensive, antioxidant, antimicrobial, antiproliferative, antitumor, anticoagulant, and immunomodulatory activities. Among these, antihypertensive peptides act as ACE inhibitors are of particular interest for prevention and treatment of hypertension (Kobayashi *et al.*, 2008).

Marine fishes are rich sources of structurally diverse bioactive compounds including polyunsaturated fatty acids, polysaccharides, minerals, vitamins, antioxidants, enzymes, and bioactive peptides (Kim *et al.*, 2008). Marine fish-derived ACE inhibitory peptides have been purified from enzymatic digestion of various fish materials from Alaska pollack (Nakajima *et al.*, 2009), bonito (Fujita *et al.*, 2000; Hideaki *et al.*, 1993; Yokoyama *et al.*, 1992), tuna (Hwang, 2010), salmon (Ohta *et al.*, 1997), shark (Wu *et al.*, 2008), and sardine (Bougatef *et al.*, 2008; Otani *et al.*, 2009). Hence, a great interest has been developed nowadays to obtain bioactive compounds, which act as ACE inhibitors from marine fishes due to their numerous health beneficial effects. This chapter discusses the marine fish-derived antihypertensive peptides and their potential applications as ingredients in functional foods and nutraceuticals to prevent hypertension in humans.

II. DEVELOPMENT OF MARINE FISH-DERIVED ANTIHYPERTENSIVE PEPTIDES

Bioactive peptides can be purified from enzymatic hydrolysis of different marine fish sources using appropriate proteolytic enzymes. Proteolytic enzymes derived from plants, animals, and microbes can be used for the hydrolysis process of marine fish products to develop bioactive peptides. The physicochemical conditions of the reaction media, such as temperature and pH of the protein solution, must then be adjusted to optimize the activity of the enzyme used (Slizyte *et al.*, 2009). Kim *et al.* (1997) used the crude proteinase which was extracted from the pyloric ceca of tuna for enzymatic hydrolysis of cod frame protein under optimal conditions in order to obtain a maximum yield. Further, α-chymotrypsin, papain, neutrase, and trypsin have been used for the hydrolysis of tuna dark muscle under optimal pH and temperature conditions of the respective enzymes. Moreover, one of the most important factors in producing bioactive peptides with desired functional properties is the molecular weight of the bioactive peptide. Therefore, for the efficient recovery and to obtain bioactive peptides with both a desired molecular size and a functional property, a suitable method is the use of an ultrafiltration membrane system. This system has the main advantage that the molecular weight distribution of the desired peptide can be controlled by adoption of an appropriate ultrafiltration membrane. In order to obtain functionally active peptides, it is a suitable method to use a three-enzyme system for sequential enzymatic digestion. Moreover, it is possible to obtain serial enzymatic digestions in a system using a multistep recycling membrane reactor combined with ultrafiltration membrane system to separate marine fish-derived bioactive peptides (Kim and Mendis, 2006). This membrane bioreactor technology equipped with ultrafiltration membranes is recently emerging for the development of bioactive compounds and considered as a potential method to utilize marine fish proteins as value-added nutraceuticals with beneficial health effects.

III. ANTIHYPERTENSIVE ACTIVITY OF BIOACTIVE PEPTIDES DERIVED FROM MARINE FISHES

Nowadays, ACE inhibitory peptides have been isolated from meat, remaining muscle proteins, skin collagen and gelatin, bone, and internal organs of fishes such as Alaska pollack, bonito, tuna, salmon, shark, and sardine. Table 16.1 provides a partial summary of ACE inhibitory peptides derived from marine fish sources, their amino acid sequence, the enzyme used for hydrolysis, and IC_{50} values. The IC_{50} value is the concentration of peptide that inhibits 50% of ACE activity.

TABLE 16.1 ACE inhibitory peptides derived from marine fish: source, enzyme used for hydrolysis, amino acid sequence and IC_{50} value

Source	Enzyme	Amino acid sequence	IC_{50} (μM)	Reference
Alaska pollack skin	Alcalase+pronase+collagenase	GPL	2.6	Byun and Kim (2002)
Alaska pollack skin		LGP	0.72	Byun and Kim (2002)
Alaska pollack skin		GLP	1.62	Byun and Kim (2002)
Alaska pollack skin		PLG	4.74	Byun and Kim (2002)
Alaska pollack skin		LPG	5.73	Byun and Kim (2002)
Alaska pollack skin		PGL	13.93	Byun and Kim (2002)
Alaska pollack frame	Pepsin	FGASTRGA	14.7	Je *et al.* (2004)
Bonito muscle	Thermolysin	IKPLNY	43	Yokoyama *et al.* (1992)
Bonito muscle	Thermolysin	DYGLYP	62	Yokoyama *et al.* (1992)
Bonito bowels	Autolysis	LRP	1	Matsumura *et al.* (1993b)
Bonito liver	Autolysis	GVYPHK	1.6	Hideaki *et al.* (1993)
Bonito intestine	Autolysis	IRPVE	1.4	Hideaki *et al.* (1993)
Bonito muscle	Thermolysin	LKP	0.32	Fujita and Yoshikawa (1999)
Bonito muscle	Thermolysin	ILP	6.9	Fujita and Yoshikawa (1999)
Bonito muscle	Thermolysin	LKPNM	2.4	Fujita and Yoshikawa (1999)
Bonito muscle	Thermolysin	IWHHT	5.1	Fujita *et al.* (2000)
Bonito meat	Pepsin	HERDPTHIKWGD	8	Hasan *et al.* (2006)
Bonito meat	Pepsin	PTHIKWGD	8	Hasan *et al.* (2006)
Bonito protein		IKW	0.4	Hasan *et al.* (2007)
Bonito protein		IKY	1	Hasan *et al.* (2007)
Bigeye tuna muscle	Pepsin	WPEAAELMMEVDP	21.6	Qian *et al.* (2007)
Bigeye tuna frame	Pepsin	GDLGKTTTVSNWSPPKYKDTP	11.28	Lee *et al.* (2010)
Salmon muscle	Thermolysin	VW	2.5	Ono *et al.* (2003)

Salmon muscle	Thermolysin	IW	4.7	Ono *et al*. (2003)
Salmon muscle	Thermolysin	MW	9.9	Ono *et al*. (2003)
Salmon muscle	Alcalase+papain	IW	1.2	Enari *et al*. (2008)
Shark meat	Protease	FE	1.45	Wu *et al*. (2008)
Shark meat	Protease	CF	1.98	Wu *et al*. (2008)
Shark meat	Protease	EY	2.68	Wu *et al*. (2008)
Shark meat	Protease	MF	0.92	Wu *et al*. (2008)
Sardine muscle	Alcalase	KW	1.63	Matsufuji *et al*. (1994)
Sardine muscle	Alcalase	AKK	3.13	Matsufuji *et al*. (1994)
Sardine muscle	Alcalase	GWAP	3.86	Matsufuji *et al*. (1994)
Sardine muscle	Alcalase	VY	26	Kawasaki *et al*. (2000)
Sardine head	Proteases	nd	1.2*	Bougatef *et al*. (2008)
Sardine viscera	Proteases	nd	0.81*	Bougatef *et al*. (2008)

nd, not detected.
* mg/ml.

The competitiveness against ACE activity of different antihypertensive peptides has been determined kinetically using Lineweaver–Burk plots (Zhao *et al.*, 2009). Generally, the mechanism of action of antihypertensive peptides is different from that of synthetic drugs. The synthetic drugs basically, indiscriminately block ACE by interfering with its action, while ACE inhibitory peptides interact much differently by competing with ACE. ACE converts angiotensin I to angiotensin II by cleaving off a small peptide. Synthetic drugs work by directly blocking the action of ACE. ACE actually reacts with the antihypertensive peptides instead of attacking angiotensin I. Antihypertensive peptides relax the arterial walls and reduce fluid volume by inhibiting the formation of angiotensin II. Therefore, antihypertensive peptides actually improve heart function and increase blood and oxygen flow to the heart, liver, and kidneys. Many studies have shown that tryptophan, tyrosine, phenylalanine, or proline at the C-terminal and branched-chain aliphatic amino acids at the N-terminal were suitable for a peptide binding to ACE as a competitive inhibitor (Ahhmed and Muguruma, 2010).

In addition, a noncompetitive mechanism has also been observed in some peptides, and this means that the peptide can combine with an enzyme molecule to produce a dead-end complex, regardless of whether a substrate molecule is bound or not. For example, LIY (Nakagomi *et al.*, 2000) and YLYEIAR (Nakagomi *et al.*, 1998) have been found to act as noncompetitive inhibitors. The hydrophobicity of the N-terminus, which is one of the common features of ACE inhibitory peptides, may contribute to the inhibitory activity. ACE inhibitory peptides are generally short-chain peptides, often carrying polar amino acid residues like proline. Further, structure–activity relationships among various peptide inhibitors of ACE indicate that binding to ACE is strongly influenced by the C-terminal tripeptide sequence of the substrate, and it is suggested that peptides, which contain hydrophobic amino acids at these positions, are potent inhibitors (Rho *et al.*, 2009).

Proteolytic digestion of gelatin extracts from Alaska pollack (*Theragra chalcogramma*) skin brought about a high ACE inhibitory activity. Gelatin extracts were hydrolyzed by serial protease treatments in the order of alcalase, pronase E, and collagenase using a three-step recycling membrane reactor. The isolated peptide was composed of GPL and showed an IC_{50} value of 2.6 μM (Byun and Kim, 2002). In addition, the peptides GLP, LGP, LPG, PGL, PLG, GP, and PL, which consisted of glycine, proline, and leucine, were synthesized by the solid-phase method from Alaska pollack skin. The IC_{50} values of each dipeptide—namely, GP and PL—were 252.6 and 337.3 μM, respectively. The IC_{50} values of each tripeptide—namely, LGP, GLP, PLG, LPG, and PGL—were 0.72, 1.62, 4.74, 5.73, and 13.93 μM, respectively. The ACE inhibitory activity of these tripeptides was higher than that of dipeptides. Among these tripeptides, LGP and GLP had higher inhibitory activity than GPL. Among the different types of tripeptides that were examined, the highest ACE inhibitory activity was observed for LGP. LGP has the highest ACE inhibitory activity among the different types of tripeptides derived from Alaska pollack skin. It had the leucine residue at the N-terminal and proline residue at the C-terminal. Further, Je *et al.* (2004) purified a novel ACE inhibitory peptide from Alaska pollack frame protein hydrolyzed with pepsin with an IC_{50} value of 14.7 μM, and the sequence of the peptide was FGASTRGA. In addition, the ACE inhibition pattern of the peptide was found to be noncompetitive.

Ten ACE inhibitory peptides were isolated from bonito bowels autolysate (Matsumura *et al.*, 1993a,b) IC_{50} of ACE inhibitory peptides—namely, YRPY, GHF, VRP, IKP, LRP, IRP, SVAKLEK, ALPHA, GVYPHK, and IRPVQ—were estimated to be 320, 1100, 2.2, 2.5, 1.0, 1.8, 82, 79, 1.6, and 1.4 μM, respectively. C-terminal amino acids were considered to be essential for their expression of ACE inhibition, while the N-terminal tripeptide IRP was presumed to inhibit ACE after the removal of a dipeptide from IRPVQ with ACE digestion. In addition, LKPNM (IC_{50}=2.4 μM) from bonito fish product has found to be hydrolyzed by ACE to produce LKP (IC_{50}=0.32 μM), which had eightfold higher ACE inhibitory activity compared with the initial peptide (Fujita and Yoshikawa, 1999). Further, two peptides HERDPTHIKWGD and PTHIKWGD from bonito muscle in an artificial gastric juice were purified and their IC_{50} values were about 8 μM (Hasan *et al.*, 2006). Tripeptides IKW and IKY derived from bonito protein showed high ACE inhibitory activities with IC_{50} of 0.4 and 1.0 μM, respectively, and acted as competitive inhibitors (Hasan *et al.*, 2007).

A novel ACE inhibitory peptide (PTHIKWGD) was purified from acid extract of tuna muscle. Further, this peptide has inhibited ACE activity by noncompetitively with K_i values of 1.7 and 5.7 μM with substrates hippuryl-L-histidyl-L-leucine and angiotensin I, respectively (Kohama *et al.*, 1989). The structure of ACE inhibitory peptide from pepsin hydrolysate of bigeye tuna dark muscle, *Thunnus obesus* was identified to be WPEAAELMMEVDP, and the IC_{50} value of the peptide was 21.6 μM (Qian *et al.*, 2007). Numerous *in vivo* studies of marine fish-derived antihypertensive peptides in spontaneously hypertensive rats (SHR) have shown potent ACE inhibitory activity, and their systolic blood pressure (SBP) has reduced significantly after oral administration of peptides. According to Lee *et al.* (2010), a single oral administration (10 mg/kg of body weight) of the peptide derived from tuna frame protein hydrolysate exhibited *in vivo* activity by lowering blood pressure in SHR, and this antihypertensive activity was similar with captopril, a commercial antihypertensive drug. Further, they have reported that no side effects observed on rats after administration of antihypertensive peptide derived from bigeye tuna.

Thermolysin hydrolysate of defatted upstream chum salmon muscle showed a high inhibitory activity against ACE, with an IC_{50} value of 27.9 μg/ml *in vitro*. After fractionation, six dipeptides containing tryptophan residue were identified as WA, WM, MW, LW, IW, and VW, with an IC_{50} value of 277.3, 96.6, 9.9, 17.4, 4.7, and 2.5 μM, respectively. When orally administrated, the hydrolysate significantly lowered blood pressure for up to 8 h after administration with a maximum decrease 4 h after administration (Ono *et al.*, 2003). Collagen extracted from Atlantic salmon (*Salmo salar* L.) skin was hydrolyzed with alcalase and papain and treated by multistage separation. After fractionation, the ACE inhibitory activities of AP (IC_{50}=0.060 mg/ml) and VR (IC_{50}=0.332 mg/ml) were found to be approximately 20-fold and 4-fold higher than that of initial salmon skin collagen peptide (1.165 mg/ml), respectively (Gu *et al.*, 2011). In addition, the antihypertensive effect of the salmon peptide on SHR was examined. After the single intravenous administration of the salmon peptide at a dose of 30 mg/kg body weight, the SBP was significantly reduced against the control. Further, a double-blind, placebo-controlled, parallel-group study determined the efficacy of the salmon peptide in mild hypertensive subjects. The SBP, after a 1.0 g of salmon peptide intake, was significantly reduced at 4 weeks after the intake, and 2 weeks after the intake finished, compared to the value before ingestion. IW had the strongest ACE inhibitory activity (IC_{50}=1.2 μM) *in vitro* (Enari *et al.*, 2008).

Shark meat hydrolyzed with protease SM98011 showed high ACE inhibitory activity, with an IC_{50} value of 0.4 mg/ml (He *et al.*, 2007). Four peptides with high ACE inhibitory activity were purified from shark meat hydrolysate. Their amino acid sequences were CF, EY, MF, and FE, and their IC_{50} values were 1.98, 2.68, 0.92, and 1.45 mM, respectively. They may have potential in the treatment of hypertension or in clinical nutrition (Wu *et al.*, 2008).

ACE inhibitory peptides derived from sardine and hair tail meat were made by Suetsuna and Osajima (1986). They reported that protease hydrolysates of sardine contained ACE inhibitory peptides with IC_{50} values *in vitro* of 3.79 and 9.01 mg/L. The ACE inhibitory activities of protein hydrolysates prepared from heads and viscera of sardine (*Sardinella aurita*) by treatment with the alkaline protease extract from the viscera of sardine were investigated. The IC_{50} values for ACE inhibitory activities of sardinelle by-products protein hydrolysates and fraction P_4 were 1.2±0.09 and 0.81±0.013 mg/ml, respectively. Fraction P_4 was rich in phenylalanine, arginine, glycine, leucine, methionine, histidine, and tyrosine. The peptide prepared from sardine muscle by *Bacillus licheniformis* alkaline protease displayed the ACE inhibitory activity with an IC_{50} of 260 μg/ml (Matsui *et al.*, 1993). This activity is about 2.4-fold higher than that of a peptic hydrolysate (620 μg/ml) of sardine muscle. The ACE inhibitory activity of an alkaline protease hydrolysate from sardine muscle did not change after being treated by gastrointestinal proteases (IC_{50}=82 μg/ml). Eleven new ACE inhibitory peptides have been isolated with IC_{50} values mostly below 100 μM; the maximal ACE inhibitory activity has been observed for KW (IC_{50}=1.63 μM) (Matsufuji *et al.*, 1994). VY with potent ACE inhibitory activity were intravenously administered to SHR, and a significant reduction of diastolic blood pressure has been determined (Matsufuji *et al.*, 1995). Further, a randomized, double-blind, placebo-controlled study has carried out on 29 volunteers. VY presented a significant antihypertensive effect on mild hypertensive subjects via ACE inhibition, as well as SHR, but no adverse effects could be detected at all (Kawasaki *et al.*, 2000). Ohba *et al.* (2003) studied the physiological functions of enzymatic hydrolysates of collagen or keratin contained in livestock and fish waste. The enzymatic hydrolysate of meat meal, a collagen-waste, showed strong ACE inhibitory activity with IC_{50} values ranging from 600 to 2800 μg/ml.

IV. CONCLUSION

Recently, much attention has been paid by consumers toward natural bioactive compounds as functional ingredients, and hence it can be suggested that marine fish-derived ACE inhibitors are alternative tools that can contribute to consumer's well-being, by being a part of novel nutraceuticals or pharmaceuticals replacing synthetic drugs. Food bioactive compounds are often effective in promoting health and lead to the reduction of disease risk. Especially, bioactive compounds derived from marine fishes have served as rich sources of health-promoting components. Among them, bioactive peptides are rich sources of natural health enhancers, and this fact implies their potential use as a functional ingredient in future nutraceutical and pharmaceutical products. Until now, most of these ACE inhibitory activities have been observed *in vitro* or in mouse model systems. Therefore, further research studies are needed in order to investigate their activity in human subjects. In conclusion, it can be suggested that marine fish-derived ACE inhibitory bioactive peptides are potential therapeutic candidates for preventing hypertension and their involvement in the future pharmaceuticals is promising.

ACKNOWLEDGMENT

This study was supported by a grant from the Marine Bioprocess Research Center of the Marine Bio 21 Project funded by the Ministry of Land, Transport, and Maritime, Republic of Korea.

REFERENCES

Ahhmed, A. M. and Muguruma, M. (2010). A review of meat protein hydrolysates and hypertension. *Meat Sci.* **86**, 110–118.

Bougatef, A., Nedjar-Arroume, N., Ravallec-Plé, R., Leroy, Y., Guillochon, D., Barkia, A., and Nasri, M. (2008). Angiotensin I-converting enzyme (ACE) inhibitory activities of sardinelle (*Sardinella aurita*) by-products protein hydrolysates obtained by treatment with microbial and visceral fish serine proteases. *Food Chem.* **111**, 350–356.

Byun, H. G. and Kim, S. K. (2002). Structure and activity of angiotensin-I converting enzyme inhibitory peptides derived from Alaskan pollack skin. *J. Biochem. Mol. Biol.* **35**, 239–243.

Enari, H., Takahashi, Y., Kawarasaki, M., Tada, M., and Tatsuta, K. (2008). Identification of angiotensin-I-converting enzyme inhibitory peptides derived from salmon muscle and their antihypertensive effect. *Fish. Sci.* **74**, 911–920.

Fujita, H. and Yoshikawa, M. (1999). LKPNM: A prodrug-type ACE-inhibitory peptide derived from fish protein. *Int. J. Immunopharmacol.* **44**, 123–127.

Fujita, H., Yokoyama, K., and Yoshikawa, M. (2000). Classification and antihypertensive activity of angiotensin I-converting enzyme inhibitory peptides derived from food proteins. *J. Food Sci.* **65**, 564–569.

Gu, R. Z., Li, C. Y., Liu, W. Y., Yi, W. X., and Cai, M. Y. (2011). Angiotensin I-converting enzyme inhibitory activity of low-molecular-weight peptides from Atlantic salmon (*Salmo salar L.*) skin. *Food Res. Int.* **44**, 1536–1540.

Hasan, F., Kitagawa, M., Kumada, Y., Hashimoto, N., Shiiba, M., Katoh, S., and Terashima, M. (2006). Production kinetics of angiotensin-I converting enzyme inhibitory peptides from bonito meat in artificial gastric juice. *Process Biochem.* **41**, 505–511.

Hasan, M. F., Kobayashi, Y., Kumada, Y., Katsuda, T., Terashima, M., and Katoh, S. (2007). ACE inhibitory activity and characteristics of tri-peptides obtained from bonito protein. *J. Chem. Eng. Japan* **40**, 59–62.

He, H. L., Chen, X. L., Sun, C. Y., Zhang, Y. Z., and Zhou, B. C. (2007). High throughput and rapid screening of marine protein hydrolysates enriched in peptides with angiotensin-I-converting enzyme inhibitory activity by capillary electrophoresis. *Bioresour. Technol.* **98**, 3499–3505.

Hideaki, K., Masayoshi, K., Shigeru, S., Chiyo, D., Kunio, D., Nobuyasu, M., and Toshio, S. (1993). Oral administration of peptides derived from bonito bowels decreases blood pressure in spontaneously hypertensive rats by inhibiting angiotensin converting enzyme. *Comp. Biochem. Physiol. C* **104**, 351–353.

Hwang, J. S. (2010). Impact of processing on stability of angiotensin I-converting enzyme (ACE) inhibitory peptides obtained from tuna cooking juice. *Food Res. Int.* **43**, 902–906.

Je, J. Y., Park, P. J., Kwon, J. Y., and Kim, S. K. (2004). A novel angiotensin-I converting enzyme inhibitory peptide from Allaska pollack (*Theragra chalcogramma*) frame protein hydrolysate. *J. Agric. Food Chem.* **52**, 7842–7845.

Kamath, V., Niketh, S., Chandrashekar, A., and Rajini, P. S. (2007). Chymotryptic hydrolysates of a-kafirin, the storage protein of sorghum (*Sorghum bicolor*) exhibited angiotensin converting enzyme inhibitory activity. *Food Chem.* **100**, 306–311.

Kawasaki, T., Seki, E., Osajima, K., Yoshida, M., Asada, K., Matsui, T., and Osajima, Y. (2000). Antihypertensive effect of valyl-tyrosine, a short chain peptide derived from sardine muscle hydrolyzate, on mild hypertensive subjects. *J. Hum. Hypertens.* **14**, 519–523.

Kim, S. K. and Mendis, E. (2006). Bioactive compounds from marine processing byproducts—A review. *Food Res. Int.* **39**, 383–393.

Kim, S. K., Jeon, Y. J., Byun, H. G., Kim, Y. T., and Lee, C. K. (1997). Enzymatic recovery of cod frame proteins with crude proteinase from tuna pyloric caeca. *Fish. Sci.* **63**, 421–427.

Kim, S. K., Ravichandran, Y. D., Khan, S. B., and Kim, Y. T. (2008). Prospective of the cosmeceuticals derived from marine organisms. *Biotechnol. Bioprocess Eng.* **13**, 511–523.

Kobayashi, Y., Yamauchi, T., Katsuda, T., Yamaji, H., and Katoh, S. (2008). Angiotensin-I converting enzyme (ACE) inhibitory mechanism of tripeptides containing aromatic residues. *J. Biosci. Bioeng.* **106**, 310–312.

Kohama, Y., Oka, H., Matsumoto, S., Nakagawa, T., Miyamoto, T., Mimura, T., Nagase, Y., Satake, M., Takane, T., and Fujita, T. (1989). Biological properties of angiotensin-converting enzyme-inhibitor derived from tuna muscle. *J. Pharmacobiodyn.* **12**, 566–571.

Lee, S. H., Qian, Z. J., and Kim, S. K. (2010). A novel angiotensin-I converting enzyme inhibitory peptide from tuna frame protein hydrolysate and its antihypertensive effect in spontaneously hypertensive rats. *Food Chem.* **118**, 96–102.

Matsufuji, H., Matsui, T., Seki, E., Osajima, K., Nakashima, M., and Osajima, Y. (1994). Angiotensin-I-converting enzyme inhibitory peptides in an alkaline protease hydrolyzate derived from sardine muscle. *Biosci. Biotechnol. Biochem.* **58**, 2244–2245.

Matsufuji, H., Matsui, T., Ohshige, S., Kawasaki, T., Osajima, K., and Osajima, Y. (1995). Antihypertensive effects of angiotensin fragments in SHR. *Biosci. Biotechnol. Biochem.* **59**, 1398–1401.

Matsui, T., Matsufuji, H., Seki, E., Osajima, K., Nakashima, M., and Osajima, Y. (1993). Inhibition of angiotensin I-converting enzyme by *Bacillus licheniformis* alkaline protease hydrolysates derived from sardine muscle. *Biosci. Biotechnol. Biochem.* **57**, 922–925.

Matsumura, N., Fujii, M., Takeda, Y., and Shimizu, T. (1993a). Isolation and characterization of angiotensin I-converting enzyme inhibitory peptides derived from bonito bowels. *Biosci. Biotechnol. Biochem.* **57**, 1743–1744.

Matsumura, N., Fujii, M., Takeda, Y., Sugita, K., and Shimizu, T. (1993b). Angiotensin I- converting enzyme inhibitory peptides derived from bonito bowels autolysate. *Biosci. Biotechnol. Biochem.* **57**, 695–697.

Nakagomi, K., Fujimura, A., Ebisu, H., Sakai, T., Sadkane, Y., Fujii, N., and Tanimura, T. (1998). Acein-1, a novel angiotensin-I-converting enzyme inhibitory peptide isolated from tryptic hydrolysate of human plasma. *FEBS Lett.* **438**, 255–257.

Nakagomi, K., Yamada, R., Ebisu, H., Sadkane, Y., Akizawa, T., and Tanimura, T. (2000). Isolation of acein-2, a novel angiotensin I-converting enzyme inhibitory peptide derived from a tryptic hydrolysate of human plasma. *FEBS Lett.* **467**, 235–238.

Nakajima, K., Yoshie-Stark, Y., and Ogushi, M. (2009). Comparison of ACE inhibitory and DPPH radical scavenging activities of fish muscle hydrolysates. *Food Chem.* **114**, 844–851.

Ohba, R., Deguchi, T., Kishikawa, M., Arsyad, F., Morimura, S., and Kida, K. (2003). Physiological functions of enzymatic hydrolysates of collagen or keratin contained in livestock and fish waste. *Food Sci. Technol. Res.* **9**, 91–93.

Ohta, T., Iwashita, A., Sasaki, S., and Kawamura, Y. (1997). Antihypertensive action of the orally administered protease hydrolysates of chum salmon head and their angiotensin I converting enzyme inhibitory peptides. *Food Sci. Technol. Int.* **4**, 339–343.

Ono, S., Hosokawa, M., Miyashita, K., and Takahashi, K. (2003). Isolation of angiotensin-I-converting enzyme inhibitory effect derived from hydrolysate of upstream chum salmon muscle. *J. Food Sci.* **68**, 1611–1614.

Otani, L., Ninomiya, T., Murakami, M., Osajima, K., Kato, H., and Murakami, T. (2009). Sardine peptide with angiotensin I-converting enzyme inhibitory activity improves glucose tolerance in stroke-prone spontaneously hypertensive rats. *Biosci. Biotechnol. Biochem.* **73**, 2203–2209.

Qian, Z. J., Je, J. Y., and Kim, S. K. (2007). Antihypertensive effect of angiotensin-I converting enzyme-inhibitory peptide from hydrolysates of bigeye tuna dark muscle, *Thunnus obesus*. *J. Agric. Food Chem.* **55**, 8398–8403.

Rho, S. J., Lee, J. S., Chung, Y. I., Kim, Y. W., and Lee, H. G. (2009). Purification and identification of an angiotensin I-converting enzyme inhibitory peptide from fermented soybean extract. *Process Biochem.* **44**, 490–493.

Shahidi, F. and Zhong, Y. (2008). Bioactive peptides. *J. AOAC Int.* **91**, 914–931.

Slizyte, R., Mozuraityte, R., Martinez-Alvarez, O., Falch, E., Fouchereau-Peron, M., and Rustad, T. (2009). Functional, bioactive and antioxidative properties of hydrolysates obtained from cod (*Gadus morhua*) backbones. *Process Biochem.* **44**, 668–677.

Suetsuna, K. and Osajima, K. (1986). The inhibitory activities against angiotensin-I-converting enzyme of basic peptides originating from sardine and hair tail meat. *Bull. Jpn. Soc. Sci. Fish.* **52**, 1981–1984.

Wijesekara, I. and Kim, S. K. (2010). Angiotensin-I-converting enzyme (ACE) inhibitors from marine resources: Prospects in the pharmaceutical industry. *Mar. Drugs* **8**, 1080–1093.

Wu, H., He, H. L., Chen, X. L., Sun, C. Y., Zhang, Y. Z., and Zhou, B. C. (2008). Purification and identification of novel angiotensin-I-converting enzyme inhibitory peptides from shark meat hydrolysate. *Process Biochem.* **43**, 457–461.

Yokoyama, K. H., Chiba, H., and Yoshikawa, M. (1992). Peptide inhibitors for angiotensin-I-converting enzyme from thermolysin digest of dried bonito. *Biosci. Biotechnol. Biochem.* **56**, 1541–1545.

Zhao, Y., Bafang, L., Dong, S., Liu, Z., Zhao, X., Wang, J., and Zeng, M. (2009). A novel ACE inhibitory peptide isolated from *Acaudina molpadioidea* hydrolysate. *Peptides* **30**, 1028–1033.

CHAPTER 17

Bioactive Sterols from Marine Resources and Their Potential Benefits for Human Health

Se-Kwon Kim[*,†,1] and **Quang Van Ta**[*]

Contents

Abstract

Bioactive agents from marine resources have shown their valuable health beneficial effects. Therefore, increase knowledge on novel functional ingredients with biological activities from marine animal and microbe has gained much attention. Sterols are recognized as potential in development functional food ingredients and pharmaceutical agents. Marine resources, with a great diversity, can be a very interesting natural resource of sterols. This chapter focuses on biological activities of marine animal and microbe sterols with potential health beneficial applications in functional foods and pharmaceuticals.

[*] Department of Chemistry, Marine Biochemistry Laboratory, Pukyong National University, Busan, Republic of Korea
[†] Marine Bioprocess Research Center, Pukyong National University, Busan, Republic of Korea
[1] Correspondence author: Se-Kwon Kim, *E-mail address:* sknkim@pknu.ac.kr

Advances in Food and Nutrition Research, Volume 65
ISSN 1043-4526, DOI: 10.1016/B978-0-12-416003-3.00017-2

I. INTRODUCTION

Marine organisms are rich sources of structurally diverse bioactive compounds with various beneficial biological activities that promote good health. The importance of marine organisms as a source of novel bioactive substances is growing rapidly. With marine species comprising approximately one-half of the total global biodiversity, the sea offers an enormous resource for novel compounds. Moreover, different kinds of substances have been procured from marine organisms because they live in an exigent, competitive, and aggressive surrounding different in many aspects from the terrestrial environment, a situation that demands the production of specific and potent active molecules. Among them, sterols have been reported as potential bioactive compound.

Sterols are an important family of lipids, present in the majority of eukaryotic cells. In addition, sterols are categorized to the steroid group, which contains the same fused four-ring core structure and has different biological roles as hormones and signaling molecules. Sterols are essential components of the membranes of all eukaryotic organisms, controlling membrane fluidity and permeability. Sterols are highly diverse in nature (Fahy *et al.*, 2005; Gaulin *et al.*, 2010; Kamenarska *et al.*, 2006). Clinical studies have showed that plant sterols could reduce the risk of heart disease by prevention and reduction of hypercholesterolemia. Several studies have indicated that phytosterols have health-promoting effects such as anticancer activity. Additionally, plant sterols have been suggested to have anti-inflammatory, antibacterial, and antifungal activities. Further, long-term studies on animal models and human did not show any toxic effect of plant sterols (Moreau *et al.*, 2002). Sterols also can be found in marine algae. It has been reported that sterols form marine algae also show several beneficial bioactivities such as antibacterial, antioxidant, anticancer, antidiabetic, and antihypercholesterolemia activities (Sánchez-Machado *et al.*, 2004). Moreover, marine animals and microbes also contain various bioactive substances.

Hence, this chapter focuses on biological activities of sterols derived from marine animals and microbes and their potential health beneficial applications in food and pharmaceutical industries.

II. BIOLOGICAL ACTIVITIES OF STEROLS

A. Anti-inflammation activity

Inflammation is the response of body to injury and danger. It is characterized by persistently activated immune cells. The course for inflammation includes infectious agents, physical and chemical agents, and dysfunction of immune system. Moreover, chronic inflammation is proposed to be at the root of approximately one-fifth of all human cancers (De Marzo *et al.*, 2007). Sustained cellular injuries can cause inflammation. Various inflammation and innate immune cells are often recruited at the site of infection or damage. In response to proinflammatory stimuli, activated immune cells generate reactive oxygen species and reactive nitrogen species, which can function as chemical effectors in inflammation (Halliday, 2005). In addition, it has been reported that cyclooxygenase 2 (COX-2) is a key mediator of inflammation. COX-2 mediates prostaglandin synthesis, which, in turn, promotes cell proliferation, cytokine synthesis, and suppress immune surveillance (Maitra *et al.*, 2002). Another factor related to inflammation is nuclear factor kappa-B (NF-κB). The NF-κB protein complex is activated in response to inflammation (Karin, 2006). There is evidence that COX-2 inhibitors reduce the incidence of carcinoma and reduce the risk of colorectal cancer in patients that overexpress COX-2 (McKay *et al.*, 2008). Hence, finding anti-inflammatory agents is required for the prevention of carcinogenesis. Fortunately, marine resources provide many bioactive compounds. Among them, sterols are abundant in marine microbes. It was reported that new sterol Bendigole F, which isolated from marine sponge *Actinomadura* sp. SBMs009, showed the translocation inhibition of NF-κB at the half maximal inhibitory concentration (IC_{50}) of 71 μM. The inhibition of NF-κB translocation suggested that Bendigole F has potential anti-inflammation activity (Simmons *et al.*, 2011). Sterol glycoside Carijoside A, which isolated from the octocorals belonging to the genus *Carijoa* (=*Telesto*), had shown the anti-inflammatory activity, significant inhibition of superoxide anion generation (IC_{50}=1.8 μg/ml), and elastase release (IC_{50}=6.8 μg/ml) by human neutrophils (Liu *et al.*, 2010). 3-Hydroxy-26-norcampest-5-en-25-oic acid from *Euryspongian* sp. was reported that it has effect on COX pathway and thus could be a potential anti-inflammatory natural compound derived from marine sponge (Mandeaua *et al.*, 2005). Huang *et al.* (2008) had reported that eight new sterols were isolated from soft coral *Nephthea chabroli* Audouin (Nephtheidae) and were named as Nebrosteroid A–H. Studies on LPS-stimulated RAW 264.7 cells showed that Nebrosteroid D, E, and G at a concentration of 10 μM significantly reduced the level of iNOS and COX-2 proteins expression. Nebrosteroid A, B, C, and H could significantly reduce the iNOS protein expression but did not inhibit the COX-2 protein expression (Huang *et al.*, 2008). In addition, fractionations of the CH_2Cl_2 extracts of the Formosan soft coral *Clavularia viridis* Quoy and Gaimard (class *Anthozoa*, subclass *Octocorallia*, order *Stolonifera*) resulted in the isolation of four new steroids, Stoloniferones R, S, T, and (25*S*)-24-methylenecholestane-3,5,6-triol-26-acetate. Among them, Stoloniferones S and (25*S*)-24-methylenecholestane-3,5,6-triol-26-acetate significantly reduced the levels of iNOS protein at concentration of 10 μM.

Stoloniferones T and (25*S*)-24-methylenecholestane-3,5,6-triol-26-acetate at concentration of 10 μM significantly reduced COX-2 protein expression (Chang *et al.*, 2008). Collectively, these results suggest that sterols derived from marine soft coral and sponge have promising potential to be used as valuable chemopreventive agents in inflammation therapy.

B. Antimicrobial activity

Tuberculosis is the second commonest cause of death worldwide. Most forms of active tuberculosis can be treated with 6 months of medication. Unfortunately, outbreaks of multidrug resistant tuberculosis have been occurring since 1990s. Natural products form one avenue in the search for new antituberculosis agents. Parguesterols A and B from Caribbean Sea sponge *Svenzea zeai* have been reported to show potent antituberculosis activity. Antituberculosis assay on *Mycobacterium tuberculosis* H_{37}Rv showed that minimum inhibition concentration for antitubercular activity of parguesterols A and B were 7.8 and 11.2 μg/ml, respectively. Moreover, they showed the low toxicity against Vero cells (IC_{50}=52 μg/ml). Therefore, they have potent in development of new antituberculosis agents (Wei *et al.*, 2007).

Moreover, sterols also showed the inhibitory effect on development of Candida fungus. Candida fungus has been reported that they can cause superficial infections such as oropharyngeal candidiasis and vulvovaginal candidiasis. In debilitated or immunocompromised patients, if introduced intravenously, candidiasis may become a systemic disease producing abscess, thrombophlebitis, endocarditis, or infections of the eyes or other organs (Darwazeh *et al.*, 1990). It has been reported that isolated sterol from sponge *Dysidea arenaria*, 9α,11α-epoxycholest-7-ene-3β,5α,6α,19-tetrol 6-acetate (ECTA), showed the enhancement in antifungal activity when combined with fluconazole, one common antifungal drug. Fluconazole alone showed IC_{50}=300 μM on drug resistance *Candida albicans*; however, the IC_{50} of fluconazole is decreased to 8.5 μM when combined with 3.8 μM of ECTA (Jacob *et al.*, 2003). Sterols from the Red Sea marine sponge *Lamellodysidea herbacea*, cholesta-8,24-dien-3β,5α,6α-triol and cholesta-8(14),24-dien-3β,5α,6α-triol, also showed the antifungal activity against *Candida tropicalis* (Sauleau and Bourguet-Kondracki, 2005). Collectively, sterols from marine resource have potential in antibacterial and antifungal application.

C. Anti-HIV activity

Human immunodeficiency virus type-1 (HIV-1) is identified as the causative agent of acquired immunodeficiency syndrome (AIDS), which is one of the most important diseases. The first generation anti-HIV drugs have been developed to treat AIDS patients after the introduction of AIDS in early 1980. However, failure in anti-AIDS treatment is observed in most of the patients infected with HIV as a result of drug-resistant strains of virus. Therefore, the search for potential drug candidates containing higher inhibitory activity against various HIV strains is increasing in pharmaceutical industry. In this regard, natural bioactive compounds and their derivatives are great sources for the development of new generation anti-HIV therapeutics, which are more effective with minor side effects.

Marine organisms are a valuable resource of anti-HIV compounds. The sulfated sterol ibisterol was isolated from the deep water Caribbean sponge *Topsentia* sp. that was investigated as anti-HIV agent (McKee *et al.*, 1993). Sun *et al.* (1991) reported that weinbersterol disulfates A was isolated from the sponge *Petrosia weinbergi*, which is active *in vitro* against HIV. Studies on sterols isolated from marine sponge have shown that ibisterol sulfate from *Topsentia* sp., halistanol sulfate from *Halichondria* cf. *moorei*, and 26-methylhalistanol sulfate and 25-demethylhalistanol sulfate from *Pseudaxinyssa digitata* showed essentially complete protection against the cytopathic effects of HIV-1 infection at half maximal effective concentration (EC_{50}) of 13, 6, 3, and 6 μM, respectively. The 10 sulfated sterols, which isolated from sea stars *Tremaster novaecaledoniae*, *Asterias amurensis*, *Styracaster caroli*, and *Echinaster brasiliensis*, showed the inactivation against HIV at the concentration of 100 μg/ml (McKee *et al.*, 1994). Moreover, clathsterol isolated from Red Sea sponge *Clathria* sp., which is inhibit HIV-1 reverse transcriptase activity at a concentration of 10 μM (Rudi *et al.*, 2001). Hence, marine sterols have potent in development of new antiviral agents.

D. Anticancer activity

Marine sponges are sources of numerous novel sterols. Their sterols have shown promising anticancer activities and hence, they are important sources to manufacture novel anticancer drugs. It has been reported that 5α,6α-epoxy-24*R**-ethylcholest-8(14)-en-3β,7α-diol and 5α,6α-epoxy-24*R**-ethylcholest-8-en-3β,7α-diol have been purified from marine sponge *Polymastia tenax*. These sterols showed antiproliferative activity toward lung (A549), colon (HT-29 and H-116), mice endothelial (MS-1), and human prostate carcinoma (PC-3) cell lines (Santafé *et al.*, 2002). Orostanal, purified sterol from marine sponge *Stellatta hiwasaensis*, inhibited proliferation in human leukemia (HL-60) cell line at IC_{50} 1.7 μM (Miyamoto *et al.*, 2001). Moreover, sterols from marine sponge *Lanthella* sp., petrosterol-3,6-dione, 5α,6α-epoxy-petrosterol, and petrosterol, showed the cytotoxicity activity against A549, HT-29, breast (MCF-7), human ovarian cancer (SK-OV-3), and two types of leukemia (HL-60 and U937) human cancer cell lines with IC_{50} values from 8.4 to 69.9 petrosterol. Petrosterol-3,6-dione showed growth-inhibitory effects with IC_{50} values of 8.4, 19.9, 17.8, 16.2, and 22.1 petrosterol against A549, HL-60, MCF-7, SK-OV-3, and U937 cell lines, respectively. 5α,6α-Epoxy-petrosterol showed cytotoxic effects with IC_{50} values of 9.8, 21.3, 19.4, 22.6, and 19.9 μM. Case of petrosterol showed growth-inhibitory activities with 11.5, 21.5, 16.4, 19.8, and 18.7 μM, respectively (Nguyen *et al.*, 2009).

In addition, sterols can also find in other marine organism such as coral, starfish. Lobophytosterol from soft coral *Lobophytum laevigatum* showed cytotoxicity effect on A549 and HL-60 with IC_{50} values of 4.5 and 5.6 μM, respectively. Treatment with lobophytosterol induced apoptosis evident by chromatin condensation in treated cells (Tran *et al.*, 2011). Sterols from Caribbean gorgonian *Plexaurella grisea* showed that cholest-3β,5α,6β,7β-tetrol and cholest-3β,5α,6β-triol have cytotoxicity activity against the A549 and HT-29 cell lines with effective dose (ED_{50}) values of 1 μg/ml. The new compounds 5β,6β-epoxyergost-24(28)-ene-3β,7β-diol, together with the known compound ergost-24(28)-ene-3α,5β,6β-triol, exhibited strong and selective cytotoxicity against the HT-29 cell line with an ED_{50}=0.1 μg/ml, and an ED_{50}=0.25 μg/ml for ergost-24(28)-ene-3β,5α,6β,7β-tetrol (Rueda *et al.*, 2001). Purified armartnol A from soft coral *Nephthea armata* exhibited cytotoxicity against A549, HT-29, and P-388 (mouse lymphocytic leukemia) cell lines with IC_{50} value of 7.6, 6.5, and 6.1 μM, respectively. Moreover, armartnol B showed cytotoxicity against P-388 and HT-29 cells with IC_{50} values of 3.2 and 3.1 μM, respectively (El-Gamal *et al.*, 2004). Wang *et al.* (2004) reported that certonardosterol Q6 from starfish *Certonardoa semiregularis* Muller & Troschel exhibited cytotoxicity effect on A549, SK-OV-3 (human ovarian cancer), SK-MEL-2 (human skin cancer), and HCT 15 cell line at ED_{50} values of 0.43, 0.22, 0.17, and 0.48 μg/ml, respectively. Collectively, sterols from marine resources have potential in development of anticancer agents.

III. CONCLUSIONS

Marine animal and microbe showed the potential role in application for human health and nutrition. Increasing knowledge on animal and microbe sterols has raised the awareness and demand for novel functional food ingredients and pharmaceuticals. Hence, sterols compounds from marine animal and microbe can be used as functional ingredients to reduce infectious diseases and cancer in human. However, there are few studies on marine animal and microbe sterols compared with those plant sterols. Therefore, the future studies on marine sterols will give more beneficial bioactive sterols compounds. Hence, the sterols, which derived from marine animal and microbe, have potential to expand its health beneficial value not only in the food industry but also in the pharmaceutical industries.

REFERENCES

Chang, C. H., Wen, Z. H., Wang, S. K., and Duh, C. Y. (2008). New anti-inflammatory steroids from the Formosan soft coral *Clavularia viridis*. *Steroids* **73**, 562–567.

Darwazeh, A., Lamey, P., Samaranayake, L., MacFarlane, T., Fisher, B., Macrury, S., and MacCuish, A. (1990). The relationship between colonisation, secretor status and in-vitro adhesion of *Candida albicans* to buccal epithelial cells from diabetics. *J. Med. Microbiol.* **33**, 43–49.

De Marzo, A. M., Platz, E. A., Sutcliffe, S., Xu, J., Gronberg, H., Drake, C. G., Nakai, Y., Isaacs, W. B., and Nelson, W. G. (2007). Inflammation in prostate carcinogenesis. *Nat. Rev. Cancer* **7**, 256–269.

El-Gamal, A. A. H., Wang, S. K., Dai, C. F., and Duh, C. Y. (2004). New nardosinanes and 19-oxygenated ergosterols from the soft coral *Nephthea armata* collected in Taiwan. *J. Nat. Prod.* **67**, 1455–1458.

Fahy, E., Subramaniam, S., Brown, H. A., Glass, C. K., Merrill, A. H., Murphy, R. C., Raetz, C. R. H., Russell, D. W., Seyama, Y., Shaw, W., Shimizu, T., Spener, F., *et al.* (2005). A comprehensive classification system for lipids. *J. Lipid Res.* **46**, 839–861.

Gaulin, E., Bottin, A., and Dumas, B. (2010). Sterol biosynthesis in oomycete pathogens. *Plant Signal. Behav.* **5**, 258–260.

Halliday, G. M. (2005). Inflammation, gene mutation and photoimmunosuppression in response to UVR-induced oxidative damage contributes to photocarcinogenesis. *Mutat. Res.* **571**, 107–120.

Huang, Y. C., Wen, Z. H., Wang, S. K., Hsua, C. H., and Duh, C. Y. (2008). New anti-inflammatory 4-methylated steroids from the Formosan soft coral *Nephthea chabroli*. *Steroids* **7**, 1181–1186.

Jacob, M. R., Hossain, C. F., Mohammed, K. A., Smillie, T. J., Clark, A. M., Walker, L. A., and Nagle, D. G. (2003). Reversal of fluconazole resistance in multidrug efflux-resistant fungi by the *Dysidea arenaria* sponge sterol 9α,11α-epoxycholest-7-ene-3β,5α,6α,19-tetrol 6-acetate. *J. Nat. Prod.* **66**, 1618–1622.

Kamenarska, Z., Stefanov, K., Stancheva, R., Dimitrova-Konaklieva, S., and Popov, S. (2006). Comparative investigation on sterols from some Black Sea red algae. *Nat. Prod. Res.* **20**, 113–118.

Karin, M. (2006). Nuclear factor-kappa B in cancer development and progression. *Nature* **441**, 431–436.

Liu, C. Y., Hwang, T. L., Lin, M. R., Chen, Y. S., Chang, Y. C., Fang, L. S., Wang, W. H., Wu, Y. C., and Sung, P. J. (2010). Carijoside A, a bioactive sterol glycoside from an octocoral *Carijoa* sp. (Clavulariidae). *Mar. Drugs* **8**, 2014–2020.

Maitra, A., Ashfaq, R., and Gunn, C. R. (2002). Cyclooxygenase 2 expression in pancreatic adenocarcinoma and pancreatic intraepithelial neoplasia—An immunohistochemical analysis with automated cellular imaging. *Am. J. Clin. Pathol.* **118**, 194–201.

Mandeaua, A., Debitus, C., Aries, M. F., and David, B. (2005). Isolation and absolute configuration of new bioactive marine steroids from *Euryspongian* sp.. *Steroids* **70**, 873–878.

McKay, C. J., Glen, P., and McMillan, D. C. (2008). Chronic inflammation and pancreatic cancer. *Best Pract. Res. Clin. Gastroenterol.* **22**, 65–73.

McKee, T. C., Cardellina, J. H., II, Tischler, M., Snader, K. M., and Boyd, M. R. (1993). Ibisterol sulfate, a novel HIV-inhibitory sulfated sterol from the deep water sponge *Topsentia* sp. *Tetrahedron Lett.* **34**, 389–392.

McKee, T. C., Cardellina, J. H., Riccio, R., D'Auria, M. V., Iorizzi, M., Minale, L., Moran, R. A., Gulakowski, R. J., McMahon, J. B., Buckheit, R. W., Snader, K. M., and Boyd, M. R. (1994). HIV-inhibitory natural products. 11. Comparative studies of sulfated sterols from marine invertebrates. *J. Med. Chem.* **37**, 793–797.

Miyamoto, T., Kodama, K., Aramaki, Y., Higuchia, R., and Van Soest, R. W. M. (2001). Orostanal, a novel abeo-sterol inducing apoptosis in leukemia cell from a marine sponge, *Stelletta hiwasaensis*. *Tetrahedron Lett.* **42**, 6349–6351.

Moreau, R. A., Whitaker, B. D., and Hicks, K. B. (2002). Phytosterols, phytostanols, and their conjugates in foods: Structural diversity, quantitative analysis, and health-promoting uses. *Prog. Lipid Res.* **41**, 457–500.

Nguyen, H. T., Chau, V. M., Tran, T. H., Phan, V. K., Hoang, T. H., Nguyen, T. D., Nguyen, X. N., Bui, H. T., Hyun, J. H., Kang, H. K., and Kim, Y. H. (2009). C29 sterols with a cyclopropane ring at C-25 and 26 from the Vietnamese marine sponge *Ianthella* sp. and their anticancer properties. *Bioorg. Med. Chem. Lett.* **19**, 4584–4588.

Rudi, A., Yosief, T., Loya, S., Hizi, A., Schleyer, M., and Kashman, Y. (2001). Clathsterol, a novel anti-HIV-1 RT sulfated sterol from the sponge Clathria species. *J. Nat. Prod.* **64**, 1451–1453.

Rueda, A., Zubía, E., Ortega, M. J., and Salva, J. (2001). Structure and cytotoxicity of new polyhydroxylated sterols from the Caribbean gorgonian *Plexaurella grisea*. *Steroids* **66**, 897–904.

Sánchez-Machado, D. I., López-Hernández, J., Paseiro-Losada, P., and López-Cervantes, J. (2004). An HPLC method for the quantification of sterols in edible seaweeds. *Biomed. Chromatogr.* **18**, 183–190.

Santafé, G., Paz, V., Rodriguez, J., and Jimenez, C. (2002). Novel cytotoxic oxygenated C29 sterols from the Colombian marine sponge *Polymastia tenax*. *J. Nat. Prod.* **65**, 1161–1164.

Sauleau, P. and Bourguet-Kondracki, M. L. (2005). Novel polyhydroxysterols from the Red Sea marine sponge *Lamellodysidea herbacea*. *Steroids* **70**, 954–959.

Simmons, L., Kaufmann, K., Garcia, R., Schwär, G., Huch, V., and Müller, R. (2011). Bendigoles D-F, bioactive sterols from the marine sponge-derived *Actinomadura* sp. SBMs009. *Bioorg. Med. Chem.* **19**, 6570–6575.

Sun, H. H., Gross, S. S., Gunasekera, M., and Koehn, F. E. (1991). Weinbersterol disulfates A and B, antiviral steroid sulfates from the sponge *Petrosia weinbergi*. *Tetrahedron* **47**, 1185–1190.

Tran, H. Q., Tran, T. H., Chau, V. M., Phan, V. K., Hoang, T. H., Nguyen, T. T. N., Nguyen, X. N., Nguyen, H. T., Nguyen, P. T., Dinh, T. T. T., Song, S. B., Boo, H. J., *et al.* (2011). Cytotoxic and PPARs transcriptional activities of sterols from the Vietnamese soft coral *Lobophytum laevigatum*. *Bioorg. Med. Chem. Lett.* **21**, 2845–2849.

Wang, W. H., Jang, H. J., Hong, J. K., Lee, C. O., Im, K. S., Bae, S. J., and Jung, J. H. (2004). Additional cytotoxic sterols and saponins from the starfish Certonardoa semiregularis. *J. Nat. Prod.* **67**, 1654–1660.

Wei, X. M., Rodriguez, A. D., Wang, Y. H., and Franzblau, S. G. (2007). Novel ring B abeo-sterols as growth inhibitors of Mycobacterium tuberculosis isolated from a Caribbean Sea sponge, Svenzea zeai. *Tetrahedron Lett.* **48**, 8851–8854.

CHAPTER 18

Therapeutic Drugs: Healing Power of Marine Fish

N.S. Sampath Kumar,[*,1] **N. Satya Vijaya Kumar,**[†] and **R. Jaiganesh**[*]

Contents

Abstract

Marine fish is a major source of high-quality protein, lipids, and a wide variety of vitamins and minerals. These macromolecules and their derivatives show different pharmacological activities, which make the fish as a therapeutic diet. Modern technology has made it easy to explore the therapeutic importance of fish-based diet on cardiovascular diseases, neurodegenerative diseases, radicals-mediated diseases, and cancer. In this review, we focus on exploration of proteins, lipids, carbohydrates, minerals, and their derivatives from marine fish as a major source for bioactive compounds and their medicinal importance.

* Department of Biotechnology, School of Bioengineering, SRM University, Kattankulathur, Tamilnadu, India
† Nanotechnology Research Center, SRM University, Kattankulathur, Tamilnadu, India
[1] Corresponding author: N.S. Sampath Kumar, *E-mail address:* hai_sampath@yahoo.co.in

Advances in Food and Nutrition Research, Volume 65
ISSN 1043-4526, DOI: 10.1016/B978-0-12-416003-3.00018-4

I. INTRODUCTION

The ocean, with its enormity and mystery, is interwoven with existence of all living organisms on earth. And during the evolution of the most dominant life forms from single-celled organisms, the ocean turned itself into the planet's best reservoir for food, medicine, and knowledge (Folmer *et al.*, 2008). Natural products isolated from marine sources proved to have great potential in drug discovery and therapy of diseases (Faulkner, 2000). Covering more than 70% of the earth, and with the uniqueness of the environmental conditions, it is easily understandable that the ocean can be considered as a very promising resource of natural drugs for the future (Newman and Cragg, 2004). With the outbreak of new diseases and the evolution of drug resistant strains of harmful bacteria, virus, etc., there is a continuous need for the development of new drugs. So, as stated by Hippocrates 2500 years ago, "let food be thy medicine and medicine be thy food," developing food with medicinal values is receiving renewed interest. Nutritious food is always associated with health benefits; thus, research efforts to identify functional and bioactive components from many natural sources including plants, animals, and microorganisms have intensified. So far, all the prescribed nutraceuticals are obtained from terrestrial plants and microorganisms (Harvey, 2000).

Marine organisms comprise approximately half of the total global biodiversity and offer a wide variety of novel compounds (McCarthy and Pomponi, 2004). Almost all marine organisms face a very exigent, competitive, aggressive, and very different surroundings in many aspects compared to the terrestrial environment, a situation that demands quite specific and potent active molecules. In recent years, many bioactive compounds have been extracted from various marine animals like tunicates, sponges, soft corals, sea hares, nudibranchs, bryozoans, sea slugs, and fishes (Donia and Hamann, 2003; Haefner, 2003). The search for new metabolites from marine organisms resulted in the isolation of nearly 10,000 metabolites (Fuesetani, 2000), many of which are endowed with pharmacodynamic properties. In this chapter, we discuss the medicinal value of marine fish procured by direct consumption in natural form or by isolating them from various biotechnology techniques.

Commercially, fish by-products are marketed as fish oil, fishmeal, fertilizer, pet food, and fish silage (Choudhury and Bublitz, 1996; Choudhury and Gogoi, 1995); however, these products have low economic value at high processing charges. Recent studies have identified a number of bioactive compounds in fish muscle, skin, and internal organs and bone (Je *et al.*, 2005a,b,c; Kittiphattanabawon *et al.*, 2005; Kumar *et al.*, 2011a,b). Therefore, in the current scenario, it is evident that it is beneficial to extract and purify these bioactive compounds by developing simple, cost-effective but efficient techniques. Further, some of these bioactive compounds such as peptides, oligosaccharides, fatty acids, enzymes, water-soluble minerals, and biopolymers have been identified to have nutraceutical potentials that are beneficial in health promotion (Defelice, 1995).

II. CHEMICAL COMPOSITION OF FISH

Fish is one of the main food constituents in our diet as it includes essential fatty acids, amino acids, and some of the principal vitamins and minerals in required quantities (Borgstrom, 1961). The medicinal importance of fish does not come exactly from the presence of fascinating metabolites or bioactive compounds, but from its balanced proximate composition. Principal composition of fish is 66–81% water, 16–21% protein, 0.2–15% fat, 1.2–1.5% mineral, and 0–0.5% carbohydrate (Love, 1970). In addition to these, fish is also a rich source of vitamins A, D, E, and K (Kinsella, 1987). The composition, however, varies greatly from species to species (Huss, 1995). Further, the variations in proximate composition of fish are based on age, sex, feed intake, environment, and season (Boran and Karacam, 2011). Irrespective of the variations observed, fish is always a good source of healthy diet with desired quantity of protein, lipid, and carbohydrate that contribute to the total energy content of an organism, while water and ash contribute mass. Not only are these macromolecules useful for consumption but also these have become a major source for the extraction of several bioactive compounds with the development of biotechnology and bioprocess research. This chapter concentrates in highlighting the role of protein, fat, carbohydrates, and minerals in making the fish a source of many therapeutic agents.

A. Carbohydrates derivatives from marine fish

Carbohydrate content in fish is generally very low and often considered zero (Anthony *et al.*, 2000; Payne *et al.*, 1999). But, carbohydrate derivatives such as heparin, a sulfated polysaccharide extracted from marine fishes, are very effective anticoagulant and antiproliferative agents. Heparin is the most widely used glycosaminoglycans because of its ability to anticoagulate blood. As such, heparin is used extensively in the treatment of myocardial infarction, unstable angina, venous thrombosis, and pulmonary embolism (Hirsh *et al.*, 2001). In addition, heparin is required to keep extracorporeal circulations during open heart surgery and renal dialysis (Bull *et al.*, 1975; Ryckelynck *et al.*, 1998). Commercially, heparin was isolated from bovine and porcine, after the outbreak of several contagious diseases and religious objections that has brought a need in search for alternative sources. And, marine fishes such as *Salmo salar* (Flengsrud *et al.*, 2010), *Dasyatis americana, Dasyatis guttata, Aetobatus narinari, Raja radula, Potamotrygon motoro* (Dellias *et al.*, 2004), *Gadus morhua*, and *Anarhichas minor* (Monica *et al.*, 2005) brought a solution for these problems. These sulfated polysaccharides can be extracted using solvents followed either by cetyl pyridinium chloride precipitation (Holick *et al.*, 1985) or by proteolytic enzymes (Walter *et al.*, 2007), and purification can be done by high-performance liquid chromatography (HPLC). So, marine fish have a carbohydrate moiety that can act as a good anticoagulant; exogenously administered heparin has proven to exert advantageous and positive effects in inflammation (Rek *et al.*, 2009) and cancer (Casu *et al.*, 2008) treatment.

B. Protein derivatives from marine fish

Marine fish is one of the major sources of animal protein (Agusa *et al.*, 2007). Fish occupies 25% of the overall animal protein supply in the world (Nurnadia *et al.*, 2011). Moreover, fish protein has been considered as the major source for therapeutic use in the form of protein, polypeptides, peptides, enzymes, and even toxin. Throughout the world, about 500 species of fish are found to be toxic (Jha and Zi-rong, 2004). The most spectacular substance of pharmacological importance extracted from fish is tetrodotoxin (TTX), the puffer or fugu poison. Other toxins isolated include ciguatoxin from electric rays, which is served as a potent antidote for pesticide poisoning (Oliviera *et al.*, 2003). TTX isolated from puffer fish and many other marine organisms have become a useful tool for studying the voltage-gated sodium channel (Kao and Levinson, 1986). The stomach extracts of dogfish shark, *Squalus acanthias*, have been the source of new class water-soluble and broad-spectrum antibiotics, squalamines (Moore *et al.*, 1993). Insulin, a hormone used in the treatment of diabetes, isolated from European spotted dogfish, hammerhead shark, tuna, and cod fish, exerts the same hormonal activity in mammals irrespective of its 40% difference in amino acid residue (Grant and Mackie, 1977; Halvey, 1990). Fish protein-based applications are very vast, and this review confines to discuss regarding collagen, gelatin, protein hydrolysates, and bioactive peptides.

C. Collagen and gelatin

Collagen is the predominant protein of animal bodies, constituting approximately 30% of the total protein content (Birk and Bruckner, 2005). Gelatin is a soluble protein obtained by partial hydrolysis of collagen, the main fibrous protein constituent in bones, cartilages, and skins (Johnston-Banks, 1990). These proteins are widely used in food, pharmaceutical, cosmetic, biomedical, and leather industries (Hassan and Sherief, 1994; Zhang *et al.*, 2006). Collagen and gelatin could be isolated from fish skin, bone, scales, and fins, which are fish processing by-products. Though there are remarkable variations with mammalian collagen regarding amino acid composition, thermal stability, and functional properties, the outbreak of several contagious diseases like bovine spongiform encephalopathy (BSE) and foot-and-mouth disease (FMD) made fish a promising alternative source (Gomez-Guillen *et al.*, 2002).Till 2002, production of gelatin from fish was very less but by the end of 2007 increased exponentially and accounted for 1.5% of total gelatin production, indicating the increasing demand for gelatin production from alternative non-mammalian species (Gómez-Guillén *et al.*, 2009). The extraction procedure includes alkali pretreatment followed by acid or enzymatic digestion (Ahmad and Benjakul, 2010; Kittiphattanabawon *et al.*, 2010). The procedure is quite effective which may yield 5–40% of the wet weight. *Lepidorhombus boscii* (Montero and Gomez-Guillen, 2000), *Solea vulgaris* (Gimenez *et al.*, 2005), *Thunnus albacares* (Cho *et al.*, 2005), *Lagocephalus gloveri* (Senaratne *et al.*, 2006), *Raja kenojei* (Cho *et al.*, 2006), *Trachurus trachurus* (Badii and Howell, 2006), *Lutjanus vitta*, and *Priacanthus macracanthus* (Jongjareonrak *et al.*, 2006a,b) are some of the good sources for extracting this extracellular matrix protein, which is being used for development of scaffolds either alone or in combination with other biomaterials for skin replacement, artificial blood vessel, bone graft, ligament replacement, etc. (Lee *et al.*, 2001). The pharmaceutical applications include production of wound dressings, vitreous implants, and carriers for drug delivery (Friess, 1998).

D. Protein hydrolysate and bioactive peptides

Hydrolysis of food proteins is widely employed for value addition through improvement of nutritional characteristics, retarding deterioration, improvement of functional properties, and removal of toxic or inhibitory ingredients (Clemente, 2000). Protein hydrolysate is produced by the use of enzymes to break down proteins into peptides, peptones, and amino acids (Vidotti *et al.*, 2003). Enzymatic digestion is the most reliable method to obtain hydrolysates with improved nutritional and functional properties (Stoilov *et al.*, 1995). The resultant product is highly functional, contributing whipping, gelling, and texturing properties when used as an ingredient in food products and had a variety of applications in pharmaceutical, food, and cosmetic industries (De Holanda and Netto, 2006). Moreover, several pharmacological actions have been reported for protein hydrolysates, viz., hypocholesterolemic effects (Korhonen and Pihlanto, 2006), antioxidant effect (Nazeer *et al.*, 2011), angiotensin

I converting enzyme (ACE) inhibitor capacity (Nam *et al.*, 2008), and immune-enhancing activity (Biziulevicius *et al.*, 2006).

A number of commercial proteases have been used for the production of these hydrolysates and peptides, including trypsin, chymotrypsin, pepsin, alcalase, properase E, and papain (Kim *et al.*, 2001; Lin and Li, 2006; Mendis *et al.*, 2005; Yang *et al.*, 2008). Besides commercial proteases, enzyme extracts from fish viscera have been used to obtain bioactive hydrolysates from skin and bones of different fish species (Jung *et al.*, 2005; Phanturat *et al.*, 2010). Protease specificity affects size, amount, free amino acid composition, and peptides and their amino acid sequences, which, in turn, influences the biological activity of the hydrolysates (Chen *et al.*, 1995). Nowadays, with the advanced purification systems like HPLC and FPLC (Fast Protein Liquid Chromatography), it is very easy to purify a specific active peptide from protein hydrolysate and the elucidation of primary structure of peptides by LC–MS/MS (Kumar *et al.*, 2011a; Qian *et al.*, 2008).

Bioactive peptides usually contain 2–20 amino acid residues per molecule (Leppala, 2001), and the lower the molecular weight, the higher their chance to cross the intestinal barrier and exert biological effects (Roberts *et al.*, 1999). Possible bioactivities reported for peptides include antioxidant, antiproliferative, antihypertensive, antiangiogenic, anticoagulant, and antiplatelet and opioid activities as well as immunomodulatory and cholesterol-lowering effects. The activity of these peptides is dependent on their amino acid composition and sequence (Shahidi and Zhong, 2008).

Antioxidants are considered as an "elixer of life," and these antiaging molecules scavenge the free radicals that avoid or neutralize the oxidative stress (Pryor and Ann, 1982; Wiseman and Halliwell, 1996). Peptides with antioxidant activity identified from fish sources have been reported to have molecular weights between 500 and 1500 Da (Jun *et al.*, 2004; Ranathunga *et al.*, 2006; Wu *et al.*, 2003). Alaska pollak (Leu-Pro-His-Ser-Gly-Tyr; 672 Da), conger eel (Leu-Gly-Leu-N-Gly-Asp-Asp-Val-Asn; 928 Da), horse mackerel (Asn-His-Arg-Tyr-Asp-Arg; 856 Da), and croaker (Gly-Asn-Arg-Gly-Phe-Ala-Cys-Arg-His-Ala; 1101 Da) (Je *et al.*, 2005a, b,c; Kumar *et al.*, 2011b; Senaratne *et al.*, 2006) but other researchers have reported <500 Da peptide purified from sardinella (431 and 263 Da) (Bougatef *et al.*, 2010) were successful in scavenging the free radicals in different oxidative systems.

Picot *et al.* (2006) assessed the antiproliferative activity of 18 fish protein hydrolysates *in vitro* and identified significant growth inhibition against human breast cancer cell lines. Cartilage-derived peptides from shark have shown antiangiogenic activity in both *in vitro* and *in vivo* models (Bukowski, 2003). Kim *et al.* (2000) has reported antihypertensive activity of fish peptides inhibiting the action of ACE these peptides exhibited even *in vivo* activities by lowering blood pressure in spontaneously hypertensive rats (Fujita and Yoshikawa, 1999; Je *et al.*, 2005a,b,c). Peptides isolated from fish muscle hydrolysate also proved to have *in vitro* anticoagulant and antiplatelet properties, and this outcome has suggested the capability of peptides to inhibit coagulation factors in the intrinsic pathway (Rajapakse *et al.*, 2005).

E. Fat derivatives from marine fish

Fish fat is considered good for health because it is rich in omega-3 fatty acids, which helps to lower cholesterol, triglycerides, blood pressure, and also prevents blood clots. And based on their fat content, fishes are divided into three groups, that is, low-fat fish (<5%), medium-fat fish (5–10%), and high-fat fish (>10%) (Silva and Chamul, 2000). Common fluctuations are recorded during periods of heavy feeding, the fat content might show a rapid increase, and gradual decrease at starvation periods due to natural or physiological reasons (spawning or migration) (Huss, 1988, 1995).

1. Omega-3 fatty acids

Fish and fish oils contain omega-3 fatty acids; in particular, eicosapentaenoic acid (C20:5; EPA) and docosahexaenoic acid (C22:6; DHA) (Burja and Radianingtyas, 2005), which originate from phytoplankton and seaweed in the food chain (Visentainer *et al.*, 2007). People all over the world have been encouraged to enhance their intake of fatty fish by at least two- to threefold. The goal behind this encouragement is to consume adequate amounts of the long-chain polyunsaturated omega-3 fatty acids, EPA, and DHA for potential health benefits (Shahidi and Miraliakbari, 2004) such as prevention of cardiovascular diseases (CVD) (Brown and Hu, 2001; Illingworth and Ullmann, 1990; Kris-Etherton *et al.*, 2002, 2003) and against specific risk factors for coronary heart disease (CHD) (Bucher *et al.*, 2002; Harris and Schacky, 2004; He *et al.*, 2004), in addition to providing protection against chronic inflammatory diseases such as arthritis, psoriasis, asthma, and inflammatory bowel disease (Simopoulos, 2002), and in improving visual function (Carlson and Werkman, 1996). Based on these findings, many health agencies now recommend an average intake of 100mg/day of omega-3 fish oils (Ervin *et al.*, 2004) and should be raised to 1000mg/week for prevention of chronic disease (Joint WHO/FAO, 2003).

Omega-3 fatty acids are a class of "polyunsaturated" fatty acids that contain two or more double bonds, one of which is located at third carbon position from the methyl terminus. The two most bioactive and extensively studied omega-3 fatty acids are EPA and DHA (Sprecher, 2002). As the body is unable to synthesize in appreciable amounts of omega-3 fatty acids with carbon chains having more than 14 carbons, they are obtained primarily from dietary sources (i.e., they are "essential" fatty acids).

The parent omega-3 fatty acid, α-linolenic acid (ALA), is derived mostly from vegetable seed oils (Kris-Etherton *et al.*, 2000). ALA is converted in the endoplasmic reticulum to long-chain omega-3 fatty acid EPA in the presence of desaturase and elongase enzymes. Elongation and desaturation of EPA to DHA subsequently occur in peroxisomes. However, enzymatic conversion of ALA to longer-chain fatty acids is limited (Hussein *et al.*, 2005), so individuals are likely to rely on dietary consumption to maintain optimal EPA and DHA levels. Many marine fatty fish species are known to be excellent dietary sources of $\sum$PUFA(EPA+DHA), such as *Macrourus berglax* (56.2±0.74), *Mora moro* (63.27±1.16), *Centroscymnus coelolepis* (48.01±0.01), *Centroscyllium fabricii* (56.10±0.28), *Centrophorus squamosus* (54.29±0.20) (Okland *et al.*, 2005), *Pampus punctatissimus* (19.3±0.48) (Zhao *et al.*, 2010), *Xiphias gladius* (56.11±1.1) (Smida *et al.*, 2009), *Pseudosciaena crocea* (28.2±0.39) (Tang *et al.*, 2009), *Clupea harengus pallasi* (43.46±1.52), *Theragra chalcogramma* (47.57±2.06), *Hypomesus pretiosus* (44.49±1.80), *Sebastes pinniger* (37.14±0.77), *Oncorhynchus gorbuscha* (38.89±2.02), *Mallotus villosus* (24.56±2.16), and *Sardinops sagax* (56.81±3.64) (Huynh and Kitts, 2009). Several methods are available for extracting them for research and industrial applications.

2. Extraction and purification

Fish oil is the main source of *n*-3 fatty acids (Shahidi and Wanasundara, 1998). They are extracted from nonconsumable parts or by-products of the fish, with these by-products subsequent to traditional methods of treatment of the fish such as head, skin, falling thread, central bones, viscera, liver, and depending on the fishing season, reproductive elements, such as eggs or milt (Justine *et al.*, 2006). Traditional fish oil production was a continuous wet reduction process, which involves cooking, pressing, and centrifugation (Chantachum *et al.*, 2000; FAO, 1986). But, with the increasing industrial application and utilization of fish oils, advanced recovery procedures that result in high yields and good quality were launched. Various methods are available for obtaining concentrated EPA and DHA from fish oil, such as urea adducts (Wanasundara and Sahidi, 1999), iodolactonization (Gaiday *et al.*, 1991), HPLC (Yamamura and Shimomura, 1997), supercritical fluid extraction (Walker *et al.*, 1999), extraction by silver nitrate (Kubota *et al.*, 1997), fractionation (Hirata *et al.*, 1993), low-temperature solvent extraction (Moffat *et al.*, 1993), solvent and enzymatic extraction (Grima *et al.*, 1994) are in practice. However, specific purification technologies have emerged to remove the unpleasant flavor, because most of fish oil has an unacceptable taste and odor. And purification mainly consists of deacidification (eliminate the free fatty acids), bleaching and oxidation products, deodorization, winterization (for PUFA concentration), and finally, addition of antioxidants to prevent oxidation (Dinamarca *et al.*, 1990).

3. Applications of omega-3 fatty acids

Most of the focus on fish oils has been related to their ability in preventing coronary artery disease. However, fish consumption generally helps in maintaining healthier lifestyle than the rest of the population. A beneficial effect of fish consumption or omega-3 fatty acid intake on heart disorders has been suggested to be related to the overall favorable effects on cardioprotective effects (Bucher *et al.*, 2002; He *et al.*, 2004; Kris-Etherton *et al.*, 2002), lipid profiles (Harris, 1997; Nestel, 1990; Phillipson *et al.*, 1985), threshold for arrhythmias (Kang and Leaf, 1996, 2000), platelet activity (Dyerberg *et al.*, 1978; Leaf and Weber, 1988), inflammation and endothelial function (De Caterina *et al.*, 1998; Pischon *et al.*, 2003), atherosclerosis (Erkkila *et al.*, 2004; Von Schacky *et al.*, 1999), and hypertension (Morris *et al.*, 1993). A systematic study summarized 47 trials and found that fish oil supplementation was effective in reducing triglyceride level by 0.34 mmol/L over an average treatment period of 24 weeks in participants with hypertriglyceridemia (Mozaffarian *et al.*, 2006). High concentration of triglycerides in blood is one of the serious problems observed in coronary heart disorders. This systematic review supported the FDA approval in 2004 for prescribing ≥500mg/dL of omega-3 fatty acids for reducing high triglycerides in adults as an adjunct to diet (Geleijnse *et al.*, 2002). Clinical trials showed positive activity of omega-3 fatty acid in men of all age groups ranging from 45 to 80 years (He *et al.*, 2008) and in postmenopausal women (Erkkila *et al.*, 2004). A meta-analysis conducted by He *et al.* (2004) has concluded that individuals, who eat fish once every week, had considerably lower CVD mortality as compared with those who eat fish once a month or less.

After the positive findings of Lukiw *et al.* (2005) that endogenous DHA-derived NPD1 is a potent regulator of an intrinsic neuroprotective agent that promotes survival in stressed human brain cells. The concept of "if fish oils protect the heart maybe it can even do same for brain" (Mazza *et al.*, 2007) and this attracted many researchers and supplementation of omega-3 fatty acids on some cognitive and physiological parameters in healthy subjects, and found that omega-3 fatty acids supplementation is associated with an improvement of physiological functions, particularly those involving complex cortical processing (Fontani *et al.*, 2005). Actually, the brain and other neural tissues are particularly rich in long-chain PUFAs, which serve as specific precursors, and particularly DHA is concentrated in membrane phospholipids at synapses (Roberts *et al.*, 1999) and is decreased in Alzheimer disease (Guan *et al.*, 1994; Lukiw *et al.*, 2005).

Vlaardingerbroek *et al.* (2006) has experimentally proved that patients with acquired or inborn error can benefit from taking preformed longer-chain versions of the omega-3 fatty acids in their diet. Studer *et al.* (2005) showed that fish oil was successful than the statin drugs in reducing mortality from cardiovascular disorders. Further, daily DHA has been found to reduce stress (Hamazaki *et al.*, 1996), and this observation gives us clear understanding about fish oils and its therapeutic nature against CHD and psychiatric disorders. Omega-3 fatty acids also play a role in fertility, cell division, and growth, suggesting that they may protect against certain types of cancer or may alter the response to cancer treatment (James *et al.*, 2000; Jones *et al.*, 2003). So, omega-3 fatty acids should be supplement with various antioxidant agents such as vitamins E, A, and C, flavonoids, polyphenols, carotenoids, lipoic acid, and coenzyme Q.

F. From minerals

Fish bone consists of 60–70% of inorganic substances and mainly composed of calcium, phosphate, and hydroxyapatite (HAP). Calcium is one of the essential elements for body metabolism (Vaghy, 1994). Generally, calcium is obtained from the regular diet, but in most of the cases, it is severely deficient. To overcome this deficiency, several commercial products, viz., milk powders, cookies, toffees, etc., are fortified with calcium to reach the growing demand (Barba and Russo, 2006). However, fishes are very good sources of calcium and could be absorbed to the body as tested *in vivo* (Larsen *et al.*, 2000). But, further studies are required to identify bioavailability of fish bone calcium and its potential applications. As per the available literature, fish bone material derived from processing of large fish is a useful calcium source, and softening of its structure by hot water and acid treatment is required for commercial usage (Kim and Mendis, 2006).

HAP is one of the inorganic components of the hard tissues such as bones, teeth, etc. (Alfar *et al.*, 2003). HAP has been widely used in dental implants (Soballe and Overgaard, 1996), alveolar bridge augmentation, orthopedics (Zablotsky, 1993), maxillofacial surgery (Cronin *et al.*, 1994), etc. In general, the high-purity HAP is very costly, but using the fish bones of Japanese sea bass (Ozawa *et al.*, 2007) and sier fish (Prabakaran and Rajeswari, 2006) as a natural source by burning gradually at high temperature, that is, at 100, 400, 700, 900, and 1200 °C, the extraction of HAP became cost-effective way in managing the environment waste in coastal areas.

III. CONCLUSION

Therapeutic importance of bioactive compounds obtained from various marine fishes exhibits different chemical and pharmacological effects on specific targets. Generally, marine fish is good source of protein diet that also contains moderate fat, minerals, and vitamins. This review highlights the medicinal properties of protein, fat, carbohydrates, and their derivates isolated from marine fish. Fish protein derivates like collagen, gelatin, and bioactive peptides are proven to be life saving metabolites. However, regular usage of omega-3 fatty acids prevents CHD, chronic inflammatory diseases, etc. Moreover, these nutraceuticals can be isolated from the fish processing by-products, which both environmental and medicinal benefits.

REFERENCES

Agusa, T., Kunito, T., Sudaryanto, A., Monirith, I., Kan Atireklap, S., Iwata, H., Ismail, A., Sanguansin, J., Muchtar, M., Tana, T. S., and Tanabe, S. (2007). Exposure assessment for trace elements from consumption of marine fish in Southeast Asia. *Environ. Pollut.* **145**(3), 766 777.

Ahmad, M. and Benjakul, S. (2010). Extraction and characterisation of pepsin-solubilised collagen from the skin of unicorn leatherjacket (*Aluterus monocerous*). *Food Chem.* **120**, 817–824.

Alfar, A., Ghorbani, M., Ehsani, N., Saeri, M. R., and Sorrell, C. C. (2003). Some important factors in the wet precipitation process of hydroxyapatite. *Mater. Design* **24**, 197–202.

Anthony, J. A., Roby, D. D., and Turco, K. R. (2000). Lipid content and energy density of forage fishes from the Northern Gulf of Alaska. *J. Exp. Mar. Biol. Ecol.* **248**, 53–78.

Badii, F. and Howell, N. K. (2006). Fish gelatin: Structure, gelling properties and interaction with egg albumen proteins. *Food Hydrocolloids* **20**, 630–640.

Barba, G. and Russo, P. (2006). Dairy foods, dietary calcium and obesity: A short review of the evidence. *Nutr. Metab. Cardiovasc. Dis.* **16**, 445–451.

Birk, D. E. and Bruckner, P. (2005). Collagen suprastructures. *Top. Curr. Chem.* **247**, 185–205.

Biziulevicius, G. A., Kislukhina, O. V., Kazlaukaitè, J., and Zukaitè, V. (2006). Food protein enzymatic hydrolysates possess both antimicrobial and immunostimulatory activities: A 'cause and effect' theory of bifunctionality. *FEMS Immunol. Med. Microbiol.* **46**, 131–138.

Boran, G. and Karacam, H. (2011). Seasonal changes in proximate composition of some fish species from the Black Sea. *Turk. J. Fish. Aquat. Sci.* **11**, 01–05.

Borgstrom, G. (1961). Fish as Food, Production, Biochemistry and Microbiology, Vol. 1. Academic Press, Inc., London (725pp).

Bougatef, A., Arroume, N. N., Manni, L., Ravallec, R., Barkia, A., Guillochon, D., and Nasri, M. (2010). Purification and identification of novel antioxidant peptides from enzymatic hydrolysates of sardinelle (*Sardinella aurita*) by-products proteins. *Food Chem.* **118**, 559–565.

Brown, A. A. and Hu, F. B. (2001). Dietary modulation of endothelial function: Implication for cardiovascular disease. *Am. J. Clin. Nutr.* **73**, 673–686.

Bucher, H. C., Hengstler, P., Schindler, C., and Meier, G. (2002). N-3 polyunsaturated fatty acids in coronary heart disease: A meta-analysis of randomized controlled trials. *Am. J. Med.* **112**, 298–304.

Bukowski, R. M. (2003). AE-941, a multifunctional antiangiogenic compound: Trials in renal cell carcinoma. *Expert Opin. Investig. Drugs* **12**(8), 1403–1411.

Bull, M. H., Huse, W. M., and Bull, B. S. (1975). Evaluation of tests used to monitor heparin therapy during extracorporeal circulation. *Anesthesiology* **43**, 346–353.

Burja, A. M. and Radianingtyas, H. (2005). Marine microbial-derived nutraceuticals biotechnology: An update. *Food Sci. Technol.* **19**, 14–16.

Carlson, S. E. and Werkman, S. H. (1996). A randomized trial of visual attention of preterm infant fed docosahexaenoic acid until two months. *Lipids* **31**, 85–90.

Casu, B., Vlodavsky, I., and Sanderson, R. D. (2008). Non-anticoagulant heparins and inhibition of cancer. *Pathophysiol. Haemost. Thromb.* **36**, 195–203.

Chantachum, S., Benjakul, S., and Sriwirat, N. (2000). Separation and quality of fish oil from precooked and non-precooked tuna heads. *Food Chem.* **69**(3), 289–294.

Chen, H. M., Muramoto, K., and Yamauchi, F. (1995). Structural analysis of antioxidative peptides from soybean b-conglycinin. *J. Agric. Food Chem.* **43**, 574–578.

Cho, S. M., Gu, Y. S., and Kim, S. B. (2005). Extraction optimization and physical properties of yellowfin tuna (*Thunnus albacares*) skin gelatin compared to mammalian gelatins. *Food Hydrocolloids* **19**, 221–229.

Cho, S. H., Jahncke, M. L., Chin, K. B., and Eun, J. B. (2006). The effect of processing conditions on the properties of gelatin from skate (*Raja kenojei*) skins. *Food Hydrocolloids* **20**, 810–816.

Choudhury, G. S. and Bublitz, C. G. (1996). Computer-based controls in fish processing industry. *In* "Computerized Control Systems in the Food Industry", (G. S. Mittal, Ed.), pp. 513–538. Marcel Dekker Inc., New York.

Choudhury, G. S. and Gogoi, B. K. (1995). Extrusion processing of fish muscle. *J. Aquat. Food Prod. Technol.* **4**, 37–67.

Clemente, A. (2000). Enzymatic protein hydrolysates in human nutrition. *Trends Food Sci. Technol.* **11**, 254–262.

Cronin, R. J., Oesterle, L. J., and Ranly, D. M. (1994). Mandibular implants and the growing patient. *Int. J. Oral Maxillofac. Implants* **9**, 55–62.

De Caterina, R., Bernini, W., Carluccio, M. A., Liao, J. K., and Libby, P. (1998). Structural requirements for inhibition of cytokine-induced endothelial activation by unsaturated fatty acids. *J. Lipid Res.* **39**, 1062–1070.

De Holanda, H. D. and Netto, F. M. (2006). Recovery of components from shrimp (*Xiphopenaeus kroyeri*) processing waste by enzymatic hydrolysis. *J. Food Sci.* **71**(5), 298–303.

Defelice, S. L. (1995). The nutritional revolution: Its impact on food industry RandD. *Trends Food Sci. Technol.* **6**, 59–61.

Dellias, J. M. M., Onofre, G. R., Werneck, C. C., Landeira-Fernandez, A. N., Melo, F. R., Farias, W. R. L., and Silva, L. C. F. (2004). Structural composition and differential anticoagulant activities of dermatan sulfates from the skin of four species of rays, *Dasyatis americana, Dasyatis guttata, Aetobattus narinari* and *Potamotrygon motoro. Biochimie* **86**, 677–683.

Dinamarca, E., Garrido, F., and Valenzuela, A. (1990). Simple high vacuum distillation equipment for deodorizing fish oil for human consumption. *Lipids* **25**(3), 170–171.

Donia, M. and Hamann, M. T. (2003). Marine natural products and their potential applications as antiinfective agents. *Lancet* **3**, 338–348.

Dyerberg, J., Bang, H. O., Stoffersen, E., Moncada, S., and Vane, J. (1978). Eicosapentaenoic acid and prevention of thrombosis and atherosclerosis. *Lancet* **2**, 117–119.

Erkkila, A. T., Lichtenstein, A. H., Mozaffarian, D., and Herrington, D. M. (2004). Fish intake is associated with a reduced progression of coronary artery atherosclerosis in postmenopausal women with coronary artery disease. *Am. J. Clin. Nutr.* **80**, 626–632.

Ervin, R. B., Wright, J. D., Wang, C. Y., and Kennedy, S. J. (2004). Dietary intake of fats and fatty acids for the United States population: 1999–2000. *Adv. Data* **8**, 1–6.

FAO (1986). Trends in capture fisheries production. *In* "Capture Production 2004". Yearbook of Fishery StatisticsVol. 98/1. FAO, Rome(560p).

Faulkner, D. J. (2000). Marine pharmacology. *Antonie Van Leeuwenhoek* **77**, 135–145.

Flengsrud, R., Larsen, M. L., and Odegaard, O. R. (2010). Purification, characterization and in vivo studies of salmon heparin. *Thromb. Res.* **126**, 409–417.

Folmer, F., Jaspars, M., Dicato, M., and Diederich, M. (2008). Marine natural products as targeted modulators of the transcription factor NF-κB. *Biochem. Pharmacol.* **75**, 603–617.

Fontani, G., Corradeschi, F., Felici, A., Alfatti, F., Migliorini, S., and Lodi, L. (2005). Cognitive and physiological effects of omega-3 polyunsaturated fatty acid supplementation in healthy subjects. *Eur. J. Clin. Invest.* **35**(11), 691–699.

Friess, W. (1998). Collagen—Biomaterial for drug delivery. *Eur. J. Pharm. Biopharm.* **45**, 113–136.

Fuesetani, N. (2000). *In* "Drugs from the Sea", (M. Fuesetani, Ed.). Karger, Basel (Chapter 1, pp. 1–5).

Fujita, H. and Yoshikawa, M. (1999). LKPNM: A prodrug-type ACE inhibitory peptide derived from fish protein. *Int. J. Immunopharmacol.* **44**, 123–127.

Gaiday, N. V., Imbs, A. B., Kuklev, D. V., and Latyshev, N. A. (1991). Separation of natural polyunsaturated fatty acids by means of iodolactonization. *J. Am. Oil Chem. Soc.* **68**, 230–233.

Geleijnse, J. M., Giltay, E. J., Grobbee, D. E., Donders, A. R., and Kok, F. J. (2002). Blood pressure response to fish oil supplementation: Metaregression analysis of randomized trials. *J. Hypertens.* **20**, 1493–1499.

Gimenez, B., Gomez-Guillen, M. C., and Montero, P. (2005). The role of salt washing of fish skins in chemical and rheological properties of gelatin extracted. *Food Hydrocolloids* **19**, 951–957.

Gomez-Guillen, M. C., Turnay, J., Fernandez-Diaz, M. D., Ulmo, N., Lizarbe, M. A., and Montero, P. (2002). Structural and physical properties of gelatin extracted from different marine species: A comparative study. *Food Hydrocolloids* **16**, 25–34.

Gómez-Guillén, M. C., Pérez-Mateos, M., Gómez-Estaca, J., López-Caballero, E., Giménez, B., and Montero, P. (2009). Fish gelatin: A renewable material for the development of active biodegradable films. *Trends Food Sci. Technol.* **20**, 3–16.

Grant, P. T. and Mackie, A. M. (1977). Drugs from the sea—Facts and fantasy. *Nature* **267**, 786–788.

Grima, E. M., Medina, A. R., Gimenez, A. G., Gutierrez, J. A. S., and Camacho, F. G. (1994). Comparison between extraction of lipids and fatty acids from microalgal biomass. *J. Am. Oil Chem. Soc.* **73**, 955–959.

Guan, Z. Z., Soderberg, M., Sindelar, P., and Edlund, C. (1994). Content and fatty acid composition of cardiolipin in the brain of patients with Alzheimer's disease. *Neurochem. Int.* **25**(3), 295–300.

Haefner, B. (2003). Drugs from the deep. *Drug Discov. Today* **8**, 536–544.

Halvey, S. (1990). *In* "Microbiology: Applications in Food Biotechnology", (B. H. Nga and Y. K. Lu, Eds), pp. 123–134. Elsevier Applied Science Press, New York.

Hamazaki, T., Sawazaki, S., and Itomura, M. (1996). The effect of docosahexaenoic acid on aggression in young adults. A placebo-controlled double-blind study. *J. Clin. Invest.* **97**, 1129–1133.

Harris, W. S. (1997). n-3 Fatty acids and serum lipoproteins: Human studies. *Am. J. Clin. Nutr.* **65**, 1645S–1654S.

Harris, W. S. and Schacky, C. V. (2004). The omega-3 index: A new risk factor for death from coronary heart disease? *Prev. Med.* **39**, 21–220.

Harvey, A. (2000). Strategies for discovering drugs from previously unexplored natural products. *Drug Discov. Today* **5**, 294–300.

Hassan, F. and Sherief, P. M. (1994). Role and application of fish collagen. *Seafood Export J.* **25**, 19–24.

He, K., Song, Y., Daviglus, M. L., Liu, K., Van Horn, L., Dyer, A. R., and Greenland, P. (2004). Accumulative evidence on fish consumption and coronary heart disease mortality: A meta-analysis of cohort studies. *Circulation* **109**, 2705–2711.

He, K., Liu, K., Daviglus, M. L., Elisabeth, M. D., Nancy, S. J., Rui, J., Pamela, O., Lyn, M. S., David, S., Colin, Wu., Graham, B., Michael, T., *et al.* (2008). Intakes of long-chain n-3 polyunsaturated fatty acids and fish in relation to measurements of subclinical atherosclerosis. *Am. J. Clin. Nutr.* **88**, 1111–1118.

Hirata, H., Saeki, H., Nonaka, M., Kawasaki, K., Ooizumi, T., and Motoe, K. (1993). Recovery of fish oil from the manufacturing process of highly nutritional fish meat for foodstuffs from sardine. *Nippon Suisan Gakkaishi* **59**, 111–116.

Hirsh, J., Anand, S. S., Halperin, J. L., and Fuster, V. (2001). American Heart Association. Guide to anticoagulant therapy. Heparin: A statement or healthcare professionals from the American Heart Association. *Circulation* **103**, 2994–3018.

Holick, M. F., Judkiewicz, A., Walworth, N., and Wang, M. H. (1985). Recovery of heparin from fish wastes. *In* "Biotechnology of Marine Polysaccharides", (R. R. Colwell, E. R. Pariser, and A. J. Sinskay, Eds), pp. 389–397. Hemisphere Publishing Corporation, New York.

Huss, H. H. (1988). Fresh Fish: Quality and Quality Changes. FAO, Rome, Italy (132pp).

Huss, H. H. (1995). Quality and Quality Changes in Fresh Fish. FAO, Rome (348pp).

Hussein, N., Ah-Sing, E., Wilkinson, P., Leach, C., Griffin, B. A., and Millward, D. J. (2005). Long-chain conversion of [13C] linoleic acid and alpha-linolenic acid in response to marked changes in their dietary intake in men. *J. Lipid Res.* **46**, 269–280.

Huynh, M. D. and David D. Kitts. (2009). Evaluating nutritional quality of pacific fish species from fatty acid signatures. *Food Chem.* **114**, 912–918.

Illingworth, R. D. and Ullmann, D. (1990). Effects of omega-3 fatty acids on risk factor for cardiovascular disease. *In* "Omega-3 Fatty Acids in Health and Disease", (R. S. Lees and M. Karel, Eds), pp. 39–70. Marcel Dekker, New York.

James, M., Gibson, R., and Cleland, L. (2000). Dietary polyunsaturated fatty acids and inflammatory mediator production. *Am. J. Clin. Nutr.* **71**(1), 343–348.

Je, J. Y., Park, P. J., and Kim, S. K. (2005a). Antioxidant activity of peptide isolated from Alaska Pollack (*Theragra chalcogramma*) frame protein hydrolysate. *Food Res. Int.* **38**, 45–50.

Je, J. Y., Park, P. J., Jung, W. K., and Kim, S. K. (2005b). Isolation of angiotensin I converting enzyme (ACE) inhibitor from fermented oyster sauce, *Crassostrea gigas*. *Food Chem.* **90**, 809–814.

Je, J. Y., Park, P. J., Kwon, J. Y., and Kim, S. K. (2005c). A novel angiotensin I converting enzyme inhibitory peptide from Alaska Pollack (*Theragra chalcogramma*) frame protein hydrolysate. *J. Agric. Food Chem.* **52**, 7842–7845.

Jha, R. K. and Zi-rong, X. (2004). Biomedical compounds from marine organisms. *Mar. Drugs* **2**, 123–146.

Johnston-Banks, F. A. (1990). Gelatin. *In* "Food Gels", (P. Harris, Ed.), pp. 233–289. Elsevier Applied Science Publishers, London.

Joint WHO/FAO Expert consultation on Diet, nutrition and the prevention of chronic diseases (2003). *WHO technical report series, 916.* the world health organization, Geneva (Switzerland).

Jones, P., Papamandjaris, A., Jones, P., and Papamandjaris, A. (2003). Lipids and cellular metabolism. Present Knowledge in Nutrition, 8th edn. International Life Sciences Institute, Washington, DC (Vol. Chapter 10).

Jongjareonrak, A., Benjakul, S., Visessanguan, W., and Tanaka, M. (2006a). Effects of plasticizers on the properties of edible films from skin gelatin of bigeye snapper and brownstripe red snapper. *Eur. Food Res. Technol.* **222**, 229–230.

Jongjareonrak, A., Benjakul, S., Visessanguan, W., Prodpran, T., and Tanaka, M. (2006b). Characterization of edible films from skin gelatin of brownstripe red snapper and bigeye snapper. *Food Hydrocolloids* **20**, 492–501.

Jun, S. Y., Park, P. J., Jung, W. K., and Kim, S. K. (2004). Purification and characterization of an antioxidative peptide from enzymatic hydrolysate of yellowfin sole (*Limanda aspera*) frame protein. *Eur. Food Res. Technol.* **219**, 20–26.

Jung, W. K., Park, P. J., Byun, H. G., Moon, S. H., and Kim, S. K. (2005). Preparation of hoki (*Johnius belengerii*) bone oligophospho peptide with a high affinity to calcium by carnivorous intestine crude proteinase. *Food Chem.* **91**, 333–340.

Justine, D., Claire, D. M., Gilles, B., Pascal, J., and Jean-Pascal, B. (2006). Improvement of lipid and phospholipid recoveries from sardine (*Sardina pilchardus*) viscera using industrial proteases. *Process Biochem.* **41**(11), 2327–2332.

Kang, J. X. and Leaf, A. (1996). Antiarrhythmic effects of polyunsaturated fatty acids. Recent studies. *Circulation* **94**, 1774–1780.

Kang, J. X. and Leaf, A. (2000). Prevention of fatal cardiac arrhythmias by polyunsaturated fatty acids. *Am. J. Clin. Nutr.* **71**, 202S–207S.

Kao, C. Y. and Levinson, S. R. (1986). Tetrodotoxin, Saxitoxin and the Molecular Biology of the Sodium Channel. New York Academy of Sciences, New York (Vol. 497, Chapter 1, pp. 1–13).

Kim, S. K. and Mendis, E. (2006). Bioactive compounds from marine processing byproducts—A review. *Food Res. Int.* **39**, 383–393.

Kim, S. K., Choi, Y. R., Park, P. J., Choi, J. H., and Moon, S. H. (2000). Screening of biofunctional peptides from cod processing wastes. *Korean Soc. Agric. Chem. Biotechnol.* **43**, 225–227.

Kim, S. K., Kim, Y. T., Byun, H. G., Park, P. J., and Ito, H. (2001). Purification and characterization of antioxidative peptides from bovine skin. *J. Biochem. Mol. Biol.* **34**, 214–219.

Kinsella, J. E. (1987). Sea Foods and Fish Oils in Human Health and Disease. Marcel Dekker, Inc., New York (pp. 231–236).

Kittiphattanabawon, P., Benjakul, S., Visessanguan, W., Nagai, T., and Tanaka, M. (2005). Characterization of acid-soluble collagen from skin and bone of bigeye snapper (*Priacanthus tayenus*). *Food Chem.* **89**, 363–372.

Kittiphattanabawon, P., Benjakul, S., Visessanguan, W., Kishimura, H., and Shahidi, F. (2010). Isolation and characterisation of collagen from the skin of brownbanded bamboo shark (*Chiloscyllium punctatum*). *Food Chem.* **119**, 1519–1526.

Korhonen, H. and Pihlanto, A. (2006). Bioactive peptides: Production and functionality. *Int. Dairy J.* **16**, 945–960.

Kris-Etherton, P. M., Taylor, D. S., Yu-Poth, S., Huth, P., Moriarty, K., Fishell, V., Hargrove, R. L., Zhao, G., and Etherton, T. D. (2000). Polyunsaturated fatty acids in the food chain in the United States. *Am. J. Clin. Nutr.* **71**, 179S–188S.

Kris-Etherton, P. M., Harris, W. S., and Appel, L. J. (2002). Fish consumption, fish oil, omega-3 fatty acids, and cardiovascular disease. *Circulation* **106**, 2747–2757.

Kris-Etherton, P. M., Harris, W. S., and Appel, L. J. (2003). Omega-3 fatty acids and cardiovascular disease: New recommendations from the American Heart Association. *Arterioscler. Thromb. Vasc. Biol.* **23**, 151–152.

Kubota, F., Goto, M., and Nakashio, F. (1997). Separation of polyunsaturated fatty acids with silver nitrate using hollow fiber membrane extractor. *Sep. Sci. Technol.* **32**, 1529–1541.

Kumar, N. S. S., Nazeer, R. A., and Ganesh, R. J. (2011a). Purification and biochemical characterization of antioxidant peptide from horse mackerel (*Magalaspis cordyla*) viscera protein. *Peptides* **32**(7), 1496–1501.

Kumar, N. S. S., Nazeer, R. A., and Ganesh, R. J. (2011b). Purification and identification of antioxidant peptide from the skin protein hydrolysate of two marine fishes horse mackerel (*Magalaspis cordyla*) and croaker (*Otolithes ruber*). *Amino Acids*. doi: 10.1007/s00726-011-0858-6.

Larsen, T., Thilsted, S. H., Konsbak, K., and Hansen, M. (2000). Whole small fish as a rich calcium source. *Br. J. Nutr.* **83**, 191–196.

Leaf, A. and Weber, P. C. (1988). Cardiovascular effects of n-3 fatty acids. *N. Engl. J. Med.* **318**, 549–557.

Lee, C. H., Singla, A., and Lee, Y. (2001). Biomedical applications of collagen. *Int. J. Pharm.* **221**, 1–22.

Leppala, A. P. (2001). Bioactive peptides derived from whey proteins: Opioid and ACE-inhibitory peptides. *Trends Food Sci. Technol.* **11**, 347–356.

Lin, L. and Li, B. (2006). Radical scavenging properties of protein hydrolysates from Jumbo flying squid (*Dosidicus eschrichitii* Steenstrup) skin gelatin. *J. Sci. Food Agric.* **86**(14), 2290–2295.

Love, R. M. (1970). The Chemical Biology of Fishes. Academic Press, Inc., London, UK (547pp).

Lukiw, W. J., Cui, J. G., Marcheselli, V. L., Bodker, M., Botkjaer, A., and Gotlinger, K. (2005). A role for docosahexaenoic acid-derived neuroprotectin D1 in neural cell survival and Alzheimer disease. *J. Clin. Invest.* **115**(10), 2774–2783.

Mazza, M., Pomponi, M., Janiri, L., Bria, P., and Mazza, S. (2007). Omega-3 fatty acids and antioxidants in neurological and psychiatric diseases: An overview. *Prog. Neuropsychopharmacol. Biol. Psychiatry* **31**, 12–26.

McCarthy, P. J. and Pomponi, S. A. (2004). A search for new pharmaceutical drugs from marine organisms. *Marine Biomed. Res.* 1–2.

Mendis, E., Rajapakse, N., and Kim, S. K. (2005). Antioxidant properties of a radicals scavenging peptide purified from enzymatically prepared fish skin gelatin hydrolysate. *J. Agric. Food Chem.* **53**, 581–587.

Moffat, C. F., McGill, A. S., Hardy, R., and Anderson, R. S. (1993). The production of fish oils enriched in polyunsaturated fatty acids-containing triglycerides. *J. Am. Oil Chem. Soc.* **70**, 133–138.

Montero, P. and Gomez-Guillen, M. C. (2000). Extracting conditions for megrim (*Lepidorhombus boscii*) skin collagen affect functional properties of the resulting gelatin. *J. Food Sci.* **65**, 434–438.

Moore, K. S., Wehrli, S., Roder, H., Rogers, M., Forrest, J. N., McCrimmon, D., and Zasloff, M. (1993). Squalamine: An aminosterol antibiotic from the shark. *Proc. Natl. Acad. Sci. USA* **90**, 1354–1358.

Morris, M. C., Sacks, F., and Rosner, B. (1993). Does fish oil lower blood pressure? A meta-analysis of controlled trials. *Circulation* **88**, 523–533.

Mozaffarian, D., Gottdiener, J. S., and Siscovick, D. S. (2006). Intake of tuna or other broiled or baked fish versus fried fish and cardiac structure, function, and hemodynamics. *Am. J. Cardiol.* **97**, 216–222.

Nam, K. A., You, S. G., and Kim, S. M. (2008). Molecular and physical characteristics of squid (*Toradores pacificus*) skin collagens and biological properties of their enzymatic hydrolysates. *J. Food Sci.* **73**(4), 249–255.

Nazeer, R. A., Prabha, K. R. D., Kumar, N. S. S., and Ganesh, R. J. (2011). Isolation of antioxidant peptides from clam (*Meretrix casta*). *J. Food Sci. Technol.* doi: 10.1007/s13197-011-0395-z.

Nestel, P. J. (1990). Effects of N-3 fatty acids on lipid metabolism. *Annu. Rev. Nutr.* **10**, 149–167.

Newman, D. J. and Cragg, G. M. (2004). Marine natural products and related compounds in clinical and advanced preclinical trials. *J. Nat. Prod.* **67**, 1216–1238.

Nurnadia, A. A., Azrina, A., and Amin, I. (2011). Proximate composition and energetic value of selected marine fish and shellfish from the West coast of Peninsular Malaysia. *Int. Food Res. J.* **18**, 137–148.

Okland, H. M. W., Stoknes, I. S., Remme, J. F., Kjerstad, M., and Synnes, M. (2005). Proximate composition, fatty acid and lipid class composition of the muscle from deep-sea teleosts and elasmobranchs. *Comp. Biochem. Physiol. B* **140**, 437–443.

Oliviera, J. S., Junior, O. R. P., Morales, R. A. V., Junior, C. B., Schwartz, C. A., and Freitas, J. S. (2003). Toxicity of Puffer fish—Two species (*Lagocephalus laevigatus*, Linnaeus 1766 and *Sphoeroides spengleri*, Bloch 1785) from the southern Brazilian coast. *J. Venom. Anim. Toxins Incl. Trop. Dis.* **9**, 76–82.

Ozawa, M., Hattori, M., and Satake, K. (2007). Waste management and application of fish bone hydroxyapatite for waste water treatment. Proceedings of International Symposium on Eco Topia Science, pp. 1–2.

Payne, S. A., Johnson, B. A., and Otto, R. S. (1999). Proximate composition of some northeastern Pacific forage fish species. *Fish Oceanogr.* **8**(3), 159–177.

Phanturat, P., Benjakul, S., Visessanguan, W., and Roytrakul, S. (2010). Use of pyloric caeca extract from bigeye snapper (*Priacanthus macracanthus*) for the production of gelatin hydrolysate with antioxidative activity. *LWT- Food Sci. Technol.* **43**(1), 86–97.

Phillipson, B. E., Rothrock, D. W., Connor, W. E., Harris, W. S., and Illingworth, D. R. (1985). Reduction of plasma lipids, lipoproteins, and apoproteins by dietary fish oils in patients with hypertriglyceridemia. *N. Engl. J. Med.* **312**, 1210–1216.

Picot, L., Bordenave, S., Didelot, S., Fruitier-Arnaudin, I., Sannier, F., Thorkelsson, G., Berge, J. P., Guerard, F., Chabeaud, A., and Piot, J. M. (2006). Antiproliferative activity of fish protein hydrolysates on human breast cancer cell lines. *Process Biochem.* **41**, 1217–1222.

Pischon, T., Hankinson, S. E., Hotamisligil, G. S., Rifai, N., Willett, W. C., and Rimm, E. B. (2003). Habitual dietary intake of n-3 and n-6 fatty acids in relation to inflammatory markers among US men and women. *Circulation* **108**, 155–160.

Prabakaran, K. and Rajeswari, S. (2006). Development of hydroxyapatite from natural fish bone through heat treatment. *Trends Biomater. Artif. Organs* **20**(1), 20–23.

Pryor, W. A. and Ann, N. Y. (1982). Free radical biology: Xenobiotics, cancer, and aging. *Acad. Sci.* **393**, 1–22.

Qian, Z. J., Jung, W. K., and Kim, S. K. (2008). Free radical scavenging activity of a novel antioxidative peptide purified from hydrolysate of bullfrog skin *Rana catesbeiana* Shaw. *Bioresour. Technol.* **99**, 1690–1698.

Rajapakse, N., Jung, W. K., Mendis, E., Moon, S. H., and Kim, S. K. (2005). A novel anticoagulant purified from fish protein hydrolysate inhibits factor XIIa and platelet aggregation. *Life Sci.* **76**, 2607–2619.

Ranathunga, S., Rajapakse, N., and Kim, S. K. (2006). Purification and characterization of antioxidative peptide derived from muscle of conger eel (*Conger myriaster*). *Eur. Food Res. Technol.* **222**, 310–315.

Rek, A., Krenn, E., and Kungl, A. J. (2009). Therapeutically targeting protein–glycan interactions. *Br. J. Pharmacol.* **157**, 686–694.

Roberts, P. R., Burney, J. D., Black, K. W., and Zaloga, G. P. (1999). Effect of chain length on absorption of biologically active peptides from the gastrointestinal tract. *Digestion* **60**, 332–337.

Ryckelynck, J. P., Chouraqui, D., and Lobbedez, T. (1998). Anticoagulation of the extracorporeal circuit in chronic hemodialysis. *Nephrologie* **19**, 223–225.

Senaratne, L. S., Park, P. J., and Kim, S. K. (2006). Isolation and characterization of collagen from brown backed toadfish (*Lagocephalus gloveri*) skin. *Bioresour. Technol.* **97**(2), 191–197.

Shahidi, F. and Miraliakbari, H. (2004). Omega-3 (n-3) fatty acids in health and diseases: Part 1—Cardiovascular disease and cancer. *J. Med. Food* **7**(4), 387–401.

Shahidi, F. and Wanasundara, U. N. (1998). Omega-3 fatty acid concentrates: Nutritional aspects and production technologies. *Food Sci. Technol.* **9**, 230–240.

Shahidi, F. and Zhong, Y. (2008). Bioactive peptides. *J. AOAC Int.* **91**(4), 914–931.

Silva, J. J. and Chamul, R. S. (2000). Composition of marine and freshwater finfish and shellfish species and their products. *In* "Marine and Freshwater Products Handbook", (R. E. Martin, E. Paine Carter, E. J. Flick, and L. M. Davis, Eds), pp. 31–46. Technomic Publishing Company, Inc., USA.

Simopoulos, A. P. (2002). Omega-3 fatty acids in inflammation and autoimmune diseases. *J. Am. Coll. Nutr.* **21**, 495–505.

Smida, M. A. B., Marzouk, B., and El Cafsi, M. (2009). The composition of fatty acids in the tissues of Tunisian swordfish (*Xiphias gladius*). *Food Chem.* **115**, 522–528.

Soballe, K. and Overgaard, S. (1996). The current status of hydroxyapatite coating of prosthesis. *J. Bone Joint Surg.* **78B**, 689–691.

Sprecher, H. (2002). The roles of anabolic and catabolic reactions in the synthesis and recycling of polyunsaturated fatty acids. *Prostaglandins Leukot. Essent. Fatty Acids* **67**, 79–83.

Stoilov, I. L., Georgiev, T. D., Taskov, M. V., and Koleva, I. D. (1995). Inventors. Oral preparation for patients with chronic renal insufficiency and other protein metabolic diseases. (WO patent 95/2952).

Studer, M., Briel, M., Leimenstoll, B., Glass, T. R., and Bucher, H. C. (2005). Effect of different antilipidemic agents and diets on mortality: A systematic review. *Arch. Intern. Med.* **165** (7), 725–730.

Tang, H. G., Chen, L. H., Xiao, C. G., and Wu, T. X. (2009). Fatty acid profiles of muscle from large yellow croaker (*Pseudosciaena crocea* R.) of different age. *J. Zhejiang Univ. Sci. B* **10**(2), 154–158.

Tingbo, M. G., Kolset, S. O., Ofstad, R., Enersen, G., and Hannesson, K. O. (2005). Sulfated glycosaminoglycans in the extracellular matrix of muscle tissue in Atlantic cod (*Gadus morhua*) and Spotted wolffish (*Anarhichas minor*). *Comp. Biochem. Physiol. B* **140**, 349–357.

Vaghy, P. L. (1994). Calcium antagonists. *In* "Human Pharmacology", (T. M. Brody, J. Larner, K. P. Minneman, and H. C. Neu, Eds), pp. 203–215. C.V. Mosby, New York.

Vidotti, R. M., Macedo-Viegas, E. M., and Carneiro, D. J. (2003). Amino acid composition of processed fish silage using different raw materials. *Anim. Feed Sci. Technol.* **105**, 199–2004.

Visentainer, J. V., Noffs, M. D., Carvalho, P. O., Almeida, V. V., Oliveira, C. C., and Souza, N. E. (2007). Lipid content and fatty acid composition of 15 marine fish species from the southeast coast of Brazil. *J. Am. Oil Chem. Soc.* **84**, 543–547.

Vlaardingerbroek, H., Hornstra, G., De Koning, T. J., Smeitink, J. A., Bakker, H. D., De Klerk, H. B., and Rubio-Gozalbo, M. E. (2006). Essential polyunsaturated fatty acids in plasma and erythrocytes of children with inborn errors of amino acid metabolism. *Mol. Genet. Metab.* **88**(2), 159–165.

Von Schacky, C., Angerer, P., Kothny, W., Theisen, K., and Mudra, H. (1999). The effect of dietary omega-3 fatty acids on coronary atherosclerosis. A randomized, double blind, placebo-controlled trial. *Ann. Intern. Med.* **130**, 554–562.

Walker, T. H., Cochran, H. D., and Hulbert, G. J. (1999). Supercritical carbon dioxide extraction of lipids from *Pythium irregular*. *J. Am. Oil Chem. Soc.* **76**, 595–602.

Walter P. Jeske, Meredith K. McDonald, Debra A. Hoppensteadt, Elaine C. Bau, Aline Mendes, Carl P. Dietrich, Jeanine M. Walenga, and Erwin Coyne (2007). Isolation and characterization of heparin from tuna skins. *Clin. Appl. Thromb. Hemost.* **13**(2), 137–145.

Wanasundara, U. N. and Sahidi, F. (1999). Concentration of omega-3 polyunsaturated fatty acids of sea blubber oil by urea complexation: Optimization of reaction conditions. *Food Chem.* **64**, 41–49.

Wiseman, H. and Halliwell, B. (1996). Damage to DNA by reactive oxygen and nitrogen species: Role in inflammatory disease and progression to cancer. *Biochem. J.* **313**, 17–29.

Wu, H. C., Chen, H. M., and Shiau, C. Y. (2003). Free amino acids and peptides as related to antioxidant properties in protein hydrolysates of markerel (*Scomber austriasicus*). *Food Res. Int.* **36**, 949–957.

Yamamura, R. and Shimomura, Y. (1997). Industrial high-performance liquid chromatography purification of docosahexaenoic acid ethyl ester from single-cell oil. *J. Am. Oil Chem. Soc.* **74**, 1435–1440.

Yang, J. I., Ho, H. Y., Chu, Y. J., and Chow, C. J. (2008). Characteristic and antioxidant activity of retorted gelatin hydrolysates from cobia (*Rachycentron canadum*) skin. *Food Chem.* **110**(1), 128–136.

Zablotsky, M. (1993). Coatings in implant dentistry: Hype, hysteria, or clinical reality? *J. Dent. Symp.* **1**, 70–72.

Zhang, Z. K., Li, G. Y., and Shi, B. (2006). Physicochemical properties of collagen, gelatin and collagen hydrolysate derived from bovine limed split wastes. *J. Soc. Leather Technol. Chem.* **90**, 23–28.

Zhao, F., Zhuang, P., Song, C., Shi, Z. H., and Zhang, L. Z. (2010). Amino acid and fatty acid compositions and nutritional quality of muscle in the pomfret, *Pampus punctatissimus*. *Food Chem.* **118**(2), 224–227.

CHAPTER 19

Beneficial Effect of Teleost Fish Bone Peptide as Calcium Supplements for Bone Mineralization

Se-Kwon Kim[*,†] and **Won-Kyo Jung**[‡,§,1]

Contents

* Department of Chemistry, Pukyong National University, Busan, Republic of Korea
† Marine Bioprocess Research Center, Pukyong National University, Busan, Republic of Korea
‡ Department of Marine Life Science, Chosun University, Gwangju, Republic of Korea
§ Marine Bio Research Center, Chosun University, Gwangju, Republic of Korea
1 Corresponding author: Won-Kyo Jung, *E-mail address:* wkjung@chosun.ac.kr

Advances in Food and Nutrition Research, Volume 65
ISSN 1043-4526, DOI: 10.1016/B978-0-12-416003-3.00019-6

Abstract A most common and trusted source of Ca is milk or other dairy products. However, some oriental people do not drink milk due to lactose indigestion and intolerance, which make them allergic to milk. There have been many studies on alternative calcium-rich diet or Ca supplements. Among them, teleost fish like anchovy and mola, which are commonly consumed in Asian countries, could be an important Ca dietary supplement, especially in population groups with low intakes of milk and dairy products. In this chapter, we summarize beneficial effects of teleost fish bone peptide (FBP) for Ca bioavailability and bone mineralization, based on our researches.

I. INTRODUCTION

Calcium is known to be an essential element required for numerous functions in our bodies including the strengthening of teeth and bones, nerve function, and many enzymatic reactions that require calcium as a cofactor. It is also necessary for muscle contraction and regulation of the permeability of sodium ion across cell membranes including those of nerve cells. The concentration of calcium in the blood plasma remains almost constant and varies only slightly over time for a given individual (Allen, 1982; Anderson and Garner, 1996).

The major source of Ca is the diet, and the most common and trusted source of Ca is milk or other dairy products (Anderson and Garner, 1996). Dairy products contain a high content of casein. Casein phosphopeptides (CPP) derived from the intestinal digestion of casein have been shown to enhance bone calcification in rats (Lee *et al.*, 1980; Tsuchita *et al.*, 1993). Such CPP have the capacity to chelate Ca and to prevent the precipitation of Ca phosphate salts (Berrocal *et al.*, 1989), thereby increasing the amount of soluble Ca availability for absorption across the mucosa (Yuan and Kitts, 1991, 1994). However, some oriental people do not drink milk due to lactose indigestion and intolerance, which make them allergic to milk. Thus, there have been many studies on various Ca supplements as alternatives (e.g., soya protein isolate fructo-oligosaccharide, fish-meal, etc.), which may affect Ca bioavailability (Brouns and Vermeer, 2000; Kumagai *et al.*, 2004; Larsen *et al.*, 2000, 2003). As reported by Larsen *et al.* (2000), the intake of small fish with bones could increase Ca bioavailability in rats, and small fish might be an important Ca dietary supplement, especially in population groups with low intakes of milk and dairy products. However, studies on the utilization of organic components or minerals in fish bone are scarce (Kim *et al.*, 1997; Larsen *et al.*, 2000, 2003). In the previous studies (Jung *et al.*, 2005, 2006), we performed isolation of fish bone phosphopeptide with the high affinity to Ca from hoki (*Johnius belengerii*) skeletons and identification of the beneficial effects of FBP for bone mineralization *in vivo*. In this review, we introduce beneficial effects and molecular mechanism of teleost fish bone peptide (FBP) from hoki (*J. belengerii*) skeletons for Ca bioavailability and bone mineralization.

II. PREPARATION, BIOCHEMICAL PROPERTIES, AND CA-BINDING ACTIVITY OF FISH BONE PEPTIDE

A. Chemical composition of hoki bone

To prepare fish bone peptide with high affinity to Ca, we use hoki (*J. belengerii*) bone frame discarded from fishery manufacturing plant. After air-drying and crushing into a powder, chemical compositions of hoki bone were obtained as shown in Table 19.1.

The organic portion of hoki bone (30.54% on dry basis) was composed of 28.0% protein, 1.94% lipid, and 0.56% carbohydrate. Collagen was present at 86.2% of total protein, and noncollageneous protein content was 13.8% in the bone. As reported by Garner *et al.* (1996), 90% of organic components in the bone matrix were a type I collagen, and the remaining

TABLE 19.1 Chemical composition of hoki bone

Chemical	Content
Protein	28.04[a]
Collageneous	86.21[b]
Noncollageneous	13.79[b]
Lipid	1.94[a]
Carbohydrate	0.56[a]
Mineral	69.46[a]
Ca	59.69[c]
P	35.81[c]
(Molar ratio of Ca/P)	1.67[c]
Na	0.62[c]
Mg	0.34[c]
Cl	0.24[c]
K	0.07[c]
Zn	60[d]
Fe	14[d]
Cu	2[d]

All data were expressed as mean values (n=3).
[a] g/100g of fish bone on dry basis.
[b] g/100g of fish bone protein on dry basis.
[c] g/100g of fish bone mineral on dry basis.
[d] ppm/100g of fish bone mineral on dry basis.

10% consisted of noncollageneous proteins, such as osteocalcin, osteopontin, osteonectin, fibronectin, thrombospondin, proteoglycan I/II, and growth factors (IGF-1, PDGF, and TGF-β). These molecules were produced by osteoblasts and their functions are related to bone formation and cell attachment.

The inorganic portion, minerals (69.46% on dry basis), was mainly composed of 59.7% calcium (Ca) and 35.8% phosphorus (P) with the mole ratio of Ca/P, 1.67. The inorganic portion of vertebrate bone is primarily composed of hydroxyapatite (HA) crystals deposited within an organic matrix of cross-linked collagen fibrils (Anderson and Garner, 1996). The HA crystals make up approximately 60–65% of bone and the HA has an extremely complicated crystalline structure. As reported by Hamada *et al.* (1995), the Ca/P mole ratio in the bone of marine teleosts (15 species) varied within the range 1.63–1.20, and these bones were organized as the combinational structure with HA and beta type $Ca_3(PO_4)_2$. The Ca/P ratio of hoki bone was significantly higher than that of other species (mackerel, sardine, flounder, anchovy, shark, tilefish, croaker, and conger eel) but was similar to mammalian (1.68) and human (1.69) ratios. These results illustrate that hoki bone was organized with abundant Ca and organic materials.

B. Isolation and biochemical properties of FBP isolated from hoki bone

Isolation of fish bone phosphopeptide with a calcium binding activity was accomplished using HA affinity chromatography and gel permeation chromatography (GPC). Phosphate content of the FBP was measured as 23.6%, and characteristic absorptions, derived from phosphate groups in the infrared spectrum (Shimadzu IR-408, Shimadzu Co., Japan) at 1250, 1150, and 1000 cm^{-1}, were assigned to P,O, P–O–C, and P–O-binding to alkyl groups, respectively. Other components such as carbohydrate and lipid were not detected in the FBP. Its amino acid sequence and molecular mass (Val-Leu-Ser-Gly-Gly-Thr-Thr-Met-Tyr-Ala-Ser-Leu-Tyr-Ala-Glu, 1.5 kDa) were elucidated by ESI–QTOF tandem mass spectroscopy. In the result of NCBInr database searching, the amino acid sequence of JFP showed high similarity to those of actin, a major component of the cytoskeleton bound to one molecule of ATP and either Ca or magnesium ions (Matsuura *et al.*, 2000). In the amino acid composition of JFP, Thr, Met, Tyr, and Ser are known as major sites of phosphorylation and combination with metal cation, and detected in the Ca-binding domain of actin and calmodulin with EF-hand motif (Houben *et al.*, 1999). However, Pro and Hyp known as typical collageneous amino acids in the organic portion of bone tissue were not found in the sequence of JFP. Nishmoto *et al.* (2003) isolated and characterized an osteocalcin from carp *Cyprinus carpio*, and carp osteocalcin consists of a high ratio of Tyr, Thr,

Gln, and Asp. The Ca-binding oligophosphopeptide prepared from hen egg yolk phosvitin by Jiang and Mine (2000) mainly consisted of Ser, Thr, and Asp, and they discussed that phosphoseryl groups in the oligophosphopeptide played an essential role in Ca^{2+}–phosphopeptide interaction. The binding of a protein for HA offers insight into the strength of protein–mineral interactions and the role of the protein in physiological calcification (Hauschka and Gallop, 1977).

C. Ca-binding activity of FBP isolated from hoki bone

Ca-binding activity of FBP was similar to that of casein oligophosphopeptide (CPP II). The solubility of Ca was dependent on the concentration of FBP, and 41.1 mg/l of Ca was obtained at a concentration of 250 mg/l. The pH of the reaction system was maintained at 7.8 because low pH could increase the solubility of insoluble calcium salt, and 28.78 mg/l of Ca was obtained at a concentration of 250 mg/l (Fig. 19.1). As reported by Jiang and Mine (2000), the solubility of 36.8 mg/l of Ca was obtained at 300 mg/l of the oligophosphopeptide from egg yolk phosvitin with 35% phosphate retention, and the solubility was higher than that of commercial CPP II.

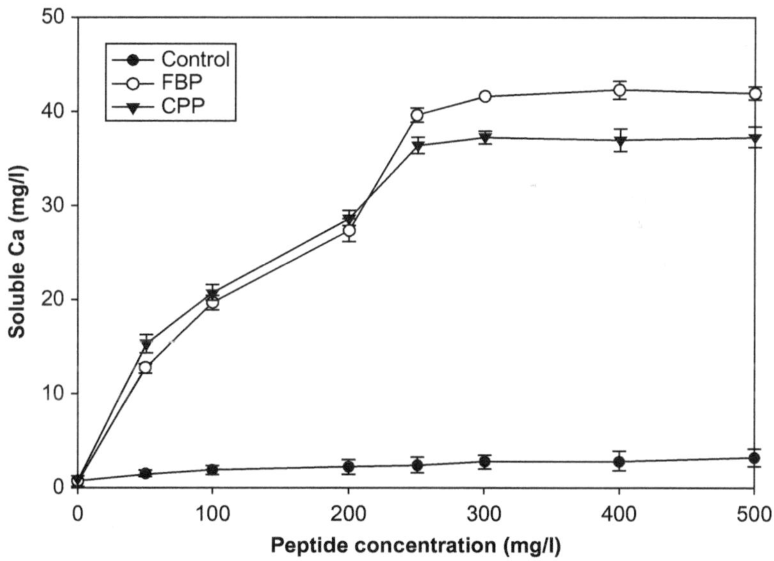

FIGURE 19.1 Calcium binding activity of FBP. Various concentrations, up to 500 mg/l, of FBP were mixed with 5 mM $CaCl_2$ and 20 mM sodium phosphate buffer (pH 7.8). The mixture was stirred at 22 °C for 30 min, and the pH was maintained at 7.8 with a pH meter. After removal of insoluble calcium phosphate salts and filtration with 0.45 m membrane, calcium contents of the supernatant were determined by FAAS. The experiments were performed in triplicate; values are expressed as means±SD.

III. *IN VITRO* AND *IN VIVO* OSTEOGENIC EFFECTS OF FISH BONE PEPTIDE

A. Beneficial effect of FBP on osteoblast-like MG63 cell differentiation

To evaluate beneficial effects of FBP on osteoblastic cell, we examined bone mineralization effects on osteoblast-like MG-63 cell. FBP could increase physiological activity for bone formation and mineralization (Fig. 19.2).

B. Beneficial effect of FBP on Ca bioavailability in ovariectomised rats (osteoporosis-modeling rats)

As illustrated in Tables 19.2 and 19.3, *in vivo* effects of FBP on Ca bioavailability were further studied in the ovariectomised rats. Menopause is a time when estrogen deficiency leads to accelerated bone resorption and

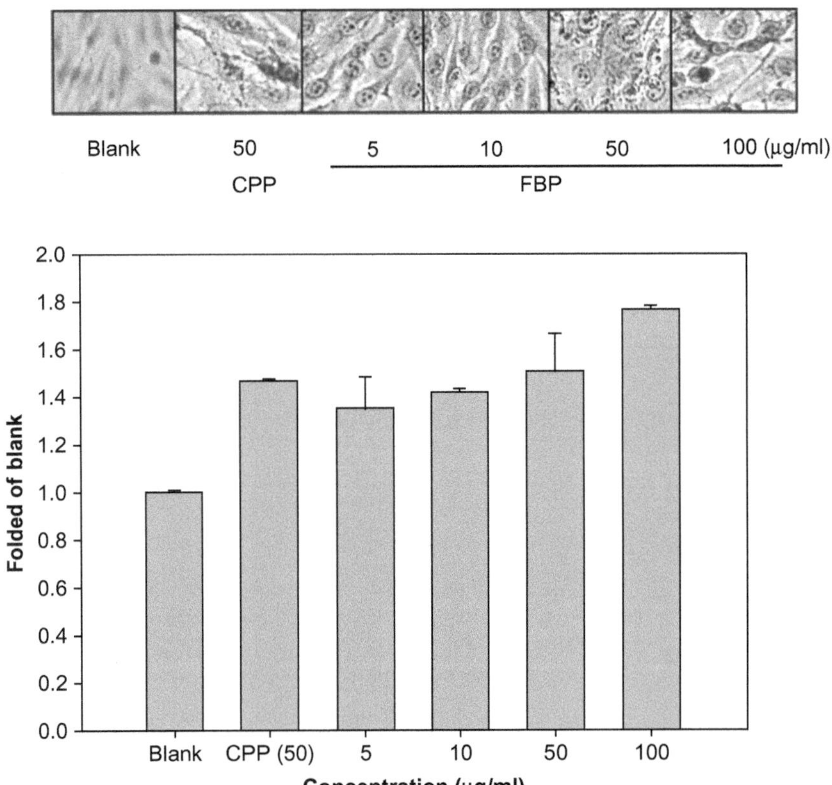

FIGURE 19.2 Bone mineralization effect of FBP on MG-63 cells in dose-dependant manner.

TABLE 19.2 Effects of FBP intake on body weight, Ca intake, and Ca retention in animals

	Control		CPP		FBP	
Experimental groups (n=8 per group)	Mean	SD	Mean	SD	Mean	SD
Body weight gain (g/d)	14.4	1.5	14.2	2.5	13.9	1.9
Food intake (g/d)	12.9	0.9	13.0	1.3	12.5	1.5
Ca intake (mg/d)	54.8	5.9	54.2	4.8	54.4	5.5
Fecal Ca (mg/d)	53.3^{a}	3.4	44.8^{b}	4.2	$46.5^{a,b}$	3.6
Urinary Ca (mg/d)	0.9	0.4	1.4	0.5	1.5	0.5
Ca retention* (mg/d)	0.2^{a}	0.5	8.0^{b}	1.2	6.4^{b}	1.7
Serum Ca (mg/d)	10.67	1.29	11.97	2.09	12.06	1.86

CPP, casein phosphopeptide.
* Calcium retention (balance) was calculated as: Ca intake−fecal Ca−urinary Ca.
[a,b] Mean values within a row with unlike superscript letters were significantly different ($P<0.05$).

TABLE 19.3 Effects of FBP intake on femur in animals

	Control		CPP		FBP	
Experimental groups (n=8 per group)	Mean	SD	Mean	SD	Mean	SD
Femoral total Ca (mg)	143^{a}	7	153^{b}	9	155^{b}	10
Femur length (mm)	33.9	0.5	34.8	0.4	34.3	0.6
Femur wet weight (g)	1.12^{a}	0.06	1.32^{b}	0.05	$1.22^{a,b}$	0.06
Bone mineral density (BMD) of the distal femur (g/cm^2)	0.161^{a}	0.018	0.229^{b}	0.029	0.213^{b}	0.025
Breaking force (kg)	3.96^{a}	0.57	8.97^{b}	1.03	8.48^{b}	0.97

CPP, casein phosphopeptide.
[a,b]Mean values within a row with unlike superscript letters were significantly different ($P<0.05$).

negative bone balance. The present study was undertaken to evaluate the beneficial effects of FBP as a Ca fortifier in osteoporosis induced by ovariectomy and a concurrent low-Ca diet. During the experimental period corresponding to the menopause with osteoporosis disease, the loss of bone mineral (Ca) was decreased by FBP supplementation in the ovariectomised rats. After the low-Ca diet, the FBP diet, including both normal levels of Ca and vitamin D, significantly decreased Ca loss in feces and increased Ca retention as compared with the control. The levels of femoral total Ca, bone mineral density, and breaking strength were also significantly increased by the FBP diet to a level similar to those of the CPP diet group (no difference; $P>0.05$). It illustrates that the increased Ca retention by FBP intake led to the prevention of mineral loss in the osteoporosis-modeling rats. As reported by Larsen *et al.* (2000), the intake of small fish with bones can increase Ca bioavailability, and the small fish may be an important source of Ca, especially in population groups with low intakes of milk and dairy products.

IV. CONCLUSION

A teleost fish can be important Ca diet source for oriental people with lactose indigestion and intolerance. The scientific results illustrate that fish bones contain potent Ca-binding peptide and the peptide could increase Ca absorption through inducing higher Ca solubility in low pH condition and bone mineralization via osteoblastic differentiation related factor activation. Further, in the osteoporosis-modeling rats, it illustrates that the increased Ca retention by FBP intake led to the prevention of mineral loss. Taken together, the intake of teleost fish diet can increase Ca bioavailability, and the fish may be an important source of Ca, especially in population groups with low intakes of milk and dairy products.

ACKNOWLEDGMENT

This research was supported by Basic Science Research Program through the National Research Foundation of Korea (NRF) funded by the Ministry of Education, Science, and Technology (2010–2003111).

REFERENCES

Allen, L. H. (1982). Calcium bioavailability and absorption: A review. *Am. J. Clin. Nutr.* **35**, 738–808.

Anderson, J. J. B. and Garner, S. C. (1996). Calcium and phosphorous nutrition in health and disease. *In* "Calcium and Phosphorous in Health and Disease", (J. J. B. Anderson and S. C. Garner, Eds), pp. 1–5. CRC Press, New York.

Berrocal, R., Chanton, S., Juillerat, M. A., Pavillard, B., Scherz, J. C., and Jost, R. (1989). Tryptic phosphopeptides from whole casein II. Physiochemical properties related to the solubilization of calcium. *J. Dairy Res.* **56**, 335–341.

Brouns, F. and Vermeer, C. (2000). Functional food ingredients for reducing the risks of osteoporosis. *Trends Food Sci. Technol.* **11**, 22–33.

Garner, S. C., Anderson, J. J. B., and Ambrose, W. W. (1996). Skeletal tissues and mineralization. *In* "Calcium and Phosphorous in Health and Disease", (J. J. B. Anderson and S. C. Garner, Eds), pp. 97–117. CRC Press, New York.

Hamada, M., Nagai, T., Kai, N., Tanoue, Y., Mae, H., Hashimoto, M., Miyoshi, K., Kumagai, H., and Saeki, K. (1995). Inorganic constituents of bone of fish. *Fish. Sci.* **63**, 517–520.

Hauschka, P. V. and Gallop, P. M. (1977). Purification and calciumbinding properties of osteocalcin the gamma-carboxyglutamate containing protein of bone. *In* "Calcium-Binding Proteins and Calcium Function", (R. H. Wasserman, R. A. Corradino, E. Carafoli, R. H. Kretsinger, D. H. MacLennan, and F. L. Siegel, Eds), pp. 338–347. Elsevier, New York.

Houben, R., Jin, D., Stafford, D., Proost, P., Ebberink, R., and Vermeer, C. (1999). Osteocalcin binds tightly to the γ-glutamylcarboxylase at a site distinct from that of the other known vitamin K-dependent proteins. *Biochem. J.* **344**, 265–269.

Jiang, B. and Mine, Y. (2000). Preparation of novel functional oligophosphopeptides from hen egg yolk phosvitin. *J. Agric. Food Chem.* **48**, 990–994.

Jung, W. K., Park, P. J., Byun, H. G., Moon, S. H., and Kim, S. K. (2005). Preparation of hoki (*Johnius belengerii*) bone oligophosphopeptide with a high affinity to calcium by carnivorous intestine crude proteinase. *Food Chem.* **91**, 333–340.

Jung, W. K., Lee, B. J., and Kim, S. K. (2006). Fish bone peptide increases Ca solubility and bioavailability in ovariectomized rats. *Br. J. Nutr.* **95**, 124–128.

Kim, S. K., Jeon, Y. J., Byun, H. G., Kim, Y. T., and Lee, C. K. (1997). Enzymatic recovery of cod frame proteins with crude proteinase from tuna pyloric caeca. *Fish. Sci.* **63**, 421–427.

Kumagai, H., Koizumi, A., Sato, N., Ishikawa, Y., Suda, A., Sakurai, H., and Kumagai, H. (2004). Effect of phytate-removal and deamidation of soybean proteins on calcium absorption in the in situ rats. *Biofactors* **22**, 21–24.

Larsen, T., Thilsted, S. H., Kongsbak, K., and Hansen, M. (2000). Whole small fish as a rich calcium source. *Br. J. Nutr.* **83**, 191–196.

Larsen, T., Thilsted, S. H., Biswas, S. K., and Tetens, I. (2003). The leafy vegetable amaranth (*Amaranthus gangeticus*) is a potent inhibitor of calcium bioavailability and retention in rice-based diets. *Br. J. Nutr.* **90**, 521–527.

Lee, Y. S., Noguchi, T., and Naito, H. (1980). Phosphopeptides and soluble calcium in the small intestine of rat given a casein diet. *Br. J. Nutr.* **43**, 457–467.

Matsuura, Y., Stewart, M., Kawamoto, M., Kamiya, N., Saeki, K., Yasunaga, T., and Wakabayashi, T. (2000). Structural basis for the higher Ca2-activation of the regulated actin-activated myosin ATPase observed with dictyostelium/tetrahymena actin chimeras. *J. Mol. Biol.* **296**, 579–595.

Nishmoto, S. K., Waite, J. H., Nishimoto, M., and Kriwacki, R. W. (2003). Structure, activity, and distribution of fish osteocalcin. *J. Biol. Chem.* **278**, 11843–11848.

Tsuchita, H., Sekiguchi, I., Kuwata, T., Igarashi, T., and Ezawa, I. (1993). The effect of casein phosphopeptides on calcium utilization in young ovariectomized rats. *Z. Ernahrungswiss.* **32**, 121–130.

Yuan, Y. V. and Kitts, D. D. (1991). Conformation of calcium absorption and femoral utilization in spontaneously hypertensive rats fed casein phosphopeptide supplemented diets. *Nutr. Res.* **11**, 1257–1272.

Yuan, Y. V. and Kitts, D. D. (1994). Calcium absorption and bone utilization in spontaneously hypertensive rats fed on native and heat-damaged casein and soybean protein. *Br. J. Nutr.* **71**, 583–603.

CHAPTER 20

Triterpene Glycosides from Sea Cucumbers and Their Biological Activities

Se-Kwon Kim[*,†,1] and **S.W.A. Himaya**[*]

Contents

Abstract

Triterpenoid glycosides are abundantly present in sea cucumbers, which are responsible for the toxicity of these echinoderms. More than 100 triterpenoid glycosides have been isolated in the past 20 years and those are grouped into four main structural categories considering their aglycone structure: 3β-hydroxyholost-9(ll)-ene aglycone skeleton, 3β-hydroxyholost-7-ene skeleton, other holostane type aglycones and nonholostane aglycone. Most of the triterpenoid glycosides are found to be possessing potential

* Department of Chemistry, Pukyong National University, Busan, Republic of Korea
† Marine Bioprocess Research Center, Pukyong National University, Busan, Republic of Korea
1 Corresponding author: Se-Kwon Kim, *E-mail address:* sknkim@pknu.ac.kr

Advances in Food and Nutrition Research, Volume 65
ISSN 1043-4526, DOI: 10.1016/B978-0-12-416003-3.00020-2

biological activities. Among the biological activities, anticancer activity and antiviral activity are the most widely studied areas. In this communication, we have presented a general view of the structural characteristics of triterpenoid glycosides and their major biological activities. The structural significance and the application limitations of triterpene glycosides are also discussed.

I. INTRODUCTION

Marine organisms have proven to be rich sources of interesting organic molecules. A great number of compounds with diverse structural features and sound biological activities have been reported and reviewed in the literature. Among these, the compounds isolated from sea cucumbers are gaining more attention in the recent past due to their structural features and bioactivities. Sea cucumbers are soft-bodied worm-like echinoderms which belong to the class *Holothuroidea* (De Moncerrat *et al.*, 2005). They have economic importance in Asian countries specifically in China where several species are used in traditional medicine or eaten as delicacies. The taxonomical distribution of sea cucumbers consists of six main orders (Apodida, Elasipodida, Aspidochirotida, Molpadiida, Dendrochirotida, and Dacthylochirotida) which include 25 families, about 200 genera, and more than 1400 species. Sea cucumbers can be found in nearly every marine environment but are most diverse on tropical shallow-water coral reefs.

Sea cucumbers gain attention of the scientific crowed in the recent past due to the presence of interesting bioactive compounds with diverse structural features. Therefore, scientists are heavily employed in searching for bioactive compounds from sea cucumbers to be used as potential drugs in the pharmaceutical industry and as neutraceuticals in the food industry. Among the compounds isolated from sea cucumbers, triterpenoid glycosides (saponins) are the major and more interesting. This chapter focuses on the structural features of the triterpenoid glycosides compounds isolated in the recent past and their biological activities.

II. THE STRUCTURAL FEATURES OF TRITERPENE GLYCOSIDES

Triterpene glycosides or triterpene saponins are the most abundant category of secondary metabolites in terrestrial plants. Interestingly, this group of compounds is found to be characteristic metabolites in sea cucumbers. And also these are the predominant secondary metabolites of holothurians, which are presumed to be responsible for their general toxicity (Zhang *et al.*, 2006a,b,c,d). More than 100 of triterpene saponins have been isolated from many species of sea cucumbers belonging to different orders from tropical Pacific, Indian, and Atlantic oceans and the Mediterranean Sea (Antonov *et al.*, 2008). Saponins of holothurians are typically composed of carbohydrate and triterpenoid moieties (Kerr and Chen, 1995). The triterpenoid moieties are composed of lanostane derivatives (Zou *et al.*, 2005) where majority belongs to holostane type (Dang *et al.*, 2007). Holostane type triterpene glycosides include a 3β,20*S*-dihydroxy-5*a*-lanostano-18,20-lactone (**1**) (Fig. 20.1) structural feature. However, recent investigations about these compounds have shown that some triterpene saponins do and those are identified as triterpene saponins with nonholostane aglycone structures. The glycon part of the sea cucumber glycosides is composed of two to six sugar units and it is linked to the C-3 position of the aglycone unit (Chiludil *et al.*, 2003; Kalinin *et al.*, 2005). Quinovose, glucose, 3-*O*-methylglucose, xylose, and 3-*O*-methylxylose are the only sugars present in the carbohydrate moieties of these glycosides (De Moncerrat *et al.*, 2005). In the structure of the oligosaccharide chain, the first monosaccharide unit is always a xylose, while 3-*O*-methylglucose and 3-*O*-methylxylose are always at the terminal. In some glycosides, sulfate groups are attached to the oligosaccharide chain. Most of them are mono-sulfated glycosides with few occurrences of di- and trisulfated glycosides (Chiludil *et al.*, 2003).

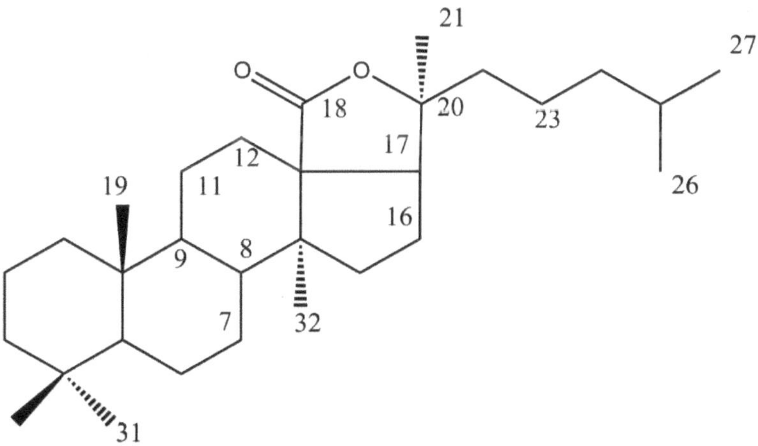

FIGURE 20.1 Structure of the holostane group which is the characteristic aglycone moiety in sea cucumber glycosides.

A. Holostane type sea cucumber glycosides

Reviewing the cited references, the holostane type glycosides were classified into three groups: saponins with 3β-hydroxyholost-9(ll)-ene aglycone skeleton, 3β-hydroxyholost-7-ene skeleton, and saponins contain a aglycone moiety different to above skeletons. Triterpene glycosides with 3β-hydroxyholost-9(ll)-ene aglycone and 3β-hydroxyholost-7-ene skeletons, isolated from different species of sea cucumbers, are listed in Tables 20.1 and 20.2 with their structures illustrated in Figs. 20.2 and 20.3, respectively (Avilov *et al.*, 2008).

B. Nonholostane type glycosides

Nonholostane type aglycone moieties are a rare feature in sea cucumber triterpene glycosides. Only few instances have been reported about nonholostane type glycosides. Some of the recently found nonholostane type sea cucumber saponins are listed in Table 20.3 with their structures in Fig. 20.4.

III. BIOACTIVITIES OF TRITERPENE GLYCOSIDES OF SEA CUCUMBERS

Sea cucumbers have long been known for its healing properties and widely used in Chinese traditional medicine rations. Extensive research on this area has revealed that the sea cucumbers gain healing power due to the presence of triterpene glycosides. These compounds exhibit wide spectrum of biological activities such as antifungal, cytotoxic, hemolytic, and immunomodulatory effects (Dang *et al.*, 2007; Han *et al.*, 2008; Zhang *et al.*, 2006c). However, most of the published data demonstrated the anticancer or the cytotoxic activity of triterpene glycosides among the other activities. This suggests that sea cucumber triterpene glycosides are more potent anticancer agents.

A. Anticancer activity

Many natural products have been characterized for the protection against some form of cancer. Sea cucumber triterpenoid glycosides have also been considered as responsible agents for such observations and much research have been conducted to explore the anticancer effect. Most of the triterpene glycosides of sea cucumbers are cytotoxic toward cancer cells. However, their mechanism of anticancer effect has not been elucidated in depth.

TABLE 20.1 Triterpenoid glycosides isolated from different species of sea cucumbers with hydroxyholost-9(ll)-ene aglycone skeleton

Compound name	Species (reference)	Yield (%)	Habitat	Reference
Parvimosides A (**2**)	*Stichopus parvimensis*	8.97×10^{-5a}	Monterey, CA	De Moncerrat *et al.* (2005)
Parvimosides B (**3**)		1.45×10^{-4a}		
Achlioniceosides A1 (**4**)	*Achlionice violaecuspidata*	0.6×10^{-2b}	Weddell sea	Antonov *et al.* (2009)
Achlioniceosides A2 (**5**)		0.8×10^{-2b}		
Achlioniceosides A3 (**6**)		0.65×10^{-2c}		
Neothyoside A (**7**)	*Neothyone gibbosa*	1.2×10^{-2a}	Cerralro, Mexico	Encarnacion *et al.* (1989)
Fuscocineroside A (**8**)	*Holothuria fuscocinera* Jaegar	4.73×10^{-3a}	South China	Zhang *et al.*, (2006b)
Fuscocineroside B (**9**)		2.02×10^{-3a}		
Fuscocineroside C (**10**)		2.14×10^{-3a}		
Pervicoside C (**11**)		6.82×10^{-4a}		
Holothurin A (**12**)		6.59×10^{-4a}		
Holothurin A3 (**13**)	*Holothuria scabra*	6×10^{-4c}	North Vietnam	Dang *et al.* (2007)
Holothurin A4 (**14**)		1.2×10^{-4c}		
Holothurins B (**15**)	*Holothuria polii*	1.54×10^{-2c}	Italy	Silchenko *et al.* (2005)
Holothurins B2 (**16**)		8.18×10^{-3c}		
Holothurins B3 (**17**)		2.73×10^{-3c}		
Holothurins B4 (**18**)		2.73×10^{-3c}		
Leucospilotaside C (**19**)	*Holothuria leucospilota*	–	South China	Han *et al.* (2008)
DS-penaustrosides C (**20**)	*Pentarta australis* Ludwig	3.16×10^{-4a}	Ariake sea	Miyamoto *et al.* (1992)
DS-penaustrosides D (**21**)		6.08×10^{-5a}		
Hemoiedemosides A (**22**)	*Hemoiedema spectabilis*	7.53×10^{-3a}	Patagonia cost of Argentina	Chiludil *et al.* (2002)

(continued)

TABLE 20.1 *(continued)*

Compound name	Species (reference)	Yield (%)	Habitat	Reference
Hemoiedemosides B (**23**)		3.53×10^{-3a}		
Marmoroside C (**24**)	*Bohadschia marmorata* Jaeger		South China	Yuan *et al.* (2008)
Fuscocineroside A (**25**)				
Leucospilotaside A (**26**)	*Holothuria leucospilota*	Minor	South China	Han *et al.* (2007)
Two triterpene glycosides (**27**)	*Telenata ananas* sp.	1.9×10^{-4a}	Andaman and Nicobar island, India	Hegde *et al.* (2002)
and (**28**)		3.5×10^{-5a}		
Arguside A (**29**)	*Bohadschia argus* Jaeger	8.27×10^{-4c}	South China	Liu *et al.* (2007)
Arguside B (**30**)	*Bohadschia argus* Jaeger	5.87×10^{-4c}	South China	Liu *et al.* (2008b)
Arguside C (**31**)		1.12×10^{-3c}		
Arguside D (**32**)	*Bohadschia argus* Jaeger	6.22×10^{-4c}	South China	Liu *et al.* (2008a)
Arguside E (**33**)		9.24×10^{-4c}		
Impatienside A (**34**)	*Holothuria impatiens*	9.11×10^{-4c}	South China	Sun *et al.* (2007)
Bivittoside D (**35**)		6.53×10^{-4c}		
17-Dehydroxyholothurinoside A (**36**)	*Holothuria grisea*	3.44×10^{-2c}	South China	Sun *et al.* (2008)
Griseaside A (**37**)		6×10^{-4c}		
Hillaside C (**38**)	*Holothuria hilla* (**113**)	9.33×10^{-4c}	Fujian, China	Wu *et al.* (2006)

[a] Wet weight basis.
[b] Ethanol extractive dry weight basis.
[c] Dry weight basis.

TABLE 20.2 Sea cucumber triterpene glycosides bearing 3β-hydroxyholost-7-ene skeleton

Compound name	Species (reference)	Yield (%)	Habitat	Reference
Eximisoside A (**39**)	*Psolus eximius* Saveljeva	1.92×10^{-3a}	Okhotsk sea	Kalinin *et al.* (1997)
Calcigeroside C2 (**40**)	*Pentamera calcigera*	1.76×10^{-4a}	Great gulf sea	Avilov *et al.* (2000a)
Okhotoside B1 (**41**)	*Cucumaria okhotensis*	3.83×10^{-4a}	Okhotsk sea	Silchenko *et al.* (2008)
Okhotoside B2 (**42**)		5.05×10^{-4a}		
Okhotoside B3 (**43**)		4.33×10^{-4a}		
Frondoside A (**44**)		3.18×10^{-4a}		
Liouvilloside B (**45**)	*Staurocucumis liouvillei*	4.28×10^{-4b}	South Georgias island	Maier *et al.* (2001)
Liouvilloside A (**46**)		1.77×10^{-3b}		
Liouvilloside A1 (**47**)	*Staurocucumis liouvillei*	2.86×10^{-2c}	Sub-Antarctic island of Bouvet	Antonov *et al.* (2008)
Liouvilloside A2 (**48**)		2.32×10^{-2c}		
Liouvilloside A3 (**49**)		5×10^{-2c}		
Liouvilloside B1 (**50**)		2.5×10^{-2c}		
Liouvilloside B2 (**51**)		5.18×10^{-2c}		
Patagonicoside A (**52**)	*Psolus patagonicus*	5.42×10^{-3c}	Argentina	Murray *et al.* (2001)
Patagonicoside B (**53**)		8.68×10^{-4c}		
Synallactoside A1 (**54**)	*Synallactes nozawai*	1.66×10^{-4a}	Japan	Silchenko *et al.* (2002)
Synallactoside A2 (**55**)		5.63×10^{-4a}		
Synallactoside B1 (**56**)		3.64×10^{-4a}		
Synallactoside B2 (**57**)		3.31×10^{-4a}		
Synallactoside C (**58**)		7.28×10^{-4a}		
Intercedenside A (**59**)	*Mensamaria intercedens*	5×10^{-5b}	South China	Zou *et al.* (2003)
Intercedenside B (**60**)		4.2×10^{-5b}		
Intercedenside C (**61**)		5.86×10^{-5b}		

(*continued*)

TABLE 20.2 *(continued)*

Compound name	Species (reference)	Yield (%)	Habitat	Reference
Frondoside B (**62**)	*Cucumaria frondosa*	7.47×10^{-5b}	New Brunswick, Canada	Findlay *et al.* (1992)
Frondecaside (**63**)		7.07×10^{-4b}		
Lefevreioside A (**64**)	*Cucumaria lefevrei*			
Calcigerosides D2 (**65**)	*Pentamera calcigera*	1.41×10^{-3b}	Japan	Avilov *et al.* (2000b)
Cucumariosides A2-5 (**66**)	*Cucumaria conicospermium*	2.18×10^{-3b}	Japan	Avilov *et al.* (2003)
Violaceuside A (**67**)	*Pseudocolochirus violaceus*	1.68×10^{-4b}	South China	Zhang *et al.* (2006c)
Violaceuside B (**68**)		3×10^{-5b}		
Violaceuside I (**69**)	*Pseudocolochirus violaceus*	2.83×10^{-5b}		Zhang *et al.* (2006d)
Violaceuside II (**70**)		7×10^{-5b}		
Violaceuside III (**71**)		4.67×10^{-5b}		
Philinopside A (**72**)	*Pentacta quadrangularis*	2.46×10^{-3a}	South China	Yi *et al.* (2006)
Philinopside B (**73**)		1.26×10^{-3b}		Zhang *et al.* (2006c)
Philinopside E (**74**)	*Pentacta quadrangularis*	5.46×10^{-4a}	South China	Zhang *et al.* (2006a)
Philinopside F (**75**)		1.62×10^{-4a}		
Pseudostichoposide B (**76**)	*Pseudostichopus trachus*		North Pacific	Silchenko *et al.* (2004)
Pseudostichoposide A (**77**)				
Intercedenside B (**78**)	*Pseudocolochirus violaceus*	9.17×10^{-5b}	South China	Zhang *et al.* (2007)
Okhotoside Ä$_1$-1 (**79**)	*Cucumaria okhotensis*	4.33×10^{-5c}	Kamchatka	Silchenko *et al.* (2007)
Okhotoside Ä$_2$-1 (**80**)		4.33×10^{-5c}		
Okhotoside Ä$_0$-1 (**81**)		7.22×10^{-5c}		

[a] Dry weight basis.
[b] Wet weight basis.
[c] Ethanol extractive dry weight basis.

The *in vitro* cytotoxicity of five triterpene glycosides, fuscocineroside A (**8**), B (**9**), and C (**10**), pervicoside C (**11**), and holothurin A (**12**) isolated from *Holothuria fuscocinerea* Jaeger on human leukemia HL-60 and human hepatoma BEL-7402 cells, was analyzed, and all compounds have shown a potent cytotoxicity in both cell lines. However, compound 10 was found to be the most potent (IC_{50}=0.88 and IC_{50}=0.58 μg/ml) in HL-60 and BEL-7402 cell lines, respectively (Zhang *et al.*, 2006d). The triterpene glycosides from the sea cucumber *Holothuria scabra*, namely, holothurin A3 (**13**) and A4 (**14**), were found to be strongly cytotoxic to cancer cell lines, human

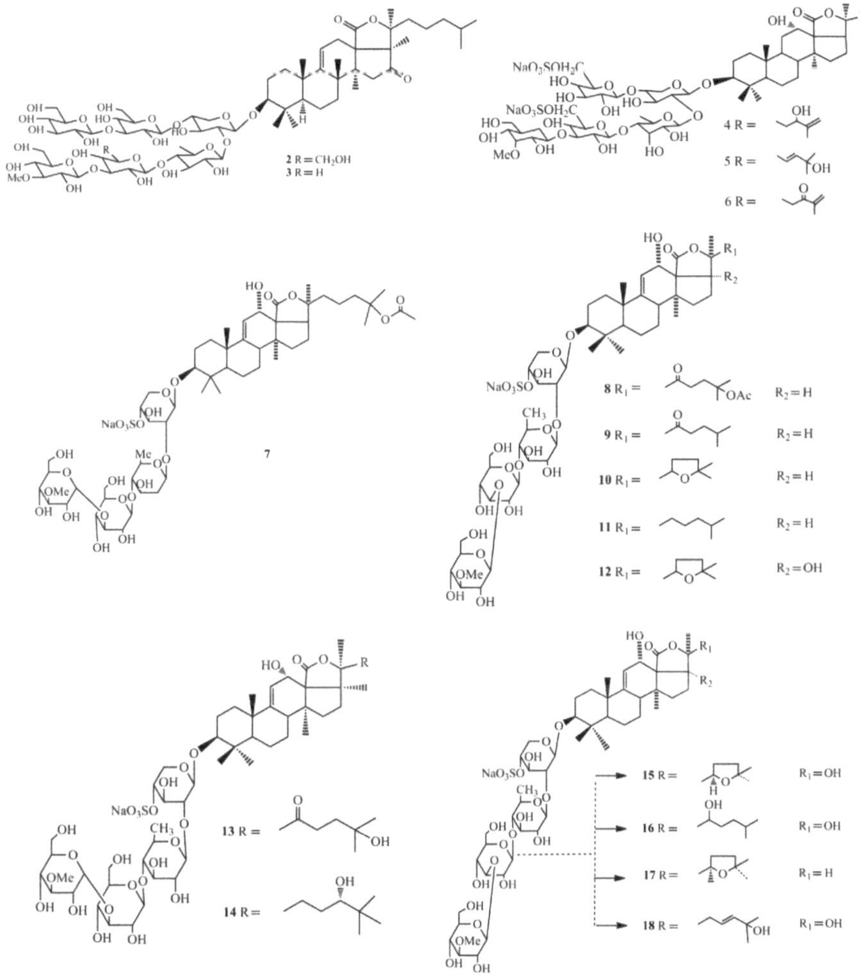

Figure 20.2 (Continued)

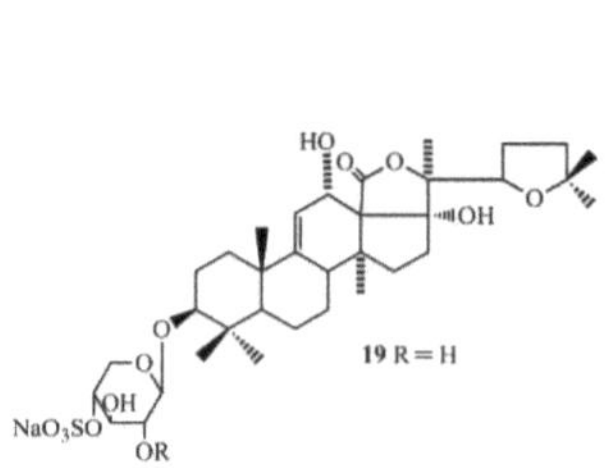

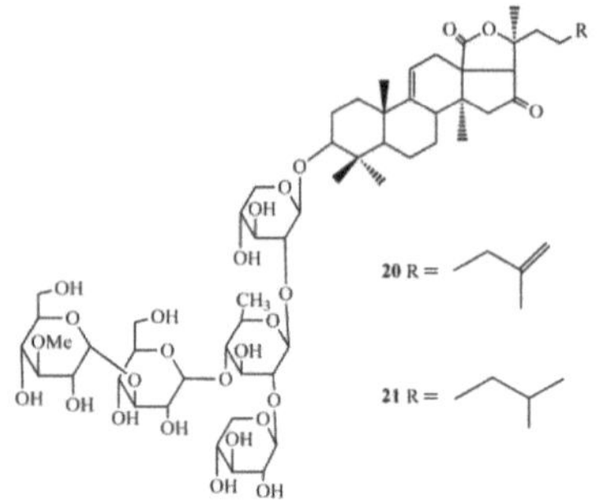

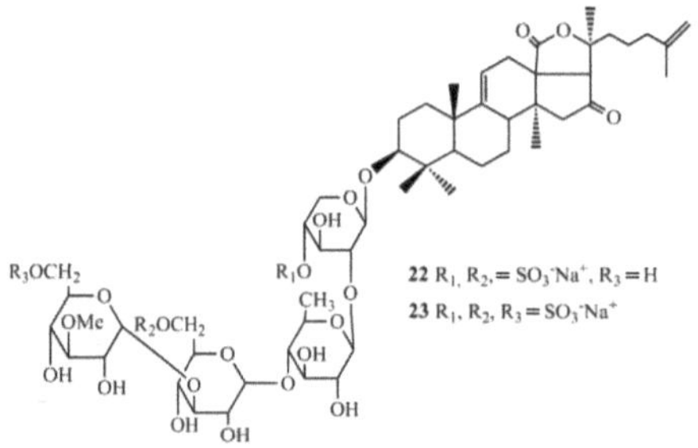

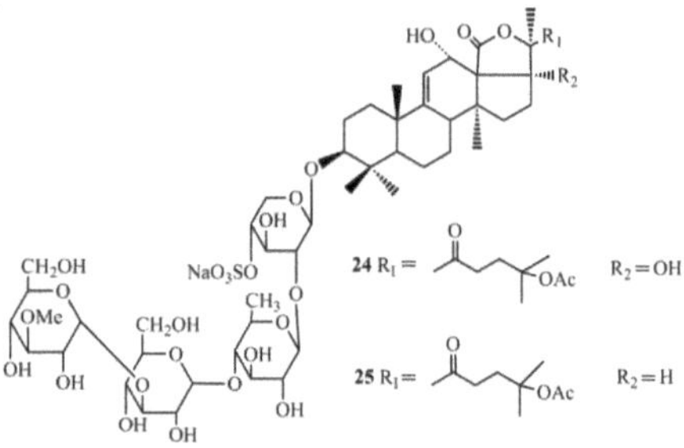

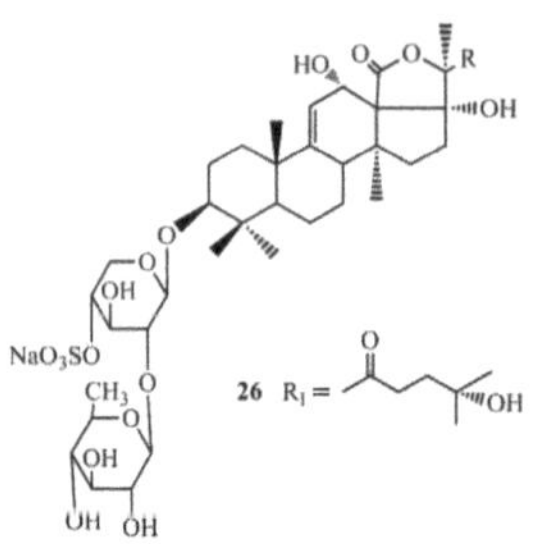

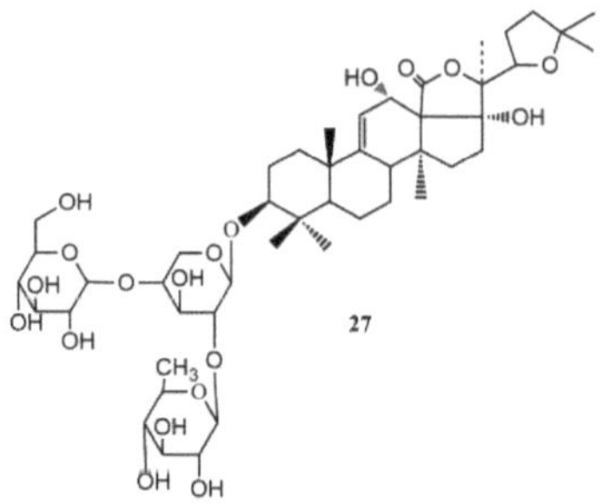

epidermoid carcinoma (KB) and human hepatocellular carcinoma (Hep-G2), with IC_{50} values of 0.87 and 0.32 μg/ml (for compound **13**) and of 1.12 and 0.57 μg/ml (for compound **14**), respectively (Dang *et al.*, 2007). Arguside A (**29**) also exhibited significant cytotoxicity against different human tumor cell lines, while showing the highest activity toward human colorectal carcinoma (HCT-116) cells (IC_{50}=0.14 μM) with more potency than the employed positive control, 10-hydroxycamptothecin (HCP) (IC_{50}=0.84 μM) (Liu *et al.*, 2007). Argusides B (**30**) and C (**31**) have also shown potent cytotoxicity against human tumor cell lines, adenocarcinomic human alveolar basal epithelial cells (A549) and HCT-116 in

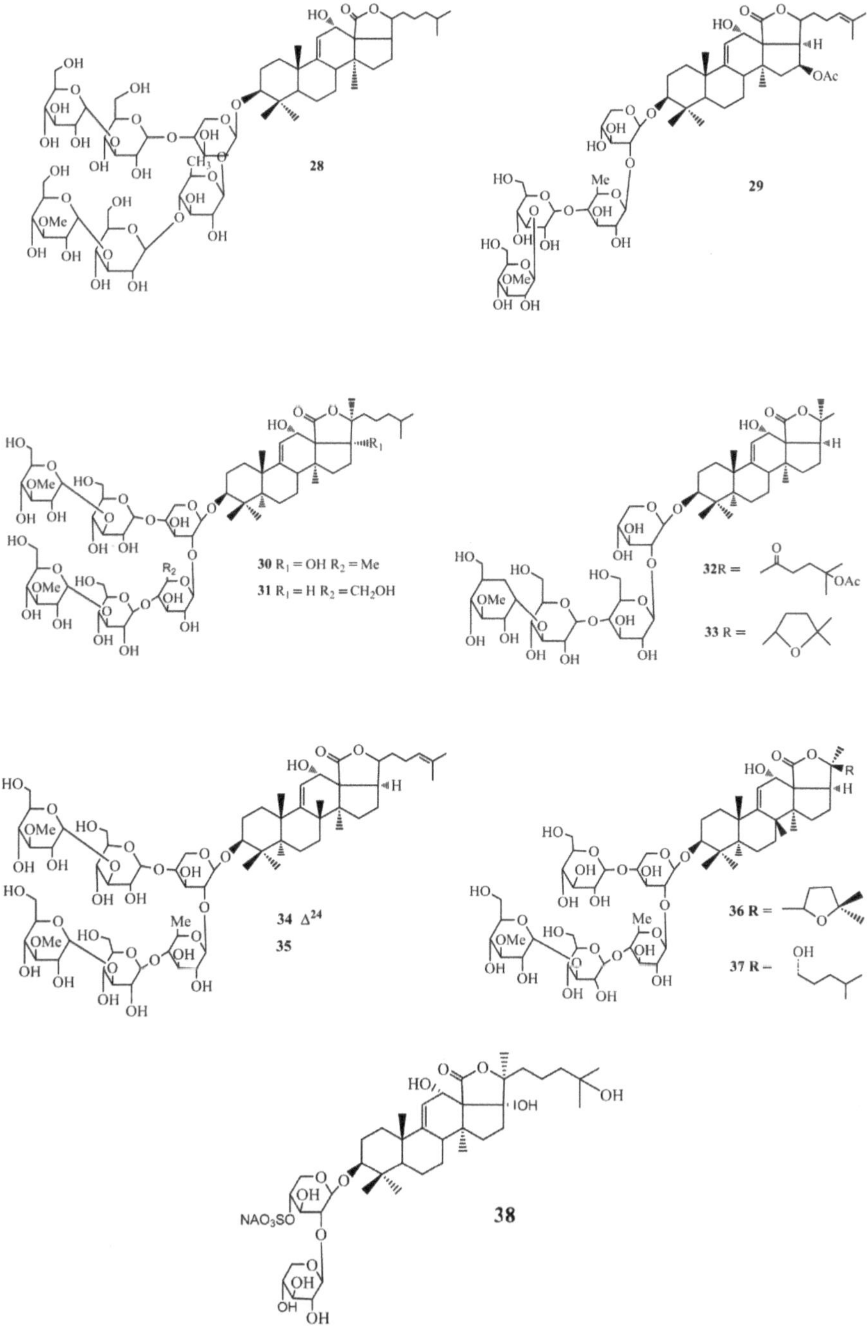

FIGURE 20.2 Triterpenoid glycosides isolated from sea cucumbers, having 3β-hydroxyholost-9(ll)-ene aglycone skeleton.

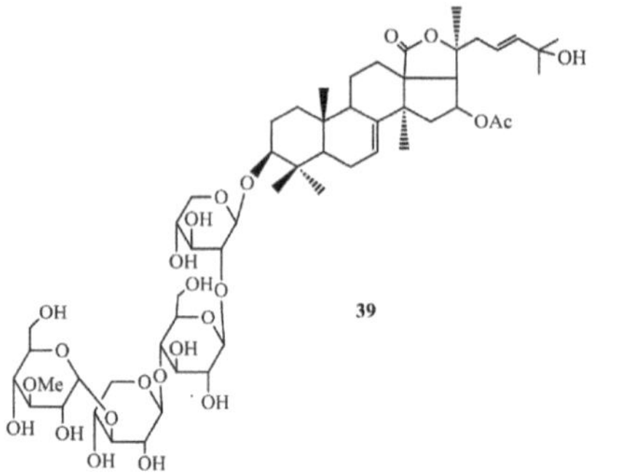

39

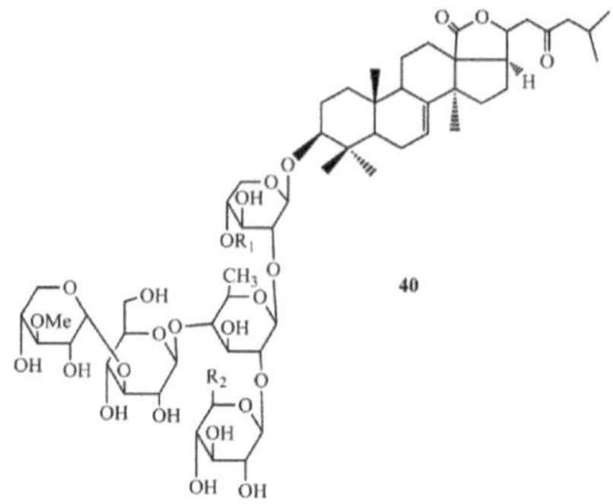

40

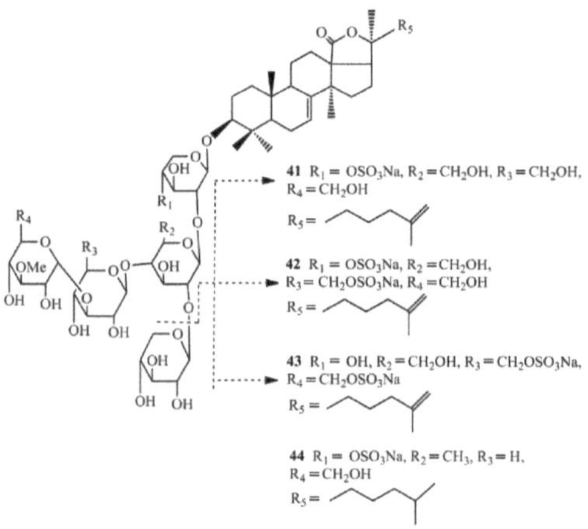

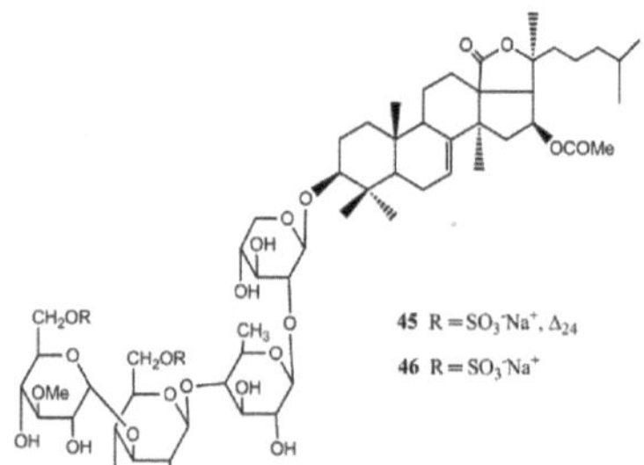

45 R = $SO_3^-Na^+$, Δ_{24}
46 R = $SO_3^-Na^+$

HepG2, and human breast adenocarcinoma (MCF-7) cell lines. The cytotoxicity of the compounds on A549 (**30**-IC_{50}: 0.48 μg/ml, **31**-IC_{50}: 0.43 μg/ml) and HCT-116 (**30**-IC_{50}: 0.46 μg/ml, **31**-IC_{50}: 0.38 μg/ml) cells was more potent than the positive control V-16. However, there was no significant difference between the cytotoxicities of 38 and 39 (Liu *et al.*, 2008b). Besides, argusides D (**32**) and E (**33**) have also been tested for their anticancer activities in above human cancer cell lines and revealed a significant activity with IC_{50} values in the range of 3.36–7.77 μg/ml (Liu *et al.*, 2008a). However, the activity is comparatively lower to the compounds **30** and **31**. This finding clearly indicates that the length and type of sugar moieties of glycosides play an important role in terms of cytotoxic activity against tumor cells. Moreover, the *in vitro* cytotoxicities of impatienside A (**34**) and bivittoside D (**35**) were evaluated extensively for

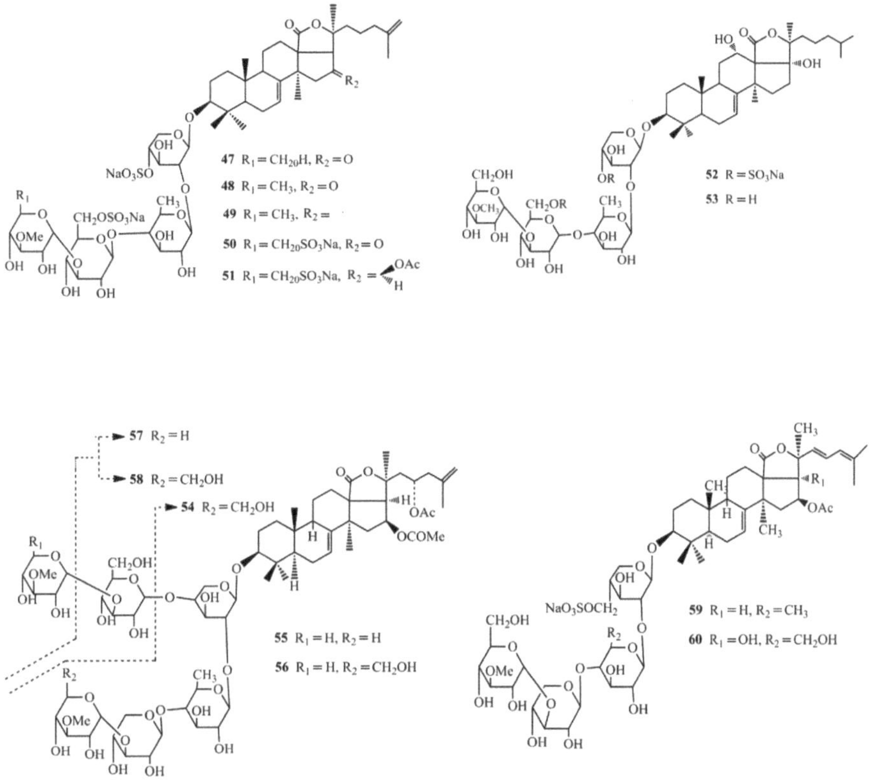

Figure 20.3 (Continued)

cytotoxicity by employing seven human cancer cell lines and the results indicate that both glycosides exhibited *in vitro* cytotoxicities similar to or better than that of the potent anticancer drug etoposide (V-16) in four human tumor cells, A549 (**34**-IC_{50}: 0.35 μg/ml, **35**-IC_{50}: 0.52 μg/ml), HCT-116 (**34**-IC_{50}: 0.45 μg/ml, **35**-IC_{50}: 0.37 μg/ml), DU-145 (**34**-IC_{50}: 1.14 μg/ml, **35**-IC_{50}: 0.937 μg/ml), and KB (**34**-IC_{50}: 1.6 μg/ml, **35**-IC_{50}: 1.42 μg/ml) (Sun *et al.*, 2007). The structural differences between glycosides **34** and **35** were limited to their holostane skeleton, and there was no significant difference in the cytotoxicities of the two glycosides. However, pervicoside C (**11**), an analogue of **35** having the same aglycone but a different sugar chain, isolated from *H. fuscocinerea* Jaeger, exhibited weak activities against HCT-116 and A549 cancer cells, with IC_{50} values of 18.7 and 28.6 μg/ml, respectively (Sun *et al.*, 2007). According to these results, it could be suggested that the length and type of sugar moieties of such glycosides play an important role in terms of cytotoxic activity against tumor cells.

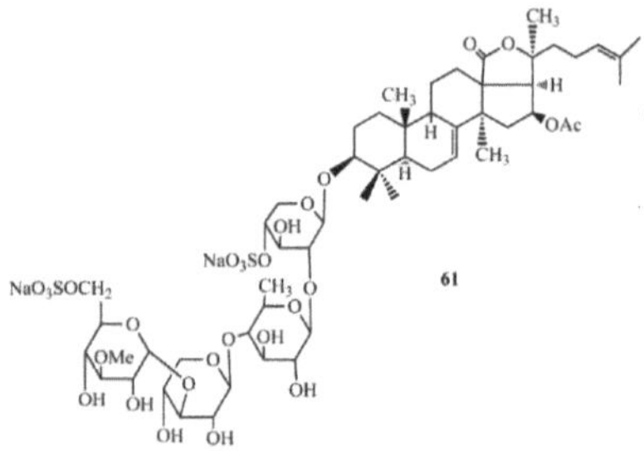

61

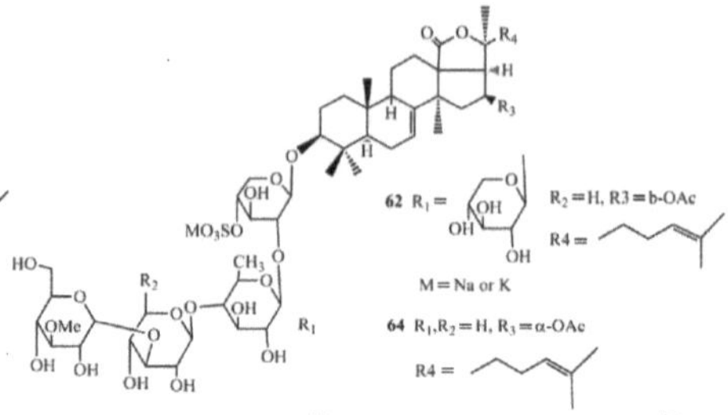

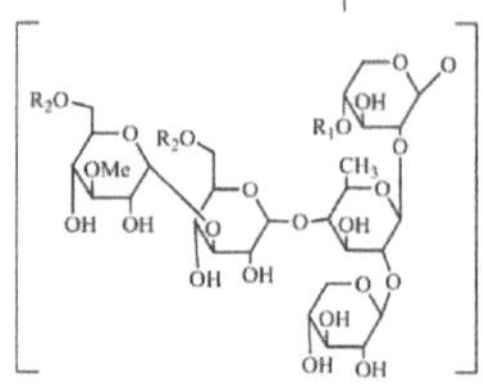

63 R_1, R_2, R_3=OSO_3M, M=Na or K

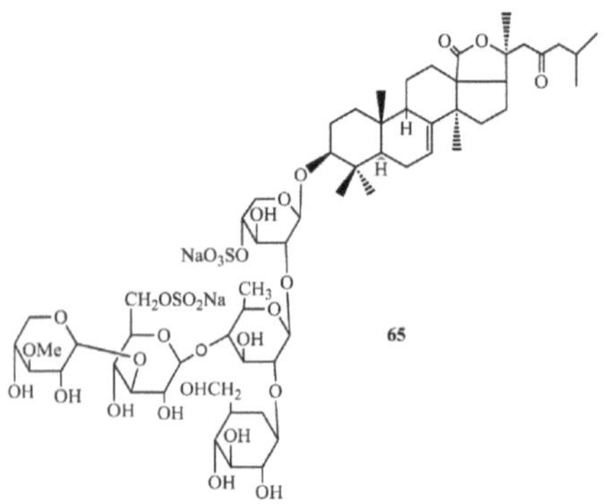

65

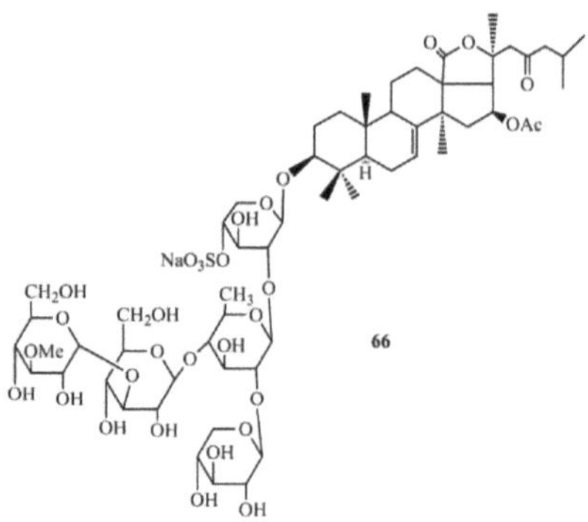

66

17-Dehydroxyholothurinoside A (**40**) and griseaside A (**41**) are identified as promising anticancer agents due to their significantly higher cytotoxicity against four human tumor cell lines, A549 (**40**-IC_{50}: 0.886μ M, **41**-IC_{50}: 1.07μM), HL-60 (**40**-IC_{50}: 0.245μM, **41**-IC_{50}: 0.427μM), BEL-7402 (**40**-IC_{50}: 0.97μM, **41**-IC_{50}: 1.114μM), and human acute lymphoblastic leukemia cell line (Molt-4) (**40**-IC_{50}: 0.34μM, **41**-IC_{50}: 0.521μM) compared to the positive control HCP (A549 IC_{50}: 2.35μM, BEL-7402 IC_{50}: 2.6μM, HL-60 IC_{50}: 1.9μM, Molt-4 IC_{50}: 2.2μM) (Sun *et al.*, 2008). Hillaside C (**38**) has also been tested for its anticancer potential against eight human tumor cell lines (A-549, MCF-7, human lung carcinoma cells—IA9, human clear cell carcinoma cells—CAKI-1, human prostate cancer cells—PC-3, KB, KB-VIN, and human colorectal adenocarcinoma cells—HCT-8) and has exhibited cytotoxicity with IC_{50} values in the range of 0.15–3.20μg/ml (Wu *et al.*, 2006). Compared to the positive control HCP, the compound **38**

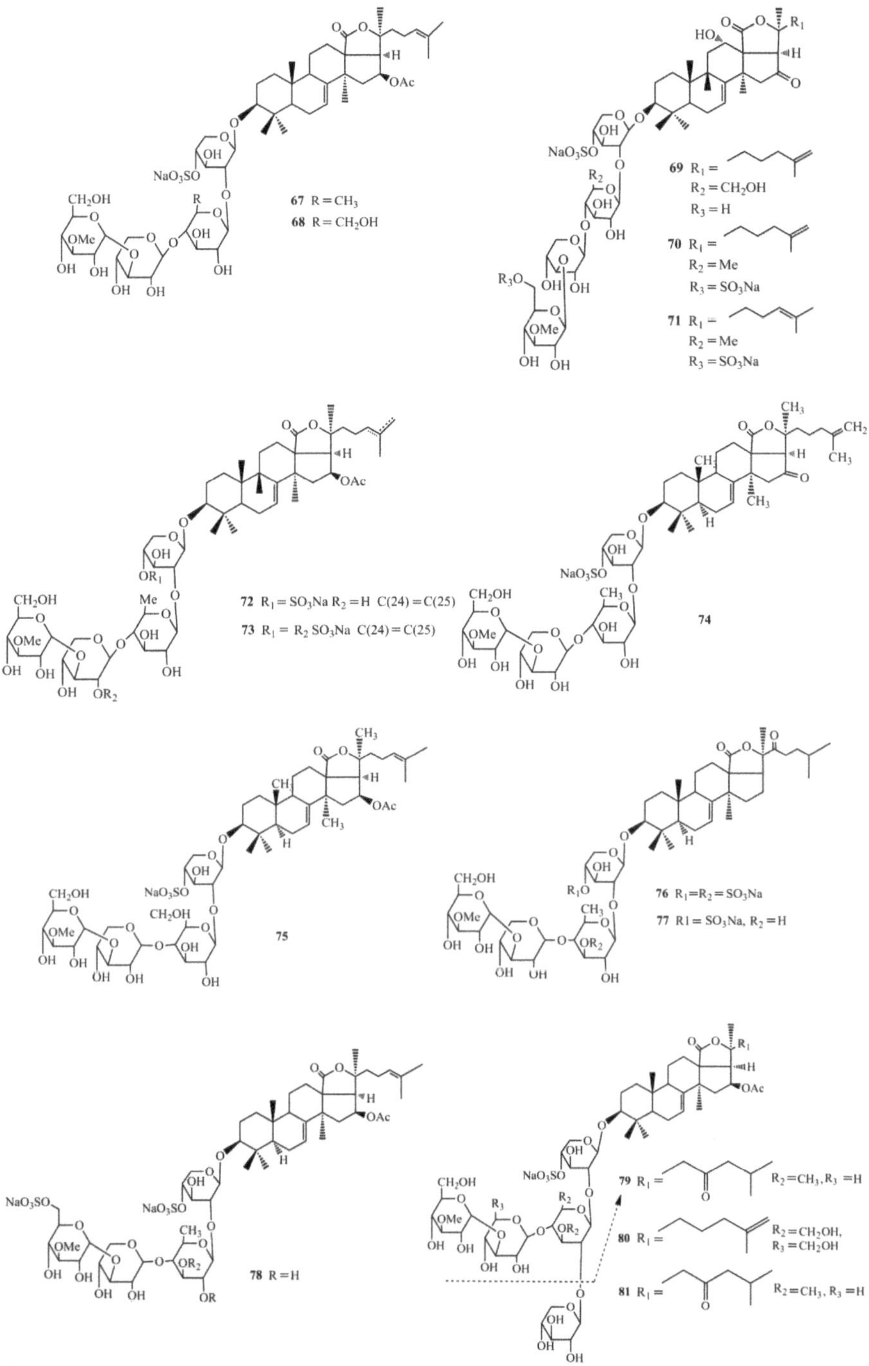

FIGURE 20.3 Triterpenoid glycosides isolated from sea cucumbers, having 3β-hydroxyholost-7(8)-ene aglycone skeleton.

TABLE 20.3 Nonholostane type triterpene glycosides isolated from sea cucumbers

Compound name	Species (reference)	Yield (%)	Habitat	Reference
Calcigerosides B (**82**)	*Pentamera calcigera*	1.76×10^{-2a}	Great gulf sea	Zou *et al.* (2006)
Calcigerosides C1 (**83**)		1.76×10^{-4a}	Japan	
Cucumarioside G2 (**84**)		6.35×10^{-2a}		
Koreoside A (**85**)	*Cucumaria koraiensis*	5×10^{-3b}	Kurile island	Avilov *et al.* (1997)
DS-penaustrosides A (**86**)	*Pentarta australis* Ludwig	4.27×10^{-4b}	Ariale sea	Miyamoto *et al.* (1992)
DS-penaustrosides B (**87**)		1.29×10^{-4b}		
Calcigerosides D1 (**88**)	*Pentamera calcigera* Stimpson	2.82×10^{-2a}	Peter the Great gulf	Avilov *et al.* (2000a)
DS-calcigeroside D1 (**89**)				

[a] Dry weight basis.
[b] Wet weight basis.

has shown more potent cytotoxicity toward CAKI-1 (**42**-IC_{50}: 0.15 μg/ml) and KB-VIN (**42**-IC_{50}: 2.81 μg/ml) cell lines. Three new triterpene glycosides, intercedensides A (**59**), B (**60**), and C (**61**) from *Mensamaria intercedens* Lampert, were widely studied for their anticancer activity employing 10 human tumor cell lines (A549, MCF-7, IA9, CAKI-1, human glioblastoma cells: U-87-MG, PC-3, KB, KB-VIN, human skin melanoma cells: SK-MEL-2, HCT-8). Interestingly, all compounds showed a significant cytotoxicity against all tumor cell lines within the IC_{50} value range of 0.7–4 μg/ml, and the compounds **59** and **61** showed similar potencies, while compound **60** was generally more potent in all cell lines. Further, **59** also exhibited significant *in vivo* antineoplastic activity against mouse Lewis lung cancer and mouse S180 sarcoma, with 48.39% and 57.48% reduction levels (Zou *et al.*, 2003).

New sulfated triterpene glycoside from *Pentacta quadrangularis*, philinopside E (**74**), showed a significant cytotoxicity (IC_{50}: 0.75–3.50 μg/ml) against 10 tumor cell lines (mouse lymphocytic leukemia cells: P388,

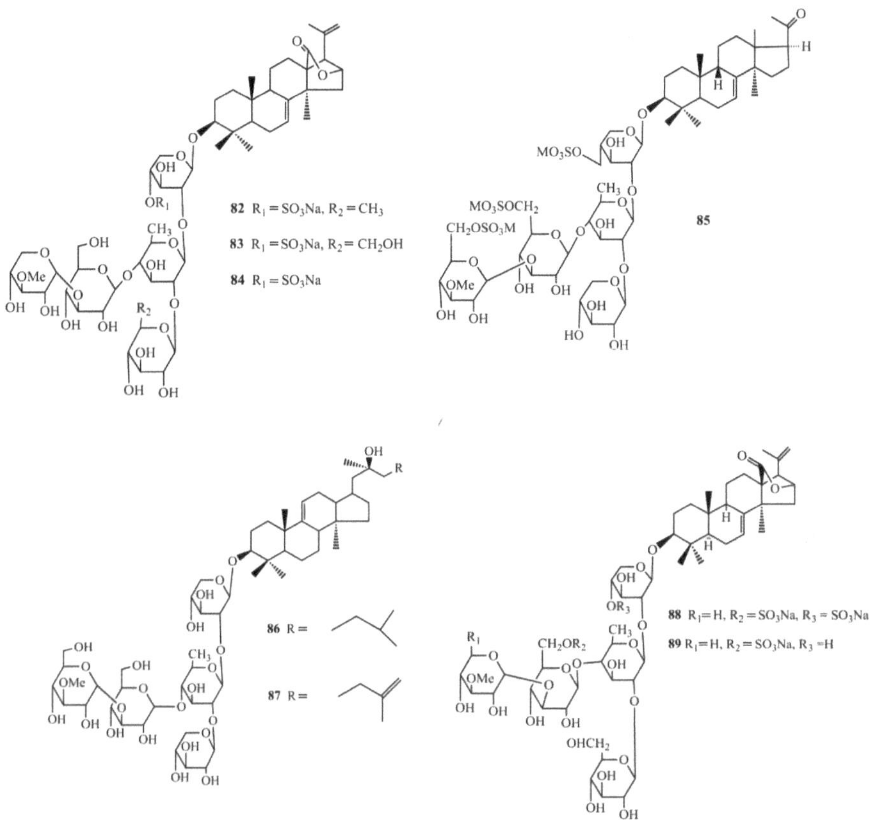

FIGURE 20.4 Triterpenoid glycosides isolated from sea cucumbers having nonholostane type aglycone moiety.

HL-60, A549, lung adenocarcinoma cells: SPC-A4, gastric carcinoma cells: MKN28, gastric carcinoma cells: SGC7901, BEL-7402, human ovarian carcinoma: HO8901, human fetal lung fibroblasts: W138, human epithelial carcinoma cells: A431) (Zhang *et al.*, 2006b). Further, sulfated triterpene glycoside, **78** from *Pseudocolochirus violaceus*, exhibited significant cytotoxicity against cancer cell lines MKN-45 (human gastric adenocarcinoma) and HCT-116 with IC_{50} values in the range of 0.052–0.442 μM, and both compounds showed significantly higher activity against HCT-116 compared to the positive control HCP (Zhang *et al.*, 2007). Moreover, the sulfated triterpene glycosides, philinopsides A (**72**) and B (**73**), showed significant cytotoxicity (IC_{50}: 0.75–3.50 μg/ml) against 10 tumor cell lines (CAKI, HOS, KB-VIN, KB, SM-MEL-2, U87-MG, HCT-8, IA9, A549, and PC3) (Yi *et al.*, 2006). Collectively, all these triterpene glycosides of sea cucumber are very potent cytotoxic agents toward a wide array of cancer types. Detailed studies on the anticancer effects of these compounds are needed to confirm their activity and use them as anticancer therapeutics.

B. Antifungal activity

Antifungal activity is the second most studied biological activity of sea cucumber triterpene glycosides and their polar extracts. Some triterpene glycosides had shown pronounced activity where they have been effective against some fungal strains than commercially available fungicides in a concentration-dependent manner. Polar fraction of the crude extracts of the sea cucumber *Holothuria polii* exhibited a significant antifungal activity in a concentration-related manner (150–300 μg/well) where the strains of *Aspergillus fumigatus* were more susceptible (Ismail *et al.*, 2008). And also the crude methanolic extract of *Actinopyga lecanora* have showed significant antifungal activity against *Candida albicans* (MIC, 62.5 μg/ml), *Candida neoformans* (MIC, 125 μg/ml), *Sporothrix schenckii* (MIC, 62.5 μg/ml), *Trichophyton mentagrophytes* (MIC, 125 μg/ml), and *A. fumigatus* (MIC, 31.2 μg/ml) (Kumar *et al.*, 2007). Sea cucumber triterpene glycoside, holothurin B (**15**) isolated from *A. lecanora*, showed an excellent antifungal activity by acting against 20 test fungal isolates. It has been reported that the activity of holothurin B is more effective than standard antifungal agent fluconazole when tested against *T. mentagrophytes* (MIC range of 1.56 μg/ml), and it has showed a comparable activity to fluconazole against *S. schenckii* and *A. fumigatus* (Kumar *et al.*, 2007). Hemoiedemosides A (**22**) and B (**23**) isolated from sea cucumber *Hemoiedema spectabilis* have been exhibited a considerable antifungal activity against the phytopathogenic fungus *Cladosporium cucumerinum*. It has been found that the disulfated glycoside (**22**) was more active than commercial antiviral agent benomyl at the higher concentrations (20–50 μg/spot) (Chiludil *et al.*, 2003). However, hemoiedemoside B (**23**), differing from hemoiedemoside A in the presence of a third sulfate group at C-6'''', is less active than hemoiedemoside A. Patagonicoside A (**52**), a disulfated tetraglycoside isolated from the holothuroid *Psolus patagonicus*, have also shown antifungal activity against the fungus, *C. cucumerinum* in a concentration-dependent manner (1.5–50 μg/spot). Also in this study, the desulfated derivative of patagonicoside A showed marked reduction in the antifungal activity (Murray *et al.*, 2001). In latter two studies, the lesser active desulfated derivatives differ only by their sulfate substitutions in the glycon part of the structure. Therefore, it can be suggested that the presence of sulfate groups in the oligosaccharide chain plays an important role with regard to the antifungal activity of the triterpene glycoside.

Moreover, the antifungal effects of patagonicoside A (**52**) and hemoiedemoside A (**23**) were compared using a phytopathogenic fungus *C. cucumerinum*. It has been found that patagonicoside A was less active than the hemoiedemoside A at the same concentrations. Both glycosides contain the same oligosaccharide chain and a triterpenoid aglycone with an 18 (20)-lactone. However, patagonicoside A contains a Δ^7 double bond and two hydroxyl functions at C-12 (R) and C-17 (R) in its aglycone part where hemoiedemoside A has Δ^{11} double bond position. Therefore, it could also be suggested that differences in the antifungal activity could be due to structural differences in the aglycone unit (Chiludil *et al.*, 2002).

IV. STRUCTURE–ACTIVITY RELATIONSHIPS

The structure–activity relationships of sea cucumber triterpene glycosides have not been extensively studied; however, several reviews have suggested the structural relationships to their activities exerted. As suggested by many authors, the bioactivity of the triterpene glycosides is a resultant of its strong membranolytic activity. And this membranolytic activity is a function of the structural feature of the glycoside (Kalinin, 2000). The presence of an 18(20)-lactone as the aglycone with at least one oxygen group near it has critical significance for biological activity of glycosides bearing 9(11)-double bond (Kitagawa, 1988). Glycosides with a 7(8)-double bond in their aglycone structure with absence of 16-ketogroup are more active than those with presence of a 16-ketogroup (Kalinin *et al.*, 1996). The characteristics of the attached glycon structure are also critical for the bioactivities of the sea cucumber triterpene glycosides. It has been found that for the actions leading to modification of the cellular membrane, a linear tetrasaccharide chain is significant (Kalinin *et al.*, 1992; Kitagawa, 1988). And also Maltsev *et al.* (1985) have reported that glycosides having quinovose as a second monosaccharide unit are more active over others. The sulfation of the sugar chain is also a significant factor related to bioactivity. A sulfate group at C-4 of the first xylose residue increases the effect against membranes. The absence of a sulfate group at C-4 of the xylose residue in biosides decreases its activity from more than onefold magnitude. However, the presence of a sulfate at C-4 of the first xylose in branched pentanosides having 3-*O*-methyl group as a terminal monosaccharide increases activity. However, the same sulfate can decrease the activity of branched pentanosides, which are having glucose as the terminal residue. And sulfate groups attached to a C-6 position of terminal glucose and 3-*O*-methylglucose residues impart a great reduction in the activity (Kalinin, 2000).

V. PROS AND CONS IN DRUG DEVELOPMENT FROM SEA CUCUMBERS

Up to date, the vast diversity of sea cucumbers paved way for natural product chemists to mine for new bioactive compounds. Among them, sea cucumber triterpene glycosides are the most studied. Similar to that in other marine invertebrates, survival demand has resulted in the evolution of these sophisticated toxic compounds and this fact is confirmed by the proved toxicity of these compounds in biochemical studies. Holothurians are used as ingredients for traditional Chinese medicine for years. And also, holothurin A is marketed in Japan as an ingredient in an antifungal medicine. Even though there are many lead compounds with promising potential to be used as drugs for cancer therapy, the cytotoxicity itself would be a constrain for this purpose, because most of the compounds could be cytotoxic toward normal cells in addition to the cancerous cells. However, in finding therapeutics from natural products, always the preference is given to the compounds having high specificity toward the cancer cells in their cytotoxic action, while minimizing the damage to normal cells. Therefore, considerable cytotoxic studies should be conducted employing the lead compounds before introducing them to the drug development phase. Moreover, the possibility of continuous supply of the product and the ecological importance of the sea cucumber are factors of importance before entering to the drug development phase. However, it is undoubtedly possible to develop new drugs from these interesting natural products of sea cucumbers with extensive research on their activity and toxicity.

VI. CONCLUDING REMARKS

Sea cucumbers can be utilized maximally in the research areas of new drug developments, specifically in the area of anticancer. Due to proven anticancer effects, several triterpenoid glycosides could be potential candidates for development of new drugs. However, the mechanisms of the activity with structure–activity relationship should be studied more, as there is considerable gap in this area compared with isolation rate of new compounds. And also adequate clinical trials are needed in drug development, since the bioactive triterpenoid compounds bares a high cytotoxic potential. And also the mode of the drug delivery into the body should be clearly studied due to the above reason.

REFERENCES

Antonov, A. S., Avilov, S. A., Kalinovsky, A. I., Anastyuk, S. D., Dmitrenok, P. S., Evtushenko, E. V., Kalinin, V. I., Smirnov, A. V., Taboada, S., Ballesteros, M., Avila, C., and Stonik, V. A. (2008). Triterpene glycosides from Antarctic sea cucumbers. 1. Structure of liouvillosides a1, a2, a3, b1, and b2 from the sea cucumber *Staurocucumis liouvillei*: New procedure for separation of highly polar glycoside fractions and taxonomic revision. *J. Nat. Prod.* **71**, 1677–1685.

Antonov, A. S., Avilov, S. A., Kalinovsky, A. I., Anastyuk, S. D., Dmitrenok, P. S., Kalinin, V. I., Taboada, S., Bosh, A., Avila, C., and Stonik, V. A. (2009). Triterpene glycosides from Antarctic sea cucumbers. 2. Structure of achlioniceosides a1, a2, and a3 from the sea cucumber *Achlionice violaecuspidata* (=*Rhipidothuria racowitzai*). *J. Nat. Prod.* **72**, 33–38.

Avilov, S. A., Kalinovsky, A. I., Kalinin, V. I., Stonik, V. A., Riguera, R., and Jimnez, C. (1997). Koreoside A, a new nonholostane triterpene glycoside from the sea cucumber *Cucumaria koraiensis*. *J. Nat. Prod.* **60**, 808–810.

Avilov, S. A., Antonov, A. S., Drozdova, O. A., Kalinin, V. I., Kalinovsky, A. I., Riguera, R., Lenis, L. A., and Jimnez, C. (2000a). Triterpene glycosides from the far eastern sea cucumber *Pentamera calcigera* II: Disulfated glycosides. *J. Nat. Prod.* **63**, 1349–1355.

Avilov, S. A., Antonov, A. S., Drozdova, O. A., Kalinin, V. I., Kalinovsky, A. I., Stonik, V. A., Riguera, R., Lenis, L. A., and Jimnez, C. (2000b). Triterpene glycosides from the far-eastern sea cucumber *Pentamera calcigera*. Monosulfated glycosides and cytotoxicity of their unsulfated derivatives. *J. Nat. Prod.* **63**, 65–71.

Avilov, S. A., Antonov, A. S., Silchenko, A. S., Kalinin, V. I., Kalinovsky, A. I., Dmitrenok, P. S., Stonik, V. A., Riguera, R., and Jimenez, C. (2003). Triterpene glycosides from the far eastern sea cucumber *Cucumaria conicospermium*. *J. Nat. Prod.* **66**, 910–916.

Avilov, S. A., Silchenko, A. S., Antonov, A. S., Kalinin, V. I., Kalinovsky, A. I., Smirnov, A. V., Dmitrenok, P. S., Evtushenko, E. V., Fedorov, S. N., Savina, A. S., Shubina, L. K., and Stonik, V. A. (2008). Synaptosides A and A1, triterpene glycosides from the sea cucumber Synapta maculata containing 3-O-methylglucuronic acid and their cytotoxic activity against tumor cells. *J. Nat. Prod.* **71**, 525–531.

Chiludil, H. D., Muniain, C. C., Seldes, A. M., and Maier, M. S. (2002). Cytotoxic and antifungal triterpene glycosides from the Patagonian sea cucumber *Hemoiedema spectabilis*. *J. Nat. Prod.* **65**, 860–865.

Chiludil, H. D., Murray, A. P., Seldes, A. M., and Maier, M. (2003). Biologically active triterpene glycosides from sea cucumbers (*Holothuroidea, Echinodermata*). *Stud. Nat. Prod. Chem.* **28**, 587–591.

Dang, N. H., Thanh, N. V., Kiem, P. V., Huong, L. M., Minh, C. V., and Kim, Y. H. (2007). Two new triterpene glycosides from the Vietnamese sea cucumber *Holothuria scabra*. *Arch. Pharm. Res.* **30**, 1387–1391.

De Moncerrat, I. M. A. M., Guerra-Rivas, G., Rios, T., and Quijano, L. (2005). Triterpenoid oligoglycosides from the sea cucumber *Stichopus parvimensis*. *J. Nat. Prod.* **68**, 1669–1673.

Encarnacion, R., Carrasco, G., Espinoza, M., Anthoni, U., Nielsen, P. H., and Carsten, C. C. (1989). Neothyoside A, proposed structure of a triterpenoid tetraglycoside from the pacific sea cucumber, *Neothyone gibbosa*. *J. Nat. Prod.* **52**, 248–251.

Findlay, J. A., Yayli, N., and Radic, L. (1992). Novel sulfated oligosaccharides from the sea cucumber *Cucumaria frondosa*. *J. Nat. Prod.* **55**, 93–101.

Han, H., Yi, Y. H., Li, L., Wang, X. H., Liu, B. S., Sun, P., and Pan, M. X. (2007). A new triterpene glycoside from sea cucumber *Holothuria leucospilota*. *Chin. Chem. Lett.* **18**, 161–164.

Han, H., Yi, Y. H., Liu, B. S., Wang, X. H., and Pan, M. X. (2008). Leucospilotaside C, a new sulfated triterpene glycoside from sea cucumber *Holothuria leucospilota*. *Chin. Chem. Lett.* **19**, 1462–1464.

Hegde, V. R., Chan, T. M., Pu, H., Gullo, V. P., Patel, M. G., Das, P., Wagner, N., Parameswaran, P. S., and Naik, C. G. (2002). Two selective novel triterpene glycosides from sea cucumber, Telenata Ananas: inhibitors of chemokine receptor-5. *Bioorg. Med. Chem. Lett.* **12**, 3203–3205.

Ismail, H., Lemriss, S., Aoun, Z. B., Mhadhebi, L., Dellai, A., Kacem, Y., Boiron, P., and Bouraoui, A. (2008). Antifungal activity of aqueous and methanolic extracts from the Mediterranean sea cucumber, *Holothuria polii*. *J. Mycol. Med.* **18**, 23–26.

Kalinin, V. I. (2000). System-theoretical (holistic) approach to the modeling of structural functional relationships of biomolecules and their evolution: An example of triterpene glycosides from sea cucumbers (*Echinodermata, Holothurioidea*). *J. Theor. Biol.* **206**, 151–168.

Kalinin, V. I., Volkova, O. V., Likhatskaya, G. N., Prokofieva, N. G., Agafonova, I. G., Anisimov, M. M., Kalinovsky, A. I., Avilov, S. A., and Stonik, V. A. (1992). Hemolytic activity of triterpene glycosides from Cucumariidae family holothurians and evolution of this group of toxins. *J. Nat. Toxins* **1**, 17–30.

Kalinin, V. I., Prkofieva, N. G., Likhatskaya, G. N., Schentsova, E. B., Agafonova, I. G., Avilov, S. A., and Drozdova, O. A. (1996). Hemolytic activities of triterpene glycosides from the holothurian order Dendrochirotida: Some trends in the evolution of this group of toxins. *Toxicon* **34**, 475–483.

Kalinin, V. I., Avilov, S. A., Kalinina, E. Y., Korolkova, O. G., Kalinovsky, A. I., Stonik, V. A., Riguera, R., and Jimnez, C. (1997). Structure of eximisoside a, a novel triterpene glycoside from the far-eastern sea cucumber *Psolus eximius*. *J. Nat. Prod.* **60**, 817–819.

Kalinin, V. I., Silchenko, A. S., Avilov, S. A., Stonik, V. A., and Smirnov, A. V. (2005). Sea cucumbers triterpene glycosides, the recent progress in structural elucidation and chemotaxonomy. *Phytochem. Rev.* **4**, 221–236.

Kerr, R. G. and Chen, Z. (1995). In vivo and in vitro biosynthesis of saponins in sea cucumbers. *J. Nat. Prod.* **58**, 172–176.

Kitagawa, I. (1988). Research of biologically active marine natural products. *Yakugaku Zasshi* **108**, 398–416.

Kumar, R., Chaturvedi, A. K., Shuklab, P. K., and Lakshm, V. (2007). Antifungal activity in triterpene glycosides from the sea cucumber Actinopyga lecanoraq. *Bioorg. Med. Chem. Lett.* **17**, 4387–4391.

Liu, B. S., Yi, Y. H., Li, L., Zhang, S. L., Han, H., Weng, Y. Y., and Pan, M. X. (2007). Arguside A: A new cytotoxic triterpene glycoside from the sea cucumber *Bohadschia argus* Jaeger. *Chem. Biodivers.* **4**, 2845–2851.

Liu, B. S., Yi, Y. H., Li, L., Sun, P., Han, H., Sun, G. Q., Wang, X. H., and Wang, Z. L. (2008a). Argusides D and E, two new cytotoxic triterpene glycosides from the sea cucumber *Bohadschia argus* Jaeger. *Chem. Biodivers.* **5**, 1425–1433.

Liu, B. S., Yi, Y. H., Li, L., Sun, P., Yuan, W. H., Sun, G. Q., Han, H., and Xue, M. (2008b). Argusides B and C, two new cytotoxic triterpene glycosides from the sea cucumber *Bohadschia argus* Jaeger. *Chem. Biodivers.* **5**, 1288–1297.

Maier, M. S., Roccatagliata, A. J., Kuriss, A., Chiludil, H., Seldes, A. M., Pujol, C. A., and Damonte, E. B. (2001). Two new cytotoxic and virucidal trisulfated triterpene glycosides from the Antarctic sea cucumber *Staurocucumis liouvillei*. *J. Nat. Prod.* **64**, 732–736.

Maltsev, I. I., Stekhova, S. I., Schentsova, E. B., Anisimov, M. M., and Stonik, V. A. (1985). Antimicrobial activities of glycosides from the sea cucumbers of family *Stichopodidae*. *Khim. Pharm. Zhurn* **19**, 54–56.

Miyamoto, T., Togawa, K., Higuchi, R., Komori, T., and Sasaki, T. (1992). Structures of four new triterpenoid oligoglycosides: DS-penaustrosides a, b, c, and d from the sea cucumber *Pentacta australis*. *J. Nat. Prod.* **55**, 940–946.

Murray, A. P., Muniain, C., Seldes, A. M., and Maier, M. S. (2001). Patagonicoside A: A novel antifungal disulfated triterpene glycoside from the sea cucumber *Psolus patagonicus*. *Tetrahedron* **57**, 9563–9568.

Silchenko, A. S., Avilov, S. A., Antonov, A. A., Kalinin, V. I., Kalinovsky, A. I., Smirnov, A. V., Riguera, R., and Jimnez, C. (2002). Triterpene glycosides from the deep-water north-pacific sea cucumber *Synallactes nozawai* Mitsukuri. *J. Nat. Prod.* **65**, 1802–1808.

Silchenko, A. S., Avilov, S. A., Kalinin, V. I., Kalinovsky, A. I., Stonik, V. A., and Smirnov, A. V. (2004). Pseudostichoposide B—New triterpene glycoside with unprecedent type of sulfatation from the deep-water North-Pacific sea cucumber *Pseudostichopus trachus*. *Nat. Prod. Res.* **18**, 565–570.

Silchenko, A. S., Stonik, V. A., Avilov, S. A., Kalinin, V. I., Kalinovsky, A. I., Zaharenko, A. M., Smirnov, A. V., Mollo, E., and Cimino, G. (2005). Holothurins B2, B3, and B4, new triterpene glycosides from Mediterranean sea cucumbers of the genus *Holothuria*. *J. Nat. Prod.* **68**, 564–567.

Silchenko, A. S., Avilov, S. A., Kalinin, V. I., Stonik, V. A., Kalinovsky, A. I., Dmitrenok, P. S., and Stepanov, V. G. (2007). Monosulfated triterpene glycosides from *Cucumaria okhotensis* Levin Et Stepanov, a new species of sea cucumbers from sea of okhotsk. *Russ. J. Bioorg. Chem.* **33**, 73–82.

Silchenko, A. S., Avilov, S. A., Kalinin, V. I., Kalinovsky, A. I., Dmitrenok, P. S., Fedorov, S. N., Stepanov, V. G., Dong, Z., and Stonik, V. A. (2008). Constituents of the sea cucumber *Cucumaria okhotensis*. Structures of okhotosides b1–b3 and cytotoxic activities of some glycosides from this species. *J. Nat. Prod.* **71**, 351–356.

Sun, P., Liu, B. S., Yi, Y. H., Li, L., Gui, M., Tang, H. F., Zhang, D. Z., and Zhang, S. L. (2007). A new cytotoxic lanostane-type triterpene glycoside from the sea cucumber *Holothuria impatiens*. *Chem. Biodivers.* **4**, 450–457.

Sun, G. Q., Li, L., Yi, Y. H., Yuan, W. H., Liua, B. S., Weng, Y. Y., Zhang, S. L., Sun, P., and Wang, Z. L. (2008). Two new cytotoxic nonsulfated pentasaccharide holostane (=20-hydroxylanostan-18-oic acid γ-lactone) glycosides from the sea cucumber *Holothuria grisea*. *Helv. Chim. Acta* **91**, 1453–1460.

Wu, J., Yi, Y. H., Tang, H. F., Zou, Z. R., and Wu, H. M. (2006). Structure and cytotoxicity of a new lanostane-type triterpene glycoside from the sea cucumber *Holothuria hilla*. *Chem. Biodivers.* **3**, 1249–1254.

Yi, Y. H., Xu, Q. Z., Li, L., Zhang, S. L., Wu, H. M., Ding, J., Tong, Y. G., Tan, W. F., Li, M. H., Tian, F., Wu, J. H., Liaw, C. C., *et al.* (2006). Philinopsides A and B, two new sulfated triterpene glycosides from the sea cucumber *Pentacta quadrangularis*. *Helv. Chim. Acta* **89**, 54–63.

Yuan, W. Y., Yi, Y. H., Li, L., Liu, B. S., Zhang, H. W., and Sun, P. (2008). Two triterpene glycosides from the sea cucumber *Bohadschia marmorata* Jaeger. *Chin. Chem. Lett.* **19**, 457–460.

Zhang, S. H., Li, L., Yi, Y. H., and Sun, P. (2006a). Philinopsides E and F, two new sulfated triterpene glycosides from the sea cucumber Pentacta quadrangularis. *Nat. Prod. Res.* **20**, 399–407.

Zhang, S. Y., Yi, Y. H., and Tang, H. F. (2006b). Bioactive triterpene glycosides from the sea cucumber *Holothuria fuscocinerea*. *J. Nat. Prod.* **69**, 1492–1495.

Zhang, S. Y., Yi, Y. H., and Tang, H. F. (2006c). Cytotoxic sulfated triterpene glycosides from the sea cucumber *Pseudocolochirus violaceus*. *Chem. Biodivers.* **3**, 807–817.

Zhang, S., Yi, Y., Tang, H., Li, L., Sun, P., and Wu, J. (2006d). Two new bioactive triterpene glycosides from the sea cucumber *Pseudocolochirus violaceus*. *J. Asian Nat. Prod. Res.* **8**, 1–8.

Zhang, S. Y., Tang, H. F., and Yi, Y. H. (2007). Cytotoxic triterpene glycosides from the sea cucumber *Pseudocolochirus violaceus*. *Fitoterapia* **78**, 283–287.

Zou, Z. R., Yi, Y. H., Wu, H. M., Wu, J. H., Liaw, C. C., and Lee, K. H. (2003). Intercedensides A–C, three new cytotoxic triterpene glycosides from the sea cucumber *Mensamaria intercedens* Lampert. *J. Nat. Prod.* **66**, 1055–1060.

Zou, Z., Yi, Y., Wu, H., Yao, X., Du, L., Jiuhong, W., Liaw, C. C., and Lee, K. H. (2005). Intercedensides D–I, cytotoxic triterpene glycosides from the sea cucumber Mensamaria intercedens Lampert. *J. Nat. Prod.* **68**, 540–546.

Zou, Z., Yi, Y., Wu, H., Yao, X., Du, L., Jiuhong, W., Liaw, C., and Lee, K. (2006). Intercedensides D–I, cytotoxic triterpene glycosides from the sea cucumber *Mensamaria intercedens* Lampert. *J. Nat. Prod.* **69**, 1530.

CHAPTER 21

Chitooligosaccharides as Potential Nutraceuticals: Production and Bioactivities

Jae-Young Je*,[1] and **Se-Kwon Kim†**

Contents

Abstract

Chitooligosaccharide (COS), a derivative of chitosan, can be produced by either chemical or enzymatic hydrolysis. Looking back through past research, several technological approaches have been taken to prepare COSs, and enzymatic approaches are favorable due to their environmentally friendly methods, safety, and a lack of toxicity. Similar to chitosan, COSs can also be considered potential nutraceuticals due to their versatile biological activities, water-soluble properties, and

* School of Food Technology and Nutrition, Chonnam National University, Yeosu, Republic of Korea
† Department of Chemistry and Marine Bioprocess Research Center, Pukyong National University, Busan, Republic of Korea
[1] Corresponding author: Jae-Young Je, *E-mail address:* jjy1915@jnu.ac.kr

Advances in Food and Nutrition Research, Volume 65
ISSN 1043-4526, DOI: 10.1016/B978-0-12-416003-3.00021-4

absorption properties in the intestine. Therefore, this chapter provides certain methods for the preparation of COSs and addresses their biological properties such as antihypertensive, antioxidant, antitumor, antimicrobial, and other select biological activities.

I. INTRODUCTION

Since the discovery of chitosan in the late 1850s, many studies have been carried out to determine its fundamental physicochemical properties. Chitosan is a nontoxic and biocompatible cationic polysaccharide produced by partial deacetylation of chitin isolated from naturally occurring crustacean shells. Therefore, chitosan has received much attention as a functional biopolymer for diverse applications, especially in biomedicine (Felt *et al.*, 1998), food (Shahidi and Synowieck, 1991), and cosmetics (Majeti and Kumar, 2000). However, in some fields such as biomedicine and the food industry, the application of chitosan is limited by its water-insoluble properties. Moreover, its biological properties are dependent on its molecular weight. Therefore, much attention has been paid to converting chitosan to oligomers or oligosaccharides. In addition, direct absorption of chitosan in the human intestine is doubtful due to its physicochemical properties, and the human gastrointestinal tract does not possess enzymes that digest chitosan to chitooligosaccharides (COSs), which have lower viscosity with relatively smaller molecular sizes and are readily absorbable in *in vivo* systems. Water-soluble COSs are beneficial for applications in various fields, especially in pharmaceuticals and food, because COSs have exhibited versatile biological activities including lowering of blood cholesterol (Macchi, 1996), lowering of high blood pressure (Giustina and Ventura, 1995), protective effects against infections (Tokoro *et al.*, 1989), controlling arthritis (Lee *et al.*, 2003), and enhancing antitumor properties (Nishimura *et al.*, 1984).

Chemical and enzymatic depolymerization processes have been used to obtain COSs, and biological properties of the obtained COSs have also been evaluated. Therefore, in this chapter, we review the production of COSs by various methods as well as some biological properties of COSs.

II. PREPARATION OF CHITOOLIGOSACCHARIDES FROM CHITOSAN

A. Chemical hydrolysis

Numerous processes have been employed for the production of COSs, which usually involve hydrolysis in acidic, alkaline, or oxidizing conditions. In general, many acids including hydrochloric acid, nitrous acid, phosphoric acid, and hydrogen fluoride have been used to obtain COSs (Defaye *et al.*, 1989; Horowitz *et al.*, 1957; Omura *et al.*, 1991; Peniston and Johnson, 1975). In early studies, COSs were produced by partial hydrolysis of chitosan with concentrated HCl. However, experimental results have shown that hydrolysis by HCl produces low yields of COSs and a larger amount of glucosamine units. In recent years, chemical depolymerization of chitosan was obtained in the solid state by means of HCl. This method allowed the production of more specific COSs without changes in structure (Belamie *et al.*, 1997). Nitrous acid depolymerization of chitosan to produce COSs has also been conducted and the obtained COSs were less than hexamers in form (Sashiwa *et al.*, 1993). The oxidative depolymerization of chitosan to produce COS using $NaNO_2$ was also performed. In this case, chitosan with a higher molecular weight was more sensitive to depolymerization and no structural changes were observed (Mao *et al.*, 2004).

There are several reports on the production of COSs from chitin or chitosan by free radicals. Nordtveit *et al.* (1994) demonstrated that the viscosity of chitosan solution decreased rapidly by random depolymerization of chitosan in the presence of hydrogen peroxide and $FeCl_3$. Tanioka *et al.* (1996) showed that Cu(II), ascorbate, and UV–H_2O_2 systems gradually reduced the molecular weight of chitosan. Chang *et al.* (2001) studied kinetic behavior during the free radical degradation of chitosan using hydrogen peroxide. The treatment resulted in breaking β-(1,4) glycosidic linkages, and the formation of glucosamine and COSs depended on the concentration of hydrogen peroxide, temperature, and the physicochemical properties of chitosan. In recent years, preparation of a series of COSs by hydrogen peroxide from chitosan has also been studied. Tian *et al.* (2010) produced COSs with molecular weights ranging from 1.7 to 3.8×10^3 by degradation of a high-molecular weight chitosan with hydrogen peroxide and obtained specific molecular weight COSs by using selective precipitation with ethanol solutions.

The chemical depolymerization of chtiosan is a very common and fast method to produce a series of COSs, but this procedure has some defects such as high cost, low yield, and acidic waste by use of concentrated HCl (Choi *et al.*, 2002). In addition, the COSs prepared by acid hydrolysis may not be suitable for human consumption because of possible production of toxic compounds during hydrolysis (Uchida *et al.*, 1989). An alternative to chemical treatment is irradiation. Radiation can provide a useful tool for the degradation of different polymers. Chitosan fibers and films were irradiated by γ rays, and the viscosity average molecular weight of the polymers decreased with increasing irradiation dose (Lim *et al.*, 1998). Choi *et al.* (2002) also studied the preparation of chitosan oligomers by irradiation. They found that a rapid decrease of viscosity was observed with up to 10kGy, and then this decrease in viscosity gradually slowed with increasing irradiation dose. In addition, chitosan oligomers from dimers to hexamers were produced by the radiational degradation of chitosan when irradiated with a 100kGy dose.

B. Enzymatic hydrolysis

Although much research has been performed regarding the production of COSs using chemical methods, there are some drawbacks in terms of commercialization due to the development of certain toxic compounds, the higher risk associated with environmental pollution, lower production yields, and changes in original structure. In contrast to chemical hydrolysis, enzymatic hydrolysis is environmental friendly and preferable to chemical hydrolysis because it can be controlled more easily, and the products can be obtained without any modifications as normally seen with acid hydrolysis. Enzymatic hydrolysis with specific and/or nonspecific enzymes has been performed to produce COSs from chitosan.

In specific enzymatic hydrolysis, various chitosanases obtained from microbes are used to produce COSs at a relatively higher proportion from chitosan. Chitosanase from *Penicillium islandicum* was able to hydrolyze 30% deacetylated chitosan to yield disaccharides (Fenton and Evelergh, 1981). Chitosanase from *Bacillus* sp. nr 7-M could produce chitodiose, chitotriose, chitotetraose, and hetero-COS (Izume *et al.*, 1992). A chitosan-degrading fungus, *Aspergillus* sp. Y2K, was isolated from soil, and chitosanase from *Aspergillus* sp. Y2K was produced using chitosan as the sole carbon source (Cheng and Li, 2000). By endo-splitting activity, the chitosanase hydrolyzes chitosan, especially higher deacetylated chitosan, to form COS with chitotriose, chitotetraose, and chitopentaose as the major products. In addition, the enzyme hydrolyzes chitohexaose to form chitotriose, while the chitopentaose and shorter oligomers remain intact. Crude enzyme with chitosanolytic activity from microbes was also used for the production of COSs from chitosan. Kuo *et al.* (2004) hydrolyzed chitosan by using crude enzyme from *Bacillus cereus* NTU-FC-4 in a membrane reactor to produce COSs, and the major COSs created in the reactor were chitobiose, chitotriose, chitotetraose, chitopentaose, and chitohexaose. Liang *et al.* (2007) obtained crude enzyme from *Bacillus amyloliquefaciens* V656, and water-soluble chitosan was hydrolyzed by a crude enzyme solution at 37°C for 1, 12, and 24h, respectively. They found that the hydrolysates with 12h of hydrolysis contained optimal hexamers. Jung *et al.* (2007) also produced crude enzyme with chitinase activity from *Paenibacillus illinoisensis* KJA-424 and investigated the production of *N*-acetyl COSs from colloidal chitin. After 24h of hydrolysis, contents of chitobiose, chitotriose, chitoheptose, and chitooctose were 0.13 (yield of 4.9%), 0.03 (yield of 1.2%), 0.01 (yield of 4.1%), and 0.24mg/mL (yield of 9.6%), respectively. However, some drawbacks using chitosanase to produce COSs from chitosan exist for industrial applications. Chitosanase, the specific enzyme for chitosanolysis, is expensive, unavailable in bulk, and results in preferential formation of chitooligomers–monomers due to its specificity, whereas nonspecific enzymes, which are inexpensive and commercially available, result mainly in the formation of low-molecular weight chitosan (LMWC; chitosan oligomers; Kittur *et al.*, 2003; Vishu Kumar *et al.*, 2004).

Several nonspecific enzymes, especially proteolytic enzymes including papain, pepsin, and pronase, were employed for the production of COSs from chitosan. Chitosan depolymerization was performed using immobilized papain. The advantage of immobilized papain was retrieval after the reaction and the retrieval enzyme could be reused, although the yield of LMWC and its molecular weight were low (Terbojevich *et al.*, 1996). Vishu Kumar *et al.* (2004) also studied chitosan depolymerization by using papain, and they achieved the production of LMWC with desired molecular weight (4.1–5.6 kDa) in higher yields at optimum conditions of pH 3.5 and 37 °C for 1–5 h. Pepsin has also been used to hydrolyze chitosan, and the optimum chitosanolysis condition was determined as pH 5.0 and 45 °C. At the optimum condition, the molecular weights of chitosans obtained using pepsin were in the range of 4.6–10 kDa depending on the reaction time (1–5 h) (Vishu Kumar and Tharanathan, 2004). Chitosan depolymerization by pronase has also been studied by the same research group. Homogeneous LMWCs of 8.5–9.5 kDa were obtained by pronase catalyzed nonspecific depolymerization at pH 3.5 and 37 °C.

Nonspecific enzymes such as cellulase, hemicellulase, lysozyme, pectinase, and α-amylase were also applied for chitosan depolymerization. Among them, many researchers have focused on cellulase for the production of COSs from chitosan due to the structural similarity of chitosan and cellulose, which are polymers of D-glucose linked by β-1,4-glycosidic linkages (in chitosan, the C-2 hydroxyl groups are replaced by amino groups), and it seems that the enzyme does not strongly recognize the group at the C-2 position in glucose or glucosamine residues when the enzyme–substrate complex is formed (Xia *et al.*, 2008). In fact, most cellulases have nonspecific hydrolytic action on chitosan, and some are even superior to specific chitosanases in degrading chitosan. Lin *et al.* (2009) prepared LMWC with the aid of cellulase from two different deacetylated chitosans (80% and 92%), and they obtained 6–8.1 kDa of LMWC depending on their degree of deacetylation (DD) and hydrolysis times. Xie *et al.* (2009) also applied cellulase derived from *Aspergillus niger* to prepare COSs, and the products were mainly COSs with degrees of depolymerization (DP) in the 3–11 range. Tasi *et al.* (2004) used cellulose to hydrolyze chitosan, producing COSs in the range of monomers to hexamers as well as LMWC (12 kDa). Qin *et al.* (2002) used commercial cellulase from *T. viride* and hemicellulase to produce water-soluble LMWC, further separated by ultrafiltration (UF) membranes based on their molecular weights. Zhang *et al.* (1999) produced COSs with 3–10 DP using cellulase complexes from 76% deacetylated chitosan. As described above, cellulase can produce COS or water-soluble LMWC by scission of glycosidic linkages, and these enzymes are relatively economical enzymes compared to chitosanases, providing an opportunity for commercial use instead of chitosanases to produce COSs or water-soluble LMWC.

C. Continuous production of COSs

Even though a set of studies regarding the production of COSs with desired molecular weights by chemical and enzymatic hydrolysis have been conducted, these methods are not applicable for large-scale production of COSs due to inefficient production of desirable chain lengths of COSs. As described above, initial enzymatic hydrolysis was carried out in batch reactors, where the enzyme was mixed with its substrate and allowed to breakdown glycosidic linkages. The main disadvantage of this system is not only a higher cost with the use of chitosanase but also low yields with desirable molecular weight COSs. To overcome these problems, the immobilization of enzymes was developed. Jeon *et al.* (1998) developed immobilized chitosanase to produce COSs with higher DP, and enzyme activity was the highest when immobilized on chitin as a carrier. However, in this method the yield of COSs was lower, because the immobilized enzyme showed a lower affinity and lower reaction rate than the free enzyme. For the production of COSs with higher DP and yield, an enzyme reactor system together with an UF membrane reactor was developed (Jeon and Kim, 2000a). This system allows the hydrolysis of 11 batches of substrate for the same amount of enzyme used in the batch reactor. However, highly viscous chitosan and the accumulation of substrate increased the fouling of the membrane and resulted in increased transmembrane pressure, which did not allow continuous production of COSs in the UF membrane system. Therefore, reducing the viscosity of chitosan prior to treatment in the UF membrane system is a requisite for a more effective continuous production system.

The continuous production of COSs from chitosan can be achieved with a dual reactor system that consists of an UF membrane reactor and a column reactor packed with an immobilized enzyme (Jeon and Kim, 2000b), as shown in Fig. 21.1. This system firstly reduces the viscosity of chitosan by the immobilized enzyme prepacked in the column reactor and the product is supplied to the UF membrane reactor for the production of COSs. The low viscosity of the hydrolyzed chitosan does not create fouling problems under the controlled conditions. This method ensures a

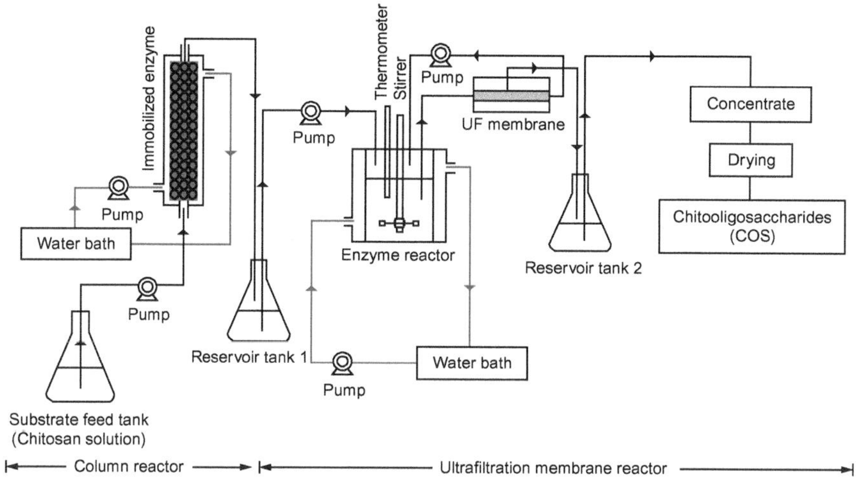

FIGURE 21.1 Schematic diagram of the dual reactor system used for continuous production of chitooligosaccharides (from Jeon and Kim, 2000b, with permission).

greater productivity per unit enzyme, the ability to control molecular weight distribution, and a more efficient continuous production process compared to those of conventional methods.

III. BIOLOGICAL PROPERTIES OF CHITOOLIGOSACCHARIDES

As for the biological activities of chitosan, LMWC or COSs show versatile biological activities, and these activities are dependent on their molecular weight and DD. Unlike high-molecular weight chitosan, COSs are easily absorbed through the intestine, quickly enter the blood flow, and have systemic biological effects in organisms. Therefore, in this chapter, the biological properties of COSs prepared from enzymatic hydrolysis are discussed.

A. Antihypertensive activity

Hypertension is one of the major independent risk factors for arteriosclerosis, stroke, myocardial infarction, and end-stage renal disease. It is a worldwide problem of epidemic proportions, which affects 15–20% of all adults, and is the most common serious chronic health problem. In our body, the renin–angiotensin system (RAS) plays a key role in the control of blood pressure. Renin hydrolyzes angiotensinogen to angiotensin I (Ang I), which is further converted by soluble or endothelial cell-associated angiotensin I converting enzyme (ACE) to angiotensin II (Ang II), a powerful vasoconstrictor that has been identified as a major factor in hypertension (Peach, 1997). Therefore, inhibition of renin and/or ACE activity by COS is a promising antihypertensive effect. Hong *et al.* (1998) studied the ACE inhibitory activity of different COSs and identified that chitotriose is more effective in lowering blood pressure compared to other oligomers. Hetero-COSs were prepared using a UF membrane reactor system and divided into three categories depending on their molecular weight. The ACE inhibitory activity of the hetero-COSs was dependent on the DD of the chitosans. 50-COSs with molecular weights of 1–5kDa, which were COSs hydrolyzed from 50% deacetylated chitosan, exhibited the highest ACE inhibitory activity with an IC_{50} value of 1.22 ± 0.13mg/mL (Park *et al.*, 2003a). However, the opposite result was observed in a renin inhibitory assay, and 90-COSs with molecular weights of 1–5kDa showed the highest renin inhibition activity with an IC_{50} value of 0.51 mg/mL (Park *et al.*, 2008). Since renin is rate-limiting enzyme in RAS and catalyzes the hydrolysis of only one naturally occurring substrate, angiotensinogen, it seems that renin inhibition is thought to be an attractive target for antihypertension strategies.

B. Antimicrobial activity

Chitosan is capable of inhibiting the growth of some microorganisms including several bacterial strains. In general, it is considered that chitosan is more effective than COS in inhibiting the growth of bacteria. However, its high-molecular weight has limited its practical applications due to its insolubility at pH values above 6.3. Therefore, the production of COS from chitosan has received much attention due to its solubility at neutral pH to utilize chitosan in its oligosaccharide form.

Like chitosan, the antibacterial effects of COS are influenced by a number of factors such as DP, DD, type of microorganism, and certain other physical–chemical properties. Three kinds of COSs with different molecular weights were produced by employing a dual reactor system and tested against Gram-positive (*Escherichia coli*, *E. coli* O-157, *Salmonella typhi*, and *Pseudomonas aeruginosa*) and Gram-negative (*Streptococcus mutans*, *Micrococcus luteus*, *Staphylococcus aureus*, *Staphylococcus epidermidis*, and *Bacillus subtilis*) bacteria (Jeon *et al.*, 2001). It was found that COSs showed considerable antimicrobial activities against Gram-positive and Gram-negative bacteria, and high-molecular weight COSs (5–10kDa) exhibited higher antimicrobial activity than those of low-molecular weight COSs. However, the antimicrobial activity of COS was lower than that of chitosan, and COSs with molecular weights higher than 10 kDa seemed to be suitable for antimicrobial activity. They also found that

COSs are more effective against pathogenic bacteria associated with human diseases in comparison to nonpathogenic bacteria (Table 21.1).

It is also well known that positively charged COS can interact with negatively charged bacterial cell walls, resulting in suppression of the metabolic activity of bacteria by blocking nutrient permeation through the cell wall. Therefore, the higher number of free amino groups in COS can make stronger interactions with bacterial cells (Tsai *et al.*, 2002). The number of primary amino groups is dependent on the DD and DP of COSs, and it has been observed that the death rate of bacterial cells tends to increase upon an increase in the DD of COSs (Tsai *et al.*, 2002). Chung *et al.* (2004) showed that 85–95% DD COSs were responsible for providing the highest antimicrobial activity. In another study, Park *et al.* (2004) prepared hetero-COSs with different molecular weights and DDs (50%, 75%, and 90%) and tested against a number of bacterial strains. They found that the hetero-COSs exhibited antibacterial activity in the same pattern as COSs in other studies, and high-molecular weight COSs showed the highest antibacterial activity. However, 75% DD COSs exhibited better antibacterial activity than those of 50% and 90% DD COSs, and the 75% DD COSs inhibited the growth of *Vibrio parahaemolyticus*,

TABLE 21.1 Antibacterial activities of chitooligosaccharides

		Antibacterial activity (%)[a]		
	Bacteria	HMWCOSs[b]	MMWCOSs[c]	LMWCOSs[d]
Gram-negative bacteria	*Escherichia coli*	98±0	62±6	51±7
	E. coli O-157	71±3	56±4	60±2
	Salmonella typhi	91±2	88±0	89±0
	Pseudomonas aeruginosa	47±5	35±5	22±8
Gram-positive bacteria	*Streptococcus mutans*	100±0	99±0	99±0
	Micrococcus luteus	97±3	95±0	93±9
	Staphylococcus aureus	82±0	57±3	23±1
	Staphylococcus epidermidis	63±5	60±5	63±7
	Bacillus subtilis	70±0	67±3	63±7

Source: Jeon *et al.* (2001).

[a] After incubation of the bacteria treated with 0.1% COS, antibacterial activity (%) was calculated by the colony formed on the medium, compared to the control.

[b] High-molecular weight COSs (5–10 kDa).

[c] Medium-molecular weight COSs (1–5 kDa).

[d] Low-molecular weight COSs (below 1 kDa).

a food-borne Gram-negative bacterial strain. These results indicate that the antibacterial activity of COSs is dependent on its molecular weight and DD as well as the tested bacteria.

The antimicrobial activity of COSs produced by nonspecific enzymes was also investigated. Homogeneous LMWCs with molecular weights of 8.5–9.5kDa were prepared by pronase depolymerization, and their antimicrobial activity against *B. cereus* and *E. coli* was investigated. The growth inhibitory effect of the LMWC was greater toward *B. cereus* than against *E. coli*. The LMWC of 9.5kDa and 81% DD caused complete growth inhibition of *B. cereus*, whereas for *E. coli*, the LMWC of 8.5kDa and 86% DD was more effective, indicating a differential role played by both the molecular weight and DD of LMWC (Vishu Kumar *et al.*, 2007). Lysozyme assisted-hydrolyzed LMWC also showed antimicrobial activity against *E. coli*, *S. aureus*, and *Campylobacter jejuni*, and the activity was dependent on their molecular weight and DD-like COSs prepared by chitosanase (Lin *et al.*, 2009; Mengíbar *et al.*, 2011).

C. Antitumor activity

It was demonstrated that COSs can inhibit the growth of tumor cells by exerting immune-enhancing effects. Suzuki *et al.* (1986) demonstrated that COSs inhibited tumor growth through an increase in immune effects. The COSs with DP4-7 showed strong inhibition of ascites cancer in BALB/c mice, while *N*-acetylchitohexaose and chitohexaose exhibited strong inhibiting effects for Sarcoma 180 (S180) and MM156 solid tumor growth in syngenic mice. The same results were also observed by Tokoro *et al.* (1998). They suggested that the observed antitumor activities were not due to direct killing of tumor cells but might have been due to increased production of lymphokines, leading to the manifestation of antitumor effects through the proliferation of cytolytic T-lymphocytes. Qin *et al.* (2002) prepared water-soluble chitosan with a molecular weight of 5kDa by hemicellulase, and intraperitoneal injection of the water-soluble chitosan inhibited the growth of S180 tumor cells in mice, and the maximum inhibitory rate reached 64.2%. Oral administration was also effective for decreasing tumor weight, but inhibitory activity was lower than that by intraperitoneal injection. They also found that the water-soluble chitosan with a higher molecular weight than hexamer forms might have better antitumor activity.

As in the results of Qin and coworkers, the antitumor activity of COSs was dependent on its structural characteristics such as DD and molecular weight. Jeon and Kim (2002) prepared COSs with different molecular weights by a UF membrane reactor system and investigated the antitumor activity of COSs against S180 or uterine cervix carcinoma No. 14 (U14) tumor-bearing BALB/c mice. They found that 1.5–5.5kDa COSs could effectively inhibit the growth of S180 or U14, and the optimum dose of these COSs for inhibiting these tumors was approximately 20mg/kg/day, which resulted in 66.6% and 73.6% tumor inhibition rates against S180 and U14-bearing mice, respectively (Table 21.2). Further, the effect was found to be caused by inducing interleukin-2 production via the development of lymphoid organs such as the spleen and thymus (Jeon and Kim, 2001). Therefore, COSs and their *N*-acetylated analogs that are soluble in basic physiologic environments may serve as good candidates for developing nutraceuticals.

D. Antioxidant activity

The formation of free radicals and reactive oxygen species (ROS) is an unavoidable consequence in aerobic organisms during respiration. Under normal conditions, ROS and free radicals are effectively eliminated by antioxidant defense systems such as antioxidant enzymes and nonenzymatic factors. However, under pathological conditions, the balance between the generation and elimination of ROS is broken, resulting in DNA, membrane lipids, and proteins becoming damaged by ROS-mediated oxidative stress. These events are strongly involved in many health disorders such as diabetes mellitus, cancer, and neurodegenerative and inflammatory diseases (Butterfield *et al.*, 2002; Pryor, 1982). Therefore, the body must be nourished with a diet that includes adequate antioxidants, and there has been growing interest to develop naturally occurring antioxidants to overcome the radical-mediated deleterious effects that occur in biological systems.

TABLE 21.2 Effects of COSs on tumor growth inhibition in BALB/c mice

		Tumor inhibition (%)	
Sample	Dose (mg/kg/day)	Sarcoma 180 (S180)	Uterine cervix carcinoma (U14)
HMWCOSs	50	–	–
	20	12.7	11.9
	10	61.7	15.3
MMWCOSs	50	66.6	73.6
	20	35.5	61.4
	10	28.4	26.6
LMWCOSs	50	12.4	27.1
	20	15.3	7.8
	10	5.7	3.5

Source: Jeon and Kim (2002).

The antioxidant properties of COSs and hetero-COSs prepared using a UF membrane reactor system were investigated, and their antioxidant effects were dependent on their molecular weight and DD values (Je *et al.*, 2004; Park *et al.*, 2003b). Low-molecular weight in a range from 1 to 5kDa and high-DD COS showed strong DPPH, hydroxyl, superoxide, and carbon-centered radical scavenging activities. Mendis *et al.* (2007) analyzed the cellular antioxidant ability of COSs in a murine melanoma cell line (B16F1), and COS below 1kDa in molecular weight exhibited appreciable suppression of intracellular ROS and elevated the intracellular glutathione (GSH) level in B16F1 cells (Fig. 21.2). Chitin oligosaccharides have also shown antioxidant activities in cell lines. Chitin oligosaccharides (NA-COS; 229.21–593.12Da) inhibited myeloperoxidase activity in HL-60 cells, as well as the oxidation of DNA and protein in RAW264.7 macrophage cells. Further, NA-COS inhibited the formation of intracellular ROS and increased GSH level (Ngo *et al.*, 2008). Therefore, COSs with specific molecular weights may be helpful in protecting body systems from oxidative stress since chitosan has shown antioxidant activity *in vivo*.

E. Other biological activities

COSs also have been reported to have protective effects against oxidative stress in various cell lines. In human umbilical vein endothelial cells, hydrogen peroxide-induced stress injuries were effectively protected against COSs by means of inhibiting intracellular ROS formation, suppressing the production of lipid peroxidation compounds such as malondialdehyde, and restoring the activities of antioxidant enzymes including superoxide dismutase and GSH peroxidase (Liu *et al.*, 2009). Karadeniz *et al.* (2010) also studied the protective effects of COSs against

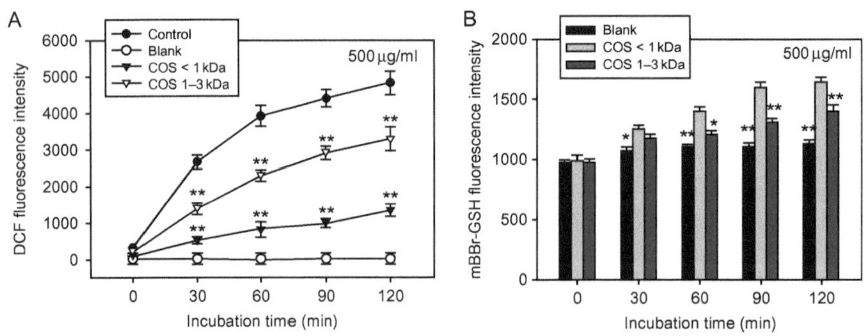

FIGURE 21.2 (A) Effects of COS on inhibition of intracellular generation of ROS and (B) regulation of GSH level in B16F1 cells (from Mendis *et al.*, 2007, with permission).

hydrogen peroxide-induced deterioration in pancreatic β-cells. COSs of 3–5kDa strongly protected pancreatic β-cells against hydrogen peroxide by inhibiting intracellular ROS formation as well as through antiapoptotic actions by decreasing the expression of caspase genes and the apoptotic Bax gene, while increasing the antiapoptotic Bcl-2 gene. In addition, neuronal damage induced by oxidative stress and glucose deprivation was significantly reduced by COSs, and these results suggest that COSs have promising potential to serve as a neuroprotective agent against neurodegenerative diseases such as Alzheimer's disease (AD) (Khodagholi *et al.*, 2010; Xu *et al.*, 2011).

In two studies, COSs were found to inhibit β-secretase activity, which is the initial step in the pathological cascade of AD, and to suppress the level of protein expression and acetylcholinesterase (AChE) activity induced by amyloid β peptide in PC12 cells (Byun *et al.*, 2005; Lee *et al.*, 2009). COSs of 90% DD and 3–5kDa showed the highest β-secretase inhibitory activity and also inhibited AChE activity and protein expression induced by amyloid β peptide in PC12 cells. These results further suggest that COSs would be helpful for preventing neurodegenerative diseases.

IV. SAFETY OF CHITOOLIGOSACCHARIDES

COSs have significant applications in the development of functional foods. Therefore, their cytotoxicity and bioavailability should be clarified. Experiments on the subacute toxicity of COSs in Sprague–Dawley rats were performed and revealed that the compounds did not induce any mortality or changes in blood chemistry, urinalysis data, or body weight in the rats (Kim *et al.*, 2001). A safety evaluation of short-term exposure to COSs was also performed in mice (Qin *et al.*, 2006). After oral administration at doses of 1–10g/kg body weight for 30days, no abnormal symptoms, clinical signs, or deaths were observed during the test, and short-term ingestion of COSs was deemed nontoxic.

REFERENCES

Belamie, E., Domard, A., and Giraud-Guille, M. M. (1997). Study of the solid-state hydrolysis of chitosan in presence of HCl. *J. Polym. Sci. A Polym. Chem.* **35**, 3181–3191.

Butterfield, D. A., Castenga, A., Pocernich, C. B., Drake, J., Scapagnini, G., and Calabrese, V. (2002). Nutritional approaches to combat oxidative stress in Alzheimer's diseases. *J. Nutr. Biochem.* **13**, 444–464.

Byun, H. G., Kim, Y. T., Park, P. J., Lin, X. L., and Kim, S. K. (2005). Chitooligosaccharides as a novel β-secretase inhibitor. *Carbohydr. Polym.* **61**, 198–202.

Chang, K. L. B., Tai, M. C., and Cheng, F. H. (2001). Kinetics and products of the degradation of chitosan by hydrogen peroxide. *J. Agric. Food Chem.* **49**, 4845–4851.

Cheng, C. Y. and Li, Y. K. (2000). An *Aspergillus* chitosanase with potential for large-scale preparation of chitosan oligosaccahrides. *Biotechnol. Appl. Biochem.* **32**, 197–203.

Choi, W. S., Ahn, K. J., Lee, D. W., Byun, M. W., and Park, H. J. (2002). Preparation of chitosan oligomers by irradiation. *Polym. Deg. Stab.* **78**, 533–538.

Chung, Y. C., Su, Y. P., Chen, C. C., Jia, G., Wang, H. L., Wu, J. C., and Lin, J. G. (2004). Relationship between antibacterial activity of chitosan and surface characteristics of cell wall. *Acta Pharmacol. Sin.* **25**, 932–936.

Defaye, J., Gadelle, A., and Pedersen, C. (1989). Chitin and chitosan oligosaccharides. *In* "Chitin and Chitosan", (G. Skjak-Braeak, T. Anthonsen, and P. Sandford, Eds), p. 415. Elsevier Applied Science, London.

Felt, O., Buri, P., and Gurny, R. (1998). Chitosan: A unique polysaccharide for drug delivery. *Drug Dev. Ind. Pharm.* **24**, 979–993.

Fenton, D. and Evelergh, D. E. (1981). Purification and mode of action of a chitosanase from *Penicillum islandicum. J. Gen. Microbiol.* **126**, 151–165.

Giustina, A. and Ventura, P. (1995). Weight-reducing regimens in obese subjects: Effects on a new dietary fiber integrator. *Acta Toxicol. Ther.* **16**, 199–214.

Hong, S. P., Kim, M. H., Oh, S. W., Han, C. H., and Kim, Y. H. (1998). ACE inhibitory and antihypertensive effect of chitosan oligosaccharides in SHR. *Korean J. Food Sci. Technol.* **30**, 1476–1479.

Horowitz, S. T., Roseman, S., and Blumenthal, H. J. (1957). The preparation of glucosamine oligosaccharides. I. Separation. *J. Am. Chem. Soc.* **709**, 5046–5049.

Izume, M., Nagae, S., Lawagishi, H., Mitsutomi, M., and Ohtakara, A. (1992). Action pattern of Bacillus sp. 7-M chitosanase on partially *N*-acetylated chitosan. *Biosci. Biotechnol. Biochem.* **56**, 448–453.

Je, J. Y., Park, P. J., and Kim, S. K. (2004). Free radical scavenging properties of hetero-chitooligosaccharides using an ESR spectroscopy. *Food Chem. Toxicol.* **42**, 381–387.

Jeon, Y. J. and Kim, S. K. (2000a). Production of chitooligosaccharides using ultrafiltration membrane reactor and their antibacterial activity. *Carbohydr. Polym.* **41**, 133–141.

Jeon, Y. J. and Kim, S. K. (2000b). Continuous production of chitooligosaccharides using a dual reactor system. *Proc. Biochem.* **35**, 623–632.

Jeon, Y. J. and Kim, S. K. (2001). Potential immuno-stimulating effect of antitumoral fraction of chitosan oligosaccharides. *J. Chtin Chitosan* **6**, 163–167.

Jeon, Y. J. and Kim, S. K. (2002). Antitumor activity of chitosan oligosaccharides produced in an ultra filtration membrane reactor system. *J. Microbiol. Biotechnol.* **12**, 503–507.

Jeon, Y. J., Park, P. J., Byun, H. G., Song, B. K., and Kim, S. K. (1998). Production of chitosan oligosaccharides using chitin-immobilized enzyme. *Korean J. Biotechnol. Bioeng.* **13**, 147–154.

Jeon, Y. J., Park, P. J., and Kim, S. K. (2001). Antimicrobial effect of chitooligosaccharides produced by bioreactor. *Carbohydr. Polym.* **44**, 71–76.

Jung, W. J., Souleimanov, A., Park, R. D., and Smith, D. L. (2007). Enzymatic production of *N*-acetyl chitooligosaccharides by crude enzyme derived from *Paenibacillus illioisensis* KJA-424. *Carbohydr. Polym.* **67**, 256–259.

Karadeniz, F., Artan, M., Kong, C. S., and Kim, S. K. (2010). Chitooligosaccharides protect pancreatic β-cells from hydrogen peroxide-induced deterioration. *Carbohydr. Polym.* **82**, 143–147.

Khodagholi, F., Eftekharzadeh, B., Maghsoudi, N., and Rezaei, P. F. (2010). Chitosan prevents oxidative stress-induced amyloid β formation and cytotoxicity in NT2 neurons: Involvement of transcription factors Nrf2 and NF-κB. *Mol. Cell. Biochem.* **337**, 39–51.

Kim, S. K., Park, P. J., Yang, H. P., and Han, S. S. (2001). Subacute toxicity of chitosan oligosaccharides in Sprague-Dawley rats. *Arzneimittelforschung* **51**, 769–774.

Kittur, F. S., Vishu Kumar, A. B., Gowda, L. R., and Tharanathan, R. N. (2003). Chitosanolysis by a pectinase isozyme of Aspergillus niger—A non specific activity. *Carbohydr. Polym.* **53**, 191–196.

Kuo, C. H., Chen, C. C., and Chiang, B. H. (2004). Process characteristics of hydrolysis of chitosan in a continuous enzymatic membrane reactor. *J. Food Sci.* **69**, 332–337.
Lee, S. H., Suh, J. S., Kim, H. S., Lee, J. D., Song, J. S., and Lee, S. K. (2003). Evaluation of radiation synovectomy of the knee by means of intraarticular injection of holmium-166-chitosan complex in patients with rheumatoid arthritis: Results at 4-months follow-up. *Korean J. Radiol.* **4**, 170–178.
Lee, S. H., Park, J. S., Kim, S. K., Ahn, C. B., and Je, J. Y. (2009). Chitooligosaccharides suppress the level of protein expression and acetylcholinesterase activity induced by $A\beta_{25-35}$ in PC12 cells. *Bioorg. Med. Chem. Lett.* **19**, 850–862.
Liang, T. W., Chen, Y. J., Yen, Y. H., and Wang, S. L. (2007). The antitumor activity of the hydrolysates of chitinous materials hydrolyzed by crude enzyme from *Bacillus amyloliquefaciens* V656. *Proc. Biochem.* **42**, 527–534.
Lim, L. Y., Khor, E., and Koo, O. (1998). Gamma irradiation of chitosan. *J. Biomed. Mater. Res.* **43**, 282–290.
Lin, S. B., Lin, Y. C., and Chen, H. H. (2009). Low molecular weight chitosan prepared with the aid of cellulase, lysozyme and chitinase: Characterization and antibacterial activity. *Food Chem.* **116**, 47–53.
Liu, H. T., Li, W. M., Xu, G., Li, X. Y., Bai, X. F., Wei, P., Yu, C., and Du, Y. G. (2009). Chitosan oligosaccharides attenuate hydrogen peroxide-induced stress injury in human umbilical vein endothelial cells. *Pharmacol. Res.* **59**, 167–175.
Macchi, G. (1996). A new approach to the treatment of obesity: Chitosan's effects on body weight reduction and plasma cholesterol levels. *Acta Toxicol. Ther.* **27**, 303–320.
Majeti, N. V. and Kumar, R. (2000). A review of chitin and chitosan. *React. Funct. Polym.* **46**, 1–27.
Mao, S., Shuai, X., Unger, F., Simon, M., Bi, D., and Kissel, T. (2004). The depolymerization of chitosan: Effects on physicochemical and biological properties. *Int. J. Pharm.* **281**, 45–54.
Mendis, E., Kim, M. M., Rajapakse, N., and Kim, S. K. (2007). An in vitro cellular analysis of the radical scavenging efficacy of chitooligosaccharides. *Life Sci.* **80**, 2118–2127.
Mengíbar, M., Ganan, M., Miralles, B., Carrascosa, A. V., Martínez-Rodriguez, A. J., Peter, M. G., and Heras, A. (2011). Antibacterial activity of products of depolymerization of chitosans with lysozyme and chitosanase against *Campylobacter jejuni*. *Carbohydr. Polym.* **84**, 844–848.
Ngo, D. N., Kim, M. M., and Kim, S. K. (2008). Chitin oligosaccharides inhibit oxidative stress in live cells. *Carbohydr. Polym.* **74**, 228–234.
Nishimura, K., Nishmura, S., Nishi, N., Saiki, I., Tokura, S., and Azuma, I. (1984). Immunological activity of chitin and its derivatives. *Vaccine* **2**, 93–99.
Nordtveit, R. J., Vårum, K. M., and Smidsrød, O. (1994). Degradation of fully water-soluble, partially N-acetylated chitosans with lysozyme. *Carbohydr. Polym.* **23**, 253–260.
Omura, H., Uehara, K., and Tanaka, Y. (1991). Manufacture of low-molecular weight chitosan. Japan patent, 03-02203.
Park, P. J., Je, J. Y., and Kim, S. K. (2003a). Angiotensin I converting enzyme (ACE) inhibitory activity of hetero-chitooligosaccharides prepared from partially different deacetylated chitosan. *J. Agric. Food Chem.* **51**, 4930–4934.
Park, P. J., Je, J. Y., and Kim, S. K. (2003b). Free radical scavenging activity of chitooligosaccharides by electron spin resonance spectrometry. *J. Agric. Food Chem.* **51**, 4624–4627.
Park, P. J., Je, J. Y., Byun, H. G., Moon, S. H., and Kim, S. K. (2004). Antimicrobial activity of hetero-chitosans and their oligosaccharides with different molecular weight. *J. Microbial. Biotechnol.* **14**, 317–323.
Park, P. J., Ahn, C. B., Jeon, Y. J., and Je, J. Y. (2008). Renin inhibition activity by chitooligosaccharides. *Bioorg. Med. Chem. Lett.* **18**, 2471–2474.
Peach, M. J. (1997). Renin-angiotensin system: Biochemistry and mechanisms of action. *Physiol. Rev.* **57**, 313–370.
Peniston, Q. P. and Johnson, E. L. (1975). Process for depolymerization of chitosan. US patent, No. 3922260.

Pryor, W. A. (1982). Free radical biology: Xenobiotics, cancer, and aging. *Ann. N. Y. Acad. Sci.* **393**, 1–22.

Qin, C., Du, Y., Xiao, L., Li, Z., and Gao, X. (2002). Enzymic preparation of water-soluble chitosan and their antitumor activity. *Int. J. Biol. Macromol.* **31**, 111–117.

Qin, C., Gao, J., Wang, L., Zeng, L., and Liu, Y. (2006). Safety evaluation of short-term exposure to chitooligomers from enzymic preparation. *Food Chem. Toxicol.* **44**, 855–861.

Sashiwa, H., Saimoto, H., Shigemasa, Y., and Tokura, S. (1993). N-Acetyl group distribution in partially deacetylated chitin prepared under homogeneous conditions. *Carbohydr. Res.* **242**, 167–172.

Shahidi, F. and Synowieck, J. (1991). Isolation and characterization of nutrients and value-added products from snow crab (*Chinoecetes opilio*) and shrimp (*Pandalus borealis*) processing discards. *J. Agric. Food Chem.* **39**, 1527–1532.

Suzuki, K., Mikami, T., Okawa, Y., Tokoro, A., Suzuki, S., and Suzuki, M. (1986). Antitumor effect of hexa-*N*-acetylchitohexaose and chitohexaose. *Carbohydr. Polym.* **151**, 403–408.

Tanioka, S., Matsui, Y., Irie, T., Tanigawa, T., Tanaka, Y., Shibata, H., Sawa, Y., and Kono, Y. (1996). Oxidative depolymerization of chitosan by hydroxyl radical. *Biosci. Biotechnol. Biochem.* **60**, 2001–2004.

Tasi, G. J., Zhang, S. L., and Shieh, P. L. (2004). Antimicrobial activity of a low molecular weight chitosan obtained from cellulose digestion of chitosan. *J. Food Prot.* **67**, 396–398.

Terbojevich, C. M., Cosani, A., and Muzzarelli, R. A. A. (1996). Molecular parameters of chitosans depolymerized with the aid of papain. *Carbohydr. Polym.* **29**, 63–68.

Tian, M., Chen, F., Ren, D., Yu, X., Zhang, X., Zhong, R., and Wan, C. (2010). Preparation of a series of chtiooligomers and their effect on hepatocytes. *Carbohydr. Polym.* **79**, 137–144.

Tokoro, A., Kobayashi, M., Tatewaki, N., Suzuki, K., Okawa, Y., Mikami, T., Suzuki, S., and Suzuki, M. (1989). Protective effect of N-acetyl chitohexaose on Listeria monocytogenes infection in mice. *Microbiol. Immunol.* **33**, 357–367.

Tokoro, A., Tatewaki, N., Suzuki, K., Mikami, T., Suzuki, S., and Suzuki, M. (1998). Growth inhibitory effect of hexa-N-acetylchitohexaose and chitohexaose and Meth-A solid tumor. *Chem. Pharm. Bull.* **36**, 784–790.

Tsai, G. J., Su, W. H., Chen, H. C., and Pan, C. L. (2002). Antimicrobial activity of shrimp chitin and chitosan from different treatments and applications of fish preservation. *Fish. Sci.* **68**, 170–177.

Uchida, Y., Izume, M., and Ohtakara, A. (1989). Preparation of chitosan oligomers with purified chitosanase and its application. *In* "Chitin and Chitosan", (G. Skjak-Braek, T. Anthonsen, and P. Standford, Eds), pp. 373–382. Elsevier Applied Science, United Kingdom.

Vishu Kumar, A. B. and Tharanathan, R. N. (2004). A comparative study on depolymerization of chitosan by proteolytic enzymes. *Carbohydr. Polym.* **58**, 275–283.

Vishu Kumar, A. B., Varadaraj, M. C., Lalitha, R. G., and Tharanathan, R. N. (2004). Low molecular weight chitosan: Preparation with the aid of papain and characterization. *Biochim. Biophys. Acta* **1670**, 137–146.

Vishu Kumar, A. B., Varadaraj, M. C., Gowda, L. R., and Tharanathan, R. N. (2007). Low molecular weight chitosans-Preparation with the aid of pronase, characterization and their bactericidal activity towards Bacillus cereus and Escherichia coli. *Biochim. Biophys. Acta* **1770**, 495–505.

Xia, W., Liu, P., and Liu, J. (2008). Advance in chitosan hydrolysis by non-specific cellulases. *Bioresour. Technol.* **99**, 6751–6762.

Xie, Y., Hu, J., Wei, Y., and Hong, X. (2009). Preparation of chitooligosaccharides by the enzymatic hydrolysis of chitosan. *Polym. Degrad. Stabil.* **94**, 1895–1899.

Xu, Y., Zhang, Q., Yu, S., Yang, Y., and Ding, F. (2011). The protective effects of chitooligosaccharides against glucose deprivation-induced cell apoptosis in cultured cortical neurons through activation of PI3K/Akt and MEK/ERK1/2 pathways. *Brain Res.* **1375**, 49–58.

Zhang, H., Du, Y., Yu, X., Mitsutomi, M., and Aiba, S. I. (1999). Preparation of chitooligosaccharides from chitosan by a complex enzyme. *Carbohydr. Res.* **320**, 257–260.

CHAPTER 22

Biological Activities of Glucosamine and Its Related Substances

Isao Nagaoka,[*,1] **Mamoru Igarashi,**[*] and **Koji Sakamoto**[†]

Contents

Abstract

Glucosamine (GlcN) has been widely used to treat osteoarthritis (OA) in humans. We revealed that among GlcN-derivatives (GlcN and *N*-acetyl-D-glucosamine) and uronic acids (D-glucuronic acid

* Department of Host Defense and Biochemical Research, Juntendo University, Graduate School of Medicine, Tokyo, Japan
† Koyo Chemical Co., Ltd., Tokyo, Japan
1 Corresponding author: Isao Nagaoka, *E-mail address:* nagaokai@juntendo.ac.jp

Advances in Food and Nutrition Research, Volume 65
ISSN 1043-4526, DOI: 10.1016/B978-0-12-416003-3.00022-6

and D-galacturonic acid), only GlcN induces the production of hyaluronic acid (HA) by synovial cells and chondrocytes, and the production level is much higher (>10-fold) in synovial cells compared with chondrocytes. Moreover, GlcN increases the expression of HA-synthesizing enzymes (HAS) in synovial cells and chondrocytes. These observations indicate that GlcN likely exhibits the chondroprotective action on OA by modulating the expression of HAS and inducing the production of HA (a major component of glycosaminoglycans contained in the synovial fluid) especially by synovial cells. The pathological change of subchondral bone is implicated in the initiation and progression of cartilage damage in OA. Thus, we further determined the effect of GlcN on the bone metabolism (osteoblastic cell differentiation). The results indicated that GlcN increases the mineralization of mature osteoblasts and the expression of middle and late stage markers (osteopontin and osteocalcin, respectively) during osteoblastic differentiation, and reduces the expression of receptor activator of NF-κB ligand (RANKL), a differentiation and activation factor for osteoclasts. These observations likely suggest that GlcN has a potential to induce the osteoblastic cell differentiation and suppress the osteoclastic cell differentiation, thereby increasing bone matrix deposition and decreasing bone resorption to modulate bone metabolism in OA.

I. INTRODUCTION

Osteoarthritis (OA) is the most common joint disorder with an immense socioeconomic impact and is characterized by qualitative and quantitative changes in the architecture and composition of all the joint structures. An altered imbalance among the biosynthesis and degradation of matrix components leads to a progressive destruction of the articular cartilage (Martel-Pelletier *et al.*, 2005). Currently, several treatments are available for OA ranging from the most conservative to more surgical extremes. Conservative measures involve lifestyle modifications, physical therapy, and pharmacologic treatment with nonsteroidal anti-inflammatory drugs and intra-articular injection of hyaluronic acid (HA) (Puhl *et al.*, 1993). Treatments for early OA are primarily based on symptomatic relief, whereas irreversible joint disability in advanced OA usually requires surgical intervention to relieve pain and improve joint function. However, current treatments are mostly targeting the symptoms but not addressing the destructed structure of articular cartilage in OA.

Glucosamine (GlcN), a naturally occurring amino monosaccharide, is present in the connective and cartilage tissues as a component of glycosaminoglycans (GAG), and contributes to maintaining the strength, flexibility, and elasticity of these tissues. Thus, GlcN has been widely used to treat OA for more than two decades in human. In this context, several short- and long-term clinical trials on OA have shown the significant symptom-modifying effect of GlcN (McAlindon *et al.*, 2000; Pavelka *et al.*, 2002; Reginster *et al.*, 2001). Moreover, the updated Osteoarthritis Research Society International recommendations for management of hip and knee OA have recently suggested that GlcN has the symptom-relieving and structure-modifying effects in knee OA (Zhang *et al.*, 2008). In this context, we revealed that GlcN is able to suppress the cytokine-induced activation of synovial cells (such as the production of nitric oxide, PGE_2, and IL-8), thereby possibly exhibiting anti-inflammatory actions in arthritis (Hua *et al.*, 2007a). However, the mechanism for the chondroprotective action of GlcN on the joint is not fully understood.

In the joints, synovial cells produce HA as a component of synovial fluid GAG, whereas chondrocytes produce chondroitin sulfate, keratan sulfate (KS), and HA as components of cartilage matrix GAG (Carrino and Caplan, 1985; Yielding *et al.*, 1957). Thus, it is reasonable to speculate that GlcN may affect the GAG production by synovial cells and chondrocytes. Thus, to elucidate the chondroprotective action of GlcN, we examined the effects of D-GlcN on the production of GAG, such as HA and KS by human synovial cells and chondrocytes, and compared with those of a GlcN-derivative (*N*-acetyl-D-glucosamine, GlcNAc) and uronic acids (D-glucuronic acid, GlcU and D-galacturonic acid, GalU) on the GAG production (Igarashi *et al.*, 2011a).

On the other hand, it is known that the pathological change of subchondral bone is associated with the initiation and progression of cartilage damage in OA (Radin and Rose, 1986). Due to nonphysiological load-bearing on the subchondral bone, the bone resorption increases, thereby enhancing the progression of cartilage damage in OA (Bettica *et al.*, 2002). However, the effect of GlcN on bone metabolism is not fully understood. To address this, we further examined the effect of GlcN on the bone metabolism using mouse osteoblastic MC3T3-E1 cells by examining the osteoblastic cell differentiation and compared with that of GlcNAc (Igarashi *et al.*, 2011b).

II. EFFECTS OF GLCN-DERIVATIVES AND URONIC ACIDS ON THE PRODUCTION OF GAG BY HUMAN SYNOVIAL CELLS AND CHONDROCYTES

GAG released from human synovial cells and chondrocytes were measured by ELISA after treatment with GlcN, GlcNAc, GlcU, and GalU. GlcN significantly increased the HA level in the culture supernatants from human synovial cells (MH7A) at 0.25–1mM (Fig. 22.1A). Similarly, Hua *et al.* (2007b) reported that GlcN dose-dependently (0.01–2mM)

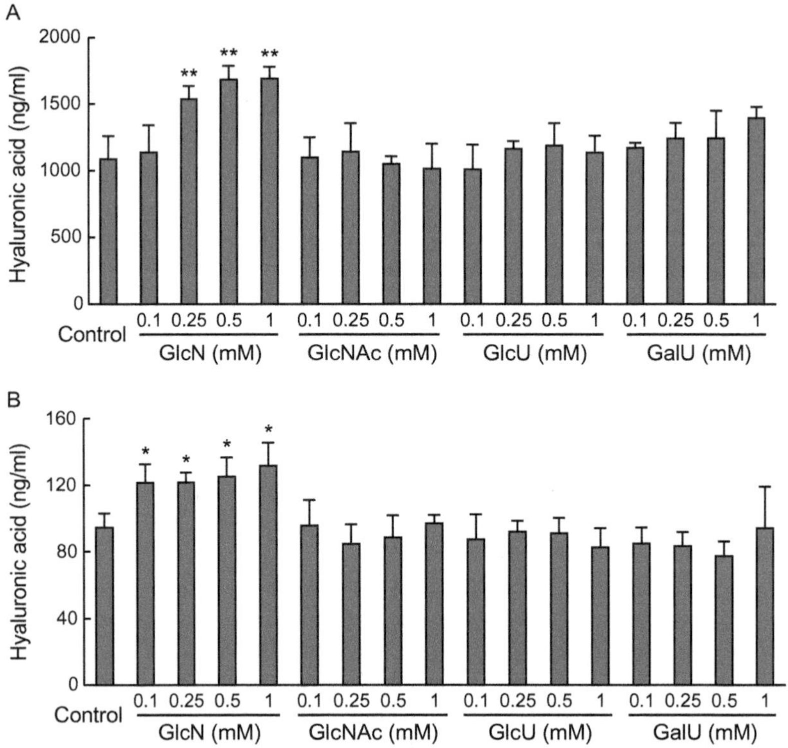

FIGURE 22.1 Effects of GlcN, GlcNAc, GlcU, and GalU on the release of HA from synovial cells and chondrocytes. Human synovial cells (MH7A) (A) and chondrocytes (SW1353) (B) were incubated without (control) or with 0.1–1mM GlcN, GlcNAc, GlcU, or GalU for 24h, and extracellularly released HA was measured by ELISA. Data are expressed as ng/ml and represent the mean±SD of three independent experiments. Values were compared between without (control) and with GlcN, GlcNAc, GlcU, or GalU. *$p<0.05$, **$p<0.01$.

increased the HA production by synovial cells. In contrast, GlcNAc, GlcU, and GalU did not essentially increase the extracellular level of HA. Of note, GlcN, GlcNAc, GlcU, and GalU do not affect the extracellular levels of KS released from human synovial cells (data not shown).

Similarly, GlcN significantly increased the HA level in the culture supernatants from human chondrocytes (SW1353) at 0.1–1mM (Fig. 22.1B). In contrast, GlcNAc, GlcU, and GalU did not increase the extracellular level of HA. As observed with synovial cells, GlcN, GlcNAc, GlcU, and GalU do not affect the extracellular levels of KS released from human chondrocytes (data not shown). Notably, the HA level in the culture supernatants from synovial cells was much higher (>10-fold) than those from chondrocytes.

In recent years, the presence of intracellular HA has been confirmed in several cell types (Evanko and Wight, 1999; Tammi *et al.*, 2001). Therefore, the effect of GlcN, GlcNAc, GlcU, and GalU on the intracellular accumulation of GAG in human synovial cells and chondrocytes was determined. In synovial cells, only GlcN significantly increased the intracellular level of HA at 0.25–1 mM (Fig. 22.2A). However, none of GlcN, GlcNAc, GlcU, and GalU affected the intracellular level of HA in chondrocytes (Fig. 22.2B). As expected, GlcN, GlcNAc, GlcU, and GalU do not affect the intracellular levels of KS in synovial cells and chondrocytes (data not shown). These observations suggest that among amino GlcN-derivatives and uronic acids examined, only GlcN is able to increase both the extracellular levels of HA released from synovial cells and chondrocytes and the intracellular levels of HA in synovial cells. Moreover, the potential to produce HA is much higher (>10-fold) in synovial cells than chondrocytes.

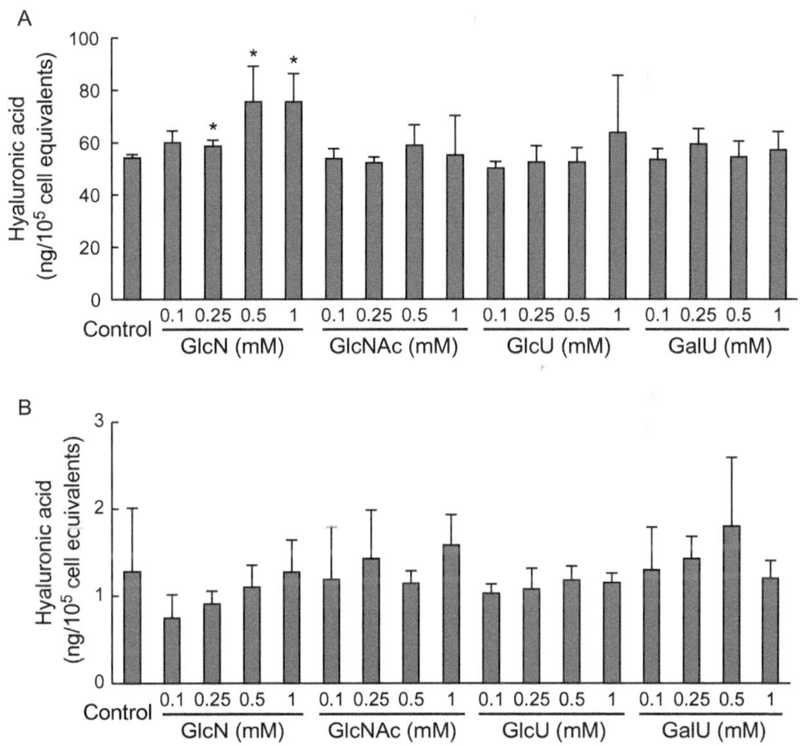

FIGURE 22.2 Effects of GlcN, GlcNAc, GlcU, and GalU on the intracellular levels of HA in synovial cells and chondrocytes. Synovial cells (A) and chondrocytes (B) were incubated without (control) or with 0.1–1 mM GlcN, GlcNAc, GlcU, or GalU for 24 h, and the intracellular levels of HA were measured by ELISA. Data are expressed as ng/10^5 cell equivalents and represent the mean±SD of three independent experiments. Values were compared between without (control) and with GlcN, GlcNAc, GlcU, or GalU. * $p<0.05$.

III. Incorporation of GlcN into GAG

As mentioned above, GlcN increased the extracellular level of HA from synovial cells and chondrocytes and the intracellular level of HA in synovial cells. Thus, it is important to clarify whether the increase of HA in the culture supernatants and within the cells is due to the production of HA utilizing GlcN as a precursor. For this purpose, synovial cells and chondrocytes were incubated with GlcN containing ^{3}H-labeled GlcN, and the radioactivities of the GAG–hexadecylpyridinium complex containing the incorporated GlcN was measured. Importantly, the incorporation of ^{3}H-labeled GlcN into the extracellular GAG as well as the intracellular GAG in synovial cells was increased depending on the concentrations of GlcN and ^{3}H-labeled GlcN (Fig. 22.3A and B). Similarly, the incorporation of ^{3}H-labeled GlcN into the extracellular and intracellular GAG was increased in chondrocytes depending on the concentrations of GlcN and ^{3}H-labeled GlcN (data not shown). These observations obviously indicate that GlcN is actually incorporated and utilized for the production of GAG (including HA) by synovial cells and chondrocytes.

IV. Effects of GlcN-Derivatives and Uronic Acids on the Expression of HA-Metabolizing Enzymes

To further clarify whether GlcN-derivatives (GlcN and GlcNAc) and uronic acids (GlcU and GalU) affect the expression of HA-metabolizing enzymes (such as HA-synthesizing enzymes, HAS), synovial cells and chondrocytes were treated with GlcN, GlcNAc, GlcU, and GalU, and then the expression levels of HAS were measured by reverse transcription polymerase chain reaction (RT-PCR). In synovial cells, the expression of HAS1 was apparently increased by GlcN treatment, but not by GlcNAc, GlcU, and GalU (Fig. 22.4A). In contrast, the expression of HAS2 and HAS3 was not essentially affected by GlcN, GlcNAc, GlcU, and GalU. In chondrocytes, the expression of HAS2 was increased not only by GlcN but also by GlcU and GalU, whereas the expression of HAS3 was increased only by GlcN (Fig. 22.4B).

V. Effects of GlcN and GlcNAc on the Mineralization of Osteoblastic MC3T3-E1 Cells

Mineralization of differentiated mouse osteoblastic MC3T3-E1 cells was measured after the incubation with GlcN or GlcNAc for 21 days. As shown in Fig. 22.5, GlcN significantly increased the mineralization at 0.1

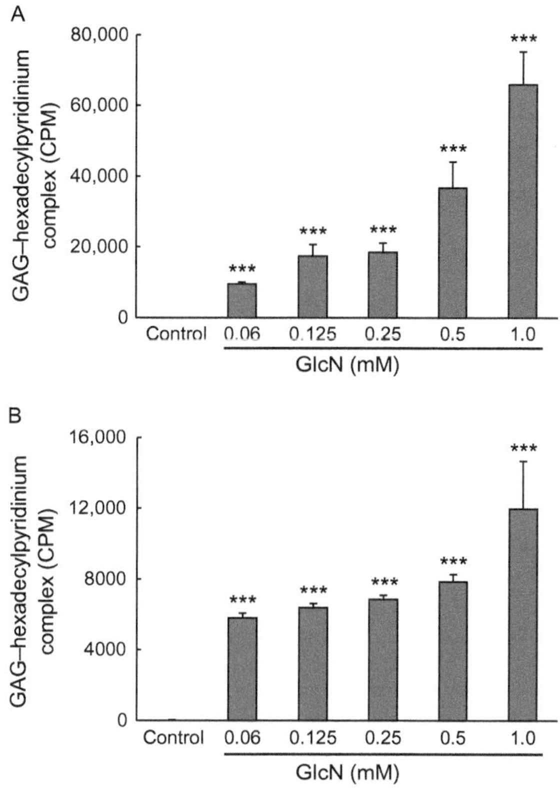

FIGURE 22.3 Incorporation of GlcN into GAG by synovial cells. Synovial cells were incubated without (control) or with 0.06–1 mM GlcN containing 1/10,000 of D-[6-^{3}H(N)]-glucosamine hydrochloride for 24 h, and the incorporation of GlcN into GAG was determined by counting the radioactivities (CPM) of GAG–hexadecylpyridinium complex recovered from the culture supernatant (A) and cell lysates (B). Data are corrected by subtracting the background count incubated with ^{3}H-labeled GlcN without cells. Data represent the mean±SD of three independent experiments. Values were compared between without (control) and with GlcN. ***$p<0.001$.

and 1 mM compared with control. Moreover, GlcNAc significantly increased the mineralization at 1 mM. Notably, GlcN was more potent than GlcNAc in the mineralization at the same concentration.

VI. EFFECTS OF GlcN AND GlcNAc ON THE OSTEOBLASTIC DIFFERENTIATION OF MC3T3-E1 CELLS

Osteoblastic cells express the differentiation markers, such as type I collagen (COLI) and alkaline phosphatase (ALP) as markers for early stage of osteoblastic differentiation, osteopontin (OPN) as a marker for

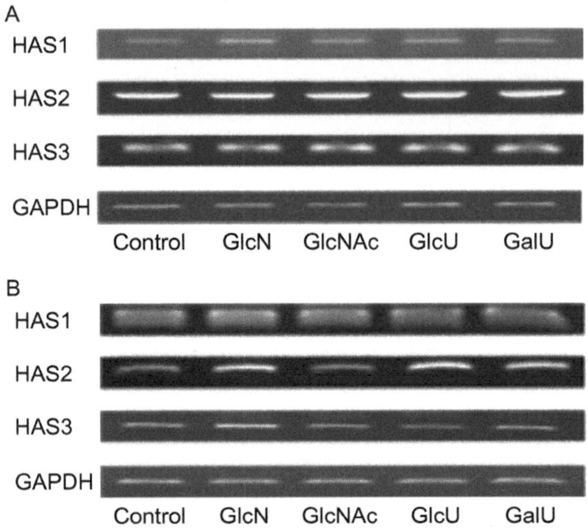

FIGURE 22.4 Effect of GlcN, GlcNAc, GlcU, and GalU on the mRNA expression of HA-metabolizing enzymes in synovial cells and chondrocytes. Synovial cells (A) and chondrocytes (B) were incubated with 1mM GlcN, GlcNAc, GlcU, or GalU for 24h. After the incubation, total RNA was purified, and the expression of mRNA for HAS1–3 and GAPDH was evaluated by RT-PCR. PCR products were resolved by 2% agarose gel electrophoresis and stained with ethidium bromide. Data represent one of two independent experiments.

middle stage differentiation, and osteocalcin (OCN) as a marker for late stage differentiation (Ibaraki *et al.*, 1992; Rodan *et al.*, 1988; Weinreb *et al.*, 1990). Thus, the effect of GlcN on osteoblastic cell differentiation was investigated by measuring the expression of differentiation markers by RT-PCR. First, the expression of the markers for early stage of osteoblastic differentiation (COLI and ALP) was evaluated. The expression of COLI was not changed by GlcN (0.1 and 1mM) after the incubation for 1–3days, compared with control. Further, the expression of ALP was not substantially changed by both GlcN and GlcNAc.

Second, the expression of a marker for middle stage of osteoblastic differentiation (OPN) was evaluated. The expression of OPN was apparently increased by both GlcN and GlcNAc (1mM) after the incubation for 21 days. Further, the expression of a marker for late stage of osteoblastic differentiation (OCN) was evaluated. Of importance, the expression of OCN was significantly increased by GlcN (1mM) after the incubation for 21days (Fig. 22.6). OCN expression was also increased by GlcNAc after 21days.

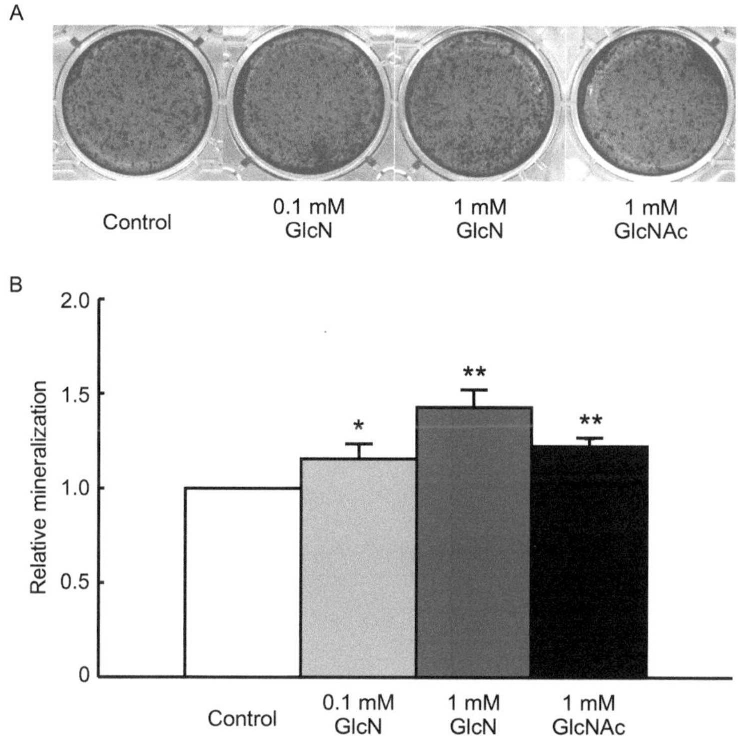

FIGURE 22.5 Effects of GlcN and GlcNAc on the mineralization of osteoblasts. Mouse osteoblastic cells (MC3T3-E1) were differentiated in the absence (control) or presence of GlcN (0.1 and 1mM) or GlcNAc (1mM) for 21days. (A) Alizarin red-S (AR-S) staining. (B) Quantification of AR-S staining. The cells were lysed in 10% hexadecylpyridinium chloride–10mM sodium phosphate buffer (pH 7.0), and the concentration of AR-S in each sample was determined by absorbance at 562nm. Concentration of AR-S was expressed as a ratio relative to control. Data represent the mean±SD of three independent experiments. Values were compared between the absence (control) and presence of GlcN or GlcNAc. *$p<0.05$, **$p<0.01$.

VII. EFFECTS OF GlcN AND GlcNAc ON THE EXPRESSION OF RECEPTOR ACTIVATOR OF NF-κB LIGAND BY MC3T3-E1 CELLS

Bone tissue is continuously remodeled through the combined activities of two cell types, osteoblasts and osteoclasts. Osteoblasts deposit bone matrix, whereas osteoclasts resorb bone matrix. The bone deposition and resorption are tightly coupled, and their balance defines both mass and quality of bone tissue (Filvaroff and Derynck, 1998). Receptor activator of NF-κB ligand (RANKL), which is expressed and secreted by mature

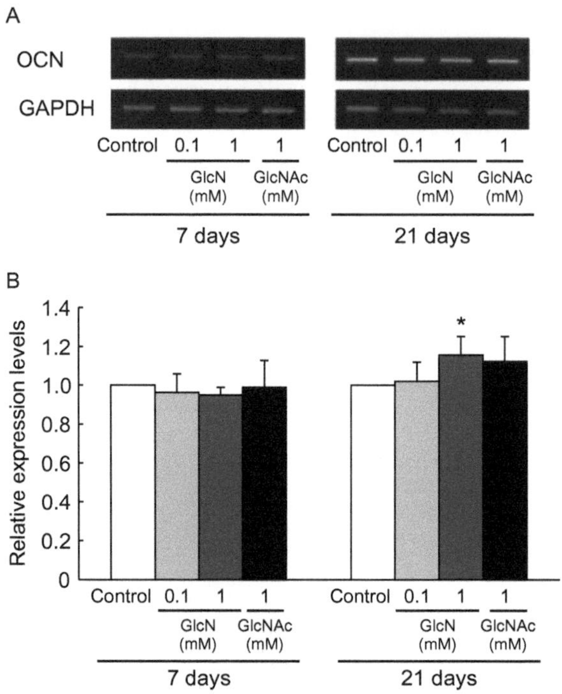

FIGURE 22.6 Effects of GlcN and GlcNAc on the mRNA expression of OCN. Osteoblastic cells were differentiated in the absence (control) or presence of GlcN (0.1 and 1 mM) or GlcNAc (1mM) for 21days. After the incubation, total RNA was purified, and the expression of mRNA for OCN was evaluated by RT-PCR. PCR products were resolved by 2% agarose gel electrophoresis and stained with ethidium bromide (A). OCN mRNA expression was expressed as a ratio relative to control (B). Data represent the mean$\pm$SD of three independent experiments. Values were compared between the absence (control) and presence of GlcN or GlcNAc. *$p<0.05$.

osteoblasts, is a key factor of osteoclastogenesis (Kobayashi *et al.*, 2009). Thus, the expression of RANKL was evaluated during osteoblastic differentiation by RT-PCR. Of note, the expression of RANKL was significantly suppressed by both GlcN and GlcNAc (1mM) after the incubation for 21 days (Figs. 22.7 and 22.8).

VIII. Other Biological Activities of GlcN

In addition to the chondroprotective action, GlcN is expected to exert an anti-inflammatory action, since GlcN suppresses inflammatory cell activation (Nagaoka *et al.*, 2011). To further extend the actions of GlcN,

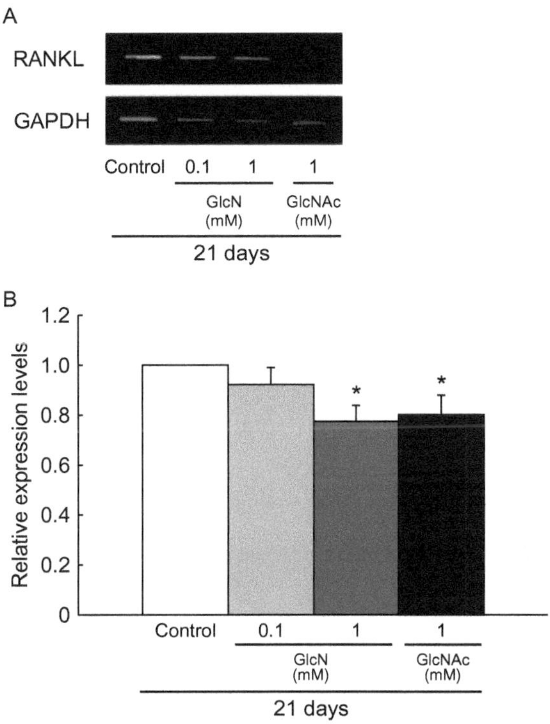

FIGURE 22.7 Effects of GlcN and GlcNAc on the mRNA expression of RANKL. Osteoblastic cells were differentiated in the absence (control) or presence of GlcN (0.1 and 1 mM) or GlcNAc (1mM) for 21days. After the incubation, the expression of mRNA for RANKL was evaluated by RT-PCR, as described in the legend for Fig. 22.6. Data represent the mean±SD of three independent experiments. Values were compared between the absence (control) and presence of GlcN or GlcNAc. *$p<0.05$.

we investigated the effects of GlcN on synovial cells, endothelial cells, and intestinal epithelial cells using *in vitro* and *in vivo* systems. First, GlcN suppressed the IL-1β-induced activation of synovial cells *in vitro* (Hua *et al.*, 2007a). Further, GlcN administration repressed synovial cell hyperplasia, cartilage destruction, and inflammatory cell infiltration in rat adjuvant arthritis (Hua *et al.*, 2005). Second, GlcN suppressed the TNF-α-induced activation of intestinal epithelial cell HT-29 *in vitro* (Yomogida *et al.*, 2008a). In addition, GlcN administration improved the clinical symptoms and colonic inflammation and tissue injury in dextran sulfate sodium-induced colitis in rats (Yomogida *et al.*, 2008b). Finally, GlcN suppressed the TNF-α-induced activation of endothelial cells *in vitro* (Ju *et al.*, 2008). Moreover, GlcN administration repressed the formation of atherosclerotic lesion and infiltration of inflammatory cells into the lesion in spontaneously hyperlipidemic mice B6 KOR *Apo*^{shl} (Yomogida

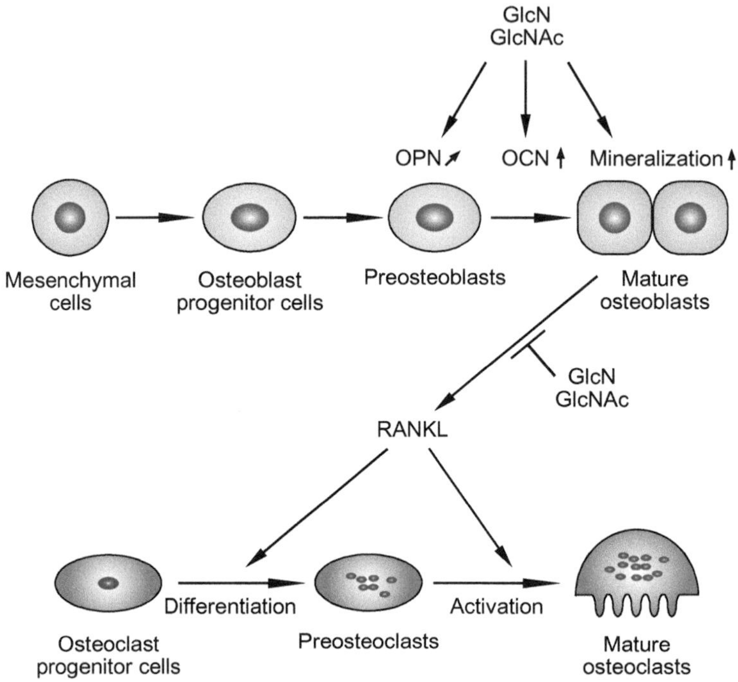

FIGURE 22.8 Schematic representation of the effects of GlcN and GlcNAc on the osteoblastic and osteoclastic cell differentiation. GlcN and GlcNAc increase the mineralization of mature osteoblasts and the expression of middle and late stage markers (OPN and OCN, respectively) during osteoblastic differentiation and reduce the expression of RANKL, a differentiation and activation factor for osteoclasts, thereby possibly increasing bone matrix deposition and decreasing bone resorption to promote the bone formation. (For color version of this figure, the reader is referred to the Web version of this chapter.)

et al., 2008c). Together these observations support the idea that GlcN can function as not only a chondroprotective agent but also an anti-inflammatory molecule in the body.

IX. CONCLUSIONS

OA is characterized by qualitative and quantitative changes in the architecture and composition of the joint structures. Importantly, the levels of HA and viscosity are decreased in the synovial fluid from OA patients (Divine *et al.*, 2006); the concentration of HA is 3.4–5mg/ml in normal synovial fluid, whereas the concentration is decreased to 2.5mg/ml in OA (Nakayama *et al.*, 2000). Thus, the intra-articular injection of HA is appreciated as a treatment of OA. Moreover, GlcN has been also used to treat OA in humans because GlcN is present in the cartilage tissues as a component of GAG (McAlindon *et al.*, 2000; Pavelka *et al.*, 2002; Reginster *et al.*, 2001). In the present study, to elucidate the chondroprotective action of GlcN, we examined the effects of GlcN-derivatives (GlcN and GlcNAc) and uronic acids (GlcU and GalU) on the production of GAG, such as HA and KS by human synovial cells and chondrocytes.

The results indicated that among GlcN-derivatives and uronic acids examined, only GlcN increases not only the extracellular levels of HA released from synovial cells and chondrocytes but also the intracellular levels of HA in synovial cells. Moreover, the experiments with ^{3}H-labeled GlcN revealed that GlcN can be incorporated and utilized for the production of GAG (including HA) by synovial cells and chondrocytes.

Moreover, RT-PCR analysis revealed that the expression of HAS1, one of HAS, was apparently increased by GlcN in synovial cells, suggesting that the upregulated HAS1 may be involved in the GlcN-induced increase of HA production by synovial cells. In contrast, the expression of HAS2 was upregulated by GlcN, GlcU, and GalU, whereas the expression of HAS3 was increased by GlcN in chondrocytes (Fig. 22.4B), suggesting that the upregulated HAS3 and possibly HAS2 may be involved in the GlcN-induced increase of HA production by chondrocytes.

Thus, among GlcN-derivatives and uronic acids, only GlcN induced the production of HA by synovial cells and chondrocytes, and the production level was much higher (>10-fold) in synovial cells compared with chondrocytes. Moreover, GlcN increased the expression of HAS in synovial cells and chondrocytes. These observations indicate that GlcN likely exhibits the chondroprotective action on the joint diseases such as OA by modulating the expression of HAS and inducing the production of HA (a major component of GAG contained in the synovial fluid) by especially synovial cells.

The pathological change of subchondral bone is implicated in the initiation and progression of cartilage damage in OA (Radin and Rose, 1986) and enhances the progression of OA (Bettica *et al.*, 2002). Thus, we next determined the effect of GlcN on the bone metabolism (osteoblastic cell differentiation) using mouse MC3T3-E1 osteoblasts by evaluating the expression of differentiation markers for early (COLI and ALP), middle (OPN), and late (OCN and mineralization) stages, and further compared with that of GlcNAc. The results indicated that mineralization of mature osteoblasts was increased by GlcN after incubation for 21 days. Further, RT-PCR analyses revealed that GlcN substantially increased the expression of a middle stage marker (OPN) and a late stage marker (OCN) after the incubation for 21 days, whereas it did not essentially affect the expression of early stage markers (COLI and ALP). In addition, GlcN suppressed the expression of RANKL, which regulates the differentiation and activation of osteoclast, a cell involved in the resorption of bone matrix. Together these observations likely suggest that GlcN not only induces the osteoblastic cell differentiation especially at middle–late stages, but also suppresses the osteoclastic cell differentiation, thereby increasing bone matrix deposition and decreasing bore resorption, and eventually promotes the bone formation.

Further, the present study has demonstrated that GlcNAc, a GlcN-derivative, also increases the mineralization of mature osteoblasts and the expression of a middle stage marker (OPN) and a late stage marker (OCN) for the osteoblastic cell differentiation and suppresses the expression of RANKL, a key factor involved in the osteoclastic cell differentiation and activation. These observations likely suggest that not only GlcN but also GlcNAc has a potential to induce the osteoblastic cell differentiation and suppresses the osteoclastic cell differentiation, thereby promoting the bone formation. Supporting this, it has been recently reported that oral administration of GlcNAc increases the bone mineral density in C57BL/6J mice, possibly by promoting the bone formation and suppressing bone resorption (Shimizu *et al.*, 2011).

Thus, GlcN and GlcNAc increase the mineralization of mature osteoblasts and the expression of middle and late stage markers (OPN and OCN, respectively) during osteoblastic differentiation and reduce the expression of RANKL, a differentiation and activation factor for osteoclasts. These observations likely suggest that both GlcN and GlcNAc have a potential to induce the osteoblastic cell differentiation and suppress the osteoclastic cell differentiation, thereby increasing bone matrix deposition and decreasing bone resorption to modulate bone metabolism in OA.

REFERENCES

Bettica, P., Cline, G., Hart, D. J., Meyer, J., and Spector, T. D. (2002). Evidence for increased bone resorption on patients with progressive knee osteoarthritis. *Arthritis Rheum.* **46**, 3178–3184.

Carrino, D. A. and Caplan, A. I. (1985). Isolation and characterization of proteoglycans synthesized in ovo by embryonic chick cartilage and new bone. *J. Biol. Chem.* **260**, 122–127.

Divine, J. G., Zazulak, B. T., and Hewett, T. E. (2006). Viscosupplementation for knee osteoarthritis. *Clin. Orthop. Relat. Res.* **455**, 113–122.

Evanko, S. P. and Wight, T. N. (1999). Intracellular localization of hyaluronan in proliferating cells. *J. Histochem. Cytochem.* **47**, 1331–1341.

Filvaroff, E. and Derynck, R. (1998). Bone remodeling: A signaling system for osteoclast regulation. *Curr. Biol.* **8**, R679–R682.

Hua, J., Suguro, S., Hirano, S., Sakamoto, K., and Nagaoka, I. (2005). Preventive actions of a high dose of glucosamine on adjuvant arthritis in rats. *Inflamm. Res.* **54**, 127–132.

Hua, J., Sakamoto, K., Kikukawa, T., Abe, C., Kurosawa, H., and Nagaoka, I. (2007a). Evaluation of the suppressive actions of glucosamine on the interleikin-1β-mediated activation of synoviocytes. *Inflamm. Res.* **56**, 432–438.

Hua, J., Miyazawa, K., Sakamoto, K., and Nagaoka, I. (2007b). Effect of glucosamine on hyaluronan production by synovial cells. *Chitin Chitosan Res.* **13**, 239–244.

Ibaraki, K., Termine, J. D., Whitson, S. W., and Young, M. F. (1992). Bone matrix mRNA expression in differentiating fetal bovine osteoblasts. *J. Bone Miner. Res.* **7**, 743–754.

Igarashi, M., Kaga, I., Takamori, Y., Sakamoto, K., Miyazawa, K., and Nagaoka, I. (2011a). Effects of glucosamine derivatives and uronic acids on the production of glycosaminoglycans by human synovial cells and chondrocytes. *Int. J. Mol. Med.* **27**, 821–827.

Igarashi, M., Sakamoto, K., and Nagaoka, I. (2011b). Effect of glucosamine, a therapeutic agent for osteoarthritis, on oseteoblastic cell differentiation. *Int. J. Mol. Med.* **28**, 373–379.

Ju, Y., Hua, J., Sakamoto, K., Ogawa, H., and Nagaoka, I. (2008). Modulation of TNF-α-induced endothelial cell activation by glucosamine, a naturally occurring amino monosaccharide. *Int. J. Mol. Med.* **22**, 809–815.

Kobayashi, Y., Udagawa, N., and Takahashi, N. (2009). Action of RANKL and OPG for osteoclastogenesis. *Crit. Rev. Eukaryot. Gene Expr.* **19**, 61–72.

Martel-Pelletier, J., Lajeunesse, D., and Pelletier, J. P. (2005). Etiopathogenesis of osteoarthritis. *In* "Arthritis and Allied Conditions: A Textbook of Rheumatology", (W. J. Koopman and L. W. Moreland, Eds), Vol. 2, pp. 2199–2226. Lippincott Williams & Wilkins, Philadelphia.

McAlindon, T. E., LaValley, M. P., Gulin, J. P., and Felson, D. T. (2000). Glucosamine and chondroitin for treatment of osteoarthritis: A systematic quality assessment and meta-analysis. *JAMA* **283**, 1469–1475.

Nagaoka, I., Igarashi, M., Hua, J., Ju, Y., Yomogida, S., and Sakamoto, K. (2011). Recent aspects of the anti-inflammatory actions of glucosamine. *Carbohydr. Polym.* **84**, 825–830.

Nakayama, Y., Shiraki, Y., Yoshihara, K., and Uesaka, S. (2000). Evaluation of glycosaminoglycans levels in normal joint fluid of the knee. *J. Nippon Med. Sch.* **67**, 92–95.

Pavelka, K., Gatterova, J., Olejarova, M., Machacek, S., Giacovelli, G., and Rovati, L. C. (2002). Glucosamine sulfate use and delay of progression of knee osteoarthrisis: A 3-year, randomized, placebo-controlled, double-blind study. *Arch. Intern. Med.* **162**, 2113–2123.

Puhl, W., Bernau, A., Greiling, H., Köpcke, W., Pförringer, W., Steck, K. J., Zacher, J., and Scharf, H. P. (1993). Intra-articular sodium hyaluronate in osteoarthritis of the knee: A multicenter, double-blind study. *Osteoarthritis Cartilage* **1**, 233–241.

Radin, E. L. and Rose, R. M. (1986). Role of subchondral bone in the initiation and progression of cartilage damage. *Clin. Orthop. Relat. Res.* **213**, 34–40.

Reginster, J. Y., Deroisy, R., Rovati, L. C., Lee, R. L., Lejeune, E., Bruyene, O., Giacovelli, G., Henrotin, Y., Dacre, J. E., and Gossett, C. (2001). Long-term effects of glucosamine sulphate on osteoarthritis progression: A randomised, placebo-controlled clinical trial. *Lancet* **357**, 251–256.

Rodan, G. A., Heath, J. K., Yoon, K., Noda, M., and Rodan, S. B. (1988). Diversity of the osteoblastic phenotype. *In* "Cell and Molecular Biology of Vertebrate Hard Tissues", (D. Everd and S. Harnett, Eds), pp. 78–91. John Wiley and Sons, Inc, Chichester Ciba Foundation Symposium 136.

Shimizu, T., Nakatani, S., Kobata, K., and Wada, M. (2011). Oral administration of *N*-acetyl-D-glucosamine and D-glucosamine hydrochloride affects bone mineral density in C57BL/6J mice. *Chitin Chitosan Res.* **17**, 74–78.

Tammi, R., Rilla, K., Pienimäki, J. P., MacCallum, D. K., Hogg, M., Luukkonen, M., Hascall, V. C., and Tammi, M. (2001). Hyaluronan enters keratinocytes by a novel endocytic route for catabolism. *J. Biol. Chem.* **276**, 35111–35122.

Weinreb, M., Shinar, D., and Rodan, G. A. (1990). Different pattern of alkaline phosphatase, osteopontin, and osteocalcin expression in developing rat bone visualized by *in situ* hybridization. *J. Bone Miner. Res.* **5**, 831–842.

Yielding, K. L., Tomkins, G. M., and Bunim, J. J. (1957). Synthesis of hyaluronic acid by human synovial tissue slices. *Science* **125**, 1300.

Yomogida, S., Hua, J., Sakamoto, K., and Nagaoka, I. (2008a). Glucosamine suppresses the interleukin-8 production and ICAM-1 expression by TNF-α-stimulated human colonic epithelial cell line HT-29. *Int. J. Mol. Med.* **22**, 205–211.

Yomogida, S., Kojima, Y., Tsutsumi-Ishii, Y., Hua, J., Sakamoto, K., and Nagaoka, I. (2008b). Glucosamine, a naturally occurring amino monosaccharide, suppresses dextran sulfate sodium-induced colitis in rats. *Int. J. Mol. Med.* **22**, 317–323.

Yomogida, S., Kojima, Y., Ju, Y., Sakamoto, K., and Nagaoka, I. (2008c). Evaluation of the effect of glucosamine on atherosclerosis using atherosclerotic B6.KOR-*Apoe*^{shl} mice. *Chitin Chitosan Res.* **14**, 64–70.

Zhang, W., Moskowitz, R. W., Nuki, G., Abramson, S., Altman, R. D., Arden, N., Bierma-Zeinstra, S., Brandt, K. D., Croft, P., Doherty, M., Dougados, M., Hochberg, M., *et al.* (2008). OARSI recommendations for the management of hip and knee osteoarthritis, part II, OARSI evidence-based, expert consensus guidelines. *Osteoarthritis Cartilage* **16**, 137–162.

CHAPTER 23

Development of Marine Probiotics: Prospects and Approach

Se-Kwon Kim,[*,†,1] **Ira Bhatnagar,**[*,†] and **Kyong-Hwa Kang**[†]

Contents

Abstract

The term probiotics stands for bacterial strain acting for the upliftment of beneficial gut microflora as well as restricting the growth of decay or disease causing bacteria and other agents. They protect us from disease causing microorganisms and are vital for the digestion of nutrients and food. Researchers are now trying to explore the possibilities of therapeutic applications of probiotics in inflammatory bowel disease, treating diarrhea, eczema prevention in children, and reducing bladder cancer recurrence and urinary tract infections. The best known probiotic strains are Bifidobacteria, *Lactobacilli*, and *Streptococcus thermophilus* and can be found in food products such as yoghurts, fermented and unfermented milk, miso, tempeh, and some juices and soy beverages. Thus, as a substitute to the use of antibiotics, the dietary supplementation

* Department of Chemistry, Pukyong National University, Busan, Republic of Korea
† Marine Bioprocess Research Center, Pukyong National University, Busan, Republic of Korea
1 Corresponding author: Se-Kwon Kim, *E-mail address:* sknkim@pknu.ac.kr

Advances in Food and Nutrition Research, Volume 65
ISSN 1043-4526, DOI: 10.1016/B978-0-12-416003-3.00023-8

with probiotic bacteria is being widely employed in the aquaculture industry. This chapter is an attempt to shift the paradigm of research in probiotic area to the marine environment and to find ways of expanding the research in marine probiotics development.

I. INTRODUCTION

The concept of probiotics (which means "for life") was introduced in early twentieth century by Metchnikoff (1907) but it remained much unknown till date and received considerable attention in recent past with advent of functional and health food market globally. Technically speaking, probiotics are "good bacteria" that are similar to the ones found in our gut. They help our gut by increasing the number of good bacteria and, in turn, inhibit the bad ones. On the other hand, *prebiotics* act as the food for probiotic bacteria where they stimulate their growth and activity. When pre- and probiotics are mixed together, they form *synbiotics*. Gatesoupe (1999) defines probiotics as "live microbial cells administered to cultured organisms to colonize the digestive tract and improve their immune response." Probiotics are useful in a number of ways as they are indispensable for the proper development of our immune system. Apart from this, probiotics protect us from disease-causing microorganisms and are vital for the digestion of nutrients and food. In fact, there have been number of studies which suggest that probiotics ease symptoms of lactose intolerance and irritable bowel syndrome. Some of them have also been shown to relieve constipation and diarrhea, as well as reduce the risk of osteoporosis as they are believed to augment the assimilation of calcium and magnesium.

Researchers are now trying to explore the possibilities of therapeutic applications of probiotics in inflammatory bowel disease, treating diarrhea, eczema prevention in children, and reducing bladder cancer recurrence and urinary tract infections. It is worth mentioning here that our knowledge of probiotics should include the fact that the beneficiary effects of probiotics vary from individual to individual depending upon how the immune system responds to it. Further, the health benefits depend on the particular strain of the organism used as a probiotic and the dose. For example, the best known probiotic strains are Bifidobacteria, *Lactobacilli*, and *Streptococcus thermophilus* and can be found in food products such as yoghurts, fermented and unfermented milk, miso, tempeh, and some juices and soy beverages. Use of probiotics is increasing day by day as they are easy to consume and can also be obtained in the form of dietary supplements such as capsules or powders. In fact, certain foods that enhance the probiotic effect, known as prebiotics, are found in whole grains, almonds, bananas, garlic, honey, leeks, onions, and artichokes (Nutrition) which makes it even more easy to lift up the health benefits of probiotic strains.

As mentioned earlier, probiotics act in a two-way fashion, either by enhancing the growth of beneficial bacteria or by reducing the growth of harmful germs and microbes. This property of probiotic strains can be explored to produce functional foods as well as in food storage industry as prolonged storage of foods under appropriate conditions leads to the development of decaying bacteria. Probiotic strains can well be used in aquaculture for proper growth and maintenance of the cultured species. The microbiota in the aquatic environment is usually in equilibrium, and it is composed by bacteria that are either beneficial or neutral to cultured animals or by harmful obligate and opportunistic pathogenic bacteria (Schulze *et al.*, 2006). Owing to inadequate maintenance, pathogenic bacteria may proliferate, disturbing the equilibrium (Karunasagar *et al.*, 1994) and incurring further financial loss due to decay of the economically important species. As a general basis, antibiotics are basically used to encounter the bacterial disease in marine shrimp culture. However, their indiscriminate use has led to the appearance of antibiotic-resistant strains (Skjermo and Vadstein, 1999), contamination of shrimp meat, and also the environment (Holmström *et al.*, 2003). Therefore, several countries have banned the use of antibiotics such as chloramphenicol (FAO, 2002). Thus, as a substitute to the use of antibiotics, the dietary supplementation with probiotic bacteria is being widely employed in the aquaculture industry (Gatesoupe, 1999; Vine *et al.*, 2006). Among the probiotic bacteria used in aquaculture, the lactic acid bacteria stand out for their easy multiplication, production of antimicrobial compounds (bacteriocins, hydrogen peroxide, organic, and lactic acids), and the stimulation of the nonspecific immune response of the host (Gatesoupe, 2008). Studies have demonstrated the beneficial effect of the addition of such bacteria in the culture of several aquatic species. Nevertheless, only a few studies have been conducted to assess the use of lactic acid bacteria strains isolated from marine shrimp (Vieira *et al.*, 2010).

This chapter is an attempt to bring forward the beneficial effects of probiotics in marine environment, the prerequisites for a strain to be considered as a probiotic and explore the possibilities of developing marine probiotics.

II. PREREQUISITES FOR PROBIOTICS

Probiotic strains should meet some standard criterion to be considered safe for use. As mentioned earlier, probiotics are bacteria that help maintain the natural balance of microflora in the intestines. The normal human digestive tract contains about 400 types of probiotic bacteria that reduce the growth of harmful bacteria and promote a healthy digestive system. A number of reports from the past suggest experimenting on the benefits of probiotic therapies and propose an array of potentially advantageous medicinal uses for probiotics. However, since it is a matter of safeguard of living creatures, various concerns exist regarding the issues of safety, efficacy, and reliability as well as labeling of the probiotic products being formulated. To ensure strict compliance to the set parameters for probiotic use and safety, it is mandatory to define certain guidelines which should be followed for a product/strain to be termed as probiotic and before a potential probiotic strain enters the market for further use. It is also necessary to follow the clinically relevant steps to move probiotic closer to being embraced by the medical community. The first and foremost step for the same is the proper identification of strain and *in vitro* screening for its probiotic characteristics. Second, before its official use, animal studies should be rigorously pursued to establish safety and *in vivo* animal and human studies to establish efficacy. Phase I, II, and III clinical trials to prove health benefits that are as good as or better than standard prevention or treatments for a particular condition or disease should be followed. The optimism associated with probiotics is, however, counterbalanced by skepticism as many "probiotic" products in the market are unreliable in content and unproven clinically (Hamilton-Miller *et al.*, 1999; Hughes and Hillier, 1990; Zhong *et al.*, 1998). Following good manufacturing practice and production of high-quality products are also very important in the probiotic field. Incongruously labeled probiotic may also lead to chaos and inapt use of the same. To avoid misleading of the consumers and further damage to the aquaculture, informative/precise labeling of the probiotic products is required which should include information about the strain, viable numbers at the end of shelf life, adequate storage conditions, etc. Apart from this, some further steps include studies to identify mechanism of action of the studied probiotic *in vivo*, development of probiotic organisms that can carry vaccines to hosts and/or antiviral probiotic, expansion of proven strains to benefit the oral cavity, nasopharynx, respiratory tract, stomach, vagina, bladder, and skin as well as for cancer, allergies, and recovery from surgery/injury (Reid, 2005). These guidelines and steps should be strictly followed to ensure a high-quality, effective probiotic with health benefits.

III. POTENTIAL OF PROBIOTICS

Probiotics find a number of applications in new functional food products, resulting in the association of more food ingredients with health claims. Marine environment is rich in certain diverse strains of microbes which can turn out to be effective and safe probiotic agents. Existing level of research has expanded the hopes of their use in several areas, including sport-related energy drinks, fortified foods and drinks, and dairy products such as yogurts, cheese, ice cream, and milks containing prebiotics and probiotics. The market for probiotics is at bloom but its sustenance greatly depends upon further research, scientific substantiation, and biotechnological intervention in exploring the marine environment. The rate of increase in resistance to antibiotics (Neu, 1992) is a major public health problem throughout the world (Bengmark, 1998). The immunostimulatory properties of probiotics (Marteau and Rambaud, 1993) offer potential in boosting the host's resistance to disease, thereby potentially reducing the frequency of antibiotic use. The use of probiotics has been proposed by several investigators as an alternative to antibiotics in animals (Canganella *et al.*, 1996; Fuller, 1989; Gardiner *et al.*, 1999; Underdahl *et al.*, 1982) and poultry (Edens *et al.*, 1997; Gardiner *et al.*, 1999). Development of novel marine probiotics for inhibition of pathogens would be a better alternative for this heightened concern of antibiotic resistance.

Some probiotic strains were shown to inhibit the growth of enteropathogens, such as *Salmonella enteritidis*, enterotoxigenic *Escherichia coli*, and *Serratia marcesens*, *in vitro* (Drago *et al.*, 1997; Gonzalez *et al.*, 1993), and in this regard may offer considerable therapeutic potential. Another report suggests that *Lactobacillus* exerts antagonist activity against *Salmonella typhimurium* C5 infection both *in vitro* and *in vivo* (Hudault *et al.*, 1997). These studies provide evidence and form a basis for the clinical use of probiotics in suppression of pathogens. Apart from the abovementioned antimicrobial activity of therapeutic importance, potential clinical applications for probiotic microorganisms include treatment of food allergy (Salminen *et al.*, 1996), reduction of hypertension (Hata *et al.*, 1996), and use as vectors for delivery of oral vaccines (Pouwels *et al.*, 1996). An effective regime for studies related to food-grade genetic markers (Allison and Klaenhammer, 1996; Lin *et al.*, 1996) would be quite beneficial for the construction of probiotics against antibiotic-resistant genes.

Probiotics not just maintain the health of living creatures but may also protect the urogenital tract against microbial infections (Reid and Bruce, 2006; Reid *et al.*, 1994, 1998). Certain probiotic strains are also shown to reduce the risk of infections associated with the use of medical devices. Another potential application of probiotic cultures is in the production of fermented food products enriched in health-promoting substances, such as conjugated linoleic acid (CLA). Beneficial effects of CLA include anticarcinogenic activity, antiatherogenic activity, the ability to reduce the catabolic effects of immune stimulation, the ability to enhance growth promotion, and the ability to reduce body fat. Foods that contain animal fat, such as beef, lamb, milk, and dairy foods, are rich sources of CLA (Stanton *et al.*, 1997). Certain strains of *Propionobacteria*, commonly used as dairy starter cultures, were shown to be able to convert free linoleic acid to CLA (Jiang, 1998), suggesting that it may be possible via fermentation to produce fermented food products enriched in CLA. Probiotics may hold a number of other potential applications which are still uncovered due to lack in technological interventions in this field.

IV. APPROACHES FOR DEVELOPMENT OF MARINE PROBIOTICS

Probiotics, especially *Lactobacillus* and *Bifidobacterium*, have been suggested to be associated with alleviation of lactose intolerance (Levri *et al.*, 2005); prevention and cure of viral, bacterial, and antibiotic or radiotherapy induced diarrheas (Guandalini, 2006; Parvez *et al.*, 2006; Szajewska and Mrukowicz, 2005); immunomodulation (Forsythe and Bienenstock, 2010); antimutagenic (Chalova *et al.*, 2008) and anticarcinogenic effects (Liong, 2008); and even blood cholesterol reduction (Ooi and Liong, 2010). Although probiotics have been employed to treat a number of diseases including certain viral and bacterial diseases, applications of probiotics in other medical arenas are still juvenile. This may be because of lack of research activities on the molecular biology of probiotic strains. Genomic studies on *Lactobacillus* have focused on the interaction with the immune system, anticancer potential, and potential as a biotherapeutic agent in cases of antibiotic-associated diarrhea, travelers' diarrhea, pediatric diarrhea, inflammatory bowel disease, and irritable bowel syndrome. Strategies for the application of biotechnological and molecular biological approaches are required for the development of marine probiotic strains to be used in aquaculture as well as other areas of day-to-day life.

Another useful probiotic strain, *Bifidobacterium*, can also be explored from marine environment for the development of latest and novel probiotic agents. The basic problem faced during the development of marine probiotics is the isolation and identification of potential strain. Some studies have used 16S rDNA-targeted genus- and species-specific PCR primers for the identification and detection of Bifidobacteria. Identification of cultured Bifidobacteria using PCR primer pairs is rapid and accurate, being based on nucleic acid sequences. Matsuki *et al.* (2002) have found that, in adult feces, the *Bifidobacterium catenulatum* group is the most commonly detected species, followed by *Bifidobacterium longum*, *Bifidobacterium adolescentis*, and *Bifidobacterium bifidum*. In breast-fed infants, *Bifidobacterium breve* was the most frequently detected species, followed by *Bifidobacterium infantis*, *B. longum*, and *B. bifidum* (Matsuki *et al.*, 2002). This knowledge may be useful in specific isolation of these species from marine environment and exploring their probiotic potential. Real-time quantitative PCR using primers targeting 16S rDNA has a promising role in the enumeration of *Bifidobacterium*. A similar approach could be applied to the oceanic environment to detect the *Bifidobacterium*, *Lactobacillus* sp., and any other related species of probable probiotic potential. This technique to target bacteria with quantitative PCR will contribute to future studies of the composition and dynamics of the intestinal microflora and thereby giving new insights into the identification of probiotic species to be sought from the marine environment. Identifying a novel marine probiotic strain and providing sufficient numbers of viable bacteria in a product are not the only tasks in this field. Efficient and strong clinical trial should be followed and industry must also provide proof of efficacy for each strain. In the early 1990s, Fonterra embarked on a programme to develop proprietary probiotic strains, and as a result, commercialized two strains, *Bifidobacterium lactis* HN019 and *Lactobacillus rhamnosus* HN001 (Crittenden *et al.*, 2005). On similar lines, marine probiotic strains can also be pursued to find a place in the market for consumption as well as utilization in aquaculture. Biomarkers have to be employed to identify strains with probiotic utility and to define the different positive health benefits of existing probiotic strains.

V. CONCLUSION

It is becoming increasingly accepted by consumers that live lactic acid bacteria do exert health benefits when eaten. In fact, many countries consume live lactic acid bacteria in the form of traditional food termed as "curd." Further, it is being strongly believed that neither do all probiotic bacteria act in the same fashion nor do their activities have the same magnitude. There is an imperative need to search for newer, safe, and effective probiotics with broad range of health benefits. Marine environment holds a plethora of bioactive compounds with different bioactivities. The marine microbes have also been reported to hold great promise as antimicrobial agents and antitumor agents (Bhatnagar and Kim, 2010a,b). It is not hard to develop methods for effective isolation and characterization of some of these marine strains to be promising probiotic strains. Although biomarker research can prove to be a very efficient tool but which biomarkers accurately reflect efficacy *in vivo* is in question as the mechanisms by which most probiotic bacteria exert their health benefits remain unclear (Dekker *et al.*, 2007). Recent technological advances have turned their wheels to unravel how probiotic bacteria work. Once we are able to define a set mechanism of action to probiotic strains, biomarker development would be easier with a promise to offer the market safe and effective probiotic medicinal foods.

REFERENCES

Allison, G. and Klaenhammer, T. (1996). Functional analysis of the gene encoding immunity to lactacin F, lafI, and its use as a Lactobacillus-specific, food-grade genetic marker. *Appl. Environ. Microbiol.* **62**, 4450–4460.

Bengmark, S. (1998). Ecological control of the gastrointestinal tract. The role of probiotic flora. *Gut* **42**, 2.

Bhatnagar, I. and Kim, S. K. (2010a). Immense essence of excellence: Marine microbial bioactive compounds. *Mar. Drugs* **8**, 2673–2701.

Bhatnagar, I. and Kim, S. K. (2010b). Marine antitumor drugs: Status, shortfalls and strategies. *Mar. Drugs* **8**, 2702–2720.

Canganella, F., Gasbarri, M., Massa, S., and Trovatelli, L. (1996). A microbiological investigation on probiotic preparations used for animal feeding. *Microbiol. Res.* **151**, 167–175. (Germany).

Chalova, V., Lingbeck, J., Kwon, Y., and Ricke, S. (2008). Extracellular antimutagenic activities of selected probiotic Bifidobacterium and Lactobacillus spp. as a function of growth phase. *J. Environ. Sci. Health B* **43**, 193–198.

Crittenden, R., Bird, A., Gopal, P., Henriksson, A., Lee, Y., and Playne, M. (2005). Probiotic research in Australia, New Zealand and the Asia-Pacific region. *Curr. Pharm. Des.* **11**, 37–53.

Dekker, J., Collett, M., Prasad, J., and Gopal, P. (2007). Functionality of probiotics—Potential for product development. *Forum Nutr.* **60**, 196–208. Karger.

Drago, L., Gismondo, M. R., Lombardi, A., Haën, C., and Gozzini, L. (1997). Inhibition of *in vitro* growth of enteropathogens by new Lactobacillus isolates of human intestinal origin. *FEMS Microbiol. Lett.* **153**, 455–463.

Edens, F., Parkhurst, C., Casas, I., and Dobrogosz, W. (1997). Principles of ex ovo competitive exclusion and in ovo administration of Lactobacillus reuteri. *Poult. Sci.* **76**, 179.

FAO (2002). The State of World Fisheries and Aquaculture. Rome.

Forsythe, P. and Bienenstock, J. (2010). Immunomodulation by commensal and probiotic bacteria. *Immunol. Invest.* **39**, 429–448.

Fuller, R. (1989). Probiotics in man and animals. *J. Appl. Bacteriol.* **66**, 365.

Gardiner, G., Stanton, C., Lynch, P., Collins, J., Fitzgerald, G., and Ross, R. (1999). Evaluation of cheddar cheese as a food carrier for delivery of a probiotic strain to the gastrointestinal tract. *J. Dairy Sci.* **82**, 1379–1387.

Gatesoupe, F. (1999). The use of probiotics in aquaculture. *Aquaculture* **180**, 147–165.

Gatesoupe, F. J. (2008). Updating the importance of lactic acid bacteria in fish farming: Natural occurrence and probiotic treatments. *J. Mol. Microbiol. Biotechnol.* **14**, 107–114.

Gonzalez, S., Apella, M., Romero, N., De Macias, M. E. N., and Oliver, G. (1993). Inhibition of enteropathogens by lactobacilli strains used in fermented milk. *J. Food Prot.* **56**, 773–776.

Guandalini, S. (2006). Probiotics for children: Use in diarrhea. *J. Clin. Gastroenterol.* **40**, 244–248.

Hamilton-Miller, J., Shah, S., and Winkler, J. (1999). Public health issues arising from microbiological and labelling quality of foods and supplements containing probiotic microorganisms. *Public Health Nutr.* **2**, 223–229.

Hata, Y., Yamamoto, M., Ohni, M., Nakajima, K., Nakamura, Y., and Takano, T. (1996). A placebo-controlled study of the effect of sour milk on blood pressure in hypertensive subjects. *Am. J. Clin. Nutr.* **64**, 767.

Holmström, K., Gräslund, S., Wahlström, A., Poungshompoo, S., Bengtsson, B. E., and Kautsky, N. (2003). Antibiotic use in shrimp farming and implications for environmental impacts and human health. *Int. J. Food Sci. Technol.* **38**, 255–266.

Hudault, S., Lievin, V., Bernet-Camard, M. F., and Servin, A. L. (1997). Antagonistic activity exerted in vitro and in vivo by Lactobacillus casei (strain GG) against Salmonella typhimurium C5 infection. *Appl. Environ. Microbiol.* **63**, 513.
Hughes, V. L. and Hillier, S. L. (1990). Microbiologic characteristics of Lactobacillus products used for colonization of the vagina. *Obstet. Gynecol.* **75**, 244.
Jiang, J. (1998). Conjugated linoleic acid-occurrence, oxidation and production by dairy starter cultures. *PhD Thesis,* Swedish University of Agricultural Sciences, Uppsala, Sweden.
Karunasagar, I., Pai, R., and Malathi, G. (1994). Mass mortality of Penaeus monodon larvae due to antibiotic-resistant Vibrio harveyi infection. *Aquaculture* **128**, 203–209.
Levri, K. M., Ketvertis, K., Deramo, M., Merenstein, J. H., and D'Amico, F. (2005). Do probiotics reduce adult lactose intolerance? A systematic review. *J. Fam. Pract.* **54**, 613.
Lin, M. Y., Harlander, S., and Savaiano, D. (1996). Construction of an integrative food-grade cloning vector for Lactobacillus acidophilus. *Appl. Microbiol. Biotechnol.* **45**, 484–489.
Liong, M. T. (2008). Roles of probiotics and prebiotics in colon cancer prevention: Postulated mechanisms and in-vivo evidence. *Int. J. Mol. Sci.* **9**, 854–863.
Marteau, P. and Rambaud, J. C. (1993). Potential of using lactic acid bacteria for therapy and immunomodulation in man. *FEMS Microbiol. Rev.* **12**, 207–220.
Matsuki, T., Watanabe, K., and Tanaka, R. (2002). Genus-and species-specific PCR primers for the detection. *In* "Probiotics and Prebiotics: Where are we Going?", (G. W. Tannock, Ed.), p. 85. Caister Academic Press, Wymondham.
Metchnikoff, E. (1907). Lactic acid as inhibiting intestinal putrefaction. *In* "The Prolongation of Life: Optimistic Studies", (P. C. Mitchell, Ed.), pp. 161–183. Heinemann, London.
Neu, H. C. (1992). The crisis in antibiotic resistance. *Science* **257**, 1064.
Nutrition. http://nutritionaz.wordpress.com/author/sandramkl3/page/3/.
Ooi, L. G. and Liong, M. T. (2010). Cholesterol-lowering effects of probiotics and prebiotics: A review of in vivo and in vitro findings. *Int. J. Mol. Sci.* **11**, 2499–2522.
Parvez, S., Malik, K., Ah Kang, S., and Kim, H. Y. (2006). Probiotics and their fermented food products are beneficial for health. *J. Appl. Microbiol.* **100**, 1171–1185.
Pouwels, P. H., Leer, R. J., and Boersma, W. J. A. (1996). The potential of Lactobacillus as a carrier for oral immunization: Development and preliminary characterization of vector systems for targeted delivery of antigens. *J. Biotechnol.* **44**, 183–192.
Reid, G. (2005). The importance of guidelines in the development and application of probiotics. *Curr. Pharm. Des.* **11**, 11–16.
Reid, G. and Bruce, A. (2006). Probiotics to prevent urinary tract infections: The rationale and evidence. *World J. Urol.* **24**, 28–32.
Reid, G., Lam, D., Bruce, A. W., van der Mei, H. C., and Busscher, H. J. (1994). Adhesion of lactobacilli to urinary catheters and diapers: Effect of surface properties. *J. Biomed. Mater. Res.* **28**, 731–734.
Reid, G., Bruce, A. W., and Smeianov, V. (1998). The role of Lactobacilli in preventing urogenital and intestinal infections. *Int. Dairy J.* **8**, 555–562.
Salminen, S., Isolauri, E., and Salminen, E. (1996). Probiotics and stabilisation of the gut mucosal barrier. *Asia Pac. J. Clin. Nutr.* **5**, 53–56.
Schulze, A. D., Alabi, A. O., Tattersall-Sheldrake, A. R., and Miller, K. M. (2006). Bacterial diversity in a marine hatchery: Balance between pathogenic and potentially probiotic bacterial strains. *Aquaculture* **256**, 50–73.
Skjermo, J. and Vadstein, O. (1999). Techniques for microbial control in the intensive rearing of marine larvae. *Aquaculture* **177**, 333–343.
Stanton, C., Lawless, F., Murphy, J. J., and Connolly, J. F. (1997). Conjugated linoleic acid—A marketing advantage for animal fats. *In* "Animal Fats: BSE and After", (K. Berger and P. J. Barnes, Eds), pp. 27–41. Barnes and Assoc, Bridgwater.
Szajewska, H. and Mrukowicz, J. Z. (2005). Use of probiotics in children with acute diarrhea. *Paediatr. Drugs* **7**, 111–122.

Underdahl, N., Torres-Medina, A., and Dosten, A. (1982). Effect of Streptococcus faecium C-68 in control of Escherichia coli-induced diarrhea in gnotobiotic pigs. *Am. J. Vet. Res.* **43**, 2227–2232.

Vieira, F. N., Buglione, C. C., Mouriño, J. P. L., Jatobá, A., Martins, M. L., Schleder, D. D., Andreatta, E. R., Barraco, M. A., and Vinatea, L. A. (2010). Effect of probiotic supplemented diet on marine shrimp survival after challenge with Vibrio harveyi. *Arq. Bras. Med. Vet. Zootec.* **62**, 631–638.

Vine, N. G., Leukes, W. D., and Kaiser, H. (2006). Probiotics in marine larviculture. *FEMS Microbiol. Rev.* **30**, 404–427.

Zhong, W., Millsap, K., Bialkowska-Hobrzanska, H., and Reid, G. (1998). Differentiation of Lactobacillus species by molecular typing. *Appl. Environ. Microbiol.* **64**, 2418.

CHAPTER **24**

Bioactive Secondary Metabolites from Marine Microbes for Drug Discovery

Chamilani Nikapitiya[1]

Contents

Abstract

The isolation and extraction of novel bioactive secondary metabolites from marine microorganisms have a biomedical potential for future drug discovery as the oceans cover 70% of the planet's surface and life on earth originates from sea. Wide range of novel

Department of Fisheries, Animal and Veterinary Science, University of Rhode Island, Kingston, Rhode Island, USA
[1] Corresponding author: Chamilani Nikapitiya, *E-mail address:* chamilaninikapitiya@yahoo.com

Advances in Food and Nutrition Research, Volume 65
ISSN 1043-4526, DOI: 10.1016/B978-0-12-416003-3.00024-X

bioactive secondary metabolites exhibiting pharmacodynamic properties has been isolated from marine microorganisms and many to be discovered. The compounds isolated from marine organisms (macro and micro) are important in their natural form and also as templates for synthetic modifications for the treatments for variety of deadly to minor diseases. Many technical issues are yet to overcome before wide-scale bioprospecting of marine microorganisms becomes a reality. This chapter focuses on some novel secondary metabolites having antitumor, antivirus, enzyme inhibitor, and other bioactive properties identified and isolated from marine microorganisms including bacteria, actinomycetes, fungi, and cyanobacteria, which could serve as potentials for drug discovery after their clinical trials.

I. INTRODUCTION

Present trend in drug discovery is to understand how the infections or diseases are controlled at the molecular and physiological level and based on that knowledge to suppress those causative factors. With advanced technology and emerging knowledge of biological systems (Anson *et al.*, 2009), still the drug discovery process is juvenile, with a slow emergence of new therapeutic discoveries. Identification of candidates, synthesis, characterization, screening, and tests for therapeutic efficacy are the initial steps in drug development, and if the compound is promising, it is followed by the clinical trials. Hence, the research and development of each new molecular entity costs approximately US$ 1.8billion (Paul *et al.*, 2010).

The mankind from inception, in their traditional medicinal practice, used natural products (terrestrial or marine origin) for prevention and curing of ailments or diseases and is continued to date. At present, though the technology is developed, with realization of their chemical diversity, natural products are better suited for successful drugs than the diversity of synthetic compound collections (Feher and Schmidt, 2003; Grabowski and Schneider, 2007), and former continues to play a major role as starting materials in drug discovery (Galm and Shen, 2007). The demand for drugs capable of controlling the existence and emergence of diseases or resistant strains of microorganisms is ever increasing. Most natural secondary metabolites show a great diversity of chemical structures, hence drugs based on these compounds are useful for diversity of disease treatments (Mayer and Lehmann, 2000; Newman and Cragg, 2007; Proksch *et al.*, 2002) such as varieties of bacterial (Nussbaum *et al.*, 2006), viral (Cragg and Newman, 2001), and fungal diseases, inflammatory conditions, immunosuppressants to aid in organ transplantation (Ruiz *et al.*, 2010) cancer, AIDS, etc. (Jain and Tiwari, 2007). As reported in 2007 in the years 1981–2006, out of 974 small-molecule new chemical entities, 63% were obtained from natural sources or semisynthetic derivatives (Newman and Cragg, 2007). According to Pharmaprojects database of March 2008 analysis by Harvey (2008), majority of the leads from natural products under development were from plants and microorganisms (Harvey, 2008). Since the discovery of penicillin in 1929, large number of natural products has been isolated from varieties of microorganisms. Biological activities were found in more than 10,000 of these natural products and over 100 are used as antibiotic drugs and for treatments of some other serious diseases like cancer (Carte, 1996). However, still only a fraction of earth's living species have been tested for bioactivity.

The ocean is complex and 70% of the earth surface is covered by the oceans. As life has originated from the ocean, there is a vast ecological, chemical, and biological diversification in the life of microorganisms to vertebrates. Thus, the rich biological and genetic diversity of marine organisms has arisen due to the harsh chemical and physical marine environmental conditions and adaptations to different habitats including stressful environments under cold, high pressure, and lightless conditions. They produce variety of compounds with unique structural and chemical properties showing various biological activities, and hence they feature a great potential to pharmaceutical industry (Blunt *et al.*, 2004). Therefore, marine biologists explore more compounds for drugs, human food, cosmetics, and other industries. Even though there were few approved and several marine organisms-derived compounds are currently under clinical trials, no descriptive reports have been published for marine microorganisms.

The field of marine microbes is a marginally explored domain, having a great potential. Up-to-date, only a limited number of marine microorganisms have been investigated for bioactive metabolites. Among them, the marine bacteria, actinomycetes (Bull and Stach, 2007; Fenical and Jensen, 2006), fungi (Bhadury *et al.*, 2006; Newman and Hill, 2006), and cyanobacteria (Tan, 2007) are attracting attentions as a new biotechnological resource and could serve as a potential source to isolate bioactive compounds.

This chapter briefly reviews the new biologically active secondary metabolites isolated, extracted or synthesized, and published recently. It highlights the chemical potential of the above marine microbes and some of the potential symbiotic microbes, with a focus on bioactive products and their mechanisms leading to pharmaceutical drug discovery. It also reviews the strategies implemented for sustainable supply and to overcome the obstacles during development of bioactive marine metabolites as potential pharmaceuticals.

II. PROBABLE CANDIDATES FOR DRUG DISCOVERY FROM MARINE MICROORGANISMS

A. Secondary metabolites from marine bacteria

It has been reported that approximately 10^6 bacteria cells/mL contains in ocean water (Ducklow, 2000). The marine organisms, especially marine bacteria comprised with unique metabolic and physiological capabilities, enable them not only to survive in extreme conditions and habitats but also to produce compounds which their terrestrial counterparts are incapable. Bioactive metabolites from marine bacteria have been increased dramatically since 1990 (Blunt *et al.*, 2004, 2005, 2006; Faulkner, 1992, 1997, 2000, 2001; Fenical, 1993a; Hill, 2003; Kobayashi and Ishibashi, 1993). As reported out of 659 marine bacterial compounds, four do not have taxonomic identification, and during 1997–2008, the compounds isolated were totally from five bacterial phyla including bacteroidetes (34 compounds), proteobacteria (256), firmicutes (35), cyanobacteria (220), and actinobacteria (256). The 85% metabolites from phyla Actinobacteria were from the genera *Streptomyces* (57%) and *Salinispora* (28%) (Williams, 2009). The secondary metabolite, *Macrolactin A* was identified from a deep-sea sediment unicellular unidentified Gram-positive bacterium (Gustafson *et al.*, 1989). This compound has antibacterial, antiviral, and cytotoxic activities. It shows significant inhibition of mammalian herpes simplex virus (type I and II) and protection of T lymphocytes against human immune-deficiency virus (HIV) replication, and inhibition of B16-F10 murine melanoma cells in *in vitro* assays (Carte, 1996). First secondary marine metabolite from the marine bacteria, *pentabromopseudilin*, is a highly brominated pyrrole antibiotic derived from *Pseudomonas bromoutilis* (Burkholder *et al.*, 1966), and active against Gram-positive bacteria. Its biosynthesis and structure–activity relationship were studied with *Alteromonas luteoviolaceus* (Laatsch *et al.*, 1995) also named as *Pseudoalteromonas luteoviolaceus*. Further, antiprotist activity has been reported for some strains of *A. luteoviolaceus* (Kamei *et al.*, 1986). The antibiotic, *3,3′,5,5′-tetrabromo-2,2′-diphenyldiol* (brominated biphenyl compound), isolated from marine bacteria *Pseudoalteromonas phenolica*, was reported to have inhibition activity against methicillin-resistant *Staphylococcus aureus* (MRSA) (Isnansetyo and Kamei, 2003). Even though the compound has the ability to permeabilize the cell membrane of MRSA, it has no lytic activity against bacterial cells or human erythrocytes (Isnansetyo and Kamei, 2003). Bioactive compounds from novel Gammaproteobacteria Gram-negative *Pseudomonas* with diverse chemical structures such as benzaldehyde, quinoline, pyrroles, phenanthren, phthalate, andrimid, moiramides, etc. have been isolated. An isolated *Pseudomonas* strain KMM 3042 was 97% sequence homology to *Pseudomonas borbori* (Romanenko *et al.*, 2008), and some of these isolated compounds can act as antimicrobial compounds with diverse mechanisms of actions where some act as cathepsin B inhibitors (dibutyl phthalate and di-(2-ethlhexyl) phthalate) (Isnansetyo and Kamei, 2009). The lysozyme from marine strain S-12-86 (LS) has exhibited *in vitro* antimicrobial activity and antiviral activity. The results of minimum inhibition concentration (MIC) (0.25–4.00 $mg\,mL^{-1}$), minimum bactericidal concentration (0.25–8.00 $mg\,mL^{-1}$) assays, and observation under the transmission electron microscope showed broad antimicrobial spectrum against several Gram-positive and Gram-negative bacteria, fungi, etc. Further it showed inhibiting effects of LS on pseudo rabies virus (PRV) in swine kidney cells (PK-15 cells) when added to cell culture medium at 0, 2, 4, 6, and 8h after PK-15 cells had been infected by PRV (Zhang *et al.*, 2008). *Tambjamines* (4-methoxypyrrole-containing bioactive compounds) have antibacterial, immunosuppressive, antiproliferative, antitumorigenic, and icthyodeterrent activities (Lindquist and Fenical, 1991). The marine bacteria *Pseudomonas aeruginosa* isolated from Sindh and Baluchistan coast of Pakistan showed antibacterial activity against several clinical isolates as well as indigenous marine bacteria (Uzair *et al.*, 2009). The bacterium *Geobacillus thermodenitrificans* exopolysaccharide, EPS-2, having antiviral and immunomodulatory effects on immunocompetent cells (Arena *et al.*, 2009) was recently discovered from marine vent of volcanic Island (Italy). The extracts of marine bacterium *Vibrio* sp. DS40M4 was shown to produce biological active molecule, *anguibactin* (Sandy *et al.*, 2010) and was found to be cytotoxic against the P388 murine leukemia cell line ($IC_{50} < 15\,\mu M$). Among two new cyclic depsipeptides, *turnagainolides A and B* isolated from laboratory cultures of marine isolate of *Bacillus* sp., the turnagainolide B showed activity in SHIP1 (SH2-containing inositol phophatase 1) activation assay (Li *et al.*, 2011b). The novel *indole alkaloids 1,1,3-tris-(3-indolyl)-butane* and several other bis- and tris-indole derivatives were isolated from *Vibrio parahaemolyticus* Bio 249 from the German North Sea (Veluri *et al.*, 2003). From the same habitat, sulfur-containing tropodithietic acid (TDA), the isolation of 3-(4′-hydroxyphenyl)-4-phenylpyrol-2,5-dicarcoxylic acid and 3.4-di(4′-hydroxyphenyl) pyrrole-2,5-dicarcoxylic from the strain RK377 (identified as γ-proteobacteria *Halomonas* sp.), and the isolation of bacteriopheophytin a_L from strain RK2207 (96% identity to *Mesorhizobium loti*) were reported (Liang, 2003). Also, the antibacterial compound TDA is produced by several marine *Roseobacter* sp. and was shown that apart from the probiotic effect of the bacteria in fish aquaculture, the compound was effective against range of human pathogenic bacteria (Porsby *et al.*, 2011). Further investigations are required to uncover whether above compounds would yield potential drug candidates.

No approved drugs by FDA, isolated from marine bacteria have been reported to date. Several potential compounds at different development stages suggest that this will change in near future. Promising anticancer agents of marine bacterial origin which are currently undergoing clinical trials are Tasidotin, Synthadotin (ILX-651) which is in Phase II at Genzyme Corporation (Cambridge, MA), and Sobidotin (TZT 1027) at Phase III at Aksa Pharmaceuticals (Tokyo, Japan) (Mayer *et al.*, 2010).

B. Secondary metabolites isolated from marine actinomycetes

Actinobacteria is one of the largest class of bacteria (Ventura *et al.*, 2007), and recent research revealed that groups of marine actinomycetes are robust sources of novel natural products for drug development (Blunt *et al.*, 2007; El-Gendy *et al.*, 2008; Maskey *et al.*, 2004; McArthur *et al.*, 2008; Olano *et al.*, 2009; Prudhomme *et al.*, 2008; Williams, 2009). One of the early identified new antibiotic named SS-228Y, isolated from actinomycete *Chainia purpurogena*, inhibits the growth of Gram-positive bacteria (Okazaki *et al.*, 1975). It has structure of peri-hydroxyquinone moiety. The potent proteasome inhibitor salinosporamide A and other distinct groups derived from the genus *Salinispora* and the members of that genus are yielding new classes of terpenoids, amino acid-derived metabolites, and polyene macrolides. *Salinosporamide A* is highly cytotoxic irreversible proteasome inhibitor (currently in Phase I clinical trials), isolated from *Salinospora tropica*, which belongs to rare obligate marine actinomycetes isolated from the ocean sediments (Feling *et al.*, 2003; Jensen *et al.*, 1991, 2005; Mincer *et al.*, 2002). This compound exhibits potent *in vitro* cytotoxic activity against human cancer (Ahn *et al.*, 2007) including human colon carcinoma, breast cancer, and non-small cell lung cancer. The unique bicyclic β-lactone γ-lactam structure salinosporamide A is a potent inhibitor, which contributes to its all proteolytic activities of the mammalian 20S proteasome (Chauhan *et al.*, 2005; Groll *et al.*, 2006) and exhibited inhibition of the chymotrypsin-like proteolytic activity of the purified 20S proteasome (Feling *et al.*, 2003). *Marinopyrroles A and B* are highly halogenated bispyrroles produced by a *Streptomyces* sp. (CNQ-418) from marine sediment samples in La Jolla, CA. Both compounds show activities against MRSA with MIC$<2\,\mu M$ and cytotoxicity against human cancer cell line (HCT-116); however, mechanisms of action are yet to be discovered in both (Hughes *et al.*, 2008). *Marineosins A and B* are two *Streptomyces*-derived diastereomer compounds with a previously unknown carbon skeleton and contain spirotetrahydropyrrole animal moiety (Boonlarppradab *et al.*, 2008). Marineosin A was more active in the HCT-116 cytotoxicity assay ($IC_{50}=0.5\,\mu M$) where marineosin B showed considerably weaker cytotoxicity ($IC_{50}=46\,\mu M$). Further, marineosine A exhibits broad cytotoxicity in the NCI 60 cell line, in addition to leukemia and melanoma cell lines. *Ammosamides A and B* are chlorinated tricyclic pyrroloquinoline alkaloids and were isolated from *Streptomyces* strain from deep-sea sediments samples from Bahamas (Hughes *et al.*, 2009). Strong *in vitro* cytotoxicity against HCT-116 colon carcinoma and selectivity against number of cancer cell lines indicates the specific mode of actions of these compounds, which has been used to identify the cellular target of these compounds (Hughes *et al.*, 2009). *Chandrananimycins A–C* were isolated from *Actinomadura* sp. exhibited antibacterial and cytotoxic activities. These are considering as anticancer compounds (Maskey *et al.*, 2003). Anticancer agent *thiocoraline* isolated from an actinomycete *Micromonospora marina* is currently undergoing preclinical evaluation as a potential future drugs (Newman and Cragg, 2004). *Mechercharmycins A and B* were isolated from the *Thermoactinomyces* sp. YM3-251 (Kanoh *et al.*, 2005). The mechercharmycin A is a cyclic peptide and shows cytotoxic activity against human lung carcinoma and human leukemia, while mechercharmycin B is a linear congener. Unprecedented *alkaloid lodopyridone,* isolated recently from a marine *Saccharomonospora* sp.; from the marine sediments collected at the mouth of the La Jolla submarine canyon, showed cytotoxicity to HCT-116 human colon cancer cells with IC_{50}=3.6 μM (Maloney *et al.*, 2009). *Echinomycin* antibiotic compound isolated from the culture broth of a marine s*treptomycete* showed antibiotic activities against *S. aureus* and *Enterococcus faecalis* with a MIC of 0.03 and 0.01 μM, respectively (Socha *et al.*, 2009). Fermentation-derived novel cytotoxic macrolide *IB-96212* was isolated from culture of *Micromonospora* sp. IB-96212 (Canedo *et al.*, 2000). Interestingly, several marine-derived actinomycetes with special reference to discovery of antituberculosis (TB) compounds for potential anti-TB drugs have been reported recently (Radhakrishan *et al.*, 2011). A novel triazolopyrimidine antibiotic, *essramycin,* was isolated from the culture broth of the marine *Streptomyces* sp., isolate Merv8102. This compound was antibacterially active against Gram-positive and Gram-negative bacteria with MIC of 2–8 μg/mL (El-Gendy *et al.*, 2008a). Among five new farnesyl-α-nitropyrroles isolated from saline culture of the marine actinomycete strain CNQ-509, the nitropyrrolin D showed cytotoxic activity toward HCT-116 human colon carcinoma cells; however, it showed a weak antibacterial activity against MRSA (Kwon *et al.*, 2010). *Fijiolide A,* a potent inhibitor of TNF-α-induced NFκB activation, has a promising lead for more advanced anticancer testing and was isolated from a marine-derived bacterium of the genus *Nocardiopsis*. Fijiolide A showed decreased TNF-α-induced NFκB activation by 70.3% (IC_{50}=0.57 μM), while fijiolide B showed less inhibition (46.5%), without dose dependence (Nam *et al.*, 2010). *Ammocidins B–D* were isolated from the culture broth of *Saccharothrix* sp. AJ9571 and ammocidins A and B showed potent antiproliferative activities against human cancer cell lines (Murakami *et al.*, 2009). Two trialkyl-substituted aromatic acids, *lorneic acid A* and *lorneic acid B*, were isolated from a marine-derived actinomyces strain (NPS554). Interestingly, lorneic acid A showed significant inhibition activities against phosphodiesterase (PDE)5 (Iwata *et al.*, 2009), and PDE5 inhibitors have pharmacological importance in pulmonary hypertension and erectile dysfunctions (Bhatnagar and Kim, 2010).

C. Secondary metabolites isolated from marine fungi

Various novel anticancer, antibacterial, antifungal, antiviral, antiplasmodial, anti-inflammatory bioactive metabolites which could beneficial for pharmaceutical therapeutics have been reported from marine-derived fungi (Bhadury *et al.*, 2006; Newman and Hill, 2006). *Ascochytatin*, a novel spirodioxynaphthalene metabolite, was isolated from marine-derived fungus *Ascochyta* sp. NGB4, which shows *B. subtilis* growth inhibition by targeting the function of the bacterial growth regulatory system TCS (YycG/YycF) (Kanoh *et al.*, 2008a). Novel sulfoalkylresorcinol isolated from marine-derived fungus *Zygosporium* sp. KNC52 showed weak antimicrobial activity against MRS. The reason for this activity may be due to inhibition of *in vitro* polymerization of a FtsZ, a protein which is a structural homologue of eukaryotic tubulin which involved in bacterial cell division (Kanoh *et al.*, 2008b). Zheng and coworkers reported that the compounds derived from marine halotolerant *Aspergillus sclerotiorum* PT06-1 have moderate antifungal, weak antibacterial, and cytotoxic activity (Zheng *et al.*, 2009, 2010). They isolated 11 novel aspochracin-type cyclic tripeptides, *Sclerotiotides A–K* from the marine-derived halo tolerant *A. sclerotiorum* PT06-1 in a nutrient-rich hypersaline medium. Ethyl acetate extracts of the fermentation broths of this organisms also produced known compounds of *JBIR-15, aspochracin*, and *penicillic acid*. Out of these 14, sclerotiotides A (MIC 7.5 μM), B (3.8 μM), F (30 μ M), and I (6.7 μM) and JBIR-15 (30 μM) exhibited antifungal activity against *Candida albicans* (Zheng *et al.*, 2010). Novel polyphenolic compounds (expansols A and B) were produced by marine-derived *Penicillium expansum* 091006 endogenous with mangrove plant *Excoecaria agallocha*. Expansol A was shown to have moderate cytotoxic activity against HL-60 cell line (IC_{50}=15.7 μM), and expansol B showed antiproliferative activity against A549 (IC_{50}=1.9 μM) and HL-60 (5.4 μM) cells (Lu *et al.*, 2010). Chrysogeside B an unsaturated C_{19}-fatty acid containing new cerebroside was isolated from fermentation broth of *Penicillium chrysogenum* PXP-55, a halotolerant fungus grown in hypersaline medium. Chrysogeside B showed antimicrobial activity against *Enterobacter aerogenes* (MIC=1.72 μM) (Peng *et al.*, 2011). Two new oxepin containing compounds were isolated from ethanol extract of marine-derived fungus *Aspergillus* sp. SF-5044, and one compound showed weak growth inhibitory activity against small panel of cell lines (Lee *et al.*, 2011). Further, cytotoxicity activity of novel triazole carboxylic acid *penipanoid A* and recently identified *quiazolinone* derivative was reported from marine sediment-derived fungus *Penicillium paneum* SD-44 (Li *et al.*, 2011a). Ethanol extracts (8 drimane sesquiterpenes, 6 isochromane derivatives, and 3 known compounds, daldinin B, 9α-hydroxy-6β-[(2*E*,4*E*,6*E*)-octa-2,4,6-trienoyloxy]-5α-drim-7-en-11,12-olide, and pergillin) of the marine-derived fungus *Aspergillus ustus* 094102 showed range of cytotoxicity from high to low (0.13–30.0 μM) against A549 and HL-60 cell lines (Lu *et al.*, 2009). Ester-substituted sesquiterpenoid related to the eremophilane class cryptosphaerolide, isolated from marine-derived ascomycete fungal strain CNL-523 (*Cryptosphaeria* sp.). Exploring the Mcl-1 pathway of this compound (an inhibitor of the protein Mcl-1, a cancer drug target involved in apoptosis) is useful for cancer chemotherapy, as isolated compound showed significant cytotoxicity against HCT-116 human colon carcinoma cell line (Oh *et al.*, 2010). Antiplasmodial activity was found from marine-derived fungus *Nodulisporium* sp. CRIF1 from Thai soft coral produced novel tetronic acid, nodlisporacid A, which showed a moderate antiplasmodial activity (IC_{50}=1–10 μM) against chloroquine-resistant *Plasmodium falciparum* strain 94 (Kasettrathat *et al.*, 2008). A potent and selective vascular disrupting agent (VDA) Plinabulin (NPI-2358), originally derived from marine fungal extract, is presently undergoing Phase II clinical trials. Preclinical trials showed that this was active to inhibit multidrug-resistant human tumor cell lines, and in variety of animal models, and enhances the efficacy when used with the existing chemo- or radiotherapy (Mayer *et al.*, 2010).

D. Secondary metabolites from marine cyanobacteria

Marine cyanobacteria (blue-green algae) are a diverse group of Gram-negative bacteria (Lopes *et al.*, 2010). They are important source of natural bioactive metabolites including toxins with pharmaceutical applications (Jaiswal *et al.*, 2008). Their uses as antibacterial, antiviral, antitumor, anti-HIV, etc., against macro and microorganisms and cell lines as well as food additives have been reported (Singh *et al.*, 2005). Being an excellent source of vitamins and proteins, some considered as health foods in worldwide markets. Chemically, secondary metabolites of cyanobacteria are macrolides, fatty acids, lipopeptides, amino acids, and amides (Jaiswal *et al.*, 2008), and majority of the molecules are nitrogen containing compounds. Between 2001 and 2006, over 120 cyanobacterial alkaloids reported and published show that there is a wide structural diversity and variety of biological actions such as antifungal, cytotoxicity, sodium channel modulation, and inhibition of proteases (Harvey, 2008). Majority of these were originated from the filamentous order Nostocales, and the genera from

Lyngbya, *Oscillatoria*, and *Symploca* (Williams, 2009). The compounds derived from most of these cyanobacteria have antitumor activities which contain common structure class of modified peptides or peptide–polyketide hybrids (Tan, 2007). Antimicrotubule agents dolastatin 10 and curacin A have pronounced antitumor activity, where preclinical and clinical trials led to the development of number of synthetic analogs using their structures. Additionally, highly cytotoxic compounds such as hectochlorin, apratoxins, aurilides, and lyngbyabellins were considered as potential lead compounds for the development of anticancer agents (Tan, 2007). Also, there is an emerging new class of molecules such as largazole, somocystinamide A, and coibamide, which are continuously identified with potent biological activities. *Largazole* was isolated from cyanobacteria *Symploca* sp. from Key Largo, Florida Keys. It was the first natural product identified from cyanobacteria containing thioester; thus, the 3-hydroxy-7-mercaptohept-4-enoic acid residue gave a novel structural feature to the marine natural product chemistry. Previous researches suggest that this compound preferentially targeting the cancer cells. It has an enhanced antiproliferative activity against fibroblastic osterosarcoma cells and human mammary epithelial cells when compared with untransformed cells (Tan, 2007). Later on, number of total synthesis has been developed (Bowers *et al.*, 2008; Nasveschuk *et al.*, 2008; Ying *et al.*, 2008) to provide sufficient material to demonstrate the mode of action as a histone deacetylase inhibitor (Bowers *et al.*, 2008; Ying *et al.*, 2008). *Coibamide A* is a highly *N*-methylated cyclic depsipeptide which was isolated from *Leptolyngbya* sp. from Coiba National Park, Panama. This unique peptidic cytotoxin showed novel selectivity profile on NCI 60 cell line, suggesting inhibition of cancer cell proliferation through a novel mechanism and leads to the drug development (Medina *et al.*, 2008). *Somocystinamide A* was first reported by Gerwick lab in 2002 (Nogle and Gerwick, 2002), and recent research demonstrated that this compound could initiate caspase-8-dependent apoptosis in many human tumor cell line (Wrasidlo *et al.*, 2008). Interestingly, modification of either lipopeptide or disulfide moiety in this structural atypical lipopeptide showed complete loss of activity. However, by integration of the compound into the nanosomes did not affect its biological activity. The compound's solubility and stability problems were solved by application of targetable nanoplatforms containing somocystinamide A, thereby facilitating its transport of the compound into required tumor cells (Torchilin, 2006). Alkynyl bromide-containing cyclic depsipeptide *Veraguamides* were isolated from marine cyanobacterium cf. *Oscillatoria margaritifera*. Out of veraguamides A–L isolated, veraguamides A, B, K, and L featured alkynyl bromide, which resembles the functionality only to other marine natural product jamaicamide A. Veraguamide A showed potent cytotoxicity to the H-460 human lung cancer cell line (LD_{50}=141 nM) (Mevers *et al.*, 2011). Seven new cyclic depsipeptides, veraguamides A–G, and known compound dolastatin 16 were isolated by cytotoxicity-directed purification of a *Symploca* cf. *hydnoides* sample from Cetti Bay, Guam. Presence of an invariant proline residue, multiple *N*-methylated amino acids, an α-hydroxy acid, and a C_8-polyketide-derived β-hydroxy acid moiety with a characteristic terminus as an alkynyl bromide, alkyne, or vinyl groups were characterized to veraguamides A–G. These compounds exhibited moderate to weak cytotoxic activity against HeLa cervical carcinoma cell lines and HT29 colorectal adenocarcinoma. Weak cytotoxic activity was observed on Dolastatin 16 toward both cell lines (Salvador *et al.*, 2011). Even though, the potent antitumor agent dolastatin 10, which originally isolated from the sea hare *Dolabella auricularia*, subsequently, reported to be isolated also from the marine cyanobacterium *Symploca* sp. VP642 from Palau (Luesch *et al.*, 2001). Symplostatin 1, a potent microtubule inhibitor, has been reisolated from *S. hydnoides* (from Guamanian and Hawaiian varieties). Antitumor activity of symplostatin 1 was tested against various murine tumors and found that it was effective against drug-insensitive colon tumor and mammary tumor. But, it was observed less effective against two multidrug-resistant tumors (Luesch *et al.*, 2001). *Caylobolide A* was previously isolated from *Lyngbya majuscula* from the Bahamas (MacMillan and Molinski, 2002). It was reported that caylobolide B, which is analogue of the known macrolactone caylobolide A, was isolated from *Phormidium* sp. collected from Key West Florida. It showed cytotoxic activity against HeLa cervical carcinoma IC_{50}=12.2 μM and HT29 colorectal adenocarcinoma cells IC_{50}=4.5 μM (Salvador *et al.*, 2010). Novel highly functionalized linear peptide, *gallinamide A*, obtained from organic extract of a marine cyanobacteria *Schizothrix* species from Panama showed potent initial antimalarial activity against the W2 chloroquine-resistant strain of *P. falciparum* (Linington *et al.*, 2009). Clark and his group investigated cytotoxic compounds from marine cyanobacterium *Blennothrix cantharidosmum* from Papua New Guinea and extracted the known compounds lyngbyastatins 1 and 3, and the first time-derived new compounds from this genus showed modest antimalarial and inhibition of quorum sensing activities (Clark *et al.*, 2008). As stated, marine cyanobacteria have a potential for large-scale production of vitamins such as vitamins of B complex group and vitamin E having a commercial interest (Plavsic *et al.*, 2004). There is a commercial value in carotenoids and phycobiliprotein pigments of cyanobacteria, as natural food coloring agents, feed additives, enhancers of the egg yolk color, drugs to improve the fertility and health of cattle, and also in cosmetic industry (Jha and Zi-rong, 2004).

E. Secondary metabolites produced by microbial symbionts

There are many interesting and useful compounds in marine invertebrates suspected to be origin from microbial symbionts (Kijjoa and sawangwong, 2004), as most metabolites isolated from marine invertebrates are structurally homologous to known metabolites of strict microbial origin. This commensally microbiota largely involved in the biosynthesis of metabolites drawn out from the macroorganisms host (Long *et al.*, 2005). *Isatin* is one of the early identified marine metabolite also called 2,3-indolinedione. The commensal *Alteromonas* sp. is responsible for production of this compound colonizing the surface of the shrimp *Palaemon macrodactylus* embryos and protects the eggs from its pathogen *Lagenidium callinectes*, the fungus (Gil-Turnes *et al.*, 1989). The yellow pigment which was later found as tambjamine like alkaloid (Franks *et al.*, 2005) and named as YPI was isolated from *Pseudoalteromonas tunicata*, was colonized on the surface of marine invertebrates, has antifungal activity (Egan *et al.*, 2002). Two new highly brominated compounds, *2,3,5,7-tetrabromobenzofuro[3,2-b] pyrrole* and *4,4′,6-tribromo-2,2′-biphenol* were identified from extracts of a marine *Pseudoalteromonas* sp. (CMMED 290) and isolated from the surface of a nudibranch collected in Kaneohe Bay, Oahu. They exhibit significant antimicrobial activity against MRSA (Feher *et al.*, 2010). Bicyclic depsipeptides, *Salinamide A and B*, were produced by the *Streptomyces* sp. CNB-091 isolated from the surface of Jellyfish *Cassiopeia xamachana*. These compounds showed moderate antibiotic activity, and chemically induced mouse ear edema assay showed that it has topical anti-inflammatory activity (Moore and Seng, 1998; Moore *et al.*, 1999; Trischman *et al.*, 1994). Previously, scientist discovered marine invertebrate *Bugula neritina* is the source of bryostatins. Bryostatin was undergone several Phase I and II clinical trials (Mann, 2002) as anticancer drug for leukemia, prostate, ovarian, and esophageal cancers and with combination of paclitaxel like other anticancer drugs (Abadi *et al.*, 2006; Clamp and Jayson, 2002; Hale *et al.*, 2002). It was reported that bacteria that live inside *B. neritina* colonies are the likely source of the anticancer compound, which passed in larvae from one generation to the next (Davidson *et al.*, 2001). Bryostatins are complex polyketides similar to bacterial secondary metabolites synthesized by modular type I polyketide synthases (PKS-I). Davidson and group identified a gene that produces this compound and showed that the gene is expressed solely in the γ-proteobacterium, called "*Candidatus* Endobugula sertula." (Davidson *et al.*, 2001; Sudek *et al.*, 2007). The compound ET-743, the DNA minor grove binder which is marketed as Yondelis by PharmaMar (http://www.pharmamar.com/en/about/), is used for the treatment of patients having soft tissue sarcoma and ovarian cancer (Butler, 2008). The drug isolated from tunicate *Ecteinascidia turbinate* in Caribbean with diverse bacterial community on the surface. It has been reported that ET-743 contains interesting structural similarity to the safracin class of DNA alkylators produced by variety of *Streptomyces* bacteria (Li *et al.*, 2008). This hypothesized that uncultured symbionts are actually responsible for the production of the drug. The fermentation of such bacteria would provide the large-scale production of compounds which could eradicate the supply problem of the drug from macroorganisms (Mendola, 2003). The commensalic bacteria harbors on marine sponge *Halichondria panicea* synthesize neuroactive compounds. Strong activation of the ionotropic glutamate receptor (iGluR) causes severe pathophysiological effects in men, for example, Alzheimer's disease (Muller *et al.*, 1997) or prion diseases (Muller *et al.*, 1993). It was proven that supernatant of the bacteria, which has highest identity to *Antarcticum vesiculatum* and *Psychroserpens burtonensis* (identified by 16S rDNA gene amplification and subsequent sequencing), has agonizing activity against subtype of the iGluR, neuronal *N*-methyl-D-aspartate receptor in cortical neurons of rats as a cell system (Perovic *et al.*, 1998). The antileukemic agent sorbicillactone-A produced by marine *P. chrysogenum* associated with sponge *Iricinia fasciculata* has shown antiviral and neuroprotective properties, now qualified for human trials. Recently, it was reported that member of marine Bacteroidetes, *Zobellia galactanivorans*, produced poryphyranases was active on the poryphyran from marine seaweed of the genus *Porphyra*. Seaweed *Porphyra* sp. (nori) gives important contribution to the daily diet in Japan (traditionally used in sushi). It has been reported that the porphyranases and agarases are common in the comparative gut metagenome analyses data in the Japanese population, but their absence in metagenome data from North American individuals. Further, it has been shown the genes coding for porphyranases, agarases and associated proteins have been transferred to the gut bacterium *Bacteroides plebeius* isolated from Japanese individuals. This suggest that these novel carbohydrate active enzymes were acquired in human gut bacteria through daily diet (Hehemann *et al.*, 2010) and would be interesting to investigate probiotic effect or health benefit to the host further.

III. LIMITATIONS IN USE OF NATURAL PRODUCTS FOR DRUG DEVELOPMENT

Restricted availability and access makes the collection of samples from marine organisms is an arduous task (Faulkner, 1992), as sampling in deep oceans and to explore diverse habitat requires specialized equipments and new technologies. There is a vast diversity of untapped microbes in the marine environment. Culturing these bacteria in marine samples using normal microbiological techniques is difficult (Konig and Wright, 1999; Ramaswamy *et al.*, 2006) and success is limited to few (Davidson, 1995; Fenical, 1993b; Kobayashi and Ishibashi, 1993); also the culturing methods are time consuming. Most compounds derived from the bacteria were partially or totally unidentified types or numbered strains, and their taxonomy of the marine bacteria is poorly defined. In marine environment, only tracer amounts of marine-based bioactive compounds could be extracted, which make it difficult to obtain even the minimum requirement needed to undergo preclinical and clinical trials. Further, very low fermentation yields are obtained even from a dense culture, sometimes limited to few milligrams per liter. Intellectual property rights and complexity of natural product chemistry are other additional restrains limiting the use of marine natural products for drug discovery (Harvey, 2008). Few compounds of marine microbial origin have so far undergone clinical trials to get approved as pharmaceuticals drugs (Mayer *et al.*, 2010).

IV. ADVANCE STRATEGIES TO EXPLOIT MARINE NATURAL PRODUCTS

Earlier symbiotic marine microorganisms in microbial environment were experimentally separated to invertebrates and microbial cells and were independently analyzed (Gulder and Moore, 2009), which was laborious and slow. Conversely, the recent progressive techniques of mass spectrometry using MALDI-TOF-MS and natural product MALDI-TOF imaging (npMALDI-I) show fast tracked the identification, isolation of microbial cells and promoted the drug discovery.

The microbes previously thought to be ''uncultivable'' were recovered and purified from the ocean both by conventional and by innovative isolation methods. Recently, detailed protocol in isolation and cultivation of marine-derived fungi associated with varieties of marine organisms for the preliminary evaluation of their pharmacological properties has been reported (Kjer *et al.*, 2010). With recent advances in fractionation and analytical techniques, it was possible to isolate and purify natural products, to determine their structures, and to screen the mixtures of natural products within similar time frame as for high throughput screening. Such identification of a compound alone does not justify its abundance for biosynthetic production and needs to know its genomic origin, as laboratory cultivation of some bacterial symbionts is yet unattainable. To overcome the difficulties and limitations associated with cultivation techniques, several molecular genetics techniques have been developed. Recombinant DNA technologies allow manipulating the biosynthetic pathways to improve the yield of bioactive metabolites and selection of lead candidate for pharmaceutical development. The cloning of genomic DNA from uncultured microbiota from marine invertebrates, heterologous expression of biosynthetic genes in a heterologous host to produce bioactive compounds will speed up the process compared to original organisms which it was produced (Olano *et al.*, 2008). Attempts are made to cultivate and grow the bacteria outside of its natural environment within *B. neritina* and to clone the genes that make the drug and transfer those into an organism which could more efficiently produced in large quantities. Moreover, genes responsible for biosynthesis of several marine metabolites have been entirely sequenced. The complete genomic sequences of *S. tropica* and of *Salinospora arenicola* have been published. *S. tropica* contains 17 distinct gene clusters encoding to diversify secondary metabolites (Udwary *et al.*, 2007), for the production of different drug classes using new strategies (conjugation of reactive compounds to a monoclonal antibody), which could serve as antitumor agents (Tang *et al.*, 2007). Symbiotic Actinomycetes, *Streptomyces* sp. JP95 isolate, have been cultured from marine tunicate *Aplidium lenticulum* and sequenced the biosynthetic gene cluster, which produces the telomerase inhibitor griseorhodin A (Long *et al.*, 2005). Metagenomics approach to marine ecosystem helps to synthesize large number of bioactive metabolites from marine bacteria and facilitate the direct access to the genomes of various uncultivable microorganisms and successful investigation of marine microbial diversity (Barrios-Llerena *et al.*, 2007; Kennedy *et al.*, 2007; Lefevre *et al.*, 2008; Li and Qin, 2005). The bioinformatics approach could be used to discover novel compounds with potential anticancer, antibacterial, and other therapeutic properties. Through this approach, it is possible to predict which microbes will produce novel chemicals by analyzing gene sequences encoding polyketide synthesis. Various genetic manipulation techniques such as combinatorial genetics (Harvey, 2000, 2007) are helpful to overcome difficulties in culturing the variety of bacterial species. Research focused on polyketide pathways of bacteria (Piel *et al.*, 2004) and fungi (Cox, 2007) to find out various drug analogues with range of activities is in progress. Molecular biological techniques have been developed to produce drug-like isoprenoid compounds, originally isolated from plants (Chang and Keasling, 2006), and now could be extended toward marine environment. Combined use of synthetic and enzymatic methods to produce complex natural products to obtain an improved activity of particular bioactive metabolite is another approach (Kennedy, 2008; Kopp and Marahiel, 2007), which could be used for marine natural products drug discovery.

V. FUTURE PROSPECTIVE OF MARINE MICROBES AND CONCLUSIONS

Natural products continue to provide a unique and diverse source of bioactive lead compounds for drug discovery, and hence, for its sustenance, it is important to keep natural products research viable and productive. However, sustainability of the source compound and changing nature of biodiversity are challenging phenomena in the pharmaceutical industry. Microbial diversity and natural product secondary metabolites in marine environment have not been explored sufficiently. Vast areas of major inhabitants (e.g., deep sea, deep-sea hydrothermal vents) (Thornburg *et al.*, 2010) remain to be discovered. Drug discovery strategies could be made more efficient and resourceful by continuous exploration of novel marine secondary natural metabolites and identification of their unique chemical structures with modern drug discovery strategies of combinatorial chemistry (Feher and Schmidt, 2003; Rester, 2008), high throughput screening, virtual screening (Rollinger *et al.*, 2008), and computer-aided design of small-molecule ligands formation.

Although this chapter focused on potential microbes producing secondary metabolites which are useful for infections and life threatening diseases, large number of microbial sources and their secondary metabolite properties still remain to be identified and discussed. Extensive and urgent attention is needed to broaden the drug discovery perspectives from marine microbes and in-depth understanding of their mechanism to use them as new therapies for AIDS, anti-inflammation, immune suppression, Alzheimer's disease, and aging processes (Schwartsmann *et al.*, 2001), also to overcome serious threats of reemerging infectious diseases and to combat multidrug resistance (Rahman *et al.*, 2010).

Novel drugs based on marine microbes could be developed by collaborative and interdisciplinary research with marine scientists, microbiologists, biotechnologists, geneticists, and medical practitioners. This will expedite the quest for innovative approaches to formulate novel classes of drugs by understanding the exact mechanism of the microbial metabolite, chemical interaction of oceanic environment, microbial biotechnology, and development of suitable bioassays. More emphasis on marine genome and microbial biotechnology approach is needed to discover optimum utilization of marine ecosystem for natural product drug discovery. A recent report showed angiotensin-converting enzyme inhibitory activity from peptide derived from single cell protein of marine yeast strain G7a (Ni *et al.*, 2009), which could open up new vistas for biomedical science. Continuous genomic libraries generation, functional proteomics, and DNA microarray analysis of gene expressed in selected surrogate hosts using metagenomic approach will provide qualitative and quantitative information of the secondary metabolites of uncultivable marine microbes. Free living cyanobacteria, yeast, and some picoplanktons also might have a greater chance to be potential host candidates for heterogeneous expression of metagenomic libraries of marine origin (Li and Qin, 2005). These will contribute toward the genetic conservation and the sustainable maintenance of marine microbial diversity. Consequently, these could serve as an alternative strategy to mitigate the complete eradication of marine microbes from threats forthcoming by disastrous environmental conditions (e.g., global warming, environmental deterioration) (Kingston, 2011).

The optimum physicochemical properties such as temperature, pH, oxygen level, and the level of nutrients will enhance the metabolic processes, chemical reactions, and secondary metabolite production of the microbes, and hence, their maintenance is important to optimize the metabolite yield (Ruiz *et al.*, 2010). Additionally, mutagenesis techniques could be used to identify strain and product yield with combination of biotechnology and other genetic engineering techniques. New advances in genetics, chemistry, and cultivation and fermentation technology could be utilized for the production of high-value novel natural lead compounds with a variety of bioactivity, for the control of human diseases, and for the protection of human health, which ultimately dictates our own survival. The bioactive compounds isolated using above strategies are equally important for in-depth understanding of the diversity of marine microbiota, ecosystem functions, and the exploitable biology.

REFERENCES

Abadi, G., Palen, W., Geddings, J., Irwin, T., Kasali, N., Colyer, J., Goodsen, F., Smith, J., Jones, K., Loftis, R., Hester, J., Noble, L., *et al.* (2006). A history of bryostatins: A prominent marine natural product, recent progress in medicinal plants. *Recent. Prog. Med. Plants* **15**, 370–411.

Ahn, K. S., Sethi, G., Chao, T. H., Neuteboom, S. T., Chaturvedi, M. M., Palladino, M. A., Younes, A., and Aggarwal, B. B. (2007). Salinosporamide A (NPI-0052) potentiates apoptosis, suppresses osteoclastogenesis and inhibits invasion through down-modulation of NF-kB-regulated gene products. *Blood* **10**, 2286–2295.

Anson, B. D., Ma, J., and He, J. Q. (2009). Identifying cardiotoxic compounds. *Genet. Eng. Biotechnol. News* **29**(9), 1–2.

Arena, A., Gugliandolo, C., Stassi, G., Pavone, B., Iannello, D., Bisignano, G., and Maugeri, T. L. (2009). An exopolysaccharide produced by *Geobacillus thermodenitrificans* strain B3-72: Antiviral activity on immunocompetent cells. *Immunol. Lett.* **123**(2), 132–137.

Barrios-Llerena, M. E., Burja, A. M., and Wright, P. C. (2007). Genetic analysis of polyketide synthase and peptide synthetase genes in cyanobacteria as a mining tool for secondary metabolites. *J. Ind. Microbiol. Biotechnol.* **34**(6), 443–456.

Bhadury, P., Mohammad, B. T., and Wright, P. C. (2006). The current status of natural products from marine fungi and their potential as anti-infective agents. *J. Ind. Microbiol. Biotechnol.* **33**, 325–337.

Bhatnagar, I. and Kim, S. K. (2010). Immense essence of excellence: Marine microbial bioactive compounds. *Mar. Drugs* **8**, 2673–2701.

Blunt, J. W., Copp, B. R., Munro, M. H. G., Northcote, P. T., and Prinsep, M. R. (2004). Marine natural products. *Nat. Prod. Rep.* **21**(1), 1–49.

Blunt, J. W., Copp, B. R., Munro, M. H. G., Northcote, P. T., and Prinsep, M. R. (2005). Marine natural products. *Nat. Prod. Rep.* **22**(1), 15–61.

Blunt, J. W., Copp, B. R., Munro, M. H. G., Northcote, P. T., and Prinsep, M. R. (2006). Marine natural products. *Nat. Prod. Rep.* **23**(1), 26–78.

Blunt, J. W., Copp, B. R., Hu, W. P., Munro, M. H., Northcote, P. T., and Prinsep, M. R. (2007). Marine natural products. *Nat. Prod. Rep.* **24**(1), 31–86.

Boonlarppradab, C., Kauffman, C. A., Jensen, P. R., and Fenical, W. (2008). Marineosins A and B, cytotoxic spiroaminals from a marine-derived actinomycete. *Org. Lett.* **10**(24), 5505–5508.

Bowers, A., West, N., Taunton, J., Schreiber, S. L., Bradner, J. E., and Williams, R. M. (2008). Total synthesis and biological mode of action of largazole: A potent class I histone deacetylase inhibitor. *J. Am. Chem. Soc.* **130**(33), 11219–11222.

Bull, A. T. and Stach, J. E. M. (2007). Marine actinobacteria: New opportunities for natural product search and discovery. *Trends Microbiol.* **15**(11), 491–499.

Burkholder, P., Pfister, R., and Leitz, F. (1966). Production of a pyrrole antibiotic by a marine bacterium. *Appl. Microbiol.* **14**(4), 649–653.

Butler, M. S. (2008). Natural products to drugs: Natural product-derived compounds in clinical trials. *Nat. Prod. Rep.* **25**, 475–516.

Canedo, L. M., Fernández-Puentes, J. L., and Baz, J. P. (2000). IB-96212, a novel cytotoxic macrolide produced by a marine Micromonospora. II. Physico-chemical properties and structure determination. *J. Antibiot. (Tokyo)* **53**(5), 479–483.

Carte, B. K. (1996). Biomedical potential of marine natural products. *Bioscience* **46**, 271–286.

Chang, M. C. Y. and Keasling, J. D. (2006). Production of isoprenoid pharmaceuticals by engineered microbes. *Nat. Chem. Biol.* **2**, 674–681.

Chauhan, D., Catley, L., Li, G., Podar, K., Hideshima, T., Velankar, M., Mitsiades, C., Mitsiades, N., Yasui, H., Letai, A., Ovaa, H., Berkers, C., *et al.* (2005). A novel orally active proteasome inhibitor induces apoptosis in multiple myeloma cells with mechanisms distinct from Bortezomib. *Cancer Cell* **8**(5), 407–419.

Clamp, A. I. and Jayson, G. C. I. (2002). The clinical development of the bryostatins. *Anti-Cancer Drugs* **13**(7), 673–683.

Clark, B. R., Engene, N., Teasdale, M. E., Rowley, D. C., Matainaho, T., Valeriote, F. A., and Gerwick, W. H. (2008). Natural products chemistry and taxonomy of the marine cyanobacterium *Blennothrix cantharidosmum*. *J. Nat. Prod.* **71**(9), 1530–1537.

Cox, R. J. (2007). Polyketides, proteins and genes in fungi: Programmed nano-machines begin to reveal their secrets. *Org. Biomol. Chem.* **5**, 2010–2026.

Cragg, G. M. and Newman, D. J. (2001). Medicinals for the millennia: The historical record. *Ann. NY Acad. Sci.* **953**, 3–25.

Davidson, B. S. (1995). New dimensions in natural products research: Cultured marine microorganisms. *Curr. Opin. Biotechnol.* **6**(3), 284–291.

Davidson, S. K., Allen, S. W., Lim, G. E., Anderson, C. M., and Haygood, M. G. (2001). Evidence for the biosynthesis of bryostatins by the bacterial symbiont "*Candidatus* Endobugula sertula" of the bryozoan *Bugula neritina*. *Appl. Environ. Microbiol.* **67**(10), 4531–4537.

Ducklow, H. (2000). Bacterial production and biomass in the oceans. *In* "Microbial Ecology of the Oceans", (D. L. Kirchman, Ed.), pp. 85–120. Wiley, New York, NY.

Egan, S., James, S., Holmström, C., and Kjelleberg, S. (2002). Correlation between pigmentation and antifouling compounds produced by *Pseudoalteromonas tunicata*. *Environ. Microbiol.* **4**(8), 433–442.

El-Gendy, M. M. A., Shaaban, M., Shaaban, K. A., El-Bondkly, A. M., and Laatsch, H. (2008). Essramycin: A first triazolopyrimidine antibiotic isolated from nature. *J. Antibiot (Tokyo)* **61**, 149–157.

El-Gendy, M. M., Hawas, U. W., and Jaspars, M. (2008a). Novel bioactive metabolites from a marine derived bacterium *Nocardia* sp. ALAA 2000. *J. Antibiot. (Tokyo)* **61**(6), 379–386.

Faulkner, D. (1992). Biomedical uses for natural marine chemicals. *J. Oceanus* **35**, 29–35.

Faulkner, D. J. (1997). Marine natural products. *Nat. Prod. Rep.* **14**(3), 259–302.

Faulkner, D. J. (2000). Marine natural products. *Nat. Prod. Rep.* **17**(1), 7–55.

Faulkner, D. J. (2001). Marine natural products. *Nat. Prod. Rep.* **18**(1), 1–49.

Feher, M. and Schmidt, J. M. (2003). Property distributions: Differences between drugs, natural products, and molecules from combinatorial chemistry. *J. Chem. Inf. Comput. Sci.* **43**(1), 218–227.

Feher, D., Barlow, R., McAtee, J., and Hemscheidt, T. K. (2010). Highly brominated antimicrobial metabolites from a marine *Pseudoalteromonas* sp. *J. Nat. Prod.* **73**(11), 1963–1966.

Feling, R. H., Buchanan, G. O., Mincer, T. J., Kauffman, C. A., Jensen, P. R., and Fenical, W. (2003). Salinosporamide A: A highly cytotoxic proteasome inhibitor from a novel microbial source, a marine bacterium of the new genus *Salinospora*. *Angew. Chem. Int. Ed Engl.* **42**(3), 355–357.

Fenical, W. (1993a). Chemical studies of marine bacteria: Developing a new resource. *Chem. Rev.* **93**(5), 1673–1683.

Fenical, W. (1993b). Marine bacteria: Developing a new chemical resource. *Chem. Rev.* **93**, 1673–1683.

Fenical, W. and Jensen, P. R. (2006). Developing a new resource for drug discovery: Marine actinomycete bacteria. *Nat. Chem. Biol.* **2**, 666–673.

Franks, A., Haywood, P., Holmström, C., Egan, S., Kjelleberg, S., and Kumar, N. (2005). Isolation and structure elucidation of a novel yellow pigment from the marine bacterium *Pseudoalteromonas tunicata*. *Molecules* **10**(10), 1286–1291.

Galm, U. and Shen, B. (2007). Natural product drug discovery: The times have never been better. *Chem. Biol.* **14**, 1098–1104.

Gil-Turnes, M. S., Hay, M. E., and Fenical, W. (1989). Symbiotic marine bacteria chemically defend crustacean embryos from a pathogenic fungus. *Science* **246**(4926), 116–118.

Grabowski, K. and Schneider, G. (2007). Properties and architecture of drugs and natural products revisited. *Curr. Chem. Biol.* **1**, 115–127.

Groll, M., Huber, R., and Potts, B. C. (2006). Crystal structures of Salinosporamide A (NPI-0052) and B (NPI-0047) in complex with the 20S proteasome reveal important consequences of beta-lactone ring opening and a mechanism for irreversible binding. *J. Am. Chem. Soc.* **128**, 5136–5141.

Gulder, A. M. G. and Moore, B. S. (2009). Chasing the treasures of the sea-bacterial marine natural products. *Curr. Opin. Microb.* **12**, 252–260.

Gustafson, K., Roman, M., and Fenical, W. (1989). The macrolactins, a novel class of antiviral and cytotoxic macrolides from a deep-sea marine bacterium. *J. Am. Chem. Soc.* **111**(19), 7519–7524.

Hale, K. J., Hummersone, M. G., Manaviazar, S., and Frigerio, M. (2002). The chemistry and biology of the bryostatin antitumour macrolides. *Nat. Prod. Rep.* **19**, 413–453.

Harvey, A. (2000). Strategies for discovering drugs from previously unexplored natural products. *Drug Discov. Today* **5**(7), 294–300.

Harvey, A. L. (2007). Natural products as a screening resource. *Curr. Opin. Chem. Biol.* **11**(5), 480–484.

Harvey, A. L. (2008). Natural products in drug discovery. *Drug Discov. Today* **13**(19/20), 894–901.

Hehemann, J. H., Correc, G., Barbeyron, T., Helbert, W., Czjzek, M., and Michel, G. (2010). Transfer of carbohydrate-active enzymes from marine bacteria to Japanese gut microbiota. *Nature* **464**, 908–912.

Hill, R. A. (2003). Marine natural products. *Annu. Rep. Prog. Chem. Sect. B.* **99**, 183–207.

Hughes, C. C., Prieto-Davo, A., Jensen, P. R., and Fenical, W. (2008). The marinopyrroles, antibiotics of an unprecedented structure class from marine *Streptomyces* sp. *Org. Lett.* **10**, 629–631.

Hughes, C. C., MacMillan, J. B., Gaudêncio, S. P., Jensen, P. R., and Fenical, W. (2009). The ammosamides: Structures of cell cycle modulators from a marine-derived *Streptomyces* species. *Angew. Chem. Int. Ed.* **48**(4), 725–727.

Isnansetyo, A. and Kamei, Y. (2003). MC21-A, a bacterial antibiotic produced by a new marine bacterium, *Pseudoalteromonas phenolica* sp. nov. O-BC30T against methicillin-resistant *Staphylococcus aureus*. *Antimicrob. Agents Chemother.* **47**, 480–488.
Isnansetyo, A. and Kamei, Y. (2009). Bioactive substances produced by marine isolates of *Pseudomonas*. *J. Ind. Microbiol. Biotechnol.* **36**, 1239–1248.
Iwata, F., Sato, S., Mukai, T., Yamada, S., Takeo, J., Abe, A., Okita, T., and Kawahara, H. (2009). Lorneic acids, trialkyl-substituted aromatic acids from a marine-derived actinomycete. *J. Nat. Prod.* **72**(11), 2046–2048.
Jain, R. and Tiwari, A. (2007). Sponges: An invertebrate of bioactive potential. *Curr. Sci.* **93**, 444–445.
Jaiswal, P., Singh, P. K., and Prasanna, R. (2008). Cyanobacterial bioactive molecules—An overview of their toxic properties. *Can. J. Microbiol.* **54**(9), 701–717.
Jensen, P. R., Dwight, R., and Fenical, W. (1991). Distribution of actinomycetes in near-shore tropical marine sediments. *Appl. Environ. Microbiol.* **57**(4), 1102–1108.
Jensen, P. R., Mincer, T. J., Williams, P. G., and Fenical, W. (2005). Marine actinomycete diversity and natural product discovery. *Antonie Van Leeuwenhoek* **87**, 43–48.
Jha, R. K. and Zi-rong, X. (2004). Biomedical compounds from marine organisms. *Mar. Drugs* **2**, 123–146.
Kamei, Y., McCarthy, S. A., Kakimoto, D., and Johnson, R. (1986). Inhibition of *Paramecium caudatum* by an *Alteromonas luteoviolacea* antibiotic. *Antimicrob. Agents Chemother.* **30**, 301–303.
Kanoh, K., Yoshihide, M., Kyoko, A., Hiroshi, I., Mugio, N., and Yoshikazu, S. (2005). Mechercharmycins A and B, cytotoxic substances from marine-derived *Thermoactinomyces* sp. YM3-251. *J. Antibiot.* **58**(4), 289–292.
Kanoh, K., Okada, A., Adachi, K., Imagawa, H., Nishizawa, M., Matsuda, S., Shizuri, Y., and Utsumi, R. (2008a). Ascochytatin, a novel bioactive spirodioxynaphthalene metabolite produced by the marine-derived fungus, *Ascochyta* sp. NGB4. *J. Antibiot. (Tokyo)* **61**, 142–148.
Kanoh, K., Adachi, K., Matsuda, S., Shizuri, Y., Yasumoto, K., Kusumi, T., Okumura, K., and Kirikae, T. (2008b). New sulfoalkylresorcinol from marine-derived fungus, *Zygosporium* sp. KNC52. *J. Antibiot. (Tokyo)* **61**(3), 192–194.
Kasettrathat, C., Ngamrojanavanich, N., Wiyakrutta, S., Mahidol, C., Ruchirawat, S., and Kittakoop, P. (2008). Cytotoxic and antiplasmodial substances from marine derived fungi, *Nodulisporium* sp. and CRI247-01. *Phytochemistry* **69**(14), 2621–2626.
Kennedy, J. (2008). Mutasynthesis, chemobiosynthesis, and back to semi-synthesis: Combining synthetic chemistry and biosynthetic engineering for diversifying natural products. *Nat. Prod. Rep.* **25**, 25–34.
Kennedy, J., Marchesi, J. R., and Dobson, A. D. W. (2007). Metagenomic approaches to exploit the biotechnological potential of the microbial consortia of marine sponges. *Appl. Microb. Biotechnol.* **75**(1), 11–20.
Kijjoa, A. and Sawangwong, P. (2004). Drugs and cosmetics from the sea. *Mar. Drugs* **2004**, 73–82.
Kingston, D. G. I. (2011). Modern natural products drug discovery and its relevance to biodiversity conservation. *J. Nat. Prod.* **74**(3), 496–511.
Kjer, J., Debbab, A., Aly, A. H., and Proksch, P. (2010). Methods for isolation of marine-derived endophytic fungi and their bioactive secondary products. *Nature Protoc.* **5**, 479–490.
Kobayashi, J. and Ishibashi, M. (1993). Bioactive metabolites of symbiotic marine microorganisms. *Chem. Rev.* **93**(5), 1753–1769.
Konig, G. M. and Wright, A. D. (1999). Trends in marine biotechnology. *In* "Drugs Discovery from Nature", (S. Grabley and R. Thiericke, Eds), pp. 180–187. Springer-Verlag, Berlin, Heidelberg.

Kopp, F. and Marahiel, M. A. (2007). Where chemistry meets biology: The chemoenzymatic synthesis of nonribosomal peptides and polyketides. *Curr. Opin. Biotechnol.* **18**(6), 513–520.

Kwon, H. C., Espindola, A. P. D. M., Park, J. S., Prieto-Dav, A., Rose, M., Jensen, P. R., and Fenical, W. (2010). Nitropyrrolins A-E, cytotoxic farnesyl-α-nitropyrroles from a marine-derived bacterium within the actinomycete family *Streptomycetaceae*. *J. Nat. Prod.* **73**(12), 2047–2052.

Laatsch, H., Rennerberg, B., Hanefeld, U., Kellner, M., Pudleine, H., Hamprecht, G., Kraemer, H., and Anke, H. (1995). Structure-activity relationship of phenyl- and benzoylpyrroles. *Chem. Pharm. Bull.* **43**(4), 537–546.

Lee, S. U., Asami, Y., Lee, D., Jang, J. H., Ahn, J. S., and Oh, H. (2011). Protuboxepins A and B and Protubonines A and B from the marine-derived fungus *Aspergillus* sp. SF-5044. *J. Nat. Prod.* **74**(5), 1284–1287.

Lefevre, F., Robe, P., Jarrin, C., Ginolhac, A., Zago, C., Auriol, D., Vogel, T. M., Simonet, P., and Nalin, R. (2008). Drugs from hidden bugs: Their discovery via untapped resources. *Res. Microbiol.* **159**(3), 153–161.

Li, X. and Qin, L. (2005). Metagenomics-based drug discovery and microbial diversity. *Trends Biotechnol.* **23**(11), 539–543.

Li, L., Deng, W., Song, J., Ding, W., Zhao, Q. F., Peng, C., Song, W. W., Tang, G. L., and Liu, W. (2008). Characterization of the saframycin A gene cluster from *Streptomyces lavendulae* NRRL 11002 revealing a nonribosomal peptide synthetase system for assembling the unusual tetrapeptidyl skeleton in an iterative manner. *J. Bacteriol.* **190**(1), 1251–1263.

Li, C. S., An, C. Y., Li, X. M., Gao, S. S., Cui, C. M., Sun, H. F., and Wang, B. G. (2011a). Triazole and dihydroimidazole alkaloids from the marine sediment-derived fungus *Penicillium paneum* SD-44. *J. Nat. Prod.* **74**(5), 1331–1334.

Li, D., Carr, G., Zhang, Y., Williams, D. E., Amlani, A., Bottriell, H., Mui, A. L. F., and Andersen, R. J. (2011b). Turnagainolides A and B, cyclic depsipeptides produced in culture by a *Bacillus* sp.: Isolation, structure elucidation, and synthesis. *J. Nat. Prod.* **74** (5), 1093–1099.

Liang, T. (2003). Investigation of secondary metabolites of North Sea bacteria: Fermentation, isolation, structure elucidation and bioactivity. PhD thesis. University of Gottingen.

Lindquist, N. and Fenical, W. (1991). New tambjamine class alkaloids from the marine ascidian *Atapozoa* sp. and its nudibranch predators—Origins of the tambjamines in atapozoa. *Experientia* **47**, 504–508.

Linington, R. G., Clark, B. R., Trimble, E. E., Almanza, A., Urena, L. D., Kyle, D. E., and Gerwick, W. H. (2009). Antimalarial peptides from marine cyanobacteria: Isolation and structural elucidation of gallinamide A. *J. Nat. Prod.* **72**(1), 14–17.

Long, P. F., Dunlap, W. C., Battershill, C. N., and Jaspars, M. (2005). Shotgun cloning and heterologous expression of the patellamide gene cluster as a strategy to achieving sustained metabolite production. *Chem. Bio. Chem.* **6**, 1760–1765.

Lopes, V. R., Fernández, N., Martins, R. F., and Vasconcelos, V. (2010). Primary screening of the bioactivity of Brackishwater cyanobacteria: Toxicity of crude extracts to *Artemia salina* larvae and *Paracentrotus lividus* embryos. *Mar. Drugs* **8**(3), 471–482.

Lu, Z., Wang, Y., Miao, C., Liu, P., Hong, K., and Zhu, W. (2009). Sesquiterpenoids and benzofuranoids from the marine-derived fungus *Aspergillus ustus* 094102. *J. Nat. Prod.* **72** (10), 1761–1767.

Lu, Z., Zhu, H., Fu, P., Wang, Y., Zhang, Z., Lin, H., Liu, P., Zhuang, Y., Hong, K., and Zhu, W. (2010). Cytotoxic polyphenols from the marine-derived fungus *Penicillium expansum*. *J. Nat. Prod.* **73**(5), 911–914.

Luesch, H., Moore, R. E., Paul, V. J., Mooberry, S. L., and Corbett, T. H. (2001). Isolation of dolastatin 10 from the marine cyanobacterium *Symploca* species VP642 and total stereochemistry and biological evaluation of its analogue symplostatin 1. *J. Nat. Prod.* **64** (7), 907–910.

MacMillan, J. B. and Molinski, T. F. (2002). Caylobolide A, a unique 36-membered macrolactone from a Bahamian *Lyngbya majuscule. Org. Lett.* **4**(9), 1535–1538.

Maloney, K. N., MacMillan, J. B., Kauffman, C. A., Jensen, P. R., DiPasquale, A. G., Rheingold, A. L., and Fenical, W. (2009). Lodopyridone, a structurally unprecedented alkaloid from a marine actinomycete. *Org. Lett.* **11**(23), 5422–5424.

Mann, J. (2002). Natural products in cancer: Chemotherapy: Past, present and future. *Nat. Rev. Cancer* **2**, 143–148.

Maskey, R. P., Li, F. C., Qin, S., Fiebig, H. H., and Laatsch, H. (2003). Chandrananimycins A-C: Production of novel anticancer antibiotics from a marine *Actinomadura* sp. isolate M048 by variation of marine composition and growth conditions. *J. Antibiot.* **56**(7), 622–629.

Maskey, R. P., Helmke, E., Kayser, O., Fiebig, H. H., Maier, A., Busche, A., and Laatsch, H. (2004). Anti-cancer and antibacterial trioxacarcins with high anti-malaria activity from a marine Streptomycete and their absolute stereochemistry. *J. Antibiot. (Tokyo)* **57**(12), 771–779.

Mayer, A. M. S. and Lehmann, V. K. B. (2000). Marine pharmacology in 1998: Marine compounds with antibacterial, anticoagulant, antifungal, anti-inflammatory, anthelminitic, antiplatelet, antiprotozoal, and antiviral activities with actions on the cardiovascular, endocrine, immune, and nervous systems, and other miscellaneous mechanisms of action. *Pharmacologist* **42**, 62–69.

Mayer, A. M., Glaser, K. B., Cuevas, C., Jacobs, R. S., Kem, W., Little, R. D., McIntosh, J. M., Newman, D. J., Potts, B. C., and Shuster, D. E. (2010). The odyssey of marine pharmaceuticals: A current pipeline perspective. *Trends Pharmacol. Sci.* **31**(6), 255–265.

McArthur, K. A., Mitchell, S. S., Tsueng, G., Rheingold, A., White, D. J., Grodberg, J., Lam, K. S., and Potts, B. C. (2008). Lynamicins A-E, chlorinated bisindole pyrrole antibiotics from a novel marine actinomycete. *J. Nat. Prod.* **71**(10), 1732–1737.

Medina, R. A., Goeger, D. E., Hills, P., Mooberry, S. L., Huang, N., Romero, L. I., Ortega-Barra, E., Gerwick, W. H., and McPhail, K. L. (2008). Coibamide A, a potent antiproliferative cyclic depsipeptide from the Panamanian marine cyanobacterium *Leptolyngbya* sp. *J. Am. Chem. Soc.* **130**(20), 6324–6325.

Mendola, D. (2003). Aquaculture of three phyla of marine invertebrates to yield bioactive metabolites: Process developments and economics. *Biomol. Eng.* **20**, 441–458.

Mevers, E., Liu, W. T., Engene, N., Mohimanis, H., Byrum, T., Pevzner, P. A., Dorrestein, P. C., Spadafora, C., and Gerwick, W. H. (2011). Cytotoxic veraguamides, alkynyl bromide-containing cyclic depsipeptides from the marine cyanobacterium cf. *Oscillatoria margaritifera. J. Nat. Prod.* **74**(5), 928–936.

Mincer, T. J., Jensen, P. R., Kauffman, C. A., and Fenical, W. (2002). Widespread and persistent populations of a major new marine actinomycete taxon in ocean sediments. *Appl. Environ. Microbiol.* **68**(10), 5005–5011.

Moore, B. S. and Seng, D. (1998). Biosynthesis of the bicyclic depsipeptide salinamide A in *Streptomyces* sp. CNB-091: Origin of the carbons. *Tetrahedron Lett.* **39**(23), 3915–3918.

Moore, B. S., Trischman, J. A., Seng, D., Kho, K., Jensen, D. P. R., and Fenical, W. (1999). Salinamides, antiinflammatory depsipeptides from a marine Streptomycete. *J. Org. Chem.* **64**(4), 1145–1150.

Muller, W. E. G., Ushijima, H., Schröder, H. C., Forrest, J., Schatton, W. H. F., Rytik, P. G., and Heffner-Lauc, M. (1993). Cytoprotective effect of NMDA receptor antagonists on prion protein (Prion^{Sc})-induced toxicity in rat cortical cell culture. *Eur. J. Pharmacol.* **246**, 261–267.

Muller, W. E. G., Romero, F. J., Perovic, S., Pergande, G., and Pialoglou, P. (1997). Protection of flupirtine on β-amyloid-induced apoptosis in neuronal cells in vitro. Prevention of amyloid-induced glutathione depletion. *J. Neurochem.* **68**, 2371–2377.

Murakami, R., Shinozaki, J., Kajiura, T., Kozone, I., Takagi, M., Shin-Ya, K., Seto, H., and Hayakawa, Y. (2009). Ammocidins B, C and D, new cytotoxic 20-membered macrolides from *Saccharothrix* sp. AJ9571 Ammocidins B, C and D from *Saccharothrix* sp. AJ9571. *J. Antibiot. (Tokyo)* **62**, 123–127.

Nam, S. J., Gaudencio, S. P., Kauffman, C. A., Jensen, P. R., Kondratyuk, T. P., Marler, L. E., Pezzuto, J. M., and Fenical, W. (2010). Fijiolides A and B, inhibitors of TNF-α-induced NFκB activation, from a marine-derived sediment bacterium of the genus *Nocardiopsis*. *J. Nat. Prod.* **73**(6), 1080–1086.

Nasveschuk, C. G., Ungermannova, D., Liu, X., and Phillips, A. J. (2008). A concise total synthesis of largazole, solution structure, and some preliminary structure activity relationships. *Org. Lett.* **10**(16), 3595–3598.

Newman, D. J. and Cragg, G. M. (2004). Advanced preclinical and clinical trials of natural products and related compounds from marine sources. *Curr. Med. Chem.* **11**, 1693–1713.

Newman, D. J. and Cragg, G. M. (2007). Natural products as sources of new drugs over the last 25 years. *J. Nat. Prod.* **70**(3), 461–477.

Newman, D. J. and Hill, R. T. (2006). New drugs from marine microbes: The tide is turning. *J. Ind. Microbiol. Biotechnol.* **33**, 539–544.

Ni, X., Youe, L., Chi, Z., Li, J., Wang, X., and Madzak, C. (2009). Alkaline protease gene cloning from the marine yeast *Aureobasidium pullulans* HN2-3 and the protease surface display on Yarrowia lipolytica for bioactive peptide production. *Mar. Biotechnol.* **11**, 81–89.

Nogle, L. M. and Gerwick, W. H. (2002). Somocystinamide A, a novel cytotoxic disulfide dimer from a Fijian marine cyanobacterial mixed assemblage. *Org. Lett.* **4**(7), 1095–1098.

Nussbaum, F., Brands, M., Hinzen, B., Weigand, S., and Häbich, D. (2006). Antibacterial natural products in medicinal chemistry-exodus or revival? *Angew. Chem. Int. Ed. Engl.* **45** (31), 5072–5129.

Oh, H., Jensen, P. R., Murphy, B. T., Fiorilla, C., Sullivan, J. F., Ramsey, T., and Fenical, W. (2010). Cryptosphaerolide, a cytotoxic Mcl-1 inhibitor from a marine-derived ascomycete related to the genus *Cryptosphaeria*. *J. Nat. Prod.* **73**(5), 998–1001.

Okazaki, T., Kitahara, T., and Okami, Y. (1975). Studies on marine microorganisms. IV. A new antibiotic SS-228 Y produced by Chainia isolated from shallow sea mud. *J. Antibiot. (Tokyo)* **28**(3), 176–184.

Olano, C., Lombó, F., Méndez, C., and Salas, J. A. (2008). Improving production of bioactive secondary metabolites in actinomycetes by metabolic engineering. *Metab. Eng.* **10**(5), 281–292.

Olano, C., Méndez, C., and Salas, J. A. (2009). Antitumor compounds from marine actinomycetes. *Mar. Drugs* **7**(2), 210–248.

Paul, S. M., Mytelka, D. S., Dunwiddie, C. T., Persinger, C. C., Munos, B. H., Lindborg, S. R., and Schacht, A. L. (2010). How to improve R&D productivity: The pharmaceutical industry's grand challenge. *Nat. Rev. Drug Dis.* **9**(3), 203–214.

Peng, X., Wang, Y., Sun, K., Liu, P., Yin, X., and Zhu, W. (2011). Cerebrosides and 2-pyridone alkaloids from the halotolerant fungus *Penicillium chrysogenum* grown in a hyperslaine medium. *J. Nat. Prod.* **74**(5), 1298–1302.

Perovic, S., Wichels, A., Schutt, C., Gunnar, G., Pahler, S., Steffen, R., and Muller, W. E. G. (1998). Neuroactive compounds produced by bacteria from the marine sponge *Halichondria panacea*: Activation of the neuronal MDA receptor. *Environ. Toxicol. Pharmacol.* **6**, 125–133.

Piel, J., Hui, D., Wen, G., Butzke, D., Platzer, M., Fusetani, N., and Matsunaga, S. (2004). Antitumor polyketide biosynthesis by an uncultivated bacterial symbiont of the marine sponge *Theonella swinhoei*. *Proc. Natl. Acad. Sci. USA* **101**(46), 16222–16227.

Plavsic, M., Terzic, S., Ahel, M., and Van den Berg, C. M. G. (2004). Folic acid in coastal waters of the Adriatic Sea. *Mar. Freshw. Res.* **53**, 1245–1252.

Porsby, C. H., Webber, M. A., Nielsen, K. F., Piddock, L. J., and Gram, L. (2011). Resistance and tolerance to tropodithietic acid, an antimicrobial in aquaculture, is hard to select. *Antimicrob. Agents Chemother.* **55**(4), 1332–1337.

Proksch, P. R., Edrada, A., and Ebel, R. (2002). Drugs from the seas—Current status and microbiological implications. *Appl. Microbiol. Biotechnol.* **59**, 25–34.

Prudhomme, J., McDaniel, E., Ponts, N., Bertani, S., Fenical, W., Jensen, P., and Le Roch, K. (2008). Marine actinomycetes: A new source of compounds against the human malaria parasite. *PLoS One* **3**(6), e2335.

Radhakrishan, M., Balagurunathan, R., Selvakumar, N., Doble, M., and Kumar, V. (2011). Bioprospecting of marine derived actinomycetes with special reference to antimicrobial activity. *Indian J. Geo-Mar. Sci.* **40**(3), 407–410.

Rahman, H., Austin, B., Mitchel, W., Morris, P. C., Jamieson, D. J., Adams, D. R., Spragg, A. M., and Schweizer, M. (2010). Novel anti-infective compounds from marine bacteria. *Mar. Drugs* **8**, 498–518.

Ramaswamy, A. V., Flatt, P. M., Edwards, D. J., Simmons, T. L., Han, B., and Gerwick, W. H. (2006). The secondary metabolites and biosynthetic gene clusters of marine cyanobacteria. Applications in biotechnology. *In* "Frontiers in Marine Biotechnology", (P. Proksch and W. E. G. Muller, Eds), pp. 175–224. Horizon Biosciences, Norfolk, England.

Rester, U. (2008). From virtuality to reality—Virtual screening in lead discovery and lead optimization: A medicinal chemistry perspective. *Curr. Opin. Drug Discov. Devel.* **11**(4), 559–568.

Rollinger, J. M., Stuppner, H., and Langer, T. (2008). Virtual screening for the discovery of bioactive natural products. *Prog. Drug Res.* **65**(211), 213–249.

Romanenko, L. A., Uchino, M., Kalinovskaya, N. I., and Mikhailov, V. V. (2008). Isolation, phylogenetic analysis and screening of marine mollusc-associated bacteria for antimicrobial, hemolytic and surface activities. *Microbiol. Res.* **163**(6), 633–644.

Ruiz, B., Chávez, A., Forero, A., García-Huante, Y., Romero, A., Sánchez, M., Rocha, D., Sánchez, B., Rodríguez-Sanoja, R., Sánchez, S., and Langley, E. (2010). Production of microbial secondary metabolites: Regulation by the carbon source. *Crit. Rev. Microbiol.* **36**(2), 146–167.

Salvador, L. A., Paul, V. J., and Luesch, H. (2010). Caylobolide B, a macrolactone from symplostatin 1-producing marine cyanobacteria *Phormidium* spp. from Florida. *J. Nat. Prod.* **73**(9), 1606–1609.

Salvador, L. A., Biggs, J. S., Paul, V. J., and Luesch, H. (2011). Veraguamides A-G, cyclic hexadepsipeptides from a dolastatin 16-producing cyanobacterium *Symploca* cf. *hydnoides* from Guam. *J. Nat. Prod.* **74**(5), 917–927.

Sandy, M., Han, A., Blunt, J., Munro, M., Haygood, M., and Butler, A. (2010). Vanchrobactin and anguibactin siderophores produced by *Vibrio* sp. DS40M4. *J. Nat. Prod.* **73**(6), 1038–1043.

Schwartsmann, G., Da Roccha, A. B., Berlink, R. G., and Jimeno, R. G. (2001). Marine organisms as a source of new anticancer agents. *Lancet Oncol.* **2**, 221–225.

Singh, S., Kate, B. N., and Banerjee, U. C. (2005). Bioactive compounds from cyanobacteria and micro algae: An overview. *Crit. Rev. Biotechnol.* **25**(3), 73–95.

Socha, A. M., Laplante, K. L., Russell, D. J., and Rowley, D. C. (2009). Structure-activity studies of echinomycin antibiotics against drug-resistant and biofilm-forming *Staphylococcus aureus* and *Enterococcus faecalis*. *Bioorg. Med. Chem. Lett.* **19**(5), 1504–1507.

Sudek, S., Lopanik, N. B., Waggoner, L. E., Hildebrand, M., Anderson, C., Liu, H., Patel, A., Sherman, D. H., and Haygood, M. G. (2007). Identification of the putative bryostatin polyketide synthase gene cluster from "*Candidatus* Endobugula sertula", the uncultivated microbial symbiont of the marine bryozoan Bugula neritina. *J. Nat. Prod.* **70**(1), 67–74.

Tan, L. T. (2007). Bioactive natural products from marine cyanobacteria for drug discovery. *Phytochemistry* **68**(7), 954–979.

Tang, R., Cohen, S., Legrand, O., and Marie, J. P. (2007). Antibody-drug conjugates in targeted therapy for cancer. *Front. Drug. Des. Discov.* **3**(1), 525–537.

Thornburg, C. C., Zabriskie, T. M., and McPhail, K. L. (2010). Deep-sea hydrothermal vents: Potential hot spots for natural products discovery? *J. Nat. Prod.* **73**(3), 489–499.

Torchilin, V. P. (2006). Recent approaches to intracellular delivery of drugs and DNA and organelle targeting. *Annu. Rev. Biomed. Eng.* **8**, 343–375.

Trischman, J. A., Tapiolas, D. M., Jensen, P. R., Dwight, R., Fenical, W., McKee, T. C., Ireland, C. M., Stout, T. J., and Clardy, J. (1994). Salinamides A and B: Anti-inflammatory depsipeptides from a marine streptomycete. *J. Am. Chem. Soc.* **116**(2), 757–758.

Udwary, D. W., Zeigler, L., Asolkar, R. N., Singan, V., Lapidus, A., Fenical, W., Jensen, P. R., and Moore, B. S. (2007). Genome sequencing reveals complex secondary metabolome in the marine actinomycete *Salinispora tropica*. *Proc. Natl. Acad. Sci. USA* **104**(25), 10376–10381.

Uzair, B., Ahmed, N., Mohammad, F. V., Ahmad, V. U., and Edwards, D. (2009). Screening of marine bacteria of Pakistan coast for drug discovery potential. *Proc. Pak. Acad. Sci.* **46**(3), 137–144.

Veluri, R., Oka, I., Wagner-Döbler, I., and Laatsch, H. (2003). New indole alkaloids from the North Sea bacterium *Vibrio parahaemolyticus* Bio249. *J. Nat. Prod.* **66**(11), 1520–1523.

Ventura, M., Canchaya, C., Tauch, A., Chandra, G., Fitzgerald, G. F., Chater, K. F., and Van Sinderen, D. (2007). Genomics of *Actinobacteria*: Tracing the evolutionary history of an ancient phylum. *Microbiol. Mol. Biol. Rev.* **71**, 495–548.

Williams, P. G. (2009). Panning for chemical gold: Marine bacteria as a source of new therapeutics. *Trends Biotechnol* **27**(1), 45–52.

Wrasidlo, W., Mielgo, A., Torres, V. A., Barbero, S., Stoletov, K., Suyama, T. L., Klemke, R. L., Gerwick, W. H., Carson, D. A., and Stupack, D. G. (2008). The marine lipopeptide somocystinamide A triggers apoptosis via caspase 8. *Proc. Natl. Acad. Sci. USA* **105**, 2313–2318.

Ying, Y., Taori, K., Kim, H., Hong, J., and Luesch, H. (2008). Total synthesis and molecular target of largazole, a histone deacetylase inhibitor. *J. Am. Chem. Soc.* **130**(26), 8455–8459.

Zhang, X., Wu, F. X., Sun, M., Wang, Q. Y., Wang, Y. J., and Yang, X. (2008). Study on antimicrobial and antiviral activities of lysozyme from marine strain S-12-86 *in vitro*. *Agric. Sci. China* **7**(1), 112–116.

Zheng, J., Zhu, H., Hong, K., Wang, Y., Liu, P., Wang, X., Peng, X., and Zhu, W. (2009). Novel cyclic hexapeptides from marine-derived fungus, *Aspergillus sclerotiorum* PT06-1. *Org. Lett.* **11**(22), 5262–5265.

Zheng, J., Xu, Z., Wang, Y., Hong, K., Liu, P., and Zhu, W. (2010). Cyclic tripeptides from the halotolerant fungus *Aspergillus sclerotiorum* PT06-1. *J. Nat. Prod.* **73**(6), 1133–1137.

CHAPTER 25

Marine Bacterial Sources of Bioactive Compounds

R. Jaiganesh[1] and **N. S. Sampath Kumar**

Contents

Abstract

Thousands of novel compounds have been isolated from various marine bacteria and tested for pharmacological properties, many of which are commercially available. Many more are being tested as potential bioactive compound at the preclinical and clinical stages. The growing interest in marine-derived antiviral compounds, along with the development of new technology in marine cultures and extraction, will significantly expedite the current exploration of the marine environment for compounds with significant pharmacological applications, which will continue to be a promising strategy

Department of Biotechnology, School of Bioengineering, SRM University, Kattankulathur, Tamil Nadu, India
[1] Corresponding author: R. Jai ganesh, *E-mail address:* jaiganesh.abi@gmail.com

Advances in Food and Nutrition Research, Volume 65
ISSN 1043-4526, DOI: 10.1016/B978-0-12-416003-3.00025-1

and new trend for modern medicine. Marine *actinomycetes* and cyanobacteria are a prolific but underexploited source for the discovery of novel secondary metabolites.

I. INTRODUCTION

The marine environment covers more than 70% of the world's surface and has proven to be a rich source of both biological and chemical diversity. In the past, this has proven to be a rich source of extremely potent compounds (Gochfeld *et al.*, 2003; Newman and Cragg, 2005), which represent a considerable number of drug candidates (Haefner, 2003). The oceans contain nearly 300,000 described species, but it is estimated that this number is only a small percentage of the total number of species that have yet to be discovered and described (Malakoff, 1997; Winston, 1988). Mankind has known for the past several thousand years that marine organisms contain substances capable of potent biological activity. However, the first serious investigation of marine organisms started only half a century ago. Since then, almost all forms of life in the marine environment (e.g., bacteria, algae, sponges, fungi, corals, ascidians) have been investigated for their natural product content.

Over the past 50years, marine organisms have provided key structures and compounds that proved their potential for industrial development as cosmetics, nutritional supplements, fine chemicals, agrochemicals, and therapeutic agents for a variety of diseases (Tziveleka *et al.*, 2003). During the past 30years, thousands of novel compounds and their metabolites, with diverse biological activities ranging from anticancer to antiviral, have been isolated from various marine sources (Arif *et al.*, 2004). Although more than 30,000 diseases have been clinically described, less than one-third of these can be treated symptomatically and only a few can be cured. New therapeutic agents are urgently needed to treat medical needs that are currently unmet.

Marine microorganisms are potentially prolific sources of highly bioactive secondary metabolites that might serve as useful leads in the development of new pharmaceutical agents. However, to date, the biodiversity of marine microbes and the versatility of their bioactive metabolites have not been fully explored. Marine microorganisms represent the greatest percentage of undescribed marine species (Colwell, 1997). Marine bacteria alone could constitute as much as 10% of the total living biomass carbon of the biosphere (Parkes *et al.*, 1994). From a relatively small number of these species that have been studied to date, thousands of chemical compounds have been isolated (Ireland *et al.*, 1993), yet only a small percentage of these compounds have been studied for their potential as useful products.

Marine bacteria are emerging as an exciting resource for the discovery of new classes of therapeutics. For example, if properly developed, marine bacteria could provide the drugs needed to sustain us for the next 100 years in our battle against drug-resistant infectious diseases. It appears that there have been only a few studies focused on finding bioactive compounds derived from marine bacteria to be used as anti-tumor agents, as well as agents against infectious organisms. Structurally and functionally diverse bioactive compounds have been isolated from prokaryotes, including members of the myxobacteria (e.g., Sorangium) and cyanobacteria (e.g., Nostoc) (Magarvey *et al.*, 2004). Most studies aimed at identifying bioactive compounds from marine bacteria have centered on bacteria of the order Actinomycetales, as well as on bacterial production of exopolysaccharides (EPSs), with cyanobacteria being considered a eukaryotic algae (Arena *et al.*, 2006, 2009; Magarvey *et al.*, 2004).

Recently, important bioactive compounds have been isolated from marine bacteria such as antibacterial agent, antifungal agent, antimalarial agent, antiprotozoan agent, antiviral agent, antitumor agent.

A. Antibacterial compounds

Today, increase in the number of drug-resistant pathogens, particularly acquired multidrug-resistant strains such as methicillin-resistant *Staphylococcus aureus* (MRSA), causes serious public health problem throughout the world. On the other hand, limited success of strategies such as combinatorial chemistry in providing new agents indicates an uncertain forecast for further antimicrobial therapy (Magarvey *et al.*, 2004). Because of the enormous biodiversity in the marine environment, in the case of microorganisms, sea water is composed of 78 million microscopic cells/ounce or the bottom (Fenical, 2009); this environment provides most prolific sources for new natural drugs. Contributing to the global search for new antimicrobials to combat antibiotic-resistant strains of pathogenic bacteria, marine ecological niches have been described recently as "particularly promising" (Fischbach and Walsh, 2009).

Several studies contributed to the search for novel antibacterial marine natural products isolated from marine bacteria, including the peptide bogorol A isolated from a culture of a marine *Bacillus* sp. (Barsby *et al.*, 2001), which was active against MRSA (MIC=2 Ag/mL) and vancomycin-resistant enterococcal strains (MIC=10 Ag/mL). Because bogorol A represents a new cationic peptide antibiotic template, the authors proposed that it might become "an attractive lead structure for structure–activity relationship (SAR) optimization." Asolkar *et al.* (2002) isolated a novel macrolide antibiotic chalcomycin B from the culture broth of a marine Streptomycete isolate which was particularly potent against *S. aureus* (MIC=0.39 Ag/mL).

Singh *et al.* (2003) isolated Pyron from *Pseudomonas* sp., the pyranyl-type kalihinols Y and X, although potent antibacterials (MIC=1.56 μg/mL) were, however, less selective inhibitors of bacterial folate biosynthesis than the furanyl type kalihinols, with the "C-10 position important for potency." Isnansetyo and Kamei (2003) reported that a bactericidal compound named MC21-A, a 3,3′,5,5′-tetrabromo-2,2′-biphenyldiol, from the new marine bacterium *Pseudoalteromonas phenolica* sp. nov. MC21-A was bactericidal (MIC=1−2 μg/mL) against 10 clinical isolates of MRSA, and displayed comparable bioactivity to vancomycin (MIC=0.25−2 μg/mL). The mechanism of action of MC21-A involved permeabilizing bacterial cell membranes and thus "might be a useful compound" because of a mode of action that differs from vancomycin.

Rondon *et al.* (2004) isolated Agrochelin from *Agrobaterium* sp., Gao and Hall (2005) isolated Tiomarinol from *Altromonas* sp., Jang *et al.* (2006) isolated Koromicin from *Pseudoalteromonas*, Sabdono and Radjasa (2006) isolated Pelagiomicin from *Pelagiobacter variabilis*, the alkaloids ambiguine isonitriles H and I isolated from a marine cyanobacterium *Fischerella* sp. (Raveh and Carmeli, 2007), lipoxazolidinones A and B isolated from a marine actinomycete *Marinispora* sp. (Macherla *et al.*, 2007), and tauramamide, a lipopeptide isolatedfrom the bacterium *Brevibacillus laterosporus* PNG276 (Desjardine *et al.*, 2007). The polyketides ariakemicins A and B isolated from a marine gliding bacterium *Rapidithrix* sp. (Oku *et al.*, 2008), ayamycin isolated from a bacterium *Nocardia* sp. ALAA 2000 (El-Gendy *et al.*, 2008a), and essramycin was obtained from the culture broth of a marine *Streptomyces* sp. Isolate Merv8102 (El-Gendy *et al.*, 2008b), the bisindole pyrroles lynamicins A–D, isolated from a novel marine actinomycete *Marinispora* sp. (McArthur *et al.*, 2008); marinopyrrole A isolated from an obligate marine *Streptomyces* strain (Hughes *et al.*, 2008); and Tuin *et al.* (2009) isolated Loloatin from *Bacillus* sp.

B. Antiviral compound

Marine antiviral compounds represent a significantly unique natural marine resource whose multipotential uses include the following applications:

1. The biological control of human enteropathogenic virus contamination and disease transmission in sewage-polluted water.

This application would be particularly important to communities that utilize their coastal waters for recreational activities and for food industries (e.g., fish, shellfish), as well as to those regions of the country, such as Hawaii, where the loss of these marine resources would have a devastating effect on the lifestyle and economy of the people.

2. The chemotherapy of viral diseases of humans and lower animals.

Only a few compounds have been identified from marine bacteria with antiviral activity: Macrolactin A is the marine bacteria-derived antiviral compound that includes 24-membered ring lactones, related glucosepyranosides, and open chain acids, and it shows protection to human T-lymphoblast cells against human HIV replication, with maximum protection at a concentration of 10g/mL (Gustafson *et al.*, 1989).

Nakao *et al.* (2001) reported that calyceramides A–C inhibited neuraminidase from bacterium *Clostridium perfringens* with IC_{50} values of 0.4, 0.2, and 0.8μg/mL, respectively, which was slightly more potent than 4-acetyl neuraminic acid (IC_{50}=1.5mg/mL). It remains to be determined if these compounds will also inhibit influenza virus neuraminidase with similar potency.

Nowadays, studies using marine bacteria as sources for antiviral compounds have focused mainly on marine bacterial EPSs (Arena *et al.*, 2006, 2009). Bacteria from deep-sea hydrothermal vents, such as *Alteromonas macleodii* spp. fijiensis, *Vibrio diabolicus*, *Alteromonas infernus*, and strain HYD 721, produce EPSs with original structures that enable them to survive in their extreme environments (Arena *et al.*, 2009). More recently, the hydrothermophilic *Thermotoga maritima* and *Thermococcus litoralis* have been reported as producers of extra polysaccharides (Rinker and Kelly, 2000). EPSs are known to interfere with the adsorption and penetration of viruses into host cells, as well as inhibit various retroviral reverse transcriptases. Two new EPSs, EPS-1 and EPS-2, were isolated from *Bacillus licheniformis* and *Geobacillus thermodenitrificans*, respectively, and two bacteria isolated from a shallow marine hot spring of Vulcano Island, Italy (Arena *et al.*, 2006, 2009).

C. Antitumor compound

Cancer is a dreadful human disease, increasing with changing lifestyle, nutrition, and global warming. Cancer treatments do not have potent medicine as the currently available drugs are causing side effects in some instances. In this context, the natural products derived from marine bacteria have gained significance in the treatment of cancer. Marine microorganisms are a source of new genes, the exploitation of which is likely to lead to the discovery of new drugs and targets. Secondary metabolites produced by marine bacteria have yielded anticancer agents (e.g., bryostatins, discodermolide, eleutherobin, and sarcodictyin).

Several studies already reported novel antitumor compounds: Kamata *et al.* (2001) extended the molecular pharmacology of cycloprodigiosin hydrochloride, a red pigment produced by the marine bacterium

Pseudoalteromonas denitrificans, previously shown to induce apoptotic cell death in various cancerous cell lines. Kamata *et al.* (2001) noted that cycloprodigiosin hydrochloride enhanced the apoptotic process induced by tumor necrosis factor-α in HeLa cells by suppressing the transcriptional activity of trancription nuclear factor κB (NF-κB), thus providing one possible explanation for the immunosuppressant and antitumor effects of this marine molecule.

Cryptophycins, a family of antimitotic depsipeptides, isolated from a marine cyanobacterium of the genus *Nostoc*, which bind with high affinity to microtubule ends at the vinca-binding domain. In a preclinical study with human non-small cell lung carcinoma cells in culture, Lu *et al.* (2001) observed that cryptophycin-52 and -55, synthetic cryptophycins produced by total chemical synthesis at Lilly Research Laboratories, induced hyperphosphorylation of the antiapoptotic protein Bcl-2, cell cycle arrest, and growth inhibition. The authors concluded that signal transduction pathways triggered by Bcl-2 phosphorylation may be involved in cryptophycin-induced lethality.

Williams *et al.* (2003a) purified xanthone antitumor compound from marine *Actinomadura* sp.; Williams *et al.* (2004) reported the cytotoxic alkaloid compound from marine cyanobacterium Symploca; Zhang *et al.* (2004) identified three new cytotoxic cyclic acyl peptides from marine *Bacillis* sp., namely, Mixirin A, B, and C; and Tasiamide B, a 4-amino-3-hydroxy-5-phenylpentanoic acid containing peptide compound, isolated from marine *Cyanobacterium Symploca* sp. and also structural elucidation of Tasiamide B was demonstrated by Williams *et al.* (2003b). The two new cytotoxic depsipeptides, Tasipeptins A and B, also isolated from the marine cyanobacterium *Symploca* sp. (Williams *et al.*, 2003c).

Macherla *et al.* (2005a,b) contributed SAR studies of salinosporamide A (NPI-0052), a novel marine bacterium-derived alkaloid shown to potently inhibit the proteasome, a multicatalytic proteolytic complex that is involved in the regulation of cellular protein degradation. With 16 analogs of salinosporamide A generated by either fermentation or derivatization, SAR studies were completed using a variety of well-characterized cytotoxicity, proteasome inhibition, and NF-κB activation assays. These studies demonstrated a marked reduction in potency resulting from the replacement of the chloroethyl group in salinosporamide A with nonhalogenated substituents.

Recently several researches identified novel antitumor compounds from marine bacteria: laxaphycins A and B, cyclic peptides (Gbankoto *et al.*, 2005), actinomycete dihydroquinone (Soria-Mercado *et al.*, 2005), ankaraholides A and B (Andrianasolo *et al.*, 2005), aurilides B and C (Han *et al.*, 2006), belamide A bacterium (Simmons *et al.*, 2006), and *N*-carboxamido-staurosporine (Wu *et al.*, 2006).

II. MARINE *ACTINOMYCETES*

Actinomycetes are the most economically and biotechnologically valuable prokaryotes. *Actinomycetes* are Gram-positive, often filamentous, bacteria known for their unsurpassed capacity for the production of secondary metabolites with diverse biological activities. Extensive screening of terrestrial *actinomycetes*, started in the early 1950s, has yielded many important drug leads, later developed into antimicrobial (amphotericin B, erythromycin, vancomycin), anticancer (daunorubicin, bleomycin, mitomycin), and immunosuppressive (rapamycin) drugs. Because of the excellent track record of *actinomycetes* in this regard, a significant amount of effort has been focused on the successful isolation of novel *actinomycetes* from terrestrial sources for drug screening programs in the past 50years.

In the recent years, *actinomycetes* isolated from the marine environment (sediments, sponges, tunicates, neuston, etc.) have attracted considerable attention (Lane and Moore, 2011; Liu *et al.*, 2010). True marine *actinomycetes* are usually considerably more difficult to culture compared to their terrestrial relatives, most likely due to the special growth requirements. However, development of both sampling and cultivation techniques allowed isolation of representatives of several true marine actinomycete genera producing novel compounds with interesting biological activities (Jensen *et al.*, 2005).

The trend of discovery of novel compounds from marine *actinomycetes* and their characterization is gaining momentum, which is not surprising considering the exhausted pipelines for discovery of new antibiotics and constant need for improved treatment of cancer, as well as chronic diseases. However, judging from the data only a small portion of discovered marine natural products with antimicrobial, anti-inflammatory, anticancer, and neuromodulating activities originate from *actinomycetes*.

Notably, most of the bioactive molecules from marine *actinomycetes* have been discovered in the representatives of *Streptomyces* spp. Most likely, this bias reflects the fact that marine *Streptomyces*, like their terrestrial relatives, are easier to culture and to induce synthesis of secondary metabolites compared to other marine *actinomycetes*.

A. Bioactive compounds from marine *actinomycetes*

Many of the marine actinomycete-derived natural products possess unique structural features rarely or never found among the compounds isolated from the terrestrial sources. It is thus not surprising that their molecular modes of action are sometimes also unique, prompting their investigation as potential leads for drug development. For example, diazepinomicin, an alkaloid isolated from a marine-derived

Micromonospora sp., is a potent inducer of apoptosis with dual mode of action-binding to the benzodiazepine receptor and inhibition of the Ras/MAP kinase pathway (Gourdeau *et al.*, 2008).

Thiocoraline is a thiodepsipeptide secondary metabolite from marine-derived *Micromonospora marina* (Perez Baz *et al.*, 1997) with unique mode of action-inhibition of DNA polymerase, leading to cell cycle arrest and induction of apoptosis. Thiocoraline derivatives are currently undergoing preclinical evaluation at PharmaMar.

Diazepinomicin (ECO-4601) is a unique farnesylated dibenzodiazepinone produced by a *Micromonospora* strain (Charan *et al.*, 2004). It possesses antibacterial, anti-inflammatory, and antitumor activity. It has a broad spectrum of *in vitro* cytotoxicity and has demonstrated *in vivo* activity against glioma, and breast and prostate cancer in mouse models. Salinosporamide A (NPI-0052) is a novel blactone-g-lactam isolated from a fermentation broth of a new obligate marine actinomycete, *Salinispora tropica* (Feling *et al.*, 2003; Maldonado *et al.*, 2005). NPI-0052 is being developed by Nereus Pharmaceuticals, Inc. and is scheduled to enter clinical studies for treatment of cancer in human in 2006.

S. tropica strains also produce two unprecedented macrolides, sporolides A and B (Buchanan *et al.*, 2005). Sporolides A and B appear to be synthesized from two different polyketides, containing a large number (23 out of 24 carbon skeleton) of oxidized carbons. This, in part, contributes to the highly unusual structures of the sporolides. Even though sporolides have not demonstrated any biological activity in the few antimicrobial and anticancer screens tested, these structures demonstrate the tremendous potential of marine *actinomycetes* for the production of novel secondary metabolites.

Abyssomicin C is a novel polycyclic polyketide antibiotic produced by a marine *Verrucosispora* strain (Riedlinger *et al.*, 2004). It is a potent inhibitor of *para*-aminobenzoic acid biosynthesis and, therefore, inhibits the folic acid biosynthesis at an earlier stage than the well-known synthetic sulfa drugs (Bister *et al.*, 2004). Abyssomicin C possesses potent activity against Gram-positive bacteria, including clinical isolates of multiple-resistant and vancomycin-resistant *S. aureus*. Abyssomicin C or its analog (Rath *et al.*, 2005) has the potential to be developed as antibacterial agent against drug-resistant pathogens.

Recent deciphering of the genetic potential for production of chemically diverse natural products by two species of *Salinispora* (Penn *et al.*, 2009) clearly suggest that what has so far been discovered using conventional screening represents only a tiny portion of what these organisms really can biosynthesize. To exploit this potential, however, better understanding of biosynthetic pathways and new tools for genome mining and gene manipulation are required. However, several researchers identified new bioactive compound from marine *actinomycetes*. Table 25.1 shows some examples of novel secondary metabolites isolated from marine *actinomycetes*. This is by no means an extensive search of all novel secondary metabolites produced by marine *actinomycetes* during this 3-year period; nevertheless this list is impressive, covering many different diverse structures with biological activities.

III. MARINE CYANOBACTERIA

Approximately half of all current drugs are derived from natural products, demonstrating their potential for drug discovery. In particular, largely unexplored marine organisms such as marine cyanobacteria are evaluated as sources of bioactive secondary metabolites. Cyanobacteria such as *Microcystis*, *Anabaena*, *Nostoc*, and *Oscittatoria* produce a great variety of secondary metabolites. Because cyanobacteria are largely unexplored, they represent a rich opportunity for discovery; the expected rate of rediscovery is far lower than for other better-studied groups of organisms (Olaizola, 2003). Cyanobacteria produce a wide variety of toxins and other bioactive compounds, which include 40% lipopeptides, 5.6% amino acids, 4.2% fatty acids, 4.2% macrolides, and 9% amides. Cyanobacterial lipopeptides include different compounds like cytotoxic (41%), antitumor (13%), antiviral (4%), antibiotics (12%), and the remaining 18% activities include antimalarial, antimycotics, multidrug-resistance reversers, antifeedant, herbicides, and immunosuppressive agents (Burja *et al.*, 2001).

The wild cyanobacteria, *Lyngbya*, grow in abundance in the shallow water bloom. The alkaloids are unexceptional; peptides are unusual, either linear type (made of L-amino acid, like the immunosuppressive microcolin) (Koehn *et al.*, 1992) or cyclic, and formed by both L- and D-amino acid.

A. Bioactive compounds from marine cyanobacteria

1. Cyanovirin-N

Cyanovirin-N (CV-N) is a unique, 101-amino-acid-long, 11-kDa protein. It was discovered as a constituent of a cultured cyanobacterium, *Nostoc ellipsosporum*. CV-N potently and irreversibly inactivates diverse primary strains of HIV-1, including M-tropic forms involved in sexual transmission of HIV. CV-N also blocks cell-to-cell transmission of HIV infection. CV-N is directly virucidal (Burja *et al.*, 2001).

TABLE 25.1 Novel bioactive compounds produced by marine *actinomycetes*

Compound	Source	Activity	Reference
Abyssomycin C	*Verrucosispora maris*	Antibacterial	Bister *et al*. (2004)
Albidopyrone (α-pyrone)	*Streptomyces* sp.	Cytotoxic	Hohmann *et al*. (2009a)
Aureoverticillactam	*Streptomyces aureoverticillatus*	Anticancer	Mitchell *et al*. (2004)
Bonactin	*Streptomyces* sp.	Antibacterial Antifungal	Schumacher *et al*. (2003)
Chandrananimycins	*Actinomadura* sp.	Antialgal Antibacterial Anticancer Antifungal	Maskey *et al*. (2003)
Chinikomycins	*Streptomyces* sp.	Anticancer	Li *et al*. (2005)
Chloro-dihydroquinones	Novel *actinomycete*	Antibacterial Anticancer	Soria-Mercado *et al*. (2005)
Dermacozines	*Dermacoccus* sp.	Cytotoxic, radical scavenging	Abdel-Mageed *et al*. (2010)
Diazepinomicin	*Micromonospora* sp.	Anticancer	Charan *et al*. (2004)
Enterocin	*Streptomyces maritimus*	Bacteriostatic	Piel *et al*. (2000)
Essramicins	*Streptomyces* sp.	Antibacterial	El-Gendy *et al*. (2008a,b)
Frigocyclinone	*Streptomyces griseus*	Antibacterial	Bruntner *et al*. (2005)
Glaciapyrroles	*Streptomyces* sp.	Antibacterial	Macherla *et al*. (2005a,b)
Gutingimycin	*Streptomyces* sp.	Antibacterial	Maskey *et al*. (2004)
Lynamicins	*Marinispora* sp.	Antibacterial	McArthur *et al*. (2008)
Mechercharmycins	*Thermoactinomyces* sp.	Anticancer	Kanoh *et al*. (2005)

Mansouramycins	*Streptomyces* sp.	Cytotoxic	Hawas *et al.* (2009)
Proximicins	*Verrucosispora* sp.	Cytostatic	Fiedler *et al.* (2008)
Salinosporamide A	*Salinispora tropica*	Anticancer	Feling *et al.* (2003)
Thiocoraline	*Micromonospora* spp.	Anticancer	Perez Baz *et al.* (1997)
Trioxacarcins	*Streptomyces* sp.	Antibacterial Anticancer Antimalarial	Maskey *et al.* (2004)
TP-1161	*Nocardiopsis* sp.	Antibacterial	Engelhardt *et al.* (2010)

2. Borophycin

Borophycin is a boron containing metabolite isolated from marine strains of cyanobacteria *Nostoc linckia* and *Nostoc spongiaeforme* var. tenue. It exhibits potent cytotoxicity against human epidermoid carcinoma and human colorectal adenocarcinoma cell lines and has been found to exhibit antimicrobial activity (Burja *et al.*, 2001).

3. Cryptophycin

Cryptophycin first isolated from *Nostoc* sp. ATCC 53789 is a potent fungicide and also isolated from *Nostoc* sp. GSV 224 and has exhibited potent cytotoxicity against human tumor cell lines. It shows good activity against broad spectrum drug-sensitive and drug-resistant murine and human solid tumors (Burja *et al.*, 2001).

4. Lipopeptides

The natural products of many marine cyanobacteria contain an amino-acid-derived fragment linked to fatty acid-derived portion, forming compounds known as lipopeptides. Lipopeptides are interesting and biochemically active, having cytotoxic, anticancer, antibiotic, enzyme inhibitor, antiviral, and antifungal activities (Burja *et al.*, 2001).

B. Anti-inflammatory activity

Cyanobacteria contain significant amounts of carotenoids (β-carotene, lycopene, lutein) having antioxidant properties. The anti-inflammatory activity of blue-green algae is also due to phycocyanin, a photo-harvesting pigment. C-phycocyanin is a free radical scavenger (Bhat and Madyastha, 2000).

C. Antiviral effect

The protection of human lymphoblastoid T cells from the cytopathic effect of HIV infection with the extract of cyanobacteria has been reported (Gustafson *et al.*, 1989). A new class of HIV inhibitors called sulfonic acid, containing glycolipid, were isolated from the extract of blue-green algae, and the compounds were found to be active against the HIV virus. Cyanoviridin-N, isolated from blue-green algae, inactivates the strains of HIV virus and inhibits cell-to-cell and virus-to-cell fusion (Yang *et al.*, 1997). Calcium spirulan (Ca-SP), a novel sulfated polysaccharide, is an antiviral agent.

TABLE 25.2 Compounds isolated from marine cyanobacteria

Compound	Source	Activity	Reference
Carazostatin	*Hyella caespitose*	Antifungal	Cardellina *et al.* (1979)
Debromoaplysia toxin	*Lyngbya gracilis*	Cytotoxic	Mynderse *et al.* (1977)
Diarrhetic toxin	*Oscillatoria* sp.	Cytotoxic	Murakami *et al.* (1991)
Acutiphycin	*Oscillatoria acutissima*	Anticancer	Barchi *et al.* (1984)
Oscillatoxin	*Oscillatoria nigroviridis*	Anticancer	Moore *et al.* (1984)
Raocyclamide	*Oscillatoria raoi*	Cytotoxic	Reshef *et al.* (1997)
30-Methyloscillatoxin	Several *Oscillatoriaceae* sp.	Cytotoxic	Harrigan *et al.* (1998a)
Hierridin	*Phormidium ectocarpi*	Antiplasmodium	Murakami *et al.* (1991)
Symplostatin	*Symploca hydnoides*	Cytotoxic	Harrigan *et al.* (1998b)
Aulosirazole	*Aulosira fertilissima*	Anticancer	Stratmann *et al.* (1994)
Calothrixin	*Calothrix* sp.	Antimalarial	Issa (1999)
Staurosporine	*Nostoc sphaericum*	Antiviral	Knubel *et al.* (1990)
Polybrominated bisindoles	*Rivularia firma*	Anti-inflammatory	Norton and Wells (1982)
Tolyporptin	*Tolypothrix hodosa*	Antibacterial	Prinsep *et al.* (1992)
Toyacamycin	*Tolypothrix tenuis*	Antifungal	Moore (1982)

D. Anticancer effect

The extracts of *Spirulina* and *Dunaliella* inhibited the chemically induced carcinogenesis in model hamster buccal pouches (Schwartz and Shklar, 1987). Studies have also showed that sulfated polysaccharide, Ca-SP, appears to inhibit tumor invasion and metastasis (Mishima *et al.*, 1998) of melanoma cells and inhibit the tumor invasion of basement membrane. Phormidium tenue contain several diacylglycerols that inhibit chemically induced tumors on mice (Tokuda *et al.*, 1996). C-phycocyanin selectively inhibits COX-2, which is overexpressed in breast cancer cells but has no effect on COX-1 (Reddy *et al.*, 2007). And also several studies reported the various bioactive compounds isolated from marine cyanobacteria, which are depicted in Table 25.2.

IV. CONCLUSION

We commented earlier on the global development of marine microbiology and the renaissance taking place in natural product research. Further, we have reiterated our belief that natural product search and discovery in marine bacteria shows exceptional promise. However, marine bacterial search and discovery is one thing, development of discoveries to end-point products is another.

REFERENCES

Abdel-Mageed, W. M., Milne, B. F., Wagner, M., Schumacher, M., Sandor, P., Pathom-aree, W., Goodfellow, M., Bull, A. T., Horikoshi, K., Ebel, R., Diederich, M., Fiedler, H. P., *et al.* (2010). Dermacozines, a new phenazine family from deep-sea *Dermacocci* isolated from a Mariana Trench sediment. *Org. Biomol. Chem.* **8**, 2352–2362.

Andrianasolo, E. H., Gross, H., Goeger, D., Musafija-Girt, M., McPhail, K., and Leal, R. M. (2005). Isolation of swinholide A and related glycosylated derivatives from two field collections of marine cyanobacteria. *Org. Lett.* **7**(7), 1375–1378.

Arena, A., Maugeri, T. L., Pavone, B., Iannello, D., Gugliandolo, C., and Bisignano, G. (2006). Antiviral and immunomodulatory effect of a novel exopolysaccharide from a marine thermotolerant *Bacillus licheniformis*. *Int. Immunopharmacol.* **6**, 8–13.

Arena, A., Gugliandolo, C., Stassi, G., Pavone, B., Iannello, D., Bisignano, G., and Maugeri, T. L. (2009). An exopolysaccharide produced by *Geobacillus thermodenitrificans* strain B3-72: Antiviral activity on immuno competent cells. *Immunol. Lett.* **123**, 132–137.

Arif, J. M., Al-Hazzani, A. A., Kunhi, M., and Al-Khodairy, F. (2004). Novel marine compounds: Anticancer or genotoxic. *J. Biomed. Biotechnol.* **2**, 93–98.

Asolkar, R. N., Maskey, R. P., Helmke, E., and Laatsch, H. (2002). Chalcomycin B, a new macrolide antibiotic from the marine isolate *Streptomyces* sp. B7064. *J. Antibiot.* **55**, 893–898.

Barchi, J. J., Jr., Moore, R. E., and Patterson, G. M. L. (1984). Acutiphycin and 20, 21-Didehydroacutiphycin, new antineoplastic agents from the Cyanophyta Oscillatoria acutissima. *J. Am. Chem. Soc.* **106**, 8193–8197.

Barsby, T., Kelly, M. T., Gagne, S. M., and Andersen, R. J. (2001). Bogorol A produced in culture by a marine *Bacillus* sp. reveals a novel template for cationic peptide antibiotics. *Org. Lett.* **3**, 437–440.

Bhat, V. B. and Madyastha, K. M. (2000). C-phycocyanin: A potent peroxyl radical scavenger in vivo and in vitro. *Biochem. Biophys. Res. Commun.* **275**(1), 20–25.

Bister, B., Bischoff, D., Strobele, M., Riedlinger, J., Reicke, A., Wolter, F., Bull, A. T., Zahner, H., Fiedler, H. P., and Sussmuth, R. D. (2004). Abyssomicin C–A polycyclic antibiotic from a marine *Verrucosispora strain* as an inhibitor of the p-aminobenzoic acid/tetrahydrofolate biosynthesis pathway. *Angew. Chem. Int. Ed Engl.* **43**, 2574–2576.

Bruntner, C., Binder, T., Pathom-aree, W., Goodfellow, M., Bull, A. T., Potterat, O., Puder, C., Horer, S., Schmid, A., and Bolek, W. (2005). Frigocyclinone, a novel angucyclinone antibiotic produced by a *Streptomyces griseus* strain from Antarctica. *J. Antibiot.* **58**, 346–349.

Buchanan, G. O., Williams, P. G., Feling, R. H., Kauffman, C. A., Jensen, P. R., and Fenical, W. (2005). Sporolides A and B: Structurally unprecedented halogenated macrolides from the marine actinomycete *Salinispora tropica*. *Org. Lett.* **7**, 2731–2734.

Burja, A. M., Banaigs, B., Abou-Mansour, E., Burgess, J. G., and Wright, P. C. (2001). Marine cyanobacteria—A prolific source of natural products. *Tetrahedron* 57, 9347–9377.

Cardellina, J. H., II, Kirkup, M. P., Moore, R. E., Mynderse, J. S., Seff, K., and Simmons, C. J. (1979). Hyellazole and chlorohyellazole, two novel carbozoles from the blue-green algae Hyella caeslitosa born. *Tetrahedron Lett.* **51**, 4915–4921.

Charan, R. D., Schlingmann, G., Janso, J., Bernan, V., Feng, X., and Carter, G. T. (2004). Diazepi-nomicin, a new antimicrobial alkaloid from a marine *Micromonospora* sp. *J. Nat. Prod.* **67**, 1431–1433.

Colwell, R. R. (1997). Microbial biodiversity and biotechnology. *In* "Biodiversity II: Understanding and Protecting Our Biological Resources", (M. E. Reaka-Kudla and E. Wilson, Eds), pp. 279–287. Joseph Henry Press, Washington, DC.

Desjardine, K., Pereira, A., Wright, H., Matainaho, T., Kelly, M., and Andersen, R. J. (2007). Tauramamide, a lipopeptide antibiotic produced in culture by *Brevibacillus laterosporus* isolated from a marine habitat: Structure elucidation and synthesis. *J. Nat. Prod.* **70**, 1850–1853.

El-Gendy, M. M., Hawas, U. W., and Jaspars, M. (2008a). Novel bioactive metabolites from a marine derived bacterium *Nocardia* sp. ALAA 2000. *J. Antibiot.* **61**, 379–386.

El-Gendy, M. M., Shaaban, M., Shaaban, K. A., El-Bondkly, A. M., and Laatsch, H. (2008b). Essramycin: A first triazolopyrimidine antibiotic isolated from nature. *J. Antibiot.* **61**, 149–157.

Engelhardt, K., Degnes, K. F., Kemmler, M., Bredholt, H., Fjaervik, E., Klinkenberg, G., Sletta, H., Ellingsen, T. E., and Zotchev, S. B. (2010). Production of a new thiopeptide antibiotic, TP-1161, by a marine *Nocardiopsis* species. *Appl. Environ. Microbiol.* **76**, 4969–4976.

Feling, R. H., Buchanan, G. O., Mincer, T. J., Kauffman, C. A., Jensen, P. R., and Fenical, W. (2003). Salinosporamide A: A highly cytotoxic proteasome inhibitor from a novel microbial source, a marine bacterium of the new genus *Salinospora*. *Angew. Chem. Int. Ed Engl.* **42**, 355–357.

Fenical, W. (2009). Marine Products, Discovery and Commercialisation. Available at: http://www.senate.gov/Senate404.htm.

Fiedler, H. P., Bruntner, C., Riedlinger, J., Bull, A. T., Knutsen, G., Goodfellow, M., Jones, A., Maldonado, L., Pathom-aree, W., Beil, W., Schneider, K., Keller, S., *et al.* (2008). Proximicin A, B and C, novel aminofuran antibiotic and anticancer compounds isolated from marine strains of the actinomycete *Verrucosispora*. *J. Antibiot.* **61**, 158–163.

Fischbach, M. A. and Walsh, C. T. (2009). Antibiotics for emerging pathogens. *Science* **325**, 1089–1093.

Gao, X. and Hall, D. G. (2005). Catalytic asymmetric synthesis of a potent thiolmarinol antibiotic. *J. Am. Chem. Soc.* **127**(6), 1628–1629.

Gbankoto, A., Vigo, J., Dramane, K., Banaigs, B., Aina, E., and Salmon, J. M. (2005). Cytotoxic effect of Laxaphycins A and B on human lymphoblastic cells (CCRF-CEM) using digitised videomicrofluorometry. *In Vivo* **19**(3), 577–582.

Gochfeld, D. J., El Sayed, K. A., Yousaf, M., Hu, J. F., Bartyzel, P., Dunbar, D. C., Wilkins, S. P., Zjawiony, J. K., Schinazi, R. F., Schlueter Wirtz, S., Tharnish, P. M., and Hamann, M. T. (2003). Marine natural products as lead anti-HIV agents. *Mini Rev. Med. Chem.* **3**, 401–424.

Gourdeau, H., McAlpine, J. B., Ranger, M., Simard, B., Berger, F., Beaudry, F., Farnet, C. M., and Falardeau, P. (2008). Identification, characterization and potent antitumor activity of ECO-4601, a novel peripheral benzodiazepine receptor ligand. *Cancer Chemother. Pharmacol.* **61**, 911–921.

Gustafson, K. R., Roman, M., and Fenical, W. (1989). The macrolactins, a novel class of antiviral and cytotoxic macrolides from a deep-sea marine bacterium. *J. Am. Chem. Soc.* **111**, 7519–7524.

Haefner, B. (2003). Drugs from the deep: Marine natural products as drug candidates. *Drug Discov. Today* **8**, 536–544.

Han, B., Gross, H., Goeger, D. E., Mooberry, S. L., and Gerwick, W. H. (2006). Aurilides B and C, cancer cell toxins from a Papua New Guinea collection of the marine cyanobacterium Lyngbya majuscula. *J. Nat. Prod.* **69**(4), 572–575.

Harrigan, G. G., Yoshida, W. Y., Moore, R. E., Nagle, D. G., Park, P. U., Biggs, J., Paul, V. J., Mooberry, S. L., Corbett, T. H., and Valeriote, F. A. (1998a). Isolation, structure determination, and biological activity of dolastatin-12 and lyngbyastatin-I from *Lyngbya majuscula Schizothrix calcicola* cyanobacterial assemblages. *J. Nat. Prod.* **61**(10), 1221–1225.

Harrigan, G. G., Luesch, H., Yoshida, W. Y., Moore, R. E., Nagle, D. G., Paul, V. J., Mooberry, S. L., Corbett, T. H., and Valeriote, F. A. (1998b). Symplostatin 1. A dolastatin 10 analogue from marine cyanobacterium *Symploca hydnoides*. *J. Nat. Prod.* **61**, 1075–1077.

Hawas, U. W., Shaaban, M., Shaaban, K. A., Speitling, M., Maier, A., Kelter, G., Fiebig, H. H., Meiners, M., Helmke, E., and Laatsch, H. (2009). Mansouramycins A–D, cytotoxic isoquinolinequinones from a marine *Streptomycete*. *J. Nat. Prod.* **72**, 2120–2124.

Hohmann, C., Schneider, K., Bruntner, C., Brown, R., Jones, A. L., Goodfellow, M., Krämer, M., Imhoff, J. F., Nicholson, G., Fiedler, H. P., and Süssmuth, R. D. (2009). Albidopyrone, a new alpha-pyrone-containing metabolite from marine-derived Streptomyces *sp*. NTK 227. *J. Antibiot.* **62**, 75–79.

Hughes, C. C., Prieto-Davo, A., Jensen, P. R., and Fenical, W. (2008). The marinopyrroles, antibiotics of an unprecedented structure class from a marine *Streptomyces sp*. *Org. Lett.* **10** (4), 629–631.

Ireland, C. M., Copp, B. R., Foster, M. D., McDonald, L. A., Radisky, D. C., and Swersey, J. C. (1993). Biomedical potential of marine natural products. *In* "Marine Biotechnology", (D. H. Attaway and O. R. Zaborsky, Eds), Pharmaceutical and Bioactive Natural Products, Vol. 1, pp. 1–43. Plenum Press, New York.

Isnansetyo, A. and Kamei, Y. (2003). MC21-A, a bactericidal antibiotic produced by a new marine bacterium, *Pseudoalteromonas phenolica sp*. nov. O-BC30 (T), against methicillin-resistant *Staphylococcus aureus*. *Antimicrob. Agents Chemother.* **47**, 480–488.

Issa, A. A. (1999). Antibiotic production by the cyanobacteria Oscillatoria angustissima and Calothrix parietina. *Environ. Toxicol. Pharmacol.* **8**, 33–37.

Jang, J. H., Kanoh, K., Adachi, K., and Shizuri, Y. (2006). New dihydrobenzofuran derivative, awajanoran, from marine-derived *Acremonium sp*. AWA16-1. *J. Antibiot.* **59**(7), 428–431.

Jensen, P. R., Mincer, T. J., Williams, P. G., and Fenical, W. (2005). Marine *Actinomycete* diversity and natural product discovery. *Antonie Van Leeuwenhoek* **87**, 43–48.

Kamata, K., Okamoto, S., and Oka, S. (2001). Cycloprodigiosin hydrochloride suppresses tumor necrosis factor (TNF) alpha-induced transcriptional activation by NF-kappaB. *FEBS Lett.* **507**, 74–80.

Kanoh, K., Matsuo, Y., Adachi, K., Imagawa, H., Nishizawa, M., and Shizuri, Y. (2005). Mechercharmycins A and B, cytotoxic substances from marine-derived *Thermoactinomyces sp*. YM3-251. *J. Antibiot.* **58**, 289–292.

Knubel, G., Larsen, L. K., Moore, R. E., Levine, I. A., and Patterson, G. M. (1990). Cytotoxic and antiviral indolocarbozoles from a blur green alga belonging to the nostocaceae. *J. Antibiot.* **43**, 1236–1239.

Koehn, F. E., Longley, R. E., and Reed, J. K. (1992). Microcolins A and B, new immunosuppressive peptide from the blue-green alga *Lyngbya majuscule*. *J. Nat. Prod.* **55**, 613–619.

Lane, A. L. and Moore, B. S. (2011). A sea of biosynthesis: Marine natural products meet the molecular age. *Nat. Prod. Rep.* **28**, 411–428.

Li, F., Maskey, R. P., Qin, S., Sattler, I., Fiebig, H. H., Maier, A., Zeeck, A., and Laatsch, H. (2005). Chinikomycins A and B: Isolation, structure elucidation, and biological activity of novel antibiotics from a marine *Streptomyces sp*. Isolate M045. *J. Nat. Prod.* **68**, 349–353.

Liu, X., Ashforth, E., Ren, B., Song, F., Dai, H., Liu, M., Wang, J., Xie, Q., and Zhang, L. (2010). Bioprospecting microbial natural product libraries from the marine environment for drug discovery. *J. Antibiot.* **63**, 415–422.

Lu, K., Dempsey, J., and Schultz, R. M. (2001). Cryptophycin-induced hyperphosphorylation of Bcl-2, cell cycle arrest and growth inhibition in human H460 NSCLC cells. *Cancer Chemother. Pharmacol.* **47**, 170–178.

Macherla, V. R., Liu, J., Bellows, C., Teisan, S., Lam, K. S., and Potts, B. C. M. (2005a). Glaciapyrroles A, B and C, pyrrolosesquiterpenes from a *Streptomyces sp.* isolated from an Alaskan marine sediment. *J. Nat. Prod.* **68**, 780–783.

Macherla, V. R., Mitchell, S. S., Manam, R. R., Reed, K. A., Chao, T. H., and Nicholson, B. (2005b). Structure–activity relationship studies of salinosporamide A (NPI-0052), a novel marine derived proteasome inhibitor. *J. Med. Chem.* **48**(11), 3684–3687.

Macherla, V. R., Liu, J., Sunga, M., White, D. J., Grodberg, J., Teisan, S., Lam, K. S., and Potts, B. C. (2007). Lipoxazolidinones A, B, and C: Antibacterial 4-oxazolidinones from a marine actinomycete isolated from a Guam marine sediment. *J. Nat. Prod.* **70**, 1454–1457.

Magarvey, N. A., Keller, J. M., Bernan, V., Dworkin, M., and Sherman, D. H. (2004). Isolation and characterization of novel marine-derived actinomycete taxa rich in bioactive metabolites. *Appl. Environ. Microbiol.* **70**(12), 7520–7529.

Malakoff, D. (1997). Extinction on the high seas. *Science* **277**, 486–488.

Maldonado, L. A., Fenical, W., Jensen, P. R., Kauffman, C. A., Mincer, T. J., Ward, A. C., Bull, A. T., and Goodfellow, M. (2005). *Salinispora arenicola* gen. nov., sp. nov. and *Salinispora tropica sp*. nov., obligate marine *actinomycetes* belonging to the family *Micromonosporaceae*. *Int. J. Syst. Evol. Microbiol.* **55**, 1759–1766.

Maskey, R. P., Helmke, E., and Laatsch, H. (2003). Himalomycin A and B: Isolation and structure elucidation of new fridamycin type antibiotics from a marine *Streptomyces* isolate. *J. Antibiot.* **56**, 942–949.

Maskey, R. P., Helmke, E., Kayser, O., Fiebig, H. H., Maier, A., Busche, A., and Laatsch, H. (2004). Anti-cancer and antibacterial trioxacarcins with high anti-malaria activity from a marine *Streptomycete* and their absolute stereochemistry. *J. Antibiot.* **57**, 771–779.

McArthur, K. A., Mitchell, S. S., Tsueng, G., Rheingold, A., White, D. J., Grodberg, J., Lam, K. S., and Potts, B. C. (2008). Lynamicins A–E, chlorinated bisindole pyrrole antibiotics from a novel marine actinomycete. *J. Nat. Prod.* **71**, 1732–1737.

Mishima, T., Murata, J., Toyoshima, M., Fujii, H., Nakajima, M., Hayashi, T., Kato, T., and Saiki, I. (1998). Inhibition of tumor invasion and metastasis by calcium spirulan (Ca-SP), a novel sulfated polysaccharide derived from a blue-green algae, Spirulina platensis. *Clin. Exp. Metastasis* **16**(6), 541–550.

Mitchell, S. S., Nicholson, B., Teisan, S., Lam, K. S., and Potts, B. C. (2004). Aureoverticillactam, a novel 22-atom macrocyclic lactam from the marine actinomycete *Streptomyces aureoverticillatus*. *J. Nat. Prod.* **67**, 1400–1402.

Moore, R. E. (1982). Toxins, anticancer agents, and tumor promoters from marine prokaryotes. *Pure Appl. Chem.* **54**, 1919–1934.

Moore, R. E., Blackman, A. J., Cheuk, C. E., Mynderse, J. S., Matsumoto, G. K., Clardy, J., Woodard, R. W., and Craig, J. C. (1984). Absolute stereochemistries of the aplysiatoxins and oscillatoxin A. *J. Org. Chem.* **49**, 2484–2489.

Murakami, N., Morimoto, T., Imamura, H., Ueda, T., Nagai, S., Sakakibara, J., and Yamada, N. (1991). Studies on glycolipids. III. Glyceroclycolipids from an axenically cultured cyanobacterium, Phormidium tenue. *Chem. Pharm. Bull.* **39**(9), 2277–2281.

Mynderse, J. S., Moore, R. E., Kashiwagi, M., and Norton, T. R. (1977). Antileukemia activity in the Oscillatoriaceae: Isolation of debromoaplysiatoxin from *Lyngbya*. *Science* **196**, 538–540.

Nakao, Y., Takada, K., Matsunaga, S., and Fusetani, N. (2001). Calyceramides A–C: Neuraminidase inhibitory sulfated ceramides from the marine sponge *Discodermia calyx*. *Tetrahedron* **57**, 3013–3017.

Newman, D. J. and Cragg, G. M. (2005). The discovery of anticancer drugs from natural sources. *In* "Natural Products: Drug Discovery and Therapeutics Medicines", (L. Zhang and A. Demain, Eds), pp. 275–294. Humana Press, New York.

Norton, R. S. and Wells, R. J. (1982). A series of chiral polybrominated biindoles from the marine blue-green alga Rivularia firma. Application of carbon-13 NMR spin–lattice relaxation data and carbon-13-proton coupling constants to structure elucidation. *J. Am. Chem. Soc.* **104**, 3628–3635.

Oku, N., Adachi, K., Matsuda, S., Kasai, H., Takatsuki, A., and Shizuri, Y. (2008). Ariakemicins A and B, novel polyketide–peptide antibiotics from a marine gliding bacterium of the genus *Rapidithrix*. *Org. Lett.* **10**, 2481–2484.

Olaizola, M. (2003). Commercial development of microalgal biotechnology: From the test tube to the market place. *Biomol. Eng.* **20**, 459–466.

Parkes, R. J., Cragg, B. A., Bale, S. J., Getliff, J. M., Goodman, K., Rochelle, P. A., Fry, J. C., Weightman, A. J., and Harvey, S. M. (1994). Deep bacterial biosphere in Pacific Ocean sediments. *Nature* **371**, 410–413.

Penn, K., Jenkins, C., Nett, M., Udwary, D. W., Gontang, E. A., McGlinchey, R. P., Foster, B., Lapidus, A., Podell, S., Allen, E. E., Moore, B. S., and Jensen, P. R. (2009). Genomic islands link secondary metabolism to functional adaptation in marine Actinobacteria. *ISME J.* **3**, 1193–1203.

Perez Baz, J., Cañedo, L. M., Fernández Puentes, J. L., and Silva Elipe, M. V. (1997). Thiocoraline, a novel depsipeptide with antitumor activity produced by a marine Micromonospora. II. Physico-chemical properties and structure determination. *J. Antibiot.* **50**, 738–741.

Piel, J., Hertweck, C., Shipley, P. R., Hunt, D. M., Newman, M. S., and Moore, B. S. (2000). Cloning, sequencing and analysis of the enterocin biosynthesis gene cluster from the marine isolate '*Streptomyces maritimus*': Evidence for the derailment of an aromatic polyketide synthase. *Chem. Biol.* **7**, 943–955.

Prinsep, M. R., Moore, R. E., Levine, I. A., and Patterson, G. M. L. (1992). Westiellamide, a bistratamide-related cyclic peptide from the blue-green alga *Westiellopsis prolifica*. *J. Nat. Prod.* **55**(1), 140–142.

Rath, J. P., Kinast, S., and Maier, M. E. (2005). Synthesis of the full functionalized core structure of the antibiotic abyssomicin C. *Org. Lett.* **7**, 3089–3092.

Raveh, A. and Carmeli, S. (2007). Antimicrobial ambiguines from the cyanobacterium *Fischerella* sp. collected in Israel. *J. Nat. Prod.* **70**, 196–201.

Reddy, C. M., Bhat, V. B., Kiranmai, G., Reddy, M. N., Reddanna, P., and Madyastha, K. M. (2007). Selective inhibition of cyclooxygenase-2 by C-phycocyanin, a biliprotein from *Spirulina platensis*. *Biochem. Biophys. Res. Commun.* **277**(3), 599–603.

Reshef, V., Mizrachi, E., Maretzki, T., Silberstein, C., Loya, S., Hizi, A., and Carmeli, S. (1997). New acylated sulfoglycolipids and digalactolipids and related known glycolipids from cyanobacteria with a potential to inhibit the reverse transcriptase of HIV-1. *J. Nat. Prod.* **60**, 1251–1260.

Riedlinger, J., Reicke, A., Zahner, H., Krismer, B., Bull, A. T., Maldonado, L. A., Ward, A. C., Goodfellow, M., Bister, B., and Bischoff, D. (2004). Abyssomicins, inhibitors of the para-aminobenzoic acid pathway produced by the marine Verrucosispora strain AB-18-032. *J. Antibiot. (Tokyo)* **57**, 271–279.

Rinker, K. D. and Kelly, R. M. (2000). Effect of carbon and nitrogen sources on growth dynamics and exopolysaccharide production for the *Hyperthermophilic archaeon, Thermococcus litoralis* and bacterium *Thermotoga maritima*. *Biotechnol. Bioeng.* **69**, 537–547.

Rondon, M. R., Ballering, K. S., and Thomas, M. G. (2004). Identification and analysis of a siderophore biosynthetic gene cluster from *Agrobacterium tumefaciens* C58. *Microbiology* **150**(11), 3857–3866.

Sabdono, A. and Radjasa, O. K. (2006). Antifouling activity of bacteria associated with soft coral *Sarcophyton sp.* against marine biofilm-forming bacteria.. *J. Coast. Dev.* **10**(1), 55–62.

Schumacher, R. W., Talmage, S. C., Miller, S. A., Sarris, K. E., Davidson, B. S., and Goldberg, A. (2003). Isolation and structure determination of an antimicrobial ester from a marine derived bacterium. *J. Nat. Prod.* **66**, 1291–1293.

Schwartz, J. and Shklar, G. (1987). Regression of experimental hamster cancer by beta carotene and algae extracts. *J. Oral Maxillofac. Surg.* **45**, 510–515.

Simmons, T. L., McPhail, K. L., Ortega-Barria, E., Mooberry, S. L., and Gerwick, W. H. (2006). Belamide A, a new antimitotic tetrapeptide from a Panamanian marine cyanobacterium. *Tetrahedron Lett.* **47**(20), 3387–3390.

Singh, M. P., Kong, F., Janso, J. E., Arias, D. A., Saurez, P. A., and Barnas, V. S. (2003). Novel-α-pyrones by a marine *Pseudomonas* sp. F92S91: Taxonomy and biological activities. *J. Antibiot.* **56**(12), 1033–1044.

Soria-Mercado, I. E., Prieto-Davo, A., Jensen, P. R., and Fenical, W. (2005). Antibiotic terpenoid chloro-dihydroquinones from a new marine actinomycete. *J. Nat. Prod.* **68**(6), 904–910.

Stratmann, K., Belli, J., Jensen, C. M., Moore, R. E., and Patterson, G. M. (1994). Aulosirazole, a novel solid tumor selective cytotoxin from the blue-green alga *Aulosire fertilissima*. *J. Org. Chem.* **59**(21), 6279–6281.

Tokuda, H., Nishino, H., Shirahashi, H., Murakami, N., Nagatsu, A., and Sakakibara, J. (1996). Inhibition of 12-O-tetradecanoylphorbol-13-acetate promoted mouse skin papilloma by digalactosyl diacylglycerols from the fresh water cyanobacterium *Phormidium tenue*. *Cancer Lett.* **104**(1), 91–95.

Tuin, A. W., Grotenbreg, G. M., Spalburg, E., de Neeling, A. J., Mars-Groenendijk, R. H., and Vander Marel, G. A. (2009). Structural and biological evaluation of some loloatin C analogues. *Bioorg. Med. Chem.* **17**(17), 6233–6240.

Tziveleka, L. A., Vagias, C., and Roussis, V. (2003). Natural products with anti HIV activity from marine organism. *Curr. Top. Med. Chem.* **3**(13), 1512–1535.

Williams, P. G., Yoshida, W. Y., Moore, R. E., and Paul, V. J. (2003a). Tasipeptins A and B: New cytotoxic depsipeptides from the marine cyanobacterium Symploca sp. *J. Nat. Prod.* **66**, 620–624.

Williams, P. G., Yoshida, W. Y., Quon, M. K., Moore, R. E., and Paul, V. J. (2003b). The structure of Palau'amide, a potent cytotoxin from a species of the marine cyanobacterium Lyngbya. *J. Nat. Prod.* **66**, 1545–1549.

Williams, P. G., Yoshida, W. Y., Moore, R. E., and Paul, V. J. (2003c). The isolation and structure elucidation of Tasiamide B, a 4-amino-3-hydroxy-5-phenylpentanoic acid containing peptide from the marine *Cyanobacterium Symploca sp.* *J. Nat. Prod.* **66**, 1006–1009.

Williams, P. G., Yoshida, W. Y., Moore, R. E., and Paul, V. J. (2004). Micromide and guamamide: Cytotoxic alkaloids from a species of the marine *cyanobacterium Symploca. J. Nat. Prod.* **67**, 49–53.

Winston, J. E. (1988). The systematists' perspective. *In* ''Biomedical Importance of Marine Organisms'', (D. G. Fautin, Ed.), pp. 1–6. California Academy of Sciences, San Francisco, CA.

Wu, S. J., Fotso, S., Li, F., Qin, S., Kelter, G., and Fiebig, H. H. (2006). N-carboxamido-staurosporine and selina-4(14), 7(11)-diene-8, 9-diol, new metabolites from a marine *Streptomyces sp. J. Antibiot. (Tokyo)* **59**(6), 331–337.

Yang, H., Lee, E., and Kim, H. (1997). *Spirulina platensis* inhibits anaphylactic reaction. *Life Sci.* **61**(13), 1237–1244.

Zhang, H. L., Hua, H. M., Pei, Y. H., and Yao, X. S. (2004). Three new cytotoxic cyclic acylpeptides from marine *Bacillus sp. Chem. Pharm. Bull. (Tokyo)* **52**, 1029–1030.

CHAPTER 26

Biological Activities and Health Effects of Terpenoids from Marine Fungi

Se-Kwon Kim[*,†,1] and **Yong-Xin Li**[*]

Contents

Abstract

Recently, a great deal of interest has been developed by the consumers toward natural bioactive compounds as functional ingredients in the nutraceutical, cosmeceutical, and pharmaceutical products due to their various health beneficial effects. Hence, it can be suggested that bioactive functional ingredients from marine bioresources and their by-products are alternative sources for synthetic ingredients that can contribute to consumer's well-being, as a part of nutraceuticals and functional foods. Marine-derived fungi produce a vast array of secondary metabolites including terpenes, steroids, polyketides, peptides, alkaloids, and polysaccharides. These secondary metabolites serve many biopharmaceutical purposes. This chapter discusses about marine fungi-derived terpenoids and presents an overview of their beneficial health effects.

* Department of Chemistry, Marine Biochemistry Laboratory, Pukyong National University, Busan, Republic of Korea
† Marine Bioprocess Research Center, Pukyong National University, Busan, Republic of Korea
1 Corresponding author: Se-Kwon Kim, *E-mail address:* sknkim@pknu.ac.kr

Advances in Food and Nutrition Research, Volume 65
ISSN 1043-4526, DOI: 10.1016/B978-0-12-416003-3.00026-3

I. INTRODUCTION

Recently, much attention has been paid by the consumers toward natural bioactive compounds as functional ingredients in the diet, pharmaceuticals, and cosmeceuticals. Especially, bioactive compounds derived from marine organisms have been served as a rich source of health-promoting components (Kim and Wijesekara, 2010; Wijesekara and Kim, 2010; Wijesekara *et al.*, 2010, 2011).

Terpenoids, including more than 35,000 well-defined molecules, probably constitute the natural product family with the greater structural diversity. All terpenoids are synthesized from two five-carbon building blocks. Based on the number of the building blocks, terpenoids are commonly classified as monoterpenes (C_{10}), sesquiterpenes (C_{15}), diterpenes (C_{20}), and sesterterpenes (C_{25}). Numerous of these compounds possess well-known applications as drugs, cosmetics, perfumes, flavors, insecticides, herbicides, phytohormones, and many others. In recent years, a number of advances in terpenoid research have led, for example, to the discovery of a second biosynthetic pathway, namely methyl-erythritol phosphate pathway, usually present in bacteria and plants.

Marine organisms produce a wide array of fascinating terpenoid structures distinguished by characteristic structural features. Certain structural classes, for example, cembrane, chamigrene, amphilectane skeletons, and unusual functional groups such as isonitrile, isothiocyanate, isocyanate, dichloroimine, and halogenated functionalities, occur predominantly in marine metabolites (Gross and Konig, 2006). In past two decades, natural product bioprospecting from the marine environment has resulted in hundreds of terpenoids with novel structures and interesting bioactivities, with more to be discovered in the future. These terpenoids display a wide range of biological activities against cancer, malaria, inflammation, and a variety of infectious diseases (viral and bacterial). This chapter focuses on terpenoids from marine-derived fungi and presents an overview of their biological activities with potential medicinal benefits.

II. DIVERSITY OF TERPENOIDS DERIVED FROM MARINE FUNGI

Marine fungi contain a pronounced degree of structurally diversified terpenoids such as monoterpenes, sesquiterpenes, diterpenes, sesterterpenes, triterpenes, steroids, and tetraterpenes. Meroterpenoids are most often isolated from fungi and marine organisms, but bacteria and higher plants produce such mixed biosynthesized products as well. Mixed polyketide–terpenoid-derived fungal natural products constitute the largest class of meroterpenoids, while those derived from orsellinic acid or its mono- and dimethyl derivatives are the most frequently isolated meroterpenoids from fungi, especially *Penicillium* and *Aspergillus* spp.

Fungi, themselves, produce a range of interesting terpenoids including trichothecanes, the protoilludane group, cyathins, fussicoccanes, and many triterpenoids. Recent research into marine natural products has shown marine organisms are a prolific source of unusual terpenoids. Detailed reviews of marine natural products appear regularly, and a comprehensive review of terpenoids of marine invertebrates concentrates on their biological activity. Marine organisms continue to provide new carbon skeletons. Many substituents that are rarely found in natural products occur in marine terpenoids, for example, bromo- and chloro-substituents in algal terpenoids and isonitrile and isothiocyanate substituents in sponge terpenoids. A chlorinated monoterpene has been found in marine-derived fungus, *Tryblidiopycnis* sp. (Huang *et al.*, 2006). Sesquiterpenes have been reported from *Dendryphiella* sp., *Penicillium* sp., and *Hansfordia* sp. (Schneider *et al.*, 1997). A marine-derived fungus *Fusarium* sp. secretes sesterterpenes as secondary metabolites (Renner *et al.*, 1998). Moreover, triterpenes have been isolated from *Phomopsis* sp. (Li *et al.*, 2008). In addition, steroids have been yielded by some marine-derived fungi, such as *Rhizopus* sp., *Gymnascella* sp., and *Penicillium* sp. (Ebel, 2010). The occurrence of tetraterpenoids or carotenoids has only rarely been studied for marine fungi, and the known carotenoids neurosporaxanthin, β-carotene, and torulene have been reported.

III. HEALTH BENEFITS AND BIOLOGICAL ACTIVITIES OF TERPENOIDS FROM MARINE FUNGI

Antioxidants may have a positive effect on human health as they can protect human body against damage by reactive oxygen species (ROS), which attack macromolecules. Moreover, deterioration of some foods has been identified due to oxidation of lipids or rancidity and formation of undesirable secondary lipid peroxidation products. Lipid oxidation by ROS such as superoxide anion, hydroxyl radicals, and H_2O_2 also causes a decrease in nutritional value of lipid foods and affects their safety and appearance. Therefore, in food and pharmaceutical industries, many synthetic commercial antioxidants such as butylated hydroxytoluene, butylated hydroxyanisole, tert-butylhydroquinone, and propyl gallate have been used to retard the oxidation and peroxidation processes. However, the use of these synthetic antioxidants must be under strict regulation due to potential health hazards (Hettiarachchy *et al.*, 1996). Hence, the search for natural antioxidants as safe alternatives is important in the food industry (Penta-Ramos and Xiong, 2001). Recently, there is a considerable interest in the food industry as well as pharmaceutical industry for the development of antioxidants from natural sources, such as marine flora and fauna.

It has been demonstrated that marine fungi-derived terpenoids such as neurosporaxanthin, β-carotene, g-carotene, and torulene have potential antioxidant activity (Ebel, 2010). Antioxidant activity of terpenoid carotenoids have been determined by various methods such as 1,1-diphenyl-2-picryl hydrazil radical scavenging, lipid peroxide inhibition, ferric reducing antioxidant power, nitric oxide scavenging, ABTS radical scavenging, and superoxide radical and hydroxyl radical scavenging assays. Marine fungi-derived terpenoids could be used as a rich source of natural antioxidants with potential application in the food industry as well as in cosmetic and pharmaceutical areas. Several studies have reported that terpenoids derived from marine fungi have antiproliferative activity in several human cancer cell lines *in vitro*, as well as inhibitory activity of tumor growth in mice. In addition, they have antimetastatic activity by blocking the interactions between cancer cells and the basement membrane. Many species of marine-derived fungi contain significant quantities of complex structural terpenoids that have been shown to inhibit the proliferation of pathogenic microorganisms.

IV. CONCLUDING REMARKS

Recent studies have provided evidence that terpenoids derived from marine fungi play a vital role in human health. The possibilities of designing new functional foods, pharmaceuticals, and cosmeceuticals reducing or regulating the chronic malfunctions are promising. Therefore, it can be suggested that due to valuable biological functions with medicinal beneficial effects, marine fungal-derived terpenoids have much potential as active ingredients for preparation of nutraceuticals and medicinal food products. Until now, most of the biological activities of marine-derived bioactive terpenoids have been observed *in vitro* or in mouse model systems. Therefore, further research studies are needed in order to investigate their activity in human subjects.

REFERENCES

Ebel, R. (2010). Terpenes from marine-derived fungi. *Mar. Drugs* **8**, 2340–2368.

Gross, H. and Konig, G. M. (2006). Terpenoids from marine organisms: Unique structures and their pharmacological potential. *Phytochem. Rev.* **5**, 115–141.

Hettiarachchy, N. S., Glenn, K. C., Gnanasambandan, R., and Johnson, M. G. (1996). Natural antioxidant extract from fenugreek (*Trigonella foenumgraecum*) for ground beef patties. *J. Food Sci.* **61**, 516–519.

Huang, H. R., Xia, X. K., She, Z. G., Lin, Y. C., Vrijmoed, L. L. P., and Jones, E. B. G. (2006). A new chloro-monoterpene from the mangrove endophytic fungus *Tryblidiopycnis* sp. (4275). *J. Asian Nat. Prod. Res.* **8**, 609–612.

Kim, S. K. and Wijesekara, I. (2010). Development and biological activities of marine-derived bioactive peptides: A review. *J. Funct. Foods* **2**, 1–9.

Li, L. Y., Sattler, I., Deng, Z. W., Groth, I., Walther, G., Menzel, K. D., Peschel, G., Grabley, S., and Lin, W. H. (2008). A seco-oleane-type triterpenes from *Phomopsis* sp. (strain HK10458) isolated from the mangrove plant *Hibiscus tiliaceus*. *Phytochemistry* **69**, 511–517.

Penta-Ramos, E. A. and Xiong, Y. L. (2001). Antioxidative activity of whey protein hydrolysates in a liposomal system. *J. Dairy Sci.* **84**, 2577–2583.

Renner, M. K., Jensen, P. R., and Fenical, W. (1998). Neomangicols: Structures and absolute stereochemistries of unprecedented halogenated sesterterpenes from a marine fungus of the genus *Fusarium*. *J. Org. Chem.* **63**, 8346–8354.

Schneider, G., Anke, H., and Sterner, O. (1997). New secondary metabolites from a mycophilic *Hansfordia* species. *Nat. Prod. Lett.* **10**, 133–138.

Wijesekara, I. and Kim, S. K. (2010). Angiotensin-I-converting enzyme (ACE) inhibitors from marine resources. *Mar. Drugs* **8**, 1080–1093.

Wijesekara, I., Yoon, N. Y., and Kim, S. K. (2010). Phlorotannins from *Ecklonia cava* (Phaeophyceae): Biological activities and potential health benefits. *Biofactors* **36**, 408–414.

Wijesekara, I., Pangestuti, R., and Kim, S. K. (2011). Biological activities and potential health benefits of sulfated polysaccharides derived from marine algae. *Carbohydr. Polym.* **84**, 14–21.

CHAPTER 27

Antimetastasis Effect of Anthraquinones from Marine Fungus, *Microsporum* sp.

Chen Zhang[*,†,‡] and **Se-Kwon Kim**[†,§,1]

Contents

Abstract

This chapter discusses about obtaining natural products which have anticancer metastasis activities from selected marine-derived fungus (*Microsporum* sp.) and investigates their biological activities such as cytotoxicity on viability cell lines, anticancer cell migration and invasion, protease inhibition, and expression of matrix metalloproteinase (MMP-2 and -9). Moreover, the correlative mechanisms behind these activities were studied.

I. INTRODUCTION

Cancer is a class of disease in which a group of cells display uncontrolled growth through division beyond normal limits, invasion that intrudes upon and destroys adjacent tissues, and sometimes metastasis, in which

* School of Medicine, Tongji University, Shanghai, China
† Department of Chemistry, Pukyong National University, Busan, Republic of Korea
‡ Key Laboratory of Molecular Enzymology and Engineering of the Ministry of Education, College of Life Science, Jilin University, Changchun, China
§ Marine Bioprocess Research Center, Pukyong National University, Busan, Republic of Korea
1 Corresponding author: Se-Kwon Kim, *E-mail address:* sknkim@pknu.ac.kr

Advances in Food and Nutrition Research, Volume 65
ISSN 1043-4526, DOI: 10.1016/B978-0-12-416003-3.00027-5

cancer cells spread to other locations in the body via lymph or blood. In 2007, cancer caused about 13% of all human deaths worldwide (7.9 million) and it is projected to be 12 million deaths per year by 2030 (Croce, 2008).

Cancer is the most dangerous enemy of human beings. Therefore, what makes the cancer so dangerous? The answer is metastasis. Metastasis is the major cause of cancer mortality (Tanigawa *et al.*, 1997). Metastasis is the spread of a disease from one organ or part to another nonadjacent organ or part. Cancer occurs after a single cell in a tissue is genetically damaged in ways that result in the formation of a cancer stem cell possessing a malignant phenotype. Some cancer cells acquire the ability to penetrate the walls of lymphatic and/or blood vessels, after which they are able to circulate through the bloodstream (circulating tumor cells) to other sites and tissues in the body (Woodhouse *et al.*, 1997).

While there is still lacking of a treatment capable of completely preventing tumorigenesis, efficiently impeding cancer metastasis will be a potentially powerful cancer therapeutic strategy. Although it has been understood to a certain degree, the molecular mechanism of cancer metastasis is still been vigorously explored and updated, which has been continuously published in the renowned international journals with high impact factors. Generally, cancer metastasis, cancer invasion, cancer cell migration, and cancer cell-mediated tissue remodeling were actually the major positive correlations with MMP levels. Numerous studies linked inhibition of MMPs by exogenous and endogenous inhibitors, like tissue inhibitors of matrix metalloproteinase (TIMPs), with a corresponding inhibition of cell invasion (Deryugina and Quigley, 2006). Conversely, upregulation of MMPs by inducers, transcriptional enhancement, or transgene constructs usually led to enhanced tumor cell invasion, in model systems by appearance of tumor cells in distant sites and *in vitro* by Matrigel or ECM invasion (Deryugina and Quigley, 2006; Fingleton, 2006).

II. MARINE FUNGI RESOURCES AND ANTHRAQUINONES

Marine microorganisms have become an important source of novel pharmacologically active metabolites. Among marine microorganisms, marine fungi are a rich source of bioactive secondary metabolites with unprecedented skeletons. Up to date, more than 4000 fungal metabolites have been reported. And 5000–7000 taxonomic species of marine fungi have been studied with respect to their chemistry. It was estimated that the probable number of existing fungi to be 1.5million with only 71,000 being described so far (Dreyfuss and Chapela, 1994). It indicates that

1 2 3

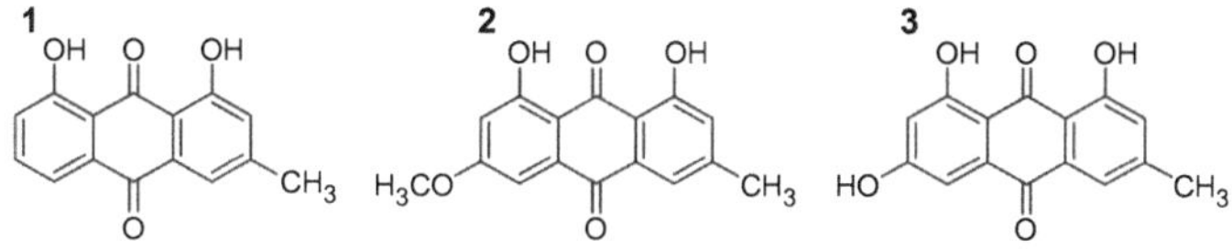

Chrysophanol Physcion Emodin

FIGURE 27.1 Chrysophanol, physcion, and emodin.

fungi are really an enormous source for natural products with unique chemical structures and interesting activities, such as, antimicrobial, antifungal, cytotoxic (Li *et al.*, 1995), antitumor (Natori *et al.*, 1994), immunostimulatory (Seki *et al.*, 2001), antiviral (Carter and Rinehar, 1978), and kinase inhibitory activities (Bugni and Ireland, 2004). Researchers have already got many lead anthraquinone compounds from marine fungi, as Xia *et al.* (2007) reported two new natural products, 1,4,5,6,7,9-hexahydroxy-2-methoxy-7-methyl-5β,9β,8aβ,6α,10aα,-hexahydroanthracen-10 (10aH)-one (1) and 1,4,6-trihydr oxy-2-methoxy-7-methylanthracene-9,10-dione (2), together with three known anthraquinones, austrocortirubin (3), demethoxyaustrocortirubin (4), and hydroxy-9,10-anthraquinone (5) from the mycelium and broth extracts of the fungus *Halorosellinia* sp. Three anthraquinones such as averufin (1), nidurufin (2), and versicolorin C (3) were found to be produced by the marine-derived endophytic fungus as the anthracycline antibiotics candidates by Zhu *et al.* (2008).

In this chapter, as part of an effort to discover novel biologically active natural products, specifically antimetastasis compounds from marine organisms, it has been investigated that the bioactive constituents of the marine-derived fungus *Microsporum* sp. yielded the chrysophanol, physcion, and emodin (Fig. 27.1).

III. EFFECT OF CHRYSOPHANOL, PHYSCION, AND EMODIN ON THE INHIBITION OF MMP-2 AND -9 EXPRESSIONS

Chrysophanol, physcion, and emodin were investigated for their cytotoxic effect on HT1080 cells at different concentrations and showed no cytotoxic effect on the tested cells. These compounds influenced the cell migration and invasion ability as measured by wound healing assay and transwell invasion assay in a concentration-dependent manner (Figs. 27.2 and 27.3), meanwhile the emodin has shown more activity.

Chrysophanol, physcion, and emodin inhibited the expressions of MMP-2 and -9 in a dose-dependent manner, according to the results of RT-PCR, Western blot, and zymography assays.

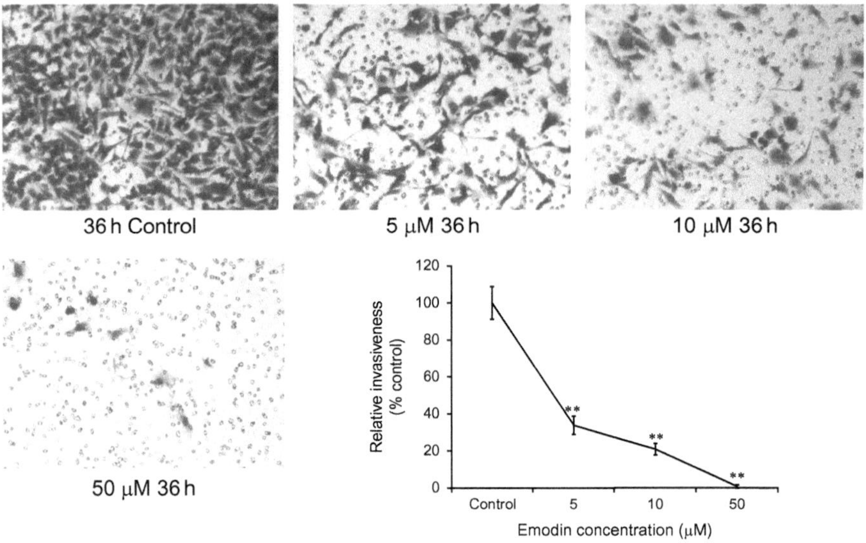

FIGURE 27.2 Effect of emodin on HT1080 cell invasion. (For color version of this figure, the reader is referred to the Web version of this chapter.)

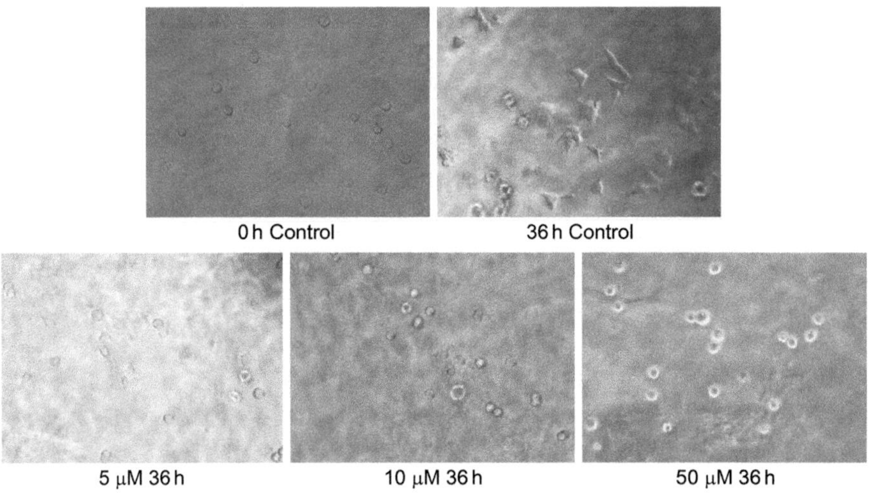

FIGURE 27.3 Effect of emodin on HT1080 cell morphology in 3D culture system. (For color version of this figure, the reader is referred to the Web version of this chapter.)

NF-κB is major transcription factor that regulates MMP-2 and -9 expressions (Huang *et al.*, 2004). Results have shown that phospho-AKT was increased by TNF-α treatment, and chrysophanol, physcion, and emodin strongly inhibited its increase in a concentration-dependent manner. In addition, the TNF-α treatment caused significant increase of phospho-

IκBα. This induction was associated with an increasing translocation of p65 into the nucleus and decreasing cytosolic p65. Pretreatment of compounds significantly inhibited both IκBα phosphorylation and p65 nuclear translocation in a concentration-dependent manner. As we know, AKT is a downstream effectors of phosphatidylinositol 3-kinase and its activation is extremely complex (PI3K; Datta *et al.*, 1999). AKT phosphorylates and activates IκB kinase (IKK), thus promoting NF-κB function (Ozes *et al.*, 1999). The inactive form of NF-κB bound by inhibitor of κB (IκB) family, and activation of NF-κB requires phosphorylation of IκB proteins. (Henkel *et al.*, 1993). A previous report demonstrated that emodin is an inhibitor of PI3K in a cell-free kinase assay system; it has been identified with IC_{50} values for emodin 3.3 μM (Frew *et al.*, 1994). Considering that three isolated compounds have highly similar chemical structure. It has been predicted that the molecular target for chrysophanol, physcion, and emodin might be PI3K. Suppression of PI3K/AKT/IKK/NF-κB pathway may be the underlying mechanisms of chrysophanol, physcion and emodin mediated MMP-2, 9 expressions. Our results are basically consistent with previous reports on inhibition of the PI3K pathway and suppression of cell migration and invasion by other anthraquinones such as aloe-emodin and rhein (Fernand *et al.*, 2011). AP-1 (c-jun) is also considered as a major transcription factor that regulates MMP-2 and -9 expressions. It is well established that c-Jun phosphorylation is mediated by MAPK pathways. The result indicates that TNF-α markedly enhances phosphorylated c-jun in nucleus, and chrysophanol, physcion, and emodin significantly decrease phosphorylated c-jun in a concentration-dependent manner. It was also investigated that isolated compounds inhibited TNF-α-induced ERK and JNK phosphorylation in a concentration-dependent manner, while having no evident effect on phosphorylation of p38 after cells stimulated by 100 ng/mL TNF-α for 1 h. MAPK super-family (ERK1/2, JNKs, and p38 kinase) is known to regulate c-jun phosphorylation and then regulate MMP expression (Moon *et al.*, 2004). Among the main downstream effectors of PI3K, the PI3K/RAS/RAF/MEK/ERK/c-Jun and PI-3K/AKT/MEK/JNK/c-Jun signaling pathway correlate with c-jun (Ding *et al.*, 2009). Overall, according to the reference and the present results, it may be discussed as follows: these isolated compounds might be targeted to the PI3K, then subsequent phosphorylation of JNK and ERK and c-Jun also was influenced. And the inhibition of ERK and JNK phosphorylation by chrysophanol, physcion, and emodin is one of the main reasons involved in inhibitory effect on c-jun transactivation and the downregulation of MMP-2 and -9.

Western blot analysis results of TIMP-1 and -2 showed that TIMPs have been downregulated by chrysophanol, physcion, and emodin in a concentration-dependent manner. Meanwhile the upregulation of RECK by chrysophanol, physcion and emodin also suppressed MMP-2 and -9 expressions.

IV. CONCLUSIONS

Anthraquinones, chrysophanol, physcion, and emodin were obtained from the fungal strain, the *Microsporum* sp. Chrysophanol, physcion, and emodin treatments have reduced the HT1080 cell migration and invasion in dose-dependent manner, and it has also observed that in high concentration (50 μM), chrysophanol, physcion, and emodin can significantly reduce HT1080 cell migration completely and block cell invasion. In 3D culture environment, the HT1080 cell branches were reduced after treating with chrysophanol, physcion, and emodin, when the concentration increased to 50 μM, these branches disappeared and cells changed to spherical shape. Among the three compounds, emodin has shown a better inhibition at the same concentration.

The molecular signaling pathways of chrysophanol, physcion, and emodin suppress MMP-2 and -9 expressions, which could be discussed as follows: (1) chrysophanol, physcion, and emodin strongly inhibited AKT phosphorylation, and significantly inhibited both IκBα phosphorylation and p65 nuclear translocation in a concentration-dependent manner in HT1080 cells after treating with TNF-α. (2) Inhibition of ERK and JNK phosphorylation by chrysophanol, physcion, and emodin is one of the main mechanisms involved in its inhibitory effect on AP-1 transactivation and the downregulation of MMP-2 and -9 in HT1080 cells after treating with TNF-α. (3) Chrysophanol, physcion, and emodin downregulated TIMP-1 and -2 expressions, conversely, upregulated RECK expressions in HT1080 cells after treating with TNF-α.

In conclusion, chrysophanol, physcion, and emodin have anti metastasis potential and may apply as nutraceutical and therapeutic agent for cancer adjunctive therapy.

REFERENCES

Bugni, T. S. and Ireland, C. M. (2004). Marine-derived fungi: A chemically and biologically diverse group of microorganisms. *Nat. Prod. Rep.* **21**, 143–163.

Carter, G. T. and Rinehar, K. L. (1978). Aplidiasphingosine, an antimicrobial and antitumor terpenoid from an Aplidium sp. (marine Tunicate). *J. Am. Chem. Soc.* **100**, 7441–7442.

Croce, C. M. (2008). Oncogenes and cancer. *N. Engl. J. Med.* **358**, 502–511.

Datta, S. R., Brunet, A., and Greenberg, M. E. (1999). Cellular survival: A play in three Akts. *Genes Dev.* **13**, 2905–2927.

Deryugina, E. I. and Quigley, J. P. (2006). Matrix metalloproteinases and tumor metastasis. *Cancer Metastasis Rev.* **25**, 9–34.

Ding, J., Ning, B., Huang, Y., Zhang, D., Li, J., Chen, C. Y., and Huang, C. (2009). PI3K/Akt/JNK/c-Jun signaling pathway is a mediator for arsenite-induced cyclin D1 expression and cell growth in human bronchial epithelial cells. *Curr. Cancer Drug Targets* **9**, 500–509.

Dreyfuss, M. M. and Chapela, I. H. (1994). Potential of fungi in the discovery of novel, low-molecular weight pharmaceuticals. *In* "The Discovery of Natural Products with Therapeutic Potential", (V. P. Gullo, Ed.), pp. 49–80. Butterworth-Heinemann, Stoneham.

Fernand, V. E., Losso, J. N., Truax, R. E., Villar, E. E., Bwambok, D. K., Fakayode, S. O., Lowry, M., and Warner, I. M. (2011). Rhein inhibits angiogenesis and the viability of hormone-dependent and -independent cancer cells under normoxic or hypoxic conditions in vitro. *Chem. Biol. Interact.* **192**, 220–232.

Fingleton, B. (2006). Matrix metalloproteinases: Roles in cancer and metastasis. *Front. Biosci.* **11**, 479–491.

Frew, T., Powis, G., Berggren, M., Abraham, R. T., Ashendel, C. L., Zalkow, L. H., Hudson, C., Qazia, S., Gruszecka-Kowalik, E., and Merriman, R. (1994). A multiwell assay for inhibitors of phosphatidylinositol-3-kinase and the identification of natural product inhibitors. *Anticancer Res.* **14**, 2425–2428.

Henkel, T., Machleidt, T., Alkalay, I., Kronke, M., Ben, N. Y., and Baeuerle, P. A. (1993). Rapid proteolysis of I kappa B-alpha is necessary for activation of transcription factor NF-kappaB. *Nature* **365**, 182–185.

Huang, Q., Shen, H. M., and Ong, C. N. (2004). Inhibitory effect of emodin on tumor invasion through suppression of activator protein-1 and nuclear factor-κB. *Biochem. Pharmacol.* **68**, 361–371.

Li, H. Y., Matsunaga, S., and Fusetani, N. (1995). Halicylindramides A-C, antifungal and cytotoxic depsipeptides from the marine sponge *Halichondria cylindrata*. *J. Med. Chem.* **51**, 2273–2279.

Moon, S. K., Kim, H. M., and Kim, C. H. (2004). PTEN induces G1 cell cycle arrest and inhibits MMP-9 expression via the regulation of NF-kappaB and AP-1 in vascular smooth muscle cells. *Arch. Biochem. Biophys.* **421**, 267–276.

Natori, T., Morita, M., Akimoto, K., and Koezuka, Y. (1994). Agelasphins, novel antitumor and immunostimulatory cerebrosides from the marine sponge Agelas mauritianus. *Tetrahedron* **50**, 2771–2778.

Ozes, O. N., Mayo, L. D., Gustin, J. A., Pfeffer, S. R., Pfeffer, L. M., and Donner, D. B. (1999). NF-kappaB activation by tumour necrosis factor requires the AKT serine-threonine kinase. *Nature* **401**, 82–85.

Seki, M., Kayo, A., and Mori, K. (2001). ChemInform abstract: Synthesis of (2S,3R,11S,12R,2'''R,11'''S,12'''R)-plakoside A, a prenylated and immunosuppressive marine galactosphingolipid with cyclopropane-containing alkyl chains. *Tetrahedron Lett* **42**, 2357–2360.

Tanigawa, N., Amaya, H., Matsumura, M., and Shimomatsuya, T. (1997). Association of tumour vasculature with tumor progression and overall survival of patients with non-early gastric carcinomas. *Br. J. Cancer* **75**, 566–571.

Woodhouse, E. C., Chuaqui, R. F., and Liotta, L. A. (1997). General mechanisms of metastasis. *Cancer* **8**, 1529–1537.

Xia, X. K., Huang, H. R., She, Z. G., Shao, C. L., Liu, F., Cai, X. L., Vrijmoed, L. L. P., and Lin, Y. C. (2007). 1H and 13C NMR assignments for five anthraquinones from the mangrove endophytic fungus *Halorosellinia* sp. (No. 1403). *Magn. Reson. Chem.* **45**, 1006–1009.

Zhu, F., Chen, G. Y., Chen, X., Yuan, Y. H., Huang, M. Z., Xiang, W. Z., and Sun, H. L. (2008). Structural elucidation of three anthraquinones from a marine-derived mangrove endophytic fungus (isolate 1850). International Conference on BioMedical Engineering and Informatics, pp. 664–667.

CHAPTER 28

Application of Marine Microbial Enzymes in the Food and Pharmaceutical Industries

Chen Zhang[*,†,‡] and **Se-Kwon Kim**[†,§,1]

Contents

Abstract Over billions of years, the ocean is regarded as the origin of life on Earth, and the ocean includes the largest habitats hosting the most life forms. Competition among microorganisms for space and nutrients in the marine environment is a powerful selective force, which has led to the evolution. The evolution prompts the marine microorganisms to generate multifarious enzyme systems to adapt to the

* School of Medicine, Tongji University, Shanghai, China
† Department of Chemistry, Pukyong National University, Busan, Republic of Korea
‡ Key Laboratory of Molecular Enzymology and Engineering of the Ministry of Education, College of Life Science, Jilin University, Changchun, China
§ Marine Bioprocess Research Center, Pukyong National University, Busan, Republic of Korea
1 Corresponding author: Se-Kwon Kim, *E-mail address:* sknkim@pknu.ac.kr

Advances in Food and Nutrition Research, Volume 65
ISSN 1043-4526, DOI: 10.1016/B978-0-12-416003-3.00028-7

complicated marine environments. Therefore, marine microbial enzymes can offer novel biocatalysts with extraordinary properties. This review deals with the research and development work done on the occurrence and bioprocessing of marine microbial enzymes.

I. INTRODUCTION

Enzymes have been isolated and purified from microorganisms, animals, and plants, and among them, microorganisms represent the most common source of enzymes because of their broad biochemical diversity, feasibility of mass culture, and ease of genetic manipulation (Niehaus *et al.*, 1999). The world's oceans cover more than 70% of our planet's surface; countless marine microorganisms contain biochemical secrets which can provide new insights and understanding of enzymes. Marine microorganisms have been attracting more and more attention as a resource for new enzymes because the microbial enzymes are relatively more stable than the corresponding enzymes derived from plants and animals (Kim, 2006). With the recent advent of biotechnology, there has been a growing interest and demand for enzymes with novel properties. When compared with the terrestrial environment, the marine environment gives marine microorganisms unique genetic structures and life habitats (Stach *et al.*, 2003). The marine environment ranges from nutrient-rich regions to nutritionally sparse locations where only a few organisms can survive; complexity of the marine environment involving high salinity, high pressure, low temperature, and special lighting conditions might make the enzymes generated by marine microorganisms significantly different from homologous enzymes generated by terrestrial microorganisms, and boosted marine microbial enzyme technology in recent years, offering valuable products. These enzymes are used as pharmaceuticals, food additives, and fine chemicals (David and Michel, 1999). In recent years, researchers have isolated a variety of enzymes with special activities from marine bacteria, actinomycetes, fungi, and other marine microorganisms, and some products have been used in industrial applications. Especially, some marine microbial enzymes have yielded a considerable number of drug candidates. Marine microorganisms, whose immense genetic and biochemical diversity is still in its infant stage, are of considerable current interest as a new promising source of enzymes with unsuspected application potentials (Ghosh *et al.*, 2005).

II. LIPASE

Lipases are ubiquitous enzymes that catalyze the breakdown of fats and oils with subsequent release of free fatty acids, diacylglycerols, monoglycerols, and glycerol. Besides this, lipases are also efficient in various reactions such as esterification, transesterification, and aminolysis (Babu *et al.*, 2008). Lipases have received increased attention recently, as evidenced by the increasing amount of information about lipases in the current literature. Also, many microbial lipases are available as commercial products, the majority of which are used in detergents, paper production, cosmetic production, food flavoring, organic synthesis, and some other industrial applications (Chi *et al.*, 2009). The enzyme detergent market share has reached 90% in current European countries, and in Japan it is around 80%. Lipases are valuable biocatalysts because they act under mild conditions and are highly stable in organic solvents, show broad substrate specificity, and usually show high regio- and/or stereoselectivity in catalysis (Amare *et al.*, 2003; Davidson, 2006).

In recent years, with marine resources exploiting, the pelagic fishes have become the primary target of fishery; these species are resourceful and have huge development potentials. However, the fat contents are high in these species; people have to face certain difficulties for fish preservation, processing, and marketing. Currently, the traditional methods for fish degreasing include expression method, extraction method, and alkaline processing, while compared with these traditional methods, using lipases has incomparable advantages, such as specific, safe, nontoxic, nonpollution, and efficient. Therefore, applications of lipases in the fish processing field are causing growing concern (Kojima and Shimizu, 2003; Seitz, 1974).

Microbial lipase was first found from *Penicillium oxalicum* in 1935 by Kirsh (David, 1935). The United States and Japan reported that the microorganisms isolated from the cold seawater can produce lipases with low temperature-resistant properties. Feller *et al.* (1990) screened four cold-resistant lipases secreted by *Moraxella* from the antarctic seawater. The optimum growth temperature for these *Moraxella* is 25°C, and the maximum secretion of lipases is supposed to happen at lower temperature condition; even the lowest secretion temperature can reach 3°C (Feller *et al.*, 1990). Wang *et al.* (2007) found nine yeast strains could produce lipase in total 427 yeast strains from seawater, sediments, mud of salterns, guts of the marine fish, and marine algae.. They belonged to *Candida intermedia* YA01a, *Pichia guilliermondii* N12c, *Candida parapsilosis* 3eA2, *Lodderomyces elongisporus* YF12c, *Candida quercitrusa* JHSb, *Candida rugosa* wl8, *Yarrowia lipolytica* N9a, *Rhodotorula mucilaginosa* L10-2, and *Aureobasidium pullulans* HN2.3, respectively. The optimal pH and temperatures of lipases produced by them were between 6.0°C and 8.5°C and between 35°C and 40°C, respectively. Some lipases from the yeast strains could actively hydrolyze different oils, indicating that they may have potential applications in industry (2007). In 2009, a novel extracellular phospholipase C (PLC) was purified from a marine (Mo *et al.*, 2009). It had a molecular mass of 28 kDa, and its enzyme activity was optimal at pH 8.0 at 45°C. The PLC hydrolyzed only phosphatidylcholine.

III. PROTEASE

Protease sales represent more than 60% of all industrial enzyme sales in the world. In modern society, the proteases are widely used in detergent industry, leather industry, and also could be used for pharmaceutical applications, such as digestive drugs, anti-inflammatory drugs (Ilona and Zdzislaw, 2007; Rajesh *et al.*, 2005; Zhang *et al.*, 2005). In 1960, Dane first purified alkaline protease from *Bacillus licheniformis*. It is believed that microorganisms are the most suitable resources for industrial protease production. In 1972, Nobuo Kato purified a new type of alkaline protease from marine Psychrobacter. Quite a few proteases continued to be obtained from marine microorganisms up till the present moment. An alkaline protease, previously isolated from a symbiotic bacterium found in the gland of deshayes of marine shipworm, was evaluated as a cleansing additive (Greene *et al.*, 1996) from the sea water, mud, and fish. Other samples selected 30 kinds of marine bacteria, and after UV mutagenesis, finally they got N1-35 strain. They found that the marine bacteria N1-35 strain produced protease activity compared with the terrestrial ones have significant advantages. In addition, extract of an extracellular collagenase from the *Vibrio alginolyticus* also showed a certain value in the industry (Greene *et al.*, 1996). A yeast strain (*A. pullulans*), which could produce the high yield of alkaline protease, was isolated from sediment of sea saltern of the Yellow sea, China, by Chi *et al.* (2007). Maximum production of enzyme was 623.1U/mg protein (7.2U/ml). The protease has the highest activity at pH 9.0 and 45°C (Chi *et al.*, 2007). *Bacillus mojavensis* A21 producing alkaline proteases was isolated from seawater by Haddar *et al.* (2009), and they purified two detergent stable alkaline serine-proteases (BM1 and BM2) from this strain. Both proteases showed high stability toward nonionic surfactants. In addition, both of them showed excellent stability and compatibility with a wide range of commercial liquid and solid detergents.

IV. POLYSACCHARIDE-DEGRADING ENZYMES

A. Chitinase and chitosanase

Chitin is widely distributed in nature as a biopolymer with nontoxic properties. After cellulose, it is the most common polysaccharide found in the nature and is the major structural component of most fungal cell wall and also quite abundant in the crust of insects and crustaceans. In nature, annual generation of chitin is about 1.0×10^{10} tons (Marguerite, 2006). Chitin and chitosan have the similar chemical structure. Chitin is made up of a linear chain of acetylglucosamine groups, and chitosan is obtained by removing enough acetylglucosamine residues. After hydrolysis, chitin and chitosan could enhance immune function, promote digestive function, and eliminate toxins from the body, even inhibit tumor cell growth, and involve in other important physiological functions (Ngo *et al.*, 2008; Je and Kim, 2006). Therefore, hydrolysis of chitin and chitosan became a hot topic recently.

As the marine zooplankton is regularly supposed to shed, there is a large number of abandoned chitin, which could be a rich source of carbon and energy for growth and reproduction of chitin-degrading microorganisms. The total production of chitin in the whole marine biocycle is at least of 2.3 million metric tons per year (Charles and Marie, 1991). Till now, researchers have found a wide range of microorganisms, which can produce chitinase or chitosanase, including *Aspergillus, Penicillium, Rhizopus, Myxobacter, Sporocytophaga, Bacillus, Enterobacter, Klebsiella, Pseudomonas, Serratia, Chromobacterium, Clostridium, Flavobacterium, Arthrobacter, Streptomyces,* and so forth. In these microorganisms, *Streptomyces plicatus, Vibrio vulnificus,* and *Beauveria bassiana* have been extensively studied (Fukamizo, 2000; Xia *et al.*, 2008). Osawa and Koga (1995) found chitinase from six species of marine bacteria *Vibrio fluvialis, Vibrio parahaemolyticus, Vibrio mimicus, V. alginolyticus, Listonella anguillarum,* and *Aeromonas hydrophila.* In addition, a variety of chitinase genes were already cloned from marine bacteria and fungi. Suolow and Jones (1988) insert two chitinase genes (*ChiA, ChiB*) into *Escherichia coli,* and subsequently these genes were transferred into Pseudomonas; finally they acquired four high-yielding chitinase strains (Suolow, 1988). Meanwhile, Roberts and Cabib (1982) have put the chitinase gene into the tobacco plant cell to develop a new tobacco plant with a strong disease resistance to the pathogen *Alternaria longipes.*

B. Alginate lyases

The brown alga is one of the largest marine biomass resources. Alginate has a wide range of applications; further, the degraded low-molecular fragment show more potential. Alginate lyases catalyze the degradation of alginate, a complex copolymer of α-L-guluronate and its C5 epimer β-D-mannuronate. They have been isolated from a wide range of organisms, including algae, marine invertebrates, and marine and terrestrial microorganisms. In recent years, the marine microorganisms' original alginate lyases have been greatly developed. Discovering and characterization of alginate lyases will enhance and expand the use of these enzymes to engineer novel alginate polymers for applications in various industrial, agricultural, and medical fields (Wong *et al.*, 2000; Xiao *et al.*, 2006).

C. Agarases

Agar is a highly heterogeneous polysaccharide. Neutral agarose is an alternating polymer of D-galactose and 3,6-anhydro-L-galactose linked by alternating β1→4 and α1→3 bonds. Agar oligosaccharides have a wide range of applications in food industry; it can be used for beverages, bread, and some low-calorie food production. Japanese use agar-oligosaccharide as a moisturizing cosmetic additive and also has good hair conditioning effects (Oren, 2004; Rasmussen and Morrissey, 2007).

Nowadays, the acid degradation of agar is being replaced by enzymatic degradation with the advantages of easy control and mild reaction. Agarase is an enzyme found in agarolytic microorganisms (Parro and Mellado, 1994), and agar-degrading microorganisms can be divided into two groups: bacteria soften the agar and the other violently liquefy the agar. In 1902, Gran isolated agar-degrading *Pseudomonas galactica* from seawater. Till now, from species within the genus *Cytophaga, Bacillus, Vibrio, Alteromonas, Pseudoalteromonas, Streptomyces,* researchers have found the presence of agarase (Hosoda *et al.*, 2003). Sugano *et al.* (1994) reported marine bacteria *Vibrio* sp. (JT0107) secrete α-Neoagaro-oligosaccharide hydrolase, which can hydrolyze the α-1,3 glycosidic bond of agar to generate Neoagaro-pentose, Neoagaro-trisaccharide, Neoagaro-disaccharide, D-galactose, etc. (Susgano *et al.*, 2002). Several agarase genes have been cloned and sequenced. In 1987, Mervyn found Streptomyces agarase gene (*dagA*), and soon after, Rosert performed gene sequence analysis of Pseudomonas agarase gene (*agrA*) (Buttner *et al.*, 1987; Belas, 1989). *Vibrio agarase* gene agaA was cloned and sequenced by Sugano *et al.* (1993). This agarase hydrolyzes not only agarose but also neoagarotetraose to yield neoagarobiose. This is a unique characteristic for a β-agarase. In 1994, from the same species of bacteria, a new β-agarase gene (*agaB*) was sequenced by the same group (Sugano *et al.*, 1994). A new β-agarase was purified from an agarolytic bacterium, *Bacillus* sp. MK03 by Suzuki *et al.* (2003); this enzyme could hydrolyze neoagarohexaose to produce neoagarotetraose and neoagarobiose. Researches also demonstrate that glucose can inhibit extracellular agarase secretion without transcription inhibition, causing it to degrade within the cell and thus limit growth of the bacteria.

D. Carrageenases

Carrageenan and carrageenin are a family of linear sulphated polysaccharides, which are extracted from red seaweeds. Eighty percent of the carrageenan is used in food and food-related industry, and it can be used as coagulant, adhesives, stabilizer, and emulsifier. In addition, it has also been widely applied in pharmaceutical and cosmetics industries.

The oligosaccharides obtained from carrageenan degradation showed a variety of specific physiological activities, such as antivirus, antitumor, anticoagulation, and so on (Roberts *et al.*, 2007). As early as 1943, Mori extracted carrageenase from the marine mollusc. Right now, *Pseudomonas, Cytophaga, Alteromonas atlantica, Alteromonas carrageenovora*, and some unidentified strains have been found to possess the carrageenan-degrading enzymes. Sarwar, *et al.* (1987), using carrageenan-containing medium, cultured cytophaga lk-C783 and obtained extracellular κ-carrageenase with a molecular weight of 10 kD. Mou *et al.* (2004) isolated an extracellular κ-carrageenase with a molecular weight of 30 kD from marine Cytophaga MCA-2. A distinct λ-carrageenan-degrading *Pseudoalteromonas* bacterium (strain CL19) was isolated from a deep-sea sediment sample by Ohta and Hatada (2006); the molecular mass of this purified enzyme was approximately 100 kD.

E. Cellulose and hemicellulose hydrolase

Cellulose is an organic compound with the formula $(C_6H_{10}O_5)_n$, a polysaccharide consisting of a linear chain of several hundred to over ten thousand β (1→4)-linked D-glucose units. Hemicellulose is a polysaccharide related to cellulose, and in contrast to cellulose, it can be derived from several sugars including glucose, xylose, mannose, galactose, rhamnose, and arabinose. Hemicellulose consists of shorter chains of around 200 sugar units. Cellulose is the maximum available saccharide in nature and is about 50% of all plant matter, and hemicellulose is about 20–30%, while the remainder is mainly lignin (Nishiyama *et al.*, 2002). Cellulolysis is the process of breaking down cellulose through hydrolysis reaction into smaller polysaccharides called cellodextrin. Because cellulose molecules strongly bind to each other, cellulolysis is relatively difficult when compared to the breakdown of other polysaccharides (Brás *et al.*, 2008). Till now, it was found that bacteria can produce cellulase, including *Cytophaga, Cellulomonas, Vibrio, Clostridium, Nocardia, Streptomyces*, and for certain fungi. *Trichoderma, Aspergillus, Fusarium, Chaetomium, Phoma, Sporotrichum, Penicillium*, etc., are also able to produce cellulase. Hemicellulase generally refers to the hydrolase, which can hydrolyze polysaccharides, for example, xylanase, galactanase, arabanase, among which xylanase has particular economic value (Brás *et al.*, 2008; Maki *et al.*, 2009).

Cellulase can be used for biotextile auxiliaries, cotton and linen products processing, and biofertilizer processing. With the rapid development of the seaweed industry, a mass of waste release into the environment leads to very serious pollution problems. Cellulases degrade seaweed processing waste to low molecular fragments, which can be easily absorbed by plants as biofertilizer.

Xylanases are hydrolases depolymerizing the plant cell component xylan, the second most abundant polysaccharide. Xylanases could be produced by fungi, bacteria, yeast, marine algae, etc., but the principal commercial source is filamentous fungi. Xylanase could be used on semi-cellulose to produce products with high economic value, such as xylitol. In paper and pulp industry, using xylanase can improve the lignin dissolution rate, reduce the usage of Cl_2 and ClO_2, and thereby reduce pollution and improve pulp properties. Xylanase also can degrade some polysaccharides in juice or beer; thus it could contribute to beverage clarification (Doi, 2008; Nishiyama *et al.*, 2002). Indian researchers obtained several fungal isolates from marine habitat that showed alkaline xylanase activity. It has been reported that the crude enzyme from NIOCC isolate # 3 (Aspergillus niger) with high xylanase activity. (Raghukumar *et al.*, 2004). Yin *et al.* (2010) purified xylanase, which had an optimal pH and temperature at 5.0 °C and 50 °C from bacterium *Bacillus* sp. YJ6. A novel cold-adaptive xylanolytic *Penicillium* strain FS010 was isolated from China Yellow sea sediments by Hou *et al.*(2006); this fungus grew well from 4 °C to 20 °C, but a lower (0 °C) or higher (37 °C) temperature limits its growth. Compared with mesophilic Penicillium chrysogenum, the cold-active xylanase shows high hydrolytic activities at low temperature (2–15 °C) and high sensitivity to high temperature (>50 °C).

F. Other polysaccharide hydrolases

Amylases were found in bread making, and they can break down complex sugars such as starch into simple sugars such as glucose, maltose, and dextrin. They can be classified into α-amylase, β-amylase, and γ-amylase. Unlike the other forms of amylase, γ-amylase is most efficient in acidic environments. To date, researchers have found some terrestrial microorganisms which can produce extracellular amylase, such as *Arxula adeninivorans, Lipomyces, Saccharomycopsis, Schwanniomyces, Candida japonica,* and *Filobasidium capsuligenum* (Gupta *et al.*, 2003). With the development of marine science and technology, researchers reported more and more microorganisms from marine habitats capable of producing amylase. The marine yeast strain *A. pullulans* N13d, producing an extracellular amylase, was isolated from the deep sea sediments of Pacific Ocean (Li *et al.*, 2007). Chakraborty *et al.* (2009) reported a novel α-amylase from marine Streptomyces sp. D1 by using media containing 2% sucrose, 0.35% peptone, and 0.15% of malt extract. Mohapatra *et al.* (1998) isolated a novel amylase from the *Mucor* sp. associated with the marine sponge *Spirastrella* sp.; this enzyme has an optimum pH of 5.0 and an optimum temperature of 60 °C.

Fucoidan is a complex sulfated polysaccharide, constituted by fucose, galactose, xylose, mannose, arabinose, and uronic acid. The fucoidan oligosaccharide, with molecular weight <1000 Da, can be used as human epidermal keratinocytes activator. Furukawa *et al.* (1992) purified fucoidanase, generated from marine *Vibrio*, and obtained three kinds enzymes, which can hydrolyze substrate to small molecule oligosaccharides. Yaphe and Morgan (1959) reported that marine bacteria *Pseudomonas atlantica* and *Pseudoalteromonas carrageenovora* were cultured in medium for 3 days by using fucoidan as the sole carbon source, with the substrate utilization as 31.5% and 29.9%, respectively.

Japanese researcher isolated *Bacillus circulans* from Tokyo Bay sea mud, which do not grow in conventional culture medium, but when the medium was properly diluted, it may grow and produce a new glucanase. This glucanase can act on α-1,3 bond and α-1,6 bond in glucan. From the marine *Bacillus*, a new glucanase has been isolated, which showed the optimum activity at 37 °C, and this property is suitable for oral and other health care (Yahata *et al.*, 1990).

Amki (1981) isolated 117 marine bacteria, belonging to *Pseudomonas, Alcaligenes, Klebsiella, Enterobacter, Vibrio, Aeromonas, Moraxella, Bacillus,* etc., which can generate β-mannanase.

V. CONCLUSION

The twenty-first century is the century of the ocean, and the ocean is a vast treasure of human life. Recently, most countries face the similar problems such as high population, resource consumption, and pollution. Meanwhile, the marine biological progress and development would give another option to human. Marine microbial enzymes, especially marine extreme microbial enzyme applications, have become more and more important.

Because enzymes have unequalled advantages, many industries are keenly interested in adapting enzymatic methods to the requirements of their processes. Clinical application of enzymes has been developing also. For example, surgeons used proteolytic enzymes for debridement of wounds, and promising clinical results have been reported by injection of certain enzymes such as streptokinase, crystalline trypsin, and chymotrypsin. Since the increased therapeutic use of enzymes, presently unpredictable, rapid advances in this field may be expected (Underkofler *et al.*, 1958; Zhang and Kim, 2010). Uses of enzymes have increased greatly during the past few years. Prospects are excellent for continuing increased usage of presently available enzymes in present applications, and in new uses, and of new enzymes for many purposes. And also some new field, like metagenomics, offers a powerful lens for viewing the microbial world that may have the potential to revolutionize understanding of the marine microbial enzymes. Japan constantly increases its support to marine microbial enzymes research, and planned to discover and clone proteins or enzymes with some special activity. In addition, Canada, Spain, Finland, Russia, and other countries have also stepped up on marine bioenzyme research.

Collectively, due to marine biological diversity and specificity of biological metabolism, the study on a global scale is still just beginning, but it has a huge potential for development and applications with industrial benefits.

ACKNOWLEDGMENTS

This study was supported by a grant from Marine Bioprocess Research Center of the Marine Biotechnology program funded by the Ministry of Land, Transport and Maritime, Republic of Korea.

REFERENCES

Amare, G., Thomas, D., Steffen, B. P., and Per, H. N. (2003). Lipase and protease extraction from activated sludge. *Water Res.* **37**, 3652–3657.

Amki, T. (1981). β-mannanse of bacteria isolated from natural habitats. *Bull. Jpn. Soc. Sci. Fish.* **47**, 753–760.

Babu, J., Pramod, W. R., and George, T. (2008). Cold active microbial lipases: Some hot issues and recent developments. *Biotechnol. Adv.* **26**, 457–470.

Belas, R. (1989). Sequence analysis of the agrA gene encoding beta-agarase from *Pseudomonas atlantica*. *J. Bacteriol.* **171**, 602–605.

Brás, N. F., Cerqueira, N. M. F. S. A., Fernandes, P. A., and Ramos, M. J. (2008). Carbohydrate binding modules from family 11: Understanding the binding mode of polysaccharides. *Int. J. Quantum Chem.* **108**, 2030–2040.

Buttner, M. J., Fearnleiy, I. M., and Bibb, M. J. (1987). The agarase gene (dagA) of Streptomyces coelicolor A3(2): Nucleotide sequence and transcriptional analysis. *Mol. Genet. Genomics* **209**, 101–109.

Chakraborty, S., Khopade, A., Kokare, C., Mahadik, K., and Chopade, B. (2009). Isolation and characterization of novel α-amylase from marine *Streptomyces* sp. D1. *J. Mol. Catal. B. Enzym.* **58**, 17–23.

Charles, J. and Marie, F. V. (1991). Chitin biomass and production in the marine environment. *Biochem. Syst. Ecol.* **19**, 347–356.

Chi, Z., Chi, Z., Zhang, T., Liu, G., Li, J., and Wang, X. (2009). Production, characterization and gene cloning of the extracellular enzymes from the marine-derived yeasts and their potential applications. *Biotechnol. Adv.* **27**, 236–255.

Chi, Z. M., Ma, C., Wang, P., and Li, H. F. (2007). Optimization of medium and cultivation conditions for alkaline protease production by the marine yeast *Aureobasidium pullulans*. *Bioresour. Technol.* **98**, 534–538.

David, K. (1935). Lipase production by *Penicillium Oxalicum* and *Aspergillus flavus*. *Bot. Gaz.* **97**, 321.

David, W. and Michel, J. (1999). Extremozymes. *Curr. Opin. Chem. Biol.* **3**, 39–46.

Davidson, M. H. (2006). Mechanisms for the hypotriglyceridemic effect of marine omega-3 fatty acids. *Am. J. Cardiol.* **98**, 27–33.

Doi, R. H. (2008). Cellulases of mesophilic microorganisms: Cellulosome and noncellulosome producers. *Ann. N. Y. Acad. Sci.* **1125**, 267–279.

Feller, G., Thiry, M., Arpigy, J. L., Mergeay, M., and Gerday, C. (1990). Lipases from psychrotrophic antarctic bacteria. *FEMS Microbiol. Lett.* **66**, 239–244.

Fukamizo, T. (2000). Chitinolytic enzymes: Catalysis, substrate binding, and their application. *Curr. Protein Pept. Sci.* **1**, 105–124.

Furukawa, S. I., Fujikawa, T., and Koga, D. (1992). Purification and some properties of exo-type fucoidanases from vibrio sp. N-5. *Biosci. Biotechnol. Biochem.* **56**, 1829–1834.

Ghosh, D., Saha, M., Sana, B., and Mukherjee, J. (2005). Marine enzymes. *Adv. Biochem. Eng. Biotechnol.* **96**, 189–218.

Greene, R. V., Grifin, H. L., and Cotta1, M. A. (1996). Utility of alkaline protease from marine shipworm bacterium in industrial cleansing applications. *Biotechnol. Lett.* **18**, 759–764.

Gupta, R., Gigras, P., Mohapatra, H., Goswami, V. K., and Chauhan, B. (2003). Microbial α-amylases: A biotechnological perspective. *Process Biochem.* **38**, 1599–1616.

Haddar, A., Agrebi, R., Bougatef, A., Hmidet, N., Sellami-Kamoun, A., and Nasri, M. (2009). Two detergent stable alkaline serine-proteases from *Bacillus mojavensis* A21: Purification, characterization and potential application as a laundry detergent additive. *Bioresour. Technol.* **100**, 3366–3373.

Hosoda, A., Sakai, M., and Kanazawa, S. (2003). Isolation and characterization of Agar-degrading *Paenibacillus* spp. associated with the rhizosphere of spinach. *Biosci. Biotechnol. Biochem.* **67**, 1048–1055.

Hou, Y. H., Wang, T. H., Long, H., and Zhu, H. Y. (2006). Novel cold-adaptive *Penicillium* strain FS010 secreting thermo-labile xylanase isolated from Yellow Sea. *Acta Biochim. Biophys. Sin.* **38**, 142–149.

Ilona, K. and Zdzislaw, E. S. (2007). Neutral and alkaline muscle proteases of marine fish and invertebrates a review. *J. Food Biochem.* **20**, 349–364.

Je, J. Y. and Kim, S. K. (2006). Antioxidant activity of novel chitin derivative. *Bioorg. Med. Chem. Lett.* **2006**(16), 1884–1887.

Kim, S. L. (2006). Discovery of novel metabolites from marine actinomycetes. *Curr. Opin. Microbiol.* **9**, 245–251.

Kojima, Y. and Shimizu, S. (2003). Purification and characterization of the lipase from *Pseudomonas fluorescens* HU380. *J. Biosci. Bioeng.* **96**, 219–226.

Li, H. F., Chi, Z. M., Wang, X. H., and Ma, C. L. (2007). Amylase production by the marine yeast *Aureobasidium pullulans* N13d. *J. Ocean Univ. Chin.* **6**, 61–66.

Maki, M., Leung, K. T., and Qin, W. (2009). The prospects of cellulase-producing bacteria for the bioconversion of lignocellulosic biomass. *Int. J. Biol. Sci.* **5**, 500–516.

Marguerite, R. (2006). Chitin and chitosan: Properties and applications. *Prog. Polym. Sci.* **2006** (31), 603–632.

Mo, S. J., Kim, J. H., and Cho, K. W. (2009). Enzymatic properties of an extracellular phospholipase C purified from a marine streptomycete. *Biosci. Biotechnol. Biochem.* **73**, 2136–2137.

Mohapatra, B. R., Banerjee, U. C., and Bapuji, M. (1998). Characterization of a fungal amylase from Mucor sp. associated with the marine sponge *Spirastrella* sp. *J. Biotechnol.* **60**, 113–117.

Mou, H. J., Jiang, X. L., Liu, Z. H., and Guan, H. S. (2004). Structural analysis of kappa-carrageenan oligosaccharides released by carrageenase from marine Cytophaga MCA-2. *J. Food Biochem.* **28**, 245–260.

Ngo, D. N., Kim, M. M., and Kim, S. K. (2008). Chitin oligosaccharides inhibit oxidative stress in live cells. *Carbohydr. Polym.* **74**, 228–234.

Niehaus, F., Bertoldo, C., Kähler, M., and Antranikian, G. (1999). Extremophiles as a source of novel enzymes for industrial application. *Appl. Microbiol. Biotechnol.* **51**, 711–729.
Nishiyama, Y., Langan, P., and Chanzy, H. (2002). Crystal structure and hydrogen-bonding system in cellulose Iβ from synchrotron X-ray and neutron fiber diffraction. *J. Am. Chem. Soc.* **124**, 9074–9082.
Ohta, Y. and Hatada, Y. (2006). A novel enzyme, lambda-carrageenase, isolated from a deep-sea bacterium. *J Biochem.* **140**, 475–481.
Oren, A. (2004). Prokaryote diversity and taxonomy: Current status and future challenges. *Philos. Trans. R. Soc. Lond. B Biol. Sci.* **359**, 623–638.
Osawa, R. and Koga, T. (1995). An investigation of aquatic bacteria capable of utilizing chitin as the sole source of nutrients. *Lett. Appl. Microbiol.* **21**, 288–291.
Parro, V. and Mellado, R. P. (1994). Effect of glucose on agarase overproduction in Streptomyces. *Gene* **145**, 49–55.
Raghukumar, C., Muraleedharan, U., Gaud, V. R., and Mishra, R. (2004). Xylanases of marine fungi of potential use for biobleaching of paper pulp. *J. Ind. Microbiol. Biotechnol.* **31**, 433–441.
Rajesh, P., Mital, D., and Satya, P. S. (2005). Extracellular alkaline protease from a newly isolated haloalkaliphilic *Bacillus* sp.: Production and optimization. *Process Biochem.* **40**, 3569–3575.
Rasmussen, R. S. and Morrissey, M. T. (2007). Marine biotechnology for production of food ingredients. *Adv. Food Nutr. Res.* **52**, 237–292.
Roberts, J. N., Christopher, B. B., Cynthia, D. T., Rhonda, K., Marcelino, B., Peter, L. C., Douglas, R. L., and John, T. S. (2007). Genital transmission of HPV in a mouse model is potentiated by nonoxynol-9 and inhibited by carrageenan. *Nat. Med.* **13**, 857–861.
Roberts, R. L. and Cabib, E. (1982). Serratia marcescens chitinase: One-step purification and use for the determination of chitin. *Anal. Biochem.* **127**, 402–412.
Sarwar, G., Matoyoshi, S., and Oda, H. (1987). Purification of a α-carrageenan from marine Cytophaga species. *Microbiol. Immunol.* **31**, 869–877.
Seitz, E. W. (1974). Industrial application of microbial lipases: A review. *J. Am. Oil Chem. Soc.* **51**, 12–16.
Stach, J. E. M., Maldonado, L. A., Ward, A. C., Goodfellow, M., and Bull, A. T. (2003). New primers for the class Actinobacteria: Application to marine and terrestrial environments. *Environ. Microbiol.* **5**, 828–841.
Sugano, Y., Matsumoto, T., and Noma, M. (1994). Sequence analysis of the agaB gene encoding a new β-agarase from *Vibrio* sp. strain JT0107. *Biochim. Biophys. Acta* **1218**, 105–108.
Suolow, T.V., and Jones, J. (1988). Chitinase-producing bacteria. US Patent No. 4751081.
Sugano, Y., Matsumoto, T., Kodama, H., and Noma, M. (1993). Cloning and sequencing of agaA, a unique agarase 0107 gene from a marine bacterium, *Vibrio* sp. strain JT0107. *Appl. Environ. Microbiol.*, **59**, 3750–3756.
Susgano, S., *et al.* (2002). US Patent, Dayhoff Accession No. BAB85019.
Suzuki, H., Sawai, Y., Suzuki, T., and Kawai, K. (2003). Purification and characterization of an extracellular β-agarase from *Bacillus* sp. MK03. *J. Biosci. Bioeng.* **95**, 328–334.
Underkofler, L. A., Barton, R. R., and Rennert, S. S. (1958). Production of microbial enzymes and their applications. *Appl. Microbiol.* **6**, 212–221.
Wang, L., Chi, Z. M., Wang, X. H., Liu, Z. Q., and Li, J. (2007). Diversity of lipase-producing yeasts from marine environments and oil hydrolysis by their crude enzymes. *Ann. Microbiol.* **57**, 495–501.
Wong, T. Y., Preston, L. A., and Schiller, N. L. (2000). Alginate lyase: Review of major sources and enzyme characteristics, structure-function analysis, biological roles, and applications. *Annu. Rev. Microbiol.* **54**, 289–340.

Xia, W., Liu, P., and Liu, J. (2008). Advance in chitosan hydrolysis by non-specific cellulases. *Bioresour. Technol.* **15**, 6751–6762.

Xiao, L., Han, F., Yang, Z., Lu, X. Z., and Yu, W. G. (2006). A novel alginate lyase with high activity on acetylated alginate of Pseudomonas aeruginosa FRD1 from *Pseudomonas* sp. QD03. *World J. Microbiol. Biotechnol.* **22**, 81–88.

Yahata, N., Watanabe, T., Nakamura, Y., Yamamoto, Y., Kamimiya, S., and Tanaka, H. (1990). Structure of the gene encoding beta-1,3-glucanase A1 of *Bacillus circulans* WL-12. *Gene* **86**, 113–117.

Yaphe, W. and Morgan, K. (1959). Enzymic hydrolysis of fucoidin by Pseudomonas atlantica and Pseudomonas carrageenovora. *Nature* **183**, 761–762.

Yin, L. J., Lin, H. H., Chiang, Y. I., and Jiang, S. T. (2010). Bioproperties and purification of xylanase from *Bacillus* sp. YJ6. *J. Agric. Food Chem.* **58**, 557–562.

Zhang, C. and Kim, S. K. (2010). Research and application of marine microbial enzymes: Status and prospects. *Mar. Drugs* **8**, 1920–1934.

Zhang, L. X., An, R., Wang, J. P., Sun, N., Zhang, S., Hu, J. C., and Kuai, J. (2005). Exploring novel bioactive compounds from marine microbes. *Curr. Opin. Microbiol.* **8**, 276–281.

CHAPTER 29

Enzymatic Properties of Starfish Phospholipase A2 and Its Application

Hideki Kishimura[1]

Contents

Abstract

Industrial phospholipase A2 (PLA2) mainly produced from porcine pancreas is used for production of lysolecithin which is well known as an excellent natural emulsifier for food, cosmetic, and pharmaceutical industries. Since the outbreak of bovine spongiform encephalopathy (BSE) or religious tradition, it is hoped that the new sources of PLA2, as well as other enzymes and proteins, will be developed instead of mammal. From these backgrounds, we studied for PLA2 from marine organisms and found that starfish *Asterina pectinifera* PLA2 possesses extremely high activity and characteristic polar-group specificity comparing with commercially available PLA2 from porcine pancreas. Therefore, it was suggested that the starfish *A. pectinifera* would be a potential source of PLA2, and the PLA2 can be utilized as alternative enzyme of mammalian PLA2.

Research Faculty of Fisheries Sciences, Hokkaido University, Minato, Hakodate, Hokkaido, Japan
[1] Corresponding author: Hideki Kishimura, *E-mail address:* kishi@fish.hokudai.ac.jp

Advances in Food and Nutrition Research, Volume 65
ISSN 1043-4526, DOI: 10.1016/B978-0-12-416003-3.00029-9

I. INTRODUCTION

Phospholipase A2 (PLA2) (EC3.1.1.4) catalyzes the selective hydrolysis of the *sn*-2-acyl group in 1,2-diacyl-*sn*-glycero-3-phospholipids (Fig. 29.1). Nikuni (1932) and Belfanti and Arnaudi (1932) independently reported the existence of PLA2 in pancreatic juice (Table 29.1). Subsequently, PLA2 obtained from various mammalian organs and from snake and bee venoms. The superfamily of PLA2 currently consists of 15 groups and many subgroups (Burke and Dennis, 2009). It is reported that this group included five distinct types of enzymes, namely, secreted PLA2s (sPLA2), cytosolic PLA2s (cPLA2), Ca^{2+}-independent PLA2s (iPLA2), platelet-activating factor acetylhydrolases (PAF-AH), and lysosomal PLA2s (lpPLA2). Among them, the sPLA2, which contains the traditional Group-I, -II, and -III PLA2s and is abundant in mammalian pancreas and snake or bee venoms, has been well characterized for functional and structural characteristics (Schaloske and Dennis, 2006; Six and Dennis, 2000). For application, the sPLA2 from mammalian pancreas and snake venom has been tried to use as diagnostic biochemical reagents. Stoll (1996) employed snake venom PLA2 to analyze the positional distribution of fatty acids in glycerophospholipids from guinea pig and pig cardiac membranes. Mine (1997) examined structural and functional changes of hen egg-yolk low density lipoproteins (LDL) as a result of modifying its phospholipids using porcine pancreatic PLA2. It was shown that LDL modified by PLA2 was removed from the circulation to liver more rapidly than unmodified LDL (Labeque *et al.*, 1993). Based on these findings, a novel therapy for hypercholesterolemia has been developed that utilizes immobilized PLA2 contained in an extracorporeal shunt. In addition, commercial PLA2 mainly produced from porcine pancreas is used for production of lysolecithin (Blasi *et al.*, 2006; Penci *et al.*, 2010). Lysolecithin, which is an excellent emulsifier, is prepared by immobilized porcine PLA2-catalyzed hydrolysis of lecithin (De Maria *et al.*, 2007; Kim *et al.*, 2001). In general, the emulsion with lysolecithin is more stable than that with lecithin under various conditions, such as high temperature, acidic solution, and high salt concentration. Lysolecithin

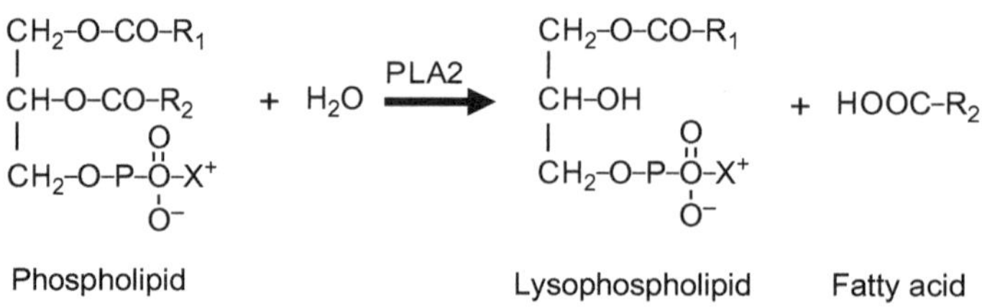

FIGURE 29.1 Scheme of phospholipid hydrolysis by PLA2 catalysis.

TABLE 29.1 Studies for phospholipase A2

Mammal	Marine invertebrates
1932 Finding in pancreas[a,b]	
1960s Purification of pancreatic PLA2[c]	
	1972 Finding in starfish *A. pectinifera*[h]
1976 3D structure of porcine PLA2[d]	Finding in marine invertebrates[i]
1977 Amino acid sequence of porcine PLA2[e]	1975 Partial purification from *A. pectinifera*[j]
	1985 Finding in starfish *A. planci*[k]
1986 cDNA cloning of dog and rat PLA2[f]	
1989 Mutation of porcine PLA2[g]	
	1995 Purification from marine snail *C. magus*[l]
	1998 Purification from starfish *A. planci* PLA2[m]
	1999 Purification from *A. pectinifera*[n]
	2000 cDNA cloning of *A. pectinifera* PLA2[o]
	2001 Bacterial expression of *A. pectinifera* PLA2[p]
	2002 cDNA cloning of sea anemone *A. carciniopados* PLA2[q]
	2004 Finding in cnidarians[r]
	Finding in poriferans[s]
	2006 cDNA cloning of *A. planci* PLA2[t]
	2007 Mutation of *A. pectinifera* PLA2[u]
	2009 Purification from sea anemone *B. caissarum*[v]
	2010 Purification from marine snail *H. trunculus*[w]

[a] Nikuni (1932).
[b] Belfanti and Arnaudi (1932).
[c] De Haas *et al.* (1968).
[d] Drenth *et al.* (1976).
[e] Puijk *et al.* (1977).
[f] Ohara *et al.* (1986).
[g] Kuipers *et al.* (1989).
[h] Hasegawa and Sasaki (1972).
[i] Vaskovsky and Suppes (1972).
[j] Okabe *et al.* (1975).
[k] Shiomi *et al.* (1985).
[l] Olivela *et al.* (1995).
[m] Shiomi *et al.* (1998).
[n] Kishimura and Hayashi (1999b).
[o] Kishimura *et al.* (2000a).
[p] Kishimura *et al.* (2001).
[q] Talvinen and Nevalainen (2002).
[r] Nevalainen *et al.* (2004a).
[s] Nevalainen *et al.* (2004b).
[t] Ota *et al.* (2006).
[u] Kishimura and Ando (2007).
[v] Martins *et al.* (2009).
[w] Zarai *et al.* (2010).

also plays a role in reduction of undesirable colorization (Tsuji and Sato, 1996), in prevention of contamination with microbial spore (Maeda and Murata, 1995), and in masking of bitter taste substances of food (Yamazawa and Katsuragi, 1996). Moreover, lysolecithin is used in protein purification as a mild detergent (Huang *et al.*, 1998), in agrochemicals as an antifungal agent (Fujita and Amano, 1995), and in cosmetic formulations as a skin moisturizing and conditioning regents (Sugino *et al.*, 1994). Dahlke *et al.* (1995) reported that PLA2 is suitable for enzymatic degumming of edible oils. The enzymatic degumming of crude edible oils reduces the amounts of acids, bases, and wastes during the refining processes.

However, similar to other enzyme and protein, the applications for food, cosmetic, and pharmaceutical industries of mammalian pancreatic PLA2 have recently been limited by the outbreak of bovine spongiform encephalopathy (BSE) or religious tradition. Therefore, it is necessary to develop new sources of PLA2. At present, fungal PLA2 from *Aspergillus niger* is available as food-grade enzyme in bulk quantities (Mustranta *et al.*, 1995). However, enzymes including PLA2 from marine organisms have hardly been used practically, because there have been few basic research for them comparing with mammalian counterparts. From these backgrounds, we are working with PLA2 from starfish viscera. Starfish belongs to the division of echinoderm similar to sea urchin and sea cucumber, but it has not been used as edible material. Approximately 300 species of starfish inhabit in the coastal area of Japan, and some of them are harmful for fisheries, especially scallop and oyster culture. In fact, huge amounts of starfish (15,000tons/year) are exterminated in the coast of Hokkaido Prefecture, Japan. Therefore, it is earnestly hoped the effective utilization of starfish. We have also performed the search and application of some other enzymes and biochemical components from starfish, such as trypsin (Kishimura and Hayashi, 2002a, 2003b), carboxypeptidase A-like enzyme (Kishimura and Hayashi, 1991), carboxypeptidase B (Kishimura and Hayashi, 2002b, 2003a; Kishimura *et al.*, 2006), free amino acids (Kishimura and Hayashi, 1990), diacyl glyceryl ethers (Hayashi and Kishimura, 1997), and collagen peptide (Aso *et al.*, 2009). In this chapter, I describe results from the investigation of the starfish PLA2 and its application.

II. PLA2 ACTIVITY OF STARFISH

Until 1995, few studies had existed on the enzymology and application of PLA2 from marine organisms (Table 29.1). For PLA2 from fish, attempts have been made to purify it from liver, hepatopancreas, and muscle of some fish, cod (*Gadus morhua*) (Aaen *et al.*, 1995; Audley *et al.*, 1978), rainbow trout (*Salmo gairdneri*) (Neas and Hazel, 1984, 1985), and red sea bream (*Pagrus major*) (Iijima *et al.*, 1990). The PLA2 from red sea bream had pH optimum 8.0 for phosphatidylethanolamine (PE) and required Ca^{2+} for the activity. These properties of red sea bream PLA2 were similar to those of mammalian pancreatic or snake venom PLA2s. In addition, it was assumed that PLA2 is expressed as a zymogen both in pyloric ceca and in hepatopancreas of adult fish (Uematsu *et al.*, 1992). However, its molecular weight was estimated to be approximately 20,000 by sodium dodecyl sulfate-polyacrylamide gel electrophoresis (SDS-PAGE) and high performance gel-filtration chromatography (HPGC) with TSK-GEL G3000SW, which was larger than that of mammalian pancreatic PLA2. On the other hand, Vaskovsky and Suppes (1972) determined PLA activity in the extracts of 46 species of marine invertebrates and showed that the majority of these species showed the activity. Nevalainen *et al.* (2004a,b) also measured PLA2 activity in 55 species of porifera and 22 species of cnidaria. Hasegawa and Sasaki (1972) found a high activity of PLA in homogenates of visceral organs of starfish *Asterina pectinifera* (*Patiria pectinifera*) that was stored at −20°C for several months (Fig. 29.2). The highest activity was observed in pyloric ceca followed by cardiac stomach and gonads. They also showed that PLA activity was caused mainly by PLA2 with minor of PLA1 (EC3.1.1.32). Okabe and Noma (1973)

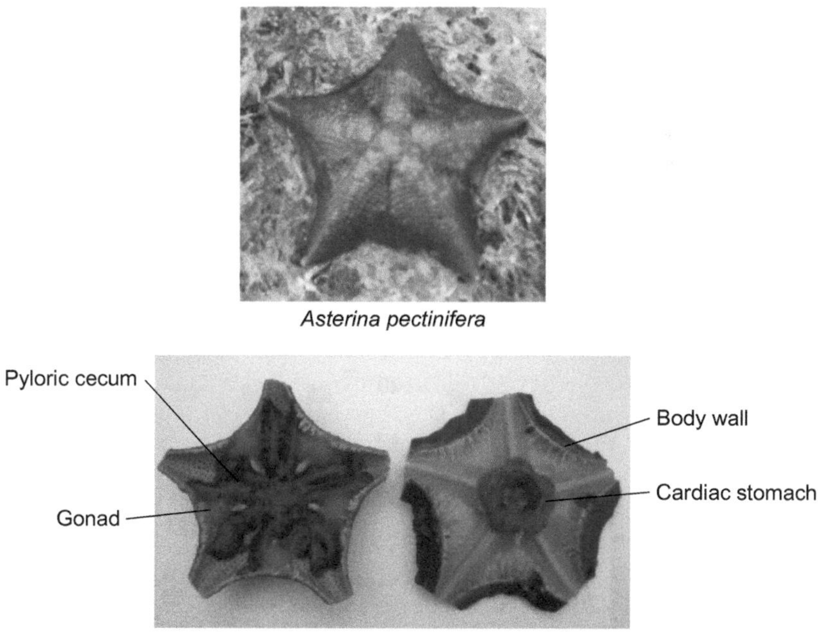

FIGURE 29.2 Starfish *Asterina pectinifera* and its organs.

investigated the properties of PLA2 in homogenate prepared from pyloric ceca of living *A. pectinifera*. The maximum activity of PLA2 exists at pH 8, and addition of Ca^{2+} and/or sodium deoxycholate enhances its activity. Further, Okabe *et al.* (1975) partially purified a PLA2 from *A. pectinifera* pyloric ceca and showed that basic properties of the enzyme, such as Ca^{2+} requirement, optimum pH, and heat stability, were similar to those of mammalian pancreatic PLA2. However, both molecular weights of 40,000 using SDS-PAGE and 72,000 using gel filtration were about three and five times higher than that of mammalian counterpart. Later, Shiomi *et al.* (1985) detected PLA2 activity in the crude toxin extracted from the spins of crown-of-thorns starfish (*Acanthaster planci*), and then McIntosh *et al.* (1995) purified a PLA2, named Conodipine-M from venom of marine snail (*Conus magus*). However, Conodipine-M comprised with two polypeptide chains linked by one or more disulfide bonds, and its amino acid sequence showed little if any homology to other previously sequenced snake venom PLA2s. Iijima *et al.* (1997) completely purified a PLA2 from red sea bream pyloric ceca, and the next year, Shiomi *et al.* (1998) isolated two PLA2s from crown-of-thorns starfish venom and determined the *N*-terminal 62 amino acid sequences.

We started a study of PLA2 from marine invertebrate in 1998 (Kishimura and Hayashi, 1998). When hydrolytic activities for egg-yolk phosphatidylcholine (PC) in pyloric ceca and viscera from several species of marine invertebrates were compared, as shown in Table 29.2, remarkably high activity was detected in starfish *A. pectinifera*, followed by starfish *Coscinasterias acutispina*, *Plazaster borealis*, *Solaster borealis*, and *Solaster paxillatus*. PLA2 activities in starfish *Aphelasterias japonica*, *Dermasterias nippon*, *Asterias amurensis*, sea urchin *Strongylocentrotus franciscanus*, and shellfish *Neptunea arthritica*, *Patinopecten yessoensis* were low (Kishimura and Hayashi, 1999a, 2005). The specific activity (1,400U/mg) of *A. pectinifera* was about three times higher than that of industrial PLA2 from porcine pancreas (440U/mg) (Amano Pharmaceutical Co., Nagoya, Japan). We also found that the crude enzyme solution prepared from fresh samples of *A. pectinifera* shows high PLA2 activity (1,427U/mg) similar to that of samples frozen at −20°C for 35 or 46 months (1,214 and 1,655U/mg, respectively) (Kishimura and Hayashi, 1999a). Therefore, these results suggest that the starfish *A. pectinifera* would be a potential source of PLA2.

III. ENZYMATIC PROPERTIES OF STARFISH PLA2

As remarkably high hydrolytic activities for egg-yolk PC were found in pyloric ceca of starfish *A. pectinifera*, a PLA2 was purified from starfish and characterized its enzymatic properties (Kishimura and Hayashi,

TABLE 29.2 Phospholipase A2 activities of various marine invertebrates[a]

Species	Organ	Activity (U[b]/g powder)	Specific activity (U/mg)
Starfish			
Asterina pectinifera	Pyloric cecum	540,000	1,400
Coscinasterias acutispina	Pyloric cecum	5,400	17
Plazaster borealis	Pyloric cecum	2,000	17
Solaster borealis	Pyloric cecum	1,900	14
Solaster paxillatus	Pyloric cecum	1,000	12
Aphelasterias japonica	Pyloric cecum	210	1.8
Distolasterias nippon	Pyloric cecum	70	0.5
Asterias amurensis	Pyloric cecum	27	0.5
Sea urchin			
Strongylocentrotus franciscanus	Viscera	120	2.0
Shellfish			
Neptunea arthritica	Viscera	110	1.9
Patinopecten yessoensis	Hepatopancreas	70	0.6

[a] Modified from Kishimura and Hayashi (2005).
[b] One unit (U) of activity was determined as one microgram of egg-yolk phosphatidylcholine hydrolyzed per minute.

1999b). At first, PLA2 was extracted from the pyloric ceca defatted with acetone. However, the extraction yield of PLA2 was very low, because phospholipids were not removed by the acetone treatment and PLA2 might combine with the remaining phospholipids. Therefore, the lipids were removed from pyloric ceca using chloroform–methanol solution. The starfish pyloric ceca were homogenized in 4 vols of chloroform–methanol (2:1, v/v) for 10 min, and the homogenate was filtrated *in vacuo* on filter paper. Similarly, the residue was homogenized in 2 vols of chloroform–methanol (2:1, v/v) and 1.3 vols of acetone for 10 min, and then the residue was air-dried at room temperature. PLA2 was extracted by stirring the delipidated powder for 3 h at 5 °C in 50 vols of 50 mM Tris–HCl buffer (pH 8.0). The extract was centrifuged at 10,000×*g* for 10 min, and then the supernatant was concentrated by lyophilization into crude enzyme solution.

A. pectinifera PLA2 was purified from the crude enzyme solution using sequential column chromatography: gel filtration on Sephacryl S-200, diethylaminoethyl (DEAE)-cellulose anion-exchange column chromatography, and gel filtration on Sephadex G-50. The final enzyme preparation, which was purified 99-fold from the crude enzyme solution in a yield of

20%, had a high activity (119,000U/mg). The enzyme was found to be nearly homogeneous using SDS-PAGE, although a minor component with brown pigment was detected at the front of gel in native-PAGE.

Enzymatic properties of *A. pectinifera* PLA2 show in Table 29.3. Molecular weight of *A. pectinifera* PLA2 was estimated as approximately 20,000 by SDS-PAGE, which was higher than that of mammalian pancreatic PLA2. Positional specificity of *A. pectinifera* PLA2 was examined using 1-palmitoyl-2-oleoyl-*sn*-glycero-3-phosphocholine (POPC). *A. pectinifera* PLA2 released mainly oleic acid at the glycerol *sn*-2 position from POPC similar to porcine pancreatic PLA2. *A. pectinifera* PLA2 hydrolyzed egg-yolk PC substrate effectively at alkaline pH with optimum activity of about pH 9, and it had optimum temperature at about 50°C. Addition of 2–4mM sodium deoxycholate enhanced *A. pectinifera* PLA2 activity, and the enzyme was also activated by 1mM or higher concentration of Ca^{2+}. The activity was stimulated by addition of Ca^{2+} followed by Mg^{2+} and Co^{2+}, while it was strongly inhibited by addition of Hg^{2+}, Zn^{2+}, and EDTA. Further, *A. pectinifera* PLA2 hydrolyzed the fatty acid ester bond exclusively at the glycerol *sn*-2 position of PC and PE regardless of its chain length and degree of unsaturation. These enzymatic properties of *A. pectinifera* PLA2 were the same as those of mammalian pancreatic PLA2 (Arni and Ward, 1996; Dennis, 1983). However, the specific activity of *A. pectinifera* PLA2 for egg-yolk PC (119,000U/mg) was about 30 times higher than that of commercial PLA2 from porcine pancreas (4,300U/mg) (Sigma Chemical Co., MO, USA). In addition, *A. pectinifera* PLA2 hydrolyzed PC more effectively than PE like snake venom PLA_2 (Ibrahim *et al.*,

TABLE 29.3 Enzymatic properties of *A. pectinifera* phospholipase A2

		Starfish PLA2		Porcine pancreatic PLA2 (Sigma)
Molecular weight		15,300		14,000
Optimal pH		pH 9		pH 9–10
Optimal temperature		50°C		50–60°C
Ca^{2+} requirement		1mM<		1mM<
Deoxycholate requirement		2–4mM		2–4mM
Fatty acid specificity	None		None	
Polar-group specificity	PC[a]>PE[b]		PC=PE	
Specific activity		119,000U/mg		4,300U/mg

[a] PC, phosphatidylcholine.
[b] PE, phosphatidylethnolamine.

1964), but not porcine pancreatic PLA2 which hydrolyzed PC almost equally to PE. These facts suggest that *A. pectinifera* PLA2 has a bright prospect for enzymology and industrial application as alternative enzyme of porcine pancreatic PLA2.

IV. STRUCTURAL PROPERTIES OF STARFISH PLA2

As described in Sections II and III, *A. pectinifera* PLA2 had extremely high activity and characteristic polar-group specificity comparing with commercially available PLA2 from porcine pancreas, and the enzyme also showed larger molecular weight than mammalian pancreatic PLA2 on SDS-PAGE. These facts suggest that *A. pectinifera* PLA2 possesses some different features in primary and/or higher order structure from the mammalian counterpart. Therefore, the primary structure of *A. pectinifera* PLA2 was determined and compared the structural properties with mammalian pancreatic PLA2.

The complete amino acid sequence of *A. pectinifera* PLA2 was determined by automated Edman degradation and cDNA cloning methods (Kishimura *et al.*, 2000a,b). *A. pectinifera* PLA2 consisted of 137 amino acid residues with an unblocked *N*-terminus. Molecular weight of *A. pectinifera* PLA2 was calculated to be 15,300.1, which was slightly larger than those of porcine and snake venom PLA2s (13,000–14,000). Amino acid sequences of *A. pectinifera* PLA2 was aligned with those of porcine pancreatic PLA2 (group IB type) (Puijk *et al.*, 1977), snake venom PLA2s from elapinae (*Naja naja atra*, group IA type) (Chang *et al.*, 1997), crotalinae (*Crotalus atrox*, group IIA type) (Randolph and Heinrikson, 1982), viperinae (*Bitis gabonica*, group IIB type) (Botes and Viljoen, 1974), and rat brain PLA2 (group IIC type) (Chen *et al.*, 1994) in Fig. 29.3. The region at positions 26–53 of *A. pectinifera* PLA2 showed fairly high sequence homology (75%) to the corresponding region of group I and group II PLA2s. The residues involved in catalytic network (His-49, Asp-111, Tyr-53, and Tyr-72) and Ca^{2+}-binding site (Tyr-29, Gly-31, Gly-33, and Asp-50) of group I and II PLA2s were completely conserved in *A. pectinifera* PLA2 (Arni and Ward, 1996; Renetseder *et al.*, 1985). Further, *A. pectinifera* PLA2 conserved 14 Cys residues at the appropriate positions, which were involved in the intramolecular disulfide bonds of group I PLA2 (Dennis, 1983). Therefore, *A. pectinifera* PLA2 is classified into group I PLA2, and the enzyme may function through a similar catalytic mechanism of mammalian pancreatic PLA2. On the other hand, the homology calculated with whole sequences between *A. pectinifera* PLA2 and the other animal PLA2s was relatively low (36–48%) since high sequence divergence exists at positions 54–107 including the pancreatic loop and the β-wing (Arni and Ward, 1996; Renetseder *et al.*, 1985). *A. pectinifera* PLA2 possessed the pancreatic

```
                              10        20        30        40        50        60 +1  +1 70
Starfish       SVYQFGKFISCYGGAGFFDGLDYNGYGCYCGYGGKGTPLDDTDRCCLVHD-NCYGKATAEAD--C-GSWD-PYII
Snake    ( IA) NLYQFKNMIQC-TVP-SRSWWDFADYGCYCGRGGSGTPVDDLDRCCQVHD-NCYNEA-EKI-SGC---W--PYFK
Porcine  ( IB) ALWQFRSMIKC-AIPGSHPLMDFNNYGCYCGLGGSGTPVDELDRCCETHD-NCYRDAKNL-DS-CKFLVDNPYTE
Snake    (IIA) SLVQFETLIM--KIAGRSGLLWYSAYGCYCGWGGHGLPQDATDRCCFVHDC-CYGKAT---D--C-----NPKTV
Snake    (IIB) DLTQFGNMIN--KM-G-QSVFDYIYYGCYCGWGGKGKPIDATDRCCFVHDC-CYGKMGT-YD---------TKWT
Rat      (IIC) SFWQFQRMVK--HITGRSAFFSYYGYGCYCGLGGRGIPVDATDRCCWAHDC-CYHKLK-EY--GC-----QPILN
Starfish 2     NLYQFGKMVMCLGNLNVLNGLEYNGYGCYCGRGGKGTPLDDTDRCCKQHD-ECYERATDEM--GC-WSI-ETYAT
Sea anemone    GAFQFSYLIK-YTGR-N-P-LDYWGYGCWCGLGGKGTPVDGVDWCCYHHD-MCFNSITQGPRPTC--SKNAPYHK
Red sea bream  ALWQFGNMIQCAQPGVN-PFL-YNDYGCWCGFGGKGAPLDDVDACCKVHD-NCY-KASR-LAPGCTAVADIPYVL
                                        Ca2+-binding loop      active site         pancreatic loop

        80 +1,2,3  90       100       110       120       130
VYDY-EQTTDASGNC---VIKCKKAADYSW-YSTNP-ECREFMCECDRAGAQCFAEKRPTYNQAYESYD--KDSC
TYSY-E--------CSQGTLTC---------KGGNN-ACAAAVCDCDRLAAICFA-GAP-YNNNNYNIDL-KARCQ
SYSYS---------CSNTEITC---------NSKNN-ACEAFICNCDRNAAICFS-KAP-YNKEHKNLDT-KKYC
SYTYSE---------ENGEIIC----------GGDD-PCGTQICECDKAAAICFRDNIPSYDNKYWLFPP-KD-CREEPEPC
SYNY-E----I----QNGGIDC----------DEDP-Q-KKELCECDRVAAICFANNRNTYNSNYFGHSSSK--CTG-TEQC
AYQFA-----I----VNGTVTC-------GCTMGGGCLCGQKACECDKLSVYCFKENLATYEKTFKQLFPTRPQCGRDKLHC
TYDYTE--SKVSGKC---TIKCKLESDYSR-FTIRK-KCKAFICECDRIGAQCFADKRSTFNRSLISY-T-KDKC
NYYFS------------G-LKC------ST-GWLTS-KCGRAICACDIAAVKCFM--RNHFNNKYQNY--KKNIC
VVDHT---------CSNQQVTC------SA---TNN-KCQAAVCECDRVAAHCFAQTQ--YNPDNKNVD-QKVHCVN
    β-wing
```

FIGURE 29.3 Alignment of the amino acid sequences of *A. pectinifera* PLA2 with the sequences of other PLA2s. Starfish: *A. pectinifera* PLA2; Snake (IA), *Naja naja atra* venom PLA2 (Chang *et al.*, 1997); porcine (IB): Porcine pancreatic PLA2 (Puijk *et al.*, 1977); Snake (IIA): *Crotalus atrox* venom PLA2 (Randolph and Heinrikson, 1982); Snake (IIB): *Bitis gabonica* venom PLA2 (Botes and Viljoen, 1974); rat (IIC): Rat brain PLA2 (Chen *et al.*, 1994); Starfish 2: *A. planci* venom PLA2 (Ota *et al.*, 2006); Sea anemone: (Talvinen and Nevalainen, 2002); Red sea bream: *P. major* PLA2 (Iijima *et al.*, 2001). Dashes indicate deletions introduced for maximizing the sequence similarity. The locations of the active site, Ca^{2+}-binding loop, pancreatic loop, and β-wing region are shown with solid bars based on the crystallographic studies of bovine pancreatic and *C. atrox* venom PLA2s (Arni and Ward, 1996; Renetseder *et al.*, 1985). The positions of deleted amino acids in pancreatic loop and β-wing regions of *A. pectinifera* PLA2 are represented as 62(+1), 66 (+1), 84(+1), 84(+2), and 84(+3).

loop-like sequence at positions 63–66; however, two amino acid deletions were required on the positions 62(+1) and 66(+1) to align with the sequence of porcine pancreatic PLA2. In addition, two insertions of each of eight residues (positions 76–83 and 89–96) and deletion of three residues (positions 84(+1)–84(+3)) were required for *A. pectinifera* PLA2 to align with the β-wing of porcine PLA2. In predicted 3D-structure, *A. pectinifera* PLA2 also showed specific structure on the pancreatic loop and the β-wing regions compared with that of bovine pancreatic PLA2 (Fig. 29.4). In order to clarify molecular evolutionary relationship between *A. pectinifera* PLA2 and group I and group II PLA2s, we made a phylogenetic tree using CLUSTAL W program (Thompson *et al.*, 1994). Consequently, *A. pectinifera* PLA2 was hardly placed in either group IA or IB

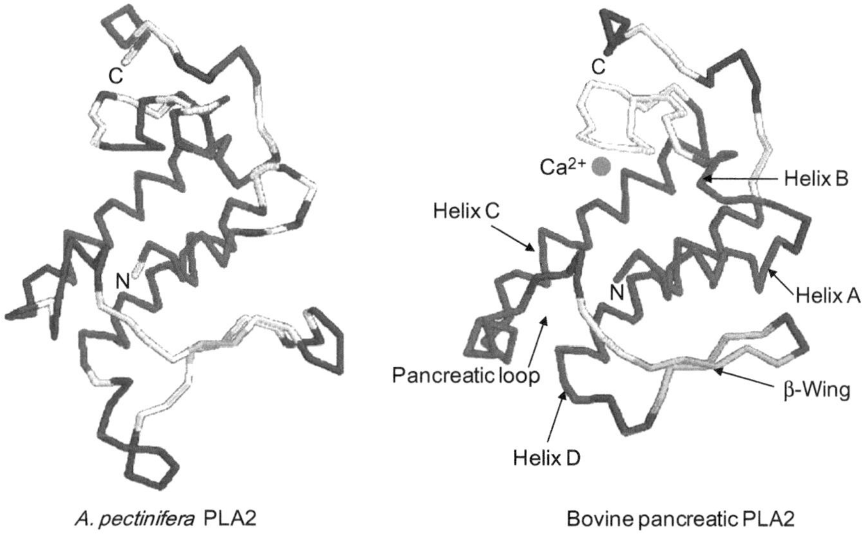

FIGURE 29.4 Predicted 3D-structure of *A. pectinifera* PLA2. *A. pectinifera* PLA2: 3D-structure of *A. pectinifera* PLA2 was predicted by CPH models (Lund *et al.*, 1997); Bovine pancreatic PLA2: Dijkstra *et al.* (1981a,b). Expression of both 3D-structures was performed using RasMac v2.6 (http://www.umass.edu/microbio/rasmol/getras.htm).

and should be classified into a new type of group I PLA2 (group IC type). Therefore, the unique structure of *A. pectinifera* PLA2 relates to its specific properties, such as the extremely high activity and the characteristic polar-group specificity.

As PLA2 exhibits enhanced activity toward lipids in lamellar and micellar aggregates both in membranes and in other lipid–water interfaces, the reaction cycle has been considered to include the interfacial binding which is distinct from the binding of a phospholipid molecule to the active site (Arni and Ward, 1996; Dennis, 1983). An earlier crystallographic study of bovine pancreatic PLA2 predicted that interfacial binding surface is composed of the residues clustered in the N- and C-termini and several other residues (Dijkstra *et al.*, 1981a). Previous mutagenesis studies indicated that Lys-62 or Arg-53 of porcine pancreatic PLA2 and Lys-53 or Lys-56 of bovine pancreatic PLA2 are involved in their specificities for anionic head group of phospholipids presumably by electrostatically repelling cationic head groups of zwitterionic phospholipids (Dua *et al.*, 1995; Kuipers *et al.*, 1989; Snitko *et al.*, 1999). Researches of mammalian pancreatic PLA2 also indicated that Arg-6, Lys-10, and lys-116 of porcine, Lys-10, Lys-56, and Lys-116 of bovine, and Arg-6, Lys-7, Lys-10, and Lys-116 of human enzymes are involved in electrostatic interactions with anionic interfaces (Lugtigheid *et al.*, 1993; Noel *et al.*, 1991). In *A. pectinifera* PLA2, most of the above positively charged residues were deleted or substituted for neutral and negatively charged residues. These facts imply that the interfacial binding mode of *A. pectinifera* PLA2 is also different from that of mammalian pancreatic PLA2.

As described in the above paragraph, *A. pectinifera* PLA2 was suggested a new type of group I PLA2 (group IC) for its characteristic structure on the pancreatic loop and the β-wing regions. Later, it was clarified that this characteristic structure on the pancreatic loop and the β-wing regions of *A. pectinifera* PLA2 was common to that of PLA2s from crown-of-thorns starfish venom, although PLA2s from fish and sea anemone did not have these properties (Fig. 29.3). McIntosh *et al.* (1995) has suggested the occurrence of a new type of PLA2 in marine snail venom, and the enzyme has been classified into group IX PLA2, as it was comprised of two polypeptide chains and showed little sequence homology to other PLA2s (Dennis, 1983). Recently, Martins *et al.* (2009) purified a PLA2 from tentacle of sea anemone (*Bunodosoma caissarum*). *N*-terminal amino acid sequence of the PLA2 showed high identity with those of group III PLA2s, and its molecular weight is about 14,000 and 30,000 in the presence or absence of dithiothreitol, respectively. Zarai *et al.* (2010) also reported a novel PLA2 from hepatopancreas of marine snail (*Hexaplex trunculus*), and it exhibits both digestive and toxic activities.

V. STRUCTURE–FUNCTION RELATIONSHIP OF STARFISH PLA2

A. pectinifera PLA2 has characteristic structures in the pancreatic loop and β-wing regions, and these may account for its specific properties. In the previous studies, it was suggested that basic amino acids in the pancreatic loop of mammalian PLA2 are involved in their polar-group specificity (Dua *et al.*, 1995; Kuipers *et al.*, 1989; Snitko *et al.*, 1999). Therefore, to probe the relationship between its polar-group specificity and structure of the pancreatic loop, site-directed mutagenesis study of *A. pectinifera* PLA2 was carried out.

First, a bacterial expression system for recombinant PLA2 (Kishimura *et al.*, 2001) was constructed. Complementary DNA of *A. pectinifera* PLA2 was inserted into expression plasmid pET-16b, and PLA2 protein was expressed in *Escherichia coli* BL21 (DE3) by induction with isopropyl-β-D-(−)-thiogalactopyranoside (IPTG). The *N*-terminal Ser in *A. pectinifera* PLA2 was substituted for Ala in recombinant PLA2 protein to express a mature enzyme (wild-type PLA2). The overnight culture of bacteria (150 ml) containing PLA2-pET-16b plasmid was transferred to a 1,000 ml culture and cultivated at 37 °C for 6 h, and then the wild-type PLA2 was expressed by induction with 1mM IPTG for 2h. Densitometric analysis showed that expression level of the recombinant protein reached a maximum (approximately 15% of total protein) in 2h of induction. Bacteria were harvested at 10,000×*g* for 5min and lysed by repeated freeze and thaw in 40ml of 50mM Tris–HCl buffer (pH 8.0) containing 50mM EDTA, 8% sucrose, 0.5% Triton X-100, and 0.5mM phenylmethanesulfonyl fluoride. Lysate was centrifuged at 10,000×*g* for 5min, and inclusion bodies precipitated were suspended in 40ml of the same buffer. Suspension was sonicated and simultaneously passed through a needle at 5°C for 10min, and then inclusion bodies were dissolved in 8M urea and 10mM 2-mercaptoethanol and incubated at room temperature for 3h. After centrifugation at 10,000×*g* for 10min, wild-type PLA2 obtained in the supernatant was renatured by successive dialysis against 6, 4, 2, and 0M urea containing 10mM Tris–HCl buffer (pH 8.0) at 5°C for 3h each. The renatured enzyme was lyophilized, dissolved in 10mM Tris–HCl buffer (pH 8.0), and centrifuged at 10,000×*g* for 10min. By this procedure, 184 mg of crude enzyme was obtained. To purify wild-type PLA2, the crude enzyme was applied to subsequent column chromatographies on DEAE-cellulose (DE-52) and Sephadex G-50. Wild-type PLA2 was purified 97-fold from the crude enzyme and had high specific activity (87,000 U/mg). Wild-type PLA2 released mainly oleic acid from POPC, and the enzyme hydrolyzed egg-yolk PC with optimum pH 9 and optimum temperature 50°C. Wild-type PLA2 was activated by 1mM or higher concentrations of Ca^{2+} and hydrolyzed PC more efficiently than PE. These properties of wild-type PLA2 were essentially the same as those of *A. pectinifera* PLA2.

Second, we constructed expression vector to make a mutant been inserted Lys residue between Cys-62 and Gly-63 in the pancreatic loop region of *A. pectinifera* PLA2 (62+K PLA2) (Kishimura and Ando, 2007). The sequence of cDNA coding *A. pectinifera* PLA2 was exchanged by the oligonucleotide-directed dual amber-long and accurate polymerase chain reaction method. 62+K PLA2 was expressed in *E. coli* Origami^{TM} B (DE3) by induction with IPTG, and then the mutant protein was renatured and purified by the same procedure as wild-type PLA2. 62+K PLA2 showed essentially the same properties as *A. pectinifera* PLA2 and wild-type PLA2 with respect to substrate positional specificity, optimum pH, optimum temperature, Ca^{2+} requirement, and sodium deoxycholate requirement. However, the specific activity of the 62+K PLA2 for egg-yolk PC (950 U/mg) was extremely lower than that of *A. pectinifera* PLA2 (119,000U/mg) and wild-type PLA2 (87,000U/mg), whereas near to that of porcine pancreatic PLA_2 (4,300U/mg). Moreover, ratio of specific activity of 62+K PLA2 for dipalmitoyl-PC to dipalmitoyl-PE (98 times) was highly lower than that of *A. pectinifera* PLA_2 (2,650 times), but similar to that of porcine pancreatic PLA2 (25 times). Now, we think that the charge and structure of pancreatic loop region of *A. pectinifera* PLA2 might carry out important role on polar-group specificity.

VI. APPLICATION OF STARFISH PLA2

In a series of studies, it was developed that *A. pectinifera* PLA2 has a bright prospect for enzymology and industrial application as alternative enzyme of porcine pancreatic PLA2. Therefore, we tried to analyze the positional distribution of fatty acids in PC and PE from marine organisms using *A. pectinifera* PLA2 instead of industrial PLA2.

To begin with, the positional distribution of fatty acids in PC and PE from scallop *P. yessoensis* and Japanese common squid *Todarodes pacificus* that are main fisheries products in Hokkaido Prefecture, Japan was analyzed. Total lipids were extracted from scallop adductor muscle and squid mantle muscle by the method of Bligh and Dyer (1959). PC and PE were prepared from the total lipids using preparative thin layer chromatography (TLC)- with chloroform–methanol–acetic acid–water (55:17:3:2, v/v/v/v) as a developing solvent. Fifteen milligram each of PC and PE were hydrolyzed at 37 °C by 9 μg (1,071 U) of *A. pectinifera* PLA2 for 3 h and by 1 mg (440 U) of porcine pancreatic PLA2 (Amano Pharmaceutical Co.) for 12 h. The released fatty acids were separated using preparative TLC with a developing solvent of hexane-diethyl ether-acetic acid (85:15:1, v/v/v) and the fatty acid compositions were analyzed by gas-liquid chromatography (GLC). As a result, the compositions of the fatty acids released from these phospholipids by *A. pectinifera* PLA2 were similar to those released by porcine PLA2 (Table 29.4). Predominant fatty acids at *sn*-2 position of scallop adductor muscle phospholipids were eicosapentaenoic acid (EPA) (35.7% for PC and 36.9% for PE) and docosahexaenoic acid (DHA) (26.8% for PC and 21.8% for PE), and those of squid mantle muscle were DHA (76.8% for PC and 44.4% for PE) and EPA (13.5% for PC and 42.3% for PE). Therefore, it was thought that *A. pectinifera* PLA2 can use as alternative enzyme of porcine pancreatic PLA2, and phospholipids from squid are rich source of functional polyunsaturated fatty acids, especially DHA.

Next, the distribution of DHA and EPA in PC and PE from different tissues of two species of squids, spear squid *Loligo bleekeri* and Japanese common squid *T. pacificus*, was analyzed using *A. pectinifera* PLA2 (Igarashi *et al.*, 2001). Mantles, fins, arms, and integuments of both squids and liver of spear squid contained small amounts of lipid (1.8–4.3% of total lipid per wet weight) that comprised almost entirely of PC (47.9–67.0%), sterols (14.2–32.7%), and PE (12.2–15.7%). Liver of spear squid contained copious amounts of total lipid (36.2%) with higher triacylglycerols (97.0%). In PC and PE from both squid tissues, DHA (24.7–37.6% for PC and 7.4–25.0% for PE), EPA (6.3–18.4% for PC and 18.9–38.8% for

TABLE 29.4 Composition of fatty acids released from phosphatidylcholine and phosphatidylethanolamine by *A. pectinifera* phospholipase A2[a] (wt%)[b]

Fatty acid	Scallop PC[c]		Scallop PE[c]		Squid PC[c]		Squid PE[c]			
	Starfish PLA2	Porcine PLA2[d]	Starfish PLA2	Porcine PLA2	Starfish PLA2	Porcine PLA2	Starfish PLA2	Porcine PLA2		
14:0	2.6	3.0	0.2		0.2	tr	tr		tr	tr
16:0	8.8	8.0	1.6		1.5	2.9	1.6		2.0	2.3
16:1n−7	1.1	1.1	0.2		0.1	0.3	0.2		0.1	0.1
18:0	1.6	1.7	1.2		1.2	1.0	1.1		1.2	1.7
18:1n−9	5.1	5.3	0.7		0.9	1.2	1.3		0.5	0.5
18:1n−7	4.2	4.0	1.4		1.3	0.5	0.4		0.4	0.4
18:2n−6	1.0	0.9	0.1		0.1	0.2	0.1		0.1	0.1
20:1n−11	0.2	0.3	3.3		3.5	0.3	0.6		0.1	0.1
20:1n−9	0.1	0.2	8.5		8.1	0.1	0.1		0.4	0.4
20:1n−7	0.4	0.4	10.5		11.4	nd	nd		nd	nd
20:4n−6	5.5	5.4	6.1		6.0	1.5	1.5		7.0	6.6
20:5n−3	35.7	35.2	39.6		39.7	13.5	14.3		42.3	40.8
21:5n−3	1.3	1.2	0.6		0.5	0.2	0.2		0.3	0.1
22:5n−3	1.6	1.7	0.5		1.4	0.3	0.2		0.5	0.5
22:6n−3	26.8	27.2	21.8		21.3	76.8	77.3		44.4	45.7
Others[e]	4.0	4.4	3.7		3.8	1.2	1.1		0.7	0.7

tr, trace (<0.05%); nd, not detected.
[a] Modified from Kishimura and Hayashi (1999b).
[b] wt% to total fatty acids.
[c] Phosphatidylcholine (PC) and phosphatidylethanolamine (PE) were prepared from scallop adductor muscle or squid mantle muscle.
[d] Phospholipase A2 from porcine pancreas (Amano Pharmaceutical Co.).
[e] Consisted of minor (<1.0%) and unknown compounds.

PE), and palmitic acid (28.6–44.0% for PC and 11.0–22.5% for PE) were characteristically present. In these squids tissues, except for liver, component fatty acids at *sn*-2 position of PC and PE were polyunsaturated for the most part with the predominant component of DHA (71.3–78.7% for PC and 17.1–45.3% for PE) and EPA (10.9–14.4% for PC and 37.2–58.0% for PE). However, component fatty acids at *sn*-1 position of squids PC and PE were more saturated with palmitic acid (66.4–72.9% for PC and 15.9–41.6% for PE) and stearic acid (4.2–6.9% for PC and 9.8–30.0% for PE).

From these results, it was clarified that *A. pectinifera* PLA2 can be utilized to determine fatty acid positional distribution of phospholipids as alternative enzyme of mammalian PLA2 in laboratory. We are hopeful of practical use of *A. pectinifera* PLA2 on industrial scale.

ACKNOWLEDGMENT

I would like to express my sincere gratitude to my advisor, Prof. Emeritus Kenji Hayashi and coadvisors, Prof. Emeritus Kiyoyoshi Nishita and Prof. Takao Ojima of Hokkaido University for accomplishment of this study. This study could not be succeeded without financial support of the Grants-in-Aid for Scientific Research from JSPS.

REFERENCES

Aaen, B., Jessen, F., and Jensen, B. (1995). Partial purification and characterization of a cellular acidic phospholipase A2 from cod (*Gadus morhua*). *Comp. Biochem. Physiol.* **110B**, 547–554.

Arni, R. K. and Ward, R. J. (1996). Phospholipase A2—A structural review. *Toxicon* **34**, 827–841.

Aso, S., Takeda, H., Kitagawa, M., Miyazaki, A., Iida, T., Ebitani, K., Tsuji, K., Sugawara, A., Kishimura, H., and Saeki, H. (2009). Blood glucose increase inhibitor for type 2 diabetes patients with insulin resistance comprises collagen peptide active ingredient contained in hydrolyzate obtained by protease treated starfishes. *Jpn. Kokai Tokkyo* JP2009235064-A.

Audley, M. A., Shetty, K. J., and Kinsella, J. E. (1978). Isolation and properties of phospholipase A from pollock muscle. *J. Food Sci.* **43**, 1771–1775.

Belfanti, S. and Arnaudi, C. (1932). A lecithinase of pancreas producing lysolecithin. *Bull. Soc. Intern. Sez. Ital.* **4**, 399–406.

Blasi, F., Cossignani, L., Simonetti, M. S., Brutti, M., Ventura, F., and Damiani, P. (2006). Enzymatic deacylation of 1,2-diacyl-*sn*-glycero-3-phosphocholines to *sn*-glycero-3-phosphocholine. *Enzyme Microb. Technol.* **39**, 1405–1408.

Bligh, E. G. and Dyer, W. J. (1959). A rapid method of total lipid extraction and purification. *Can. J. Biochem. Physiol.* **37**, 911–917.

Botes, D. P. and Viljoen, C. C. (1974). *Bitis gabonica* venom: The amino acid sequence of phospholipase A. *J. Biol. Chem.* **249**, 3827–3835.

Burke, J. E. and Dennis, E. A. (2009). Phospholipase A2 structure/function, mechanism, and signaling. *J. Lipid Res.* **50**, s237–s242.

Chang, L., Lin, S., and Chang, C. (1997). The structural variations of ε-amino groups in phospholipase A2 enzymes from *Naja naja atra* and *Bungarus multicinctus* venoms. *J. Protein Chem.* **16**, 133–137.

Chen, J., Engle, S. J., Seilhamer, J. J., and Tischfield, J. A. (1994). Cloning and characterization of novel rat and mouse low molecular weight Ca^{2+}-dependent phohpolipase A2s containing 16 cysteines. *J. Biol. Chem.* **269**, 23018–23024.

Dahlke, K., Buchold, H., Munch, E. M., and Paulitz, B. (1995). First experiences with enzymatic oil refining. *Inform* **6**, 1284–1291.

De Haas, G. H., Postema, N. M., Nieuwenhuizen, W., and Van Deenen, L. L. M. (1968). Purification and properties of phospholipase A from porcine pancreas. *Biochim. Biophys. Acta* **159**, 103–117.

De Maria, L., Vind, J., Oxenboll, K. M., Svendsen, A., and Patkar, S. (2007). Phospholipases and their industrial applications. *Appl. Microbiol. Biotechnol.* **74**, 290–300.

Dennis, E. A. (1983). Phospholipases. *In* "The Enzymes", (P. Boyer, Ed.), 3rd edn, Vol. XV, pp. 307–353. Academic Press, New York.

Dijkstra, B. W., Drenth, J., and Kalk, K. H. (1981a). Active site and catalytic mechanism of phospholipase A2. *Nature* **289**, 604–606.

Dijkstra, B. W., Kalk, K. H., Hol, W. G. J., and Drenth, J. (1981b). Structure of bovine pancreatic phospholipase A2 at 1.7 A resolution. *J. Mol. Biol.* **147**, 97–123.

Drenth, J., Enzing, C. M., Kalk, K. H., and Vessies, J. C. A. (1976). Structure of porcine pancreatic prephospholipase A2. *Nature* **264**, 373–377.

Dua, R., Wu, S. K., and Cho, W. (1995). A structure-function study of bovine pancreatic phospholipase A2 using polymerized mixed liposomes. *J. Biol. Chem.* **270**, 263–268.

Fujita, S. and Amano, S. (1995). Control of fungal infections in plants. *Jpn. Kokai Tokkyo* JP199748218.

Hasegawa, H. and Sasaki, T. (1972). Studies on visceral lipids of starfish II: Phospholipase A in starfish *Asterina pectinifera*. *Sapporo Med. J.* **40**, 1–6.

Hayashi, K. and Kishimura, H. (1997). Content and composition of diacyl glyceryl ethers in the pyloric ceca and ovaries of the asteroids *Solaster paxillatus* and *Asterias amurensis*. *Fish. Sci.* **63**, 945–949.

Huang, P., Liu, Q., and Scarborough, G. A. (1998). Lysophosphatidylglycerol: A novel effective detergent for solubilizing and purifying the cystic fibrosis transmembrane conductance regulator. *Anal. Biochem.* **259**, 89–97.

Ibrahim, S. A., Sanders, H., and Thompson, R. H. (1964). The action of phospholipase A on purified phospholipids, plasma and tissue preparations. *Biochem. J.* **93**, 588–594.

Igarashi, D., Hayashi, K., and Kishimura, H. (2001). Positional distribution of DHA and EPA in phosphatidylcholine and phosphatidylethanolamine from different tissues of squids. *J. Oleo Sci.* **50**, 729–734.

Iijima, N., Nakamura, M., Uematsu, K., and Kayama, M. (1990). Partial purification and characterization of phospholipase A2 from the hepatopancreas of red sea bream. *Nippon Suisan Gakkaishi* **56**, 1331–1339.

Iijima, N., Chosa, K., Uematsu, K., Goto, T., Hoshita, T., and Kayama, M. (1997). Purification and characterization of phospholipase A2 from the pyloric caeca of red sea bream Pagrus major. *Fish Physiol. Biochem.* **16**, 487–498.

Iijima, N., Fujikawa, Y., Tateishi, Y., Takashima, Y., Uchiyama, S., and Esaka, M. (2001). Cloning and expression of group IB phospholipase A2 isoforms in the red sea bream, *Pagrus major*. *Lipids* **36**, 499–506.

Kim, J., Lee, C. S., Oh, J., and Kim, B. G. (2001). Production of egg yolk lysolecithin with immobilized phospholipase A2. *Enzyme Microb. Technol.* **29**, 587–592.

Kishimura, H. and Ando, S. (2007). Characteristics of phospholipase A2 mutant of the starfish *Asterina pectinifera*. *Enzyme Microb. Technol.* **40**, 461–465.

Kishimura, H. and Hayashi, K. (1990). Free amino acids of starfish. *Nippon Suisan Gakkaishi* **56**, 1693.

Kishimura, H. and Hayashi, K. (1991). Purification and properties of carboxypeptidase A-like enzyme from the starfish *Asterias amurensis*. *Nippon Suisan Gakkaishi* **57**, 1939–1944.

Kishimura, H. and Hayashi, K. (1998). Purification and properties of phospholipase A2-like enzyme from the pyloric ceca of the starfish *Solaster paxillatus*. *Nippon Suisan Gakkaishi* **64**, 264–269.

Kishimura, H. and Hayashi, K. (1999a). Phospholipase A activity in the pyloric ceca of starfish. *Nippon Suisan Gakkaishi* **65**, 110–111.

Kishimura, H. and Hayashi, K. (1999b). Isolation and characteristics of phospholipase A2 from the pyloric ceca of the starfish *Asterina pectinifera*. *Comp. Biochem. Physiol.* **124B**, 483–488.

Kishimura, H. and Hayashi, K. (2002a). Isolation and characteristics of trypsin from pyloric ceca of the starfish *Asterina pectinifera*. *Comp. Biochem. Physiol.* **132B**, 485–490.

Kishimura, H. and Hayashi, K. (2002b). Isolation and characteristics of carboxypeptidase B from pyloric ceca of the starfish *Asterias amurensis*. *Comp. Biochem. Physiol.* **133B**, 183–189.

Kishimura, H. and Hayashi, K. (2003a). Simple method for purification of carboxypeptidase B from starfish pyloric ceca. *Nippon Suisan Gakkaishi* **69**, 646–648.

Kishimura, H. and Hayashi, K. (2003b). N-terminal amino acid sequence of trypsin from the pyloric ceca of starfish *Asterias amurensis*. *Fish. Sci.* **69**, 867–869.

Kishimura, H. and Hayashi, K. (2005). Characterization of phospholipase A2 from the pyloric ceca of two species of starfish *Coscinasterias acutispina* and *Plazaster borealis*. *Food Chem.* **92**, 407–411.

Kishimura, H., Ojima, T., Hayashi, K., and Nishita, K. (2000a). cDNA cloning and sequencing of phospholipase A2 from the pyloric ceca of the starfish *Asterina pectinifera*. *Comp. Biochem. Physiol.* **126B**, 579–586.

Kishimura, H., Ojima, T., Tanaka, H., Hayashi, K., and Nishita, K. (2000b). Amino acid sequence of phospholipase A2 from the pyloric ceca of starfish *Asterina pectinifera*. *Fish. Sci.* **66**, 104–109.

Kishimura, H., Ojima, T., Hayashi, K., and Nishita, K. (2001). Bacterial expression and characterization of starfish phospholipase A2. *Comp. Biochem. Physiol.* **128B**, 565–573.

Kishimura, H., Hayashi, K., and Ando, S. (2006). Characteristics of carboxypeptidase B from pyloric ceca of the starfish *Asterina pectinifera*. *Food Chem.* **95**, 264–269.

Kuipers, O. P., Thunnissen, M. M. G. M., De Geus, P., Dijkstra, B. W., Drenth, J., Verheij, H. M., and De Haas, G. H. (1989). Enhanced activity and altered specificity of phospholipase A2 by deletion of a surface loop. *Science* **244**, 82–85.

Labeque, R., Mullon, C. J. P., Ferreira, J. P. M., Lees, R. S., and Langer, R. (1993). Enzymatic modification of plasma low density lipoproteins in rabbits: A potential treatment for hypercholesterolemia. *Proc. Natl. Acad. Sci. USA* **90**, 3476–3480.

Lugtigheid, R. B., Otten-Kuipers, M. A., Verheij, H. M., and De Haas, G. H. (1993). Arginine 53 is involved in head-group specificity of the active site of porcine pancreatic phospholipase A2. *Eur. J. Biochem.* **213**, 517–522.

Lund, O., Frimand, K., Gorodkin, J., Bohr, H., Bohr, J., Hansen, J., and Brunak, S. (1997). Protein distance constraints predicted by neural networks and probability density functions. *Protein Eng.* **11**, 1241–1248.

Maeda, Y. and Murata, M. (1995). Lysolecithin in preparation of heat-sterilized food packages. *Jpn. Kokai Tokkyo* JP19977123923.

Martins, R. D., Alves, R. S., Martins, A. M. C., Barbosa, P. S. F., Evangelista, J. S. A. M., Evangelista, J. J. F., Ximenes, R. M., Toyama, M. H., Toyama, D. O., Souza, A. J. F., Orts, D. J. B., Marangoni, S., *et al.* (2009). Purification and characterization of the biological effects of phospholipase A2 from sea anemone *Bunodosoma caissarum*. *Toxicon* **54**, 413–420.

McIntosh, J. M., Ghomashchi, F., Gelb, M. H., Dooley, D. J., Stoehr, S. J., Giordani, A. B., Naisbitt, S. R., and Olivera, B. M. (1995). Conodipine-M, a novel phospholipase A2 isolated from the venom of the marine snail *Conus magus*. *J. Biol. Chem.* **270**, 3518–3526.

Mine, Y. (1997). Structural and functional changes of hen's egg yolk low-density lipoproteins with phospholipase A2. *J. Agric. Food Chem.* **45**, 4558–4563.

Mustranta, A., Forssel, P., and Poutanen, K. (1995). Comparison of lipases and phospholipases in the hydrolysis of phospholipids. *Process Biochem.* **30**, 393–401.

Neas, N. P. and Hazel, J. R. (1984). Temperature-dependent deacylation of molecular species of phosphatidylcholine by microsomal phospholipase A2 of thermally acclimated rainbow trout, *Salmo gairdneri*. *Lipids* **19**, 258–263.

Neas, N. P. and Hazel, J. R. (1985). Partial purification and kinetic characterization of the microsomal phospholipase A2 of thermally acclimated rainbow trout (*Salmo gairdneri*). *J. Comp. Physiol.* **155B**, 461–469.

Nevalainen, T. J., Peuravuori, H. J., Quinn, R. J., Llewellyn, L. E., Benzie, J. A. H., Fenner, P. J., and Winkel, K. D. (2004a). Phospholipase A2 in cnidaria original research article. *Comp. Biochem. Physiol. B Biochem. Mol. Biol.* **139**, 731–735.

Nevalainen, T. J., Quinn, R. J., and Hooper, J. N. A. (2004b). Phospholipase A2 in porifera original research article. *Comp. Biochem. Physiol. B Biochem. Mol. Biol.* **137**, 413–420.

Nikuni, J. (1932). The formation of lysolecithin from egg-yolk lecithin by pancreas extract. *Proc. Imp. Acad. (Tokyo)* **8**, 300–303.

Noel, J. P., Bingman, C. A., Deng, T., Dupureur, C. M., Hamilton, K. J., Jiang, R. T., Kwak, J. G., Sekharudu, C., Sundaralingam, M., and Tsai, M. D. (1991). Phospholipase A2 engineering: X-ray structural and functional evidence for the interaction of lysine-56 with substrates. *Biochemistry* **30**, 11801–11811.

Ohara, O., Tamaki, M., Nakamura, E., Tsuruta, Y., Fujii, Y., Shin, M., Teraoka, H., and Okamoto, M. (1986). Dog and rat pancreatic phospholipase A2: Complete amino acid sequences deduced from complementary DNAs. *J. Biochem.* **99**, 733–739.

Okabe, H. and Noma, A. (1973). Studies on the phospholipase A2 in pyloric ceca of the starfish *Asterina pectinifera*. *Tohoku J. Exp. Med.* **110**, 263–271.

Okabe, H., Noma, A., and Ohno, K. (1975). Partial purification and properties of phospholipase A2 from the starfish, *Asterina pectinifera*. *Biochim. Biophys. Acta* **398**, 149–158.

Olivela, B. M., McIntosh, J. M., Ghomashchi, F., Gelb, M. H., Dooley, D. J., Stoehr, S. J., Giordani, A. B., and Baisbitt, S. R. (1995). Conodipine-M, a novel phospholipase A2 isolated from the venom of the marine snail Conus magaus. *J. Biol. Chem.* **270**, 3518–3526.

Ota, E., Nagai, H., Nagashima, Y., and Shiomi, K. (2006). Molecular cloning of two toxic phospholipases A2 from the crown-of-thorns starfish Acanthaster planci venom. *Comp. Biochem. Physiol. B* **143**, 54–60.

Penci, M. C., Constenla, D. T., and Carelli, A. A. (2010). Free-fatty acid profile obtained by enzymatic solvent-free hydrolysis of sunflower and soybean lecithin. *Food Chem.* **120**, 332–338.

Puijk, W. C., Verheij, H. M., and De Haas, G. H. (1977). The primary structure of phospholipase A2 from porcine pancreas—A reinvestigation. *Biochim. Biophys. Acta* **492**, 254–259.

Randolph, A. and Heinrikson, R. L. (1982). *Crotalus atrox* phospholipase A2: Amino acid sequence and studies on the function of the NH2-terminal region. *J. Biol. Chem.* **257**, 2155–2161.

Renetseder, R., Brunie, S., Dijkstra, B. W., Drenth, J., and Sigler, P. B. (1985). A comparison of the crystal structures of phospholipase A2 from bovine pancreas and *Crotalus atrox* venom. *J. Biol. Chem.* **260**, 11627–11634.

Schaloske, R. H. and Dennis, E. A. (2006). The phospholipase A2 superfamily and its group numbering system. *Biochim. Biophys. Acta* **1761**, 1246–1259.

Shiomi, K., Itoh, K., Yamanaka, H., and Kikuchi, T. (1985). Biological activity of crude venom from the crown-of-thorns starfish *Acanthaster planci*. *Nippon Suisan Gakkaishi* **51**, 1151–1154.

Shiomi, K., Kazama, A., Shimakura, K., and Nagashima, Y. (1998). Purification and properties of phospholipase A2 from the crown-of-thorns starfish (*Acanthaster planci*) venom. *Toxicon* **36**, 589–599.

Six, D. A. and Dennis, E. A. (2000). The expanding superfamily of phospholipase A2. *Biochim. Biophys. Acta* **1488**, 1–19.

Snitko, Y., Han, S. K., Lee, B. I., and Cho, W. (1999). Differential interfacial and substrate binding modes of mammalian pancreatic phospholipase A2: A comparison among human, bovine, and porcine enzymes. *Biochemistry* **38**, 7803–7810.

Stoll, U. (1996). The analysis of phospholipids from cardiac membranes by phospholipase A2. *Fett/Lipid* **98**, 26–30.

Sugino, T., Reka, R. J., and Aoi, N. (1994). Cosmetics containing sphingomyelin and lysophosphatidylcholine. *Jpn. Kokai Tokkyo* JP1996199627.

Talvinen, K. A. and Nevalainen, T. J. (2002). Cloning of a novel phospholipase A2 from the cnidarian *Adamsia carciniopados. Comp. Biochem. Physiol.* **132B**, 571–578.

Thompson, J. D., Higgins, D. G., and Gibson, T. J. (1994). CLUSTAL W: Improving the sensitivity of progressive multiple sequence alignment through sequence weighting, positions-specific gap penalties and weight matrix choice. *Nucleic Acids Res.* **22**, 4673–4680.

Tsuji, H. and Sato, H. (1996). Lysolecithin-containing compositions as additives for rice. *Jpn. Kokai Tokkyo* JP1998266234.

Uematsu, K., Kitano, M., Morita, M., and Iijima, N. (1992). Presence and ontogeny of intestinal and pancreatic phospholipase A2-like proteins in the red sea bream *Pagrus major*: An immunocytochemical study. *Fish Physiol. Biochem.* **9**, 427–438.

Vaskovsky, V. E. and Suppes, Z. S. (1972). Phospholipases of marine invertebrates-I: Distribution of phospholipase A. *Comp. Biochem. Physiol.* **43B**, 601–609.

Yamazawa, S. and Katsuragi, Y. (1996). Making compositions for bitter food and pharmaceuticals. *Jpn. Kokai Tokkyo* JP1998009896.

Zarai, Z., Bacha, A. B., Horchani, H., Bezzine, S., Zouari, N., Gargouri, Y., and Mejdoub, H. (2010). A novel hepatopancreatic phospholipase A2 from *Hexaplex trunculus* with digestive and toxic activities. *Arch. Biochem. Biophys.* **494**, 121–129.

CHAPTER 30

Applications of Marine Nutraceuticals in Dairy Products

Janak K. Vidanarachchi,[1] Maheshika S. Kurukulasuriya, A. Malshani Samaraweera, and **K. F. S. T. Silva**

Contents

Department of Animal Science, Faculty of Agriculture, University of Peradeniya, Peradeniya, Sri Lanka
[1] Corresponding author: Janak K. Vidanarachchi, *E-mail address:* janakvid@pdn.ac.lk

Advances in Food and Nutrition Research, Volume 65
ISSN 1043-4526, DOI: 10.1016/B978-0-12-416003-3.00030-5

Abstract

The concept of nutraceutical has been derived by coining the terms "nutrition" and "pharmaceutical". In this context, active substances with pharmaceutical properties are delivered to the humans through food-based approaches to prevent or treat certain disease conditions. Since the natural sources are recognized as safe for human consumption, the active substances produced in the diverse group of marine organisms have a wide role in the nutraceutical industry. These marine-derived active ingredients include certain polysaccharides, polyphenols, bioactive peptides, polyunsaturated fatty acids, and carotenoids which are known to have anticancer, anti-inflammatory, antioxidant, antiobese, hypocholesteroleic, antimicrobial, prebiotic, and probiotic activity enabling them to be applied as nutraceuticals. As the dairy products are widely accepted by the consumers, the delivering of nutraceuticals through dairy products have received a greater attention of the dairy industry. Since the incorporation of marine-derived active ingredients into the dairy products have caused minimal changes in the physico-chemical properties of the final product, marine-derived substances have been widely applied and have the potential to be applied as nutraceuticals in the dairy industry.

I. INTRODUCTION

The concept of "nutraceutical" has been introduced by Stephen DeFelice in 1989 combining the terms "nutrition" and "pharmaceutical". The nutraceuticals are substances that can be considered as a food or part of a food which provides medical or human health benefits including prevention and treatment of diseases (Rajasekaran *et al.*, 2008). Hence, food-based approaches for the improvement of the human health have been popularized during the recent past, and the nutraceuticals have received a greater attention from the global food industry.

Among the different food formulations, dairy products either fermented or nonfermented can play a major role in delivering the active ingredients of the nutraceuticals to the people due to the wide consumer acceptability of the dairy products (Ozer and Kirmaci, 2010). These nutraceutical active ingredients include certain polysaccharides, peptides, phytochemicals, vitamins, and fatty acids (FAs) which can be obtained either by natural extraction or chemical and biotechnological synthesis. Since the natural sources are recognized as safe for human consumption, the active ingredients extracted from the natural sources have attracted the attention of the food industry for the development of novel nutraceuticals. In this context, plant-derived active ingredients have been identified as an important source of nutraceutical, while usage of marine-derived secondary metabolites is a blooming aspect in the global nutraceutical industry.

The massive area of the earth surface (71%) is covered by oceans with enormous biological diversity (Baharum *et al.*, 2010), and these heterogenic group of organisms including macroalgae, microalgae, bacteria, cyanobacteria, certain fish species, and crustaceans produce secondary metabolites as an adaptation for their hostile marine environment. Though these marine-derived secondary metabolites have many human health benefits which enable them to be applied as nutraceuticals, most of these marine sources remain unexploited, limiting their beneficial effects for the humans. As the marine sources have received a greater attention during the recent past, the research on marine-derived substances (MDSs) may have discovered novel bioactive substances with important biological properties which increase their applicability as nutraceuticals in food industry including the dairy products. Therefore, this chapter was written with the objectives of reviewing the potential applications of MDSs as nutraceuticals in the global dairy industry.

II. MARINE-DERIVED COMPOUNDS APPLIED AS NUTRACEUTICALS

The bioactive compounds which result from the metabolism as well as certain structural compounds found in the marine organisms are having wide range of biological properties which allow them to thrive under stressful marine environment. Similarly, these biological properties can be used to overcome certain stressful disease conditions in the humans contributing to many human health benefits. Therefore, the MDSs such as certain polysaccharides, bioactive peptides (BPs), polyphenols, polyunsaturated fatty acids (PUFAs), and carotenoids have a greater potential to be applied as nutraceuticals, especially in the dairy industry.

A. Polysaccharides

Usually, polysaccharides play a structural role in the marine organisms protecting them from the surrounding environment. Carrageenans and alginates are linear biopolymers that have been identified as the most abundant polysaccharides found in the red and brown algae, respectively (Davis *et al.*, 2003; Kim *et al.*, 2008). Apart from the alginate, marine brown algae also comprise highly complex, sulfated matrix polysaccharides called fucoidans. The complex structure of the fucoidans extracted from different marine species varies in sugar composition, sulfate content, position of sulfate groups, molecular weight, linkage mode, and sequence of sugar residue (Li *et al.*, 2008). The presence of sulfate groups in the structure improves the biological properties of fucoidans which enables their application as nutraceuticals in the dairy industry.

Similar to the matrix polysaccharides found in algae, the exoskeleton of the crustaceans and cell wall of marine fungi comprise a linear amino polysaccharide, chitin, which has many industrial applications. Since chitin is the major structural component found in the crustacean exoskeleton, the sufficient amounts of this polysaccharide for the industrial applications result from the seafood processing industry. The complex structure of chitin is composed of poly β(1→4)2-acetamido-2-deoxy-D-glucose polysaccharide chains which are developed from chitin nanofibrils (Morganti and Morganti, 2008). The polysaccharide structure of chitin is deacetylated into different degrees for the production of chitosan which has three functional groups: an amino group at C-2 position and primary and secondary hydroxyl groups at C-3 and C-6 positions, respectively (Furusaki *et al.*, 1996). The chemical or enzymatic hydrolysis of chitin and chitosan has resulted in the production of chitooligosaccharides (COSs) with a hydroxylate of glucosamines or *N*-acetyl glucosamines which have been identified as an important constituent of nutraceuticals (Il'ina and Varlamov, 2004; Verum *et al.*, 2001).

More recently, the *N*-acetyl glucosamines have been also found as a constituent in the exopolysaccharides secreted from the marine bacteria. The exopolysaccharides from marine bacteria consist of different combinations of neutral sugars, proteins, amino acids, *N*-acetyl glucosamines, *N*-acetyl galactosamines, sulfate groups, and uronic acids. The species-dependant amount of uronic acid present in the exopolysaccharide highly contributes to the property of binding capacity which increases their industrial applications including in the nutraceutical industry (Vidanarachchi and Kurukulasuriya, 2011).

B. Bioactive peptides

As polysaccharides, proteins also mainly play a structural role in the marine organisms. The enzymatic hydrolysis of the marine proteins such as collagen as well as gelatin results in BPs which may have the potential to be used as nutraceuticals (Kim and Wijesekara, 2010). The BPs are health-beneficial compounds with 3–20 amino acid residues, the beneficial effects of which depend on the amino acid composition and the sequence present in these molecules (Pihalanto-Leppala, 2001). Since these favorable BPs are released from the proteins during the gastrointestinal digestion (Ngo *et al.*, 2011), the proteins such as collagen and gelatin which contains the sequences of important BPs have a greater potential to be used as nutraceuticals in the industrial scenario. Further, the BP, nisin, has been isolated from the marine *Lactobacillus* species that have a wide role as a nutraceutical in the dairy products (Ra *et al.*, 1996).

C. Polyunsaturated fatty acids

Certain marine fish species and algae have been identified as sources for PUFAs which are rich in ω-3 or ω-6 FAs and the precursors of ω-3 FAs, α-linoleic acid (ALA). Eicosapentaenoic acid (EPA) and docosahexaenoic acid (DHA) are considered as the main ω-3 FAs which are involved in the human health benefits. The presence of EPA and DHA in the marine-derived foods increases the applicability of PUFAs as nutraceuticals in food industry (Harris *et al.*, 2010).

D. Phenolic compounds

Phenolic compounds are found in the marine algae mainly as a mechanism of adaptation for the oxidative stresses. Usually, phlorotannins are the most abundant polyphenols found in the marine brown algae whereas flavonoids contribute to considerable proportion of total phenolic content in the green algae. The brown algal phlorotannin profile mainly consists of phloroglucinol, eckol, and dieckol (Ahn *et al.*, 2007; Shibata *et al.*, 2008), while the flavonoids found in the green algae are highly complex molecules with multiactive properties such as antiangiogenesis and antioxidant ability. Antioxidant activity has been also reported from the phlorotannins, allowing both these phenolic compounds to be used as active ingredients in the nutraceuticals (Arct and Pytkowska, 2008).

E. Carotenoids

Similar to polyphenols, carotenoids, synthesized in certain marine bacteria and algal species, also have antioxidant properties which increase their applicability as nutraceuticals. Carotenoids are lipid-soluble, natural pigments with 40-carbon structures (Lee *et al.*, 2007). Among the different carotenoids synthesized within the marine organisms, β-carotene, astaxanthin, and fucoxanthin are known to have a high antioxidant capacity. Moreover, the presence of an additional hydroxyl and ketone groups in each β-ionine ring of astaxanthin molecule leads to a high antioxidant capacity, increasing their importance as nutraceuticals (Capelli and Cysewski, 2006). Currently, commercial preparation of β-carotene and astaxanthin is available using *Dunaliella* species and *Haematococcus* species, respectively (Vidanarachchi and Kurukulasuriya, 2011).

Apart from the antioxidant property, brown algal fucoxanthins are known to have a thermogenic property which increases the applicability of fucoxanthin as a nutraceutical. Therefore, β-carotene, astaxanthin, and fucoxanthin have a high potential to be applied in the food industry as nutraceuticals.

III. APPLICATIONS OF MARINE-DERIVED NUTRACEUTICALS IN DAIRY INDUSTRY

The marine-derived polysaccharides, BPs, PUFAs, polyphenols, and carotenoids have been identified as human health-beneficial compounds due to the presence of important biological properties including anticancer, antioxidative, anti-inflammatory, antiangiogenesis, thermogenic, antimicrobial, hypocholesterolemic, prebiotic, and probiotic activities. These biological properties enable the MDSs to be applied as nutraceuticals which improve the health status of the people.

A. Antitumor/anticancer marine nutraceuticals

The global cancer incidence remains as an unresolved problem, and thus the nutraceuticals with anticancer effects have a wide role in treatment and prevention of cancer in suspected people. In this context, many studies have reported the anticancer activity of marine-derived fucoidans, chitosan, COS, β-carotene, astaxanthin, and phlorotannins (Table 30.1). However, the mode of action of these marine-derived anticancer compounds varies, and the cancer inhibitory pathways include activation of P53 antiproliferative gene, induction of cell apoptosis through various pathways, inhibition of angiogenesis, and inhibition of matrix metalloproteinases (MMPs; Fig. 30.1).

Usually, the inhibition of cell apoptosis is a characteristic of cancers, and the induction of apoptosis through nutraceuticals can play a major role in destruction of cancer cells. Marine-derived fucoidans and phlorotannins enhance cell apoptosis through activation of caspases which are enzymes involved in the apoptotic pathway (Kim *et al.*, 2010; Kong *et al.*, 2009), while κ-carrageenan oligosaccharides and carotenoids stimulate the host immune system for the cancer cell apoptosis. Though κ-carrageenan oligosaccharides stimulate both TNF-α- and lymphocyte-mediated cytotoxicity (Yuan *et al.*, 2006), β-carotene stimulates the host immune system by enhancing the expression of tumor necrosis factors (Jawanda, 2009) whereas astaxanthin destroys the cancer cells via T cytotoxic cells (Jyonouchi *et al.*, 2000). Further, Shan *et al.* (1999) have also reported an increased antitumor activity through cytotoxic T lymphocytes from an algal polysaccharide.

Apart from the stimulation of host immune system, β-carotene is also responsible for the cancer cell apoptosis via expression of P53 antiproliferative gene in cancer cells and inhibition of angiogenesis which prevents the blood supply to the developing cancer cells (Jawanda, 2009). Antiangiogenesis has been also reported from the fucoidans extracted from certain seaweeds favoring the apoptosis of cancer cells (Ye *et al.*, 2005).

TABLE 30.1 *In vitro* anticancer activity of marine-derived fucoidans, chitosan, COS, β-carotene, astaxanthin, and phlorotannins in different human cell lines

Type of organism	Active ingredient	Cell line	Dose of inhibition	Mode of action	Reference
Brown algae	Fucoidans	HT-29 and HCT116 human colon cancer cells	5–20 μg/mL	Enhance caspases	Kim *et al.* (2010)
		PC-3 (prostate cancer), HeLa (cervical cancer), A549 (alveolar carcinoma), HepG2 (hepatocellular carcinoma)	N/A	N/A	Synytsya *et al.* (2010)
		Breast cancer MCF-7 cells	N/A	Enhance caspases	Miyamoto *et al.* (2009)
		Human uterine carcinoma HeLa cells	N/A	Antiangiogenesis	Ye *et al.* (2005)
		Human fibrosarcoma HT1080 cells	N/A	Inhibit MMPs	Ye *et al.* (2005)
	κ-Carrageenan oligosaccharides	S180 cell suspension	N/A	Stimulate TNF-α and lymphocyte-mediated cytotoxicity	Yuan *et al.* (2006)
Red algae	Phloroglucinol	Breast cancer MCF-7 cells	N/A	Enhance caspases	Kong *et al.* (2009)

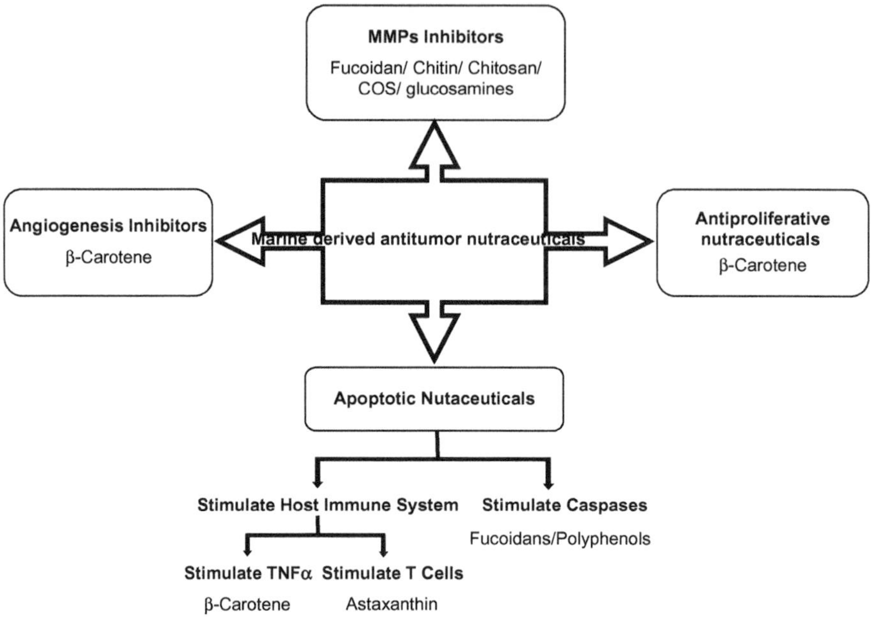

FIGURE 30.1 Mode of action of marine-derived antitumor nutraceuticals.

Though β-carotene exerts the anticancer property through several mechanisms, Chew *et al.* (1999) have reported a higher anticancer activity from astaxanthin which increases the potential application of astaxanthin as a nutraceutical in the food industry.

Moreover, the nutraceuticals which have the ability to inhibit MMPs are important for the prevention of tumor invasion to the surrounding tissues. Fucoidans as well as chitin and chitin derivatives such as chitosan, COS, and glucosamines (Kim and Kim, 2006; Rajapakse *et al.*, 2007) have been identified as important for the inhibition of MMPs which degrade the extracellular matrix and allow the tissue cells to get invaded. Since marine-derived fucoidans inhibit the cancer development as well as the tumor invasion, these complex polysaccharides have been identified as an important source of anticancer nutraceutical (Ye *et al.*, 2005). Further, the anticancer activity of fucoidan molecules increases with sulfate content in the structure (Cho *et al.*, 2011) which favors the potential application of highly sulfated fucoidans as nutraceuticals in the dairy products. Since fucoidans as well as all the other marine-derived anticancer compounds exert their anticancer activity in a dose-dependant manner (Kim *et al.*, 2010), incorporation of appropriate concentrations of these anticancer compounds according to the dietary recommendations is essential for the application of these marine-derived nutraceuticals in the dairy products.

B. Anti-inflammatory marine-derived nutraceuticals

Inflammation is an important biological process which is expressed against the disturbance of normal body functions. However, the inflammatory reaction can attack the host's own tissues, and chronic inflammation can lead to certain diseases such as arthritis, hepatitis, gastritis, periodontal disease, colitis, atherosclerosis, pneumonia, and neuroinflammatory diseases (Kim and Kim, 2010; Pangestuti and Kim, 2004). Further, activation of MMP enzyme systems leads to tissue destruction aggravating the inflammatory condition. In this context, potential anti-inflammatory nutraceuticals such as fucoidan, chitin, chitosan, COS, and glucosamine play a significant role in prevention of chronic inflammation either by inhibiting the production of nitric oxide (NO), prostaglandin E_2, inducible NO synthase, cyclooxygenase-2, or proinflammatory cytokines (Kim and Kim, 2010; Park *et al.*, 2011; Peerapornpisal *et al.*, 2010; Rajapakse *et al.*, 2008; Yoon *et al.*, 2007) or by inhibition of MMPs (Moon *et al.*, 2009; Zhang and Kim, 2009) in a dose-dependant manner. However, the ω-3 FAs such as EPA, which has been identified as a main active ingredient in the nutraceuticals, exert the anti-inflammatory action by inhibiting the metabolism of arachidonic acid by downregulation, thereby preventing the production of proinflammatory metabolites (Geering *et al.*, 2006).

Since the inflammatory diseases have a high public health significance, many studies have been conducted on anti-inflammatory effects of MDSs in human and animal models (Ananthi *et al.*, 2010; Bruyere *et al.*, 2004; Myers *et al.*, 2010), especially for osteoarthritis. Bruyere *et al.* (2004) have observed an anti-inflammatory effect of glucosamines in prevention of osteoarthritis while Myers *et al.* (2010) also have reported decreased symptoms of osteoarthritis from fucoidans found in the seaweed extract nutrient complex given orally which will be more applicable as an anti-inflammatory nutraceutical.

C. Antioxidant marine-derived nutraceuticals

The antioxidant properties of certain marine-derived nutraceuticals are believed to be linked to the ability of these compounds to relieve the inflammation (Guerin *et al.*, 2003). Similarly, the marine-derived antioxidant compounds have a major function in prevention of tissue damage in aging, arthrosclerosis, and cancer through neutralizing reactive oxygen species (ROS). These ROS, hydrogen peroxide, hydroxyl radical (OH^-), superoxide anion, and NO are synthesized in the body due to the various stresses caused to the body tissues (Lee *et al.*, 2004). Therefore, antioxidant properties of MDSs have greater potential to be applied as nutraceuticals in the food industry.

The MDSs, which are known to have high antioxidant activity, include β-carotene, astaxanthin, fucoxanthin, polyphenols, and fucoidans. The different functional groups present in the structure of these MDSs involve in their mode of action which can be mainly categorized as energy-absorbing antioxidants and electron-donating antioxidants (Table 30.2). According to the review of Guerin *et al.* (2003), carotenoids can absorb the excited energy of singlet oxygen onto the double bonds present in their chains, preventing the possible tissue damage through neutralizing the singlet oxygen. Further, carotenoids also act as radical scavengers in the oxidative stresses where singlet oxygen is not formed (Murthy *et al.*, 2005; Tsuchihashi *et al.*, 1995). Moreover, fucoxanthin molecules can play an important role in 2,2-diphenyl-1-picryhydrasyl radical scavenging activity which is unique to the fucoxanthin among the different marine-derived carotenoids (Nomura *et al.*, 1997).

Apart from the energy-absorbing marine-derived antioxidants, certain electron-rich MDSs such as polyphenols, chitosan, COS, and sulfated polysaccharides act as electron donors to free radicals including superoxide radical, NO, OH^-, and hydrogen peroxide (Ahn *et al.*, 2007; Shibata

TABLE 30.2 Different functional groups responsible for the antioxidant activity of marine-derived antioxidant compounds

Antioxidant compound	Species	Functional group	Reference
β-Carotene	*Dunaliella salina* *Dunaliella bardawil*	Double bonds in the carotenoid chain	Guerin *et al.* (2003)
Astaxanthin	*Haematococcus* species	Hydroxyl and ketone group on the each β-ionine ring	Capelli and Cysewski (2006); Lyons and O'Brien (2002)
Fucoxanthin	*Sargassum horneri* *Sargassum thunbergii*	Double allenic carbon at C-7′ position and two hydroxyl groups	Yan *et al.* (1999); Sachindra *et al.* (2007); Miyashita *et al.* (2011)
Phlorotannins		Phenolic hydroxyl groups	Shibata *et al.* (2008)
Fucoidans	*Laminaria japonica* *Fucus vesiculosus*	Sulfate group (low molecular weight)	Souza *et al.* (2007)

et al., 2008; Vijayaval and Martinez, 2010). The study of Vijayaval and Martinez (2010) has suggested the potential application of MDSs with antioxidant properties as nutraceuticals in the food industry.

D. Antiobese marine nutraceuticals

The increasing incidence of obesity is becoming a medicinal problem around the world. Thus, the prevention of accumulation of fat in the body or increased usage of adipose tissue is essential for the prevention of obesity.

The increased usage of adipose tissue has been reported from the marine carotenoid, fucoxanthin. Usually, the mitochondrial uncoupling protein 1 (UCP1) is a key molecule for the metabolic thermogenesis and fucoxanthin that reduces the abdominal fat accumulation through enhancing the expression of UCP1 in the human body (Maeda *et al.*, 2005; Miyashita *et al.*, 2011). Further, fucoxanthin has been responsible for the suppression of adipocyte differentiation which may also increase the applicability of fucoxanthin as antiobese nutraceutical (Maeda *et al.*, 2006). Since the fucoxanthine is proven to be safe for human consumption without any toxicities (Beppu *et al.*, 2009), it can be applied as a nutraceutical in the food industry. However, the hypercholesterolemic effects of fucoxanthin that has been observed by Beppu *et al.* (2009) limit the applicability of fucoxanthin and need further investigation prior to the application of fucoxanthin as a nutraceutical.

Similar to the fat catabolism, reduced fat absorption also plays a significant role in the human antiobesity. The reduced fat absorption that has been observed from the chitosan reflects the possibility of application of chitosan as a nutraceutical for the human obesity (Kaats *et al.*, 2006). Therefore, the fat catabolism enhancers such as fucoxanthin and the fat absorption inhibitors such as chitosan can be used to improve the human health by dietary inclusion of such compounds.

E. Marine-derived nutraceuticals against cardiovascular diseases

The term cardiovascular disease (CVD) is used for a group of disorders of the heart and blood vessels which includes hypertension, coronary heart disease, heart failure, and peripheral vascular disease (Rajasekaran *et al.*, 2008). The hypercholesterolemia is known to have highly contributed for the development of CVD, and thus the hypocholesterolemic nutraceuticals have received a greater attention during the recent past. The hypocholesterolemic effects of chitosan are mainly due to the decreased fat absorption from the diets whereas Xu *et al.* (2007) have reported a mechanism which increases the lipid metabolism through upregulation of hepatic LDL-receptor mRNA expression. The recent review of Kong and

Kim (2010) has suggested several possible mechanisms of chitosan for the reduced fat absorption through the formation of chitosan–cholesterol complexes in the digestive system. According to Bokura and Kobayashi (2003), chitosan entraps the lipid molecules and forms a gel using gastric HCl which makes them unavailable for the intestinal absorption. Further, the chitosan–cholesterol complexes can be formed either by interactions between positively charged chitosan and negatively charged cholesterol molecules (Tai *et al.*, 2000) or by hydrogen bonding between the OH^- of cholesterol and amine groups of chitosan (Pavinatto *et al.*, 2008).

Similar to chitosan, a cholesterol-lowering ability of hydrophilic COS has been observed by Kim *et al.* (1998). In contrast to the above study, an early study has reported poor hypocholesterolemic effects from COSs (Sugano *et al.*, 1988). Since the controversial results of the two studies may have contributed from the difference of composition of the COS, the appropriate COS composition needed to be selected prior to applying them as nutraceuticals. However, the presence of hypocholesterolemic effects increases the potential application of COSs in the food industry (Vidanarachchi *et al.*, 2010), especially in the dairy products.

The hypocholesterolemic effects as well as the antihypertensive effects have a similar importance in prevention of CVD. The flavonoids can block the angiotensin-converting enzyme which is important for the rise of blood pressure through inhibiting cyclooxygenase enzyme. Further, the inhibition of cyclooxygenase prevents the prostaglandin breakdown leading to reduced platelet stickiness and aggregation which is important for the maintenance of a healthy vascular system (Rajasekaran *et al.*, 2008). Apart from the chitin derivatives and flavonoids, the marine-derived algal polysaccharides, carrageenans, alginate, and fucoidans are also reported to be effective in lowering cholesterol and blood pressure than the other fiber sources (Jiménez-Escrig and Sánchez-Muniz, 2000).

Moreover, the marine-derived ω-3 FAs are reported to be highly beneficial for the prevention of the CVDs, exerting their action in many ways including hypocholesterolemia, antihypertention, reduction of heart rate, and platelet aggregation (Levitan *et al.*, 2010; Marinangeli and Jones, 2010). Therefore, the chitin derivatives, flavonoids, carrageenan, alginate, fucoidans, and the ω-3 FAs have a greater potential for application as nutraceuticals in the dairy industry.

F. Marine-derived antimicrobial nutraceuticals

Milk is a favorable media for the pathogenic microbes such as *Listeria monocytogenes* and *Staphylococcus aureus*, and these organisms can cause diseases in humans, especially in pregnant women, neonates, and immune-compromised adults. In this context, marine-derived antimicrobial compounds can play a significant role as nutraceuticals. Further, the antimicrobial properties can improve the shelf life of foods. The extracts from the marine algae, chitin derivatives, and certain bacteriocin have been identified as important antimicrobial compounds which have the potential of preventing the disease risk in the people (Table 30.3).

The antimicrobial activities of chitosan and COS depend mainly on the degree of polymerization and the degree of deacetylation of their molecular structure (Vidanarachchi *et al.*, 2010). The same authors have reported two distinct mechanisms of antibacterial activity of chitin derivatives toward Gram-positive and Gram-negative bacteria which depend on the cell wall constituents of these bacteria.

Similarly, many marine bacterial bacteriocins such as nisin are known to have antibacterial activities against the foodborne pathogens, especially which can be found in the dairy products (Arques *et al.*, 2008; Rajaram *et al.*, 2010). Thus, the incorporation of bacteriocins in the dairy products can be used as a tool for the maintenance of human health, preventing the incidence of foodborne diseases. Therefore, marine-derived antimicrobials have a high potential to be used as nutraceuticals in the dairy industry.

G. Marine nutraceuticals as probiotics

The antimicrobial properties of MDSs also enable them to be used as probiotics in the food products. The nisin from the *Lactobacillus* species has a higher commercial utility in the dairy products. Since the marine

TABLE 30.3 Minimum inhibitory concentrations of marine-derived antimicrobial compounds against different pathogenic bacteria

Type of bacteria	Bacterial strain	MDS	Minimum inhibitory concentration	Reference
Gram-positive	*Listeria monocytogenes*	COS	880 ± 438^a	Lin *et al.* (2009)
			2.5^b	Benabbou *et al.* (2009)
		Nisin	N/A	Arques *et al.* (2008)
	Staphylococcus aureus	COS	0.12 ± 0.01^c	Wang *et al.* (2007)
		Nisin	N/A	Arques *et al.* (2008)
Gram-negative	*Escherichia coli*	COS	0.08 ± 0.00^c	Wang *et al.* (2007)
			55 ± 27^a	Lin *et al.* (2009)
	Pseudomonas aeruginosa	COS	200 ± 122^a	Lin *et al.* (2009)

[a] MIC as μg/mL.
[b] MIC as mg/mL.
[c] MIC as %.

Lactobacillus species are known to have more biological properties than the terrestrial species (Itoi *et al.*, 2009), the nisin of marine origin has a higher potential to be applied as a nutraceutical.

Apart from the direct probiotic activities, marine seaweed-derived κ-carrageenan, alginate, and chitosans help to deliver the probiotic organisms to the target sites. The encapsulation of probiotic organism with MDSs prevents the gastrointestinal digestion of the probiotic organism, enabling them to be present at the large intestine which is the target site of probiotics (Samaraweera *et al.*, 2011). Therefore, MDSs are having direct probiotic effects as well as the probiotic delivering ability which increases their applicability as nutraceuticals in the dairy industry.

H. Marine nutraceuticals as prebiotics

Prebiotics are nondigestible, selectively fermented compounds that stimulate the growth/activity of beneficial gut microbiota which, in turn, confers the health benefits to the host. Usually, prebiotics are oligosaccharides such as COSs, while certain other algal polysaccharides are also known to have a prebiotic activity (O'Sullivan *et al.*, 2010). The highly deacetylated COSs have the ability to suppress the pathogenic bacteria (Pan *et al.*, 2009) while stimulating the growth of *Lactobacilli* and *Bifidobacteria* at a low concentration, reflecting the importance of COS as a prebiotic (Lee *et al.*, 2002).

Bifidogenic benefits have been also reported from the exopolysaccharide produced by marine lactic acid bacteria (Honypattarakere *et al.*, 2010). Further, the cyanobacterial biomass of *Spirulina platensis* is able to stimulate both *Lactobacillus* and *Bifidobacterium* species, favoring their prebiotic effect. Moreover, the *S. platensis* is rich in other nutrients, increasing the human health benefits and the applicability of MDSs as nutraceuticals in the dairy industry (Varga *et al.*, 1999a,b).

IV. APPLICABILITY OF THE MARINE-DERIVED NUTRACEUTICALS IN THE DAIRY INDUSTRY

The MDSs have many biological properties which are beneficial for the human health improvement. As dairy products are widely accepted by the consumers and the global dairy sector is expanding continuously, the possibility of delivering the biological properties through dairy products has many advantages. Among the functional foods, dairy-based functional foods account for nearly 43% of the market which is mainly contributed from the fermented dairy products (Ozer and Kirmaci, 2010). According to the same review, there is a high possibility of incorporating nutraceuticals into the dairy beverages for the benefits of human health

(Fig. 30.2). However, the other nonbeverage dairy products are also being identified as potential vehicles for the marine nutraceuticals.

As the addition of compounds can alter the physico-chemical properties of the final product, the incorporation of marine nutraceuticals into the dairy products is a challenge. However, many dairy products with marine nutraceuticals have been developed, maintaining the acceptable qualities of the final product (Table 30.4).

β-Carotene has been widely used as a food colorant in many food products including the dairy products such as yoghurt and cheese (Dufosse *et al.*, 2005). Therefore, the health benefits of β-carotene can be easily delivered to the humans through functional dairy products. Further, dairy products have been identified as the main vehicle for the probiotic supplements. The addition of specific probiotic strain at a level of 10^6–10^7CFU/g improves the health status of the humans without changing the quality of the dairy product (Granato *et al.*, 2010). Since the digestive enzymes can destroy the active substances of the nutraceuticals, especially the probiotics, the delivering of probiotics to the target site has received a greater attention in development of nutraceutical dairy products. Choi *et al.* (2006) have used microencapsulated COSs in milk for the production of healthy milk with minimum adverse effects on the physical and sensory properties.

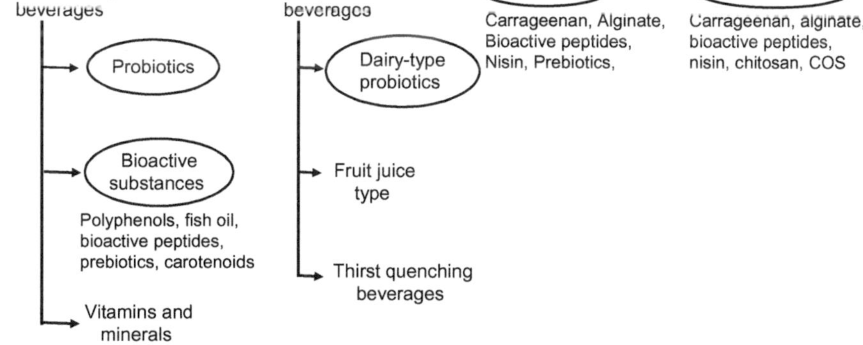

FIGURE 30.2 Schematic diagram of the marine-derived nutraceutical applications in the dairy products adapted from Ozer and Kirmaci, 2010.

TABLE 30.4 Dairy products with different marine-derived nutraceuticals

Dairy product	Marine-derived nutraceutical	Reference
Yoghurt	ω-3 FAs	Awaisheh *et al.* (2005) Let *et al.* (2007a,b)
	Alginate	Harris *et al.* (2010) Kailasapathy (2006)
Cheese	ω-3 FAs	Ye *et al.* (2009) Mele *et al.* (2011)
Liquid milk	COS	Choi *et al.* (2006)
	ω-3 FAs and ALA	Chilliard *et al.* (2001) Harris *et al.* (2010)
Fortified dairy beverages	Polyphenols, fish oil, bioactive peptides, prebiotics, carotenoids, probiotics	Ozer and Kirmaci (2010)
Whey-based dairy products	Probiotics	Ozer and Kirmaci (2010)

The addition of certain nutraceuticals such as ω-3 FAs to the dairy products has led to development of oxidative rancidity in the final product (Lal *et al.*, 2006). This problem of low oxidative stability of the ω-3 FAs-added dairy products can be overcome by the use of appropriate combinations of marine-derived emulsifiers and stabilizers such as carrageenan and alginate into the dairy products (Let *et al.*, 2007a) or by homogenization of milk (Let *et al.*, 2007b). Since carrageenan and alginates are also considered as important nutraceutical compounds, the dairy products with these marine polysaccharides may have additional human health benefits. In contrast to Lal *et al.* (2006), Awaisheh *et al.* (2005) and Mele *et al.* (2011) have produced yoghurt and cheese with ω-3 FAs, respectively, without the problem of lipid oxidation in the final product.

Usually, the ω-3 FAs or ALA-incorporated dairy products are produced by the synthesis of ALA through dairy starter culture, addition of the powder of ω-3 FAs into the dairy product, or enrichment of the cow milk through animal feed supplement (Harris *et al.*, 2010; Jiang *et al.*, 1998). Since the ω-3 FAs extracted from the marine fish species can alter the sensory properties of the dairy product, the enrichment of the cow milk with ω-3 FAs has received a greater attention. The consumer-benefited milk fat composition has been achieved through the addition of marine algae as well as the marine oils into the diet of the dairy animals (Chilliard *et al.*, 2001; Franklin *et al.*, 1999). The dairy products produced from the favorable FA-enriched milk can deliver the health benefits to the humans. Therefore, the dairy products can be identified as an important marine-derived nutraceutical delivering method which favors the improvement of the human health.

V. CONCLUSION

Many MDSs such as polysaccharides, COSs, BPs such as nisin, PUFAs, polyphenols, and carotenoids can be identified as marine-derived nutraceuticals which have many human health benefits including anticancer, anti-inflammatory, antioxidative, antiobese, hypocholesterolemic, antimicrobial, probiotic, and prebiotic activity. Therefore, MDSs have been widely applied as well as have a higher potential to be applied as nutraceuticals in the dairy industry.

REFERENCES

Ahn, G. N., Kim, K. N., Cha, S. H., Song, C. B., Lee, J., Heo, M. S., Yeo, I. K., Lee, N. H., Jee, Y. H., Kim, J. S., Heu, M. S., and Jeon, Y. J. (2007). Antioxidant activities of phlorotannins purified from Ecklonia cava on free radical scavenging using ESR and H_2O_2-mediated DNA damage. *Eur. Food Res. Technol.* **226**, 71–79.

Ananthi, S., Raghavendran, H. R. B., Sunil, A. G., Gayathri, V., Ramakrishnan, G., and Vasanthi, H. R. (2010). *In vitro* antioxidant and *in vivo* anti-inflammatory potential of crude polysaccharide from *Turbinaria ornata* (Marine Brown Alga). *Food Chem. Toxicol.* **48**, 187–192.

Arct, J. and Pytkowska, K. (2008). Flavonoids as components of biologically active cosmeceuticals. *Clin. Dermatol.* **26**, 347–357.

Arques, J. L., Rodriguez, E., Nunez, M., and Medina, M. (2008). Antimicrobial activity of nisin, reuterin, and the lactoperoxidase system on *Listeria monocutogenes* and *Staphylococcus aureus* in Caujada a semisolid dairy product manufactured in Spain. *J. Dairy Sci.* **91**, 70–75.

Awaisheh, S. S., Haddadin, M. S. Y., and Robinson, R. K. (2005). Incorporation of selected nutraceuticals and probiotic bacteria into a fermented milk. *Int. Dairy J.* **15**, 1184–1190.

Baharum, S. N., Beng, E. K., and Mokhtar, M. A. A. (2010). Marine microorganisms: Potential application and challenges. *J. Biol. Sci.* **10**, 555–564.

Benabbou, R., Zihler, A., Desbiens, M., Kheadr, E., Subirade, M., and Fliss, I. (2009). Inhibition of *Listeria monocytogenes* by a combination of chitosan and divergicin M35. *Can. J. Microbiol.* **55**, 347–355.

Beppu, F., Niwano, Y., Tsukui, T., Hosokawa, M., and Miyashita, K. (2009). Single and repeated oral dose toxicity study of fucoxanthin (FX), a marine carotenoid, in mice. *J. Toxicol. Sci.* **34**, 501–510.

Bokura, H. and Kobayashi, S. (2003). Chitosan decreases total cholesterol in women: A randomized double blind placebo-controlled trial. *Eur. J. Clin. Nutr.* **57**, 721–725.

Bruyere, O., Pavelka, K., Rovati, L. C., Deroisy, R., Olegarova, M., Gatterova, J., Giacovelli, G., and Reginster, J. V. (2004). Glucosamine sulfate reduces osteoarthritis progression in ostmenopausal women with knee osteoarthritis: Evidence from two 3-year studies. *Menopause* **11**, 138–143.

Capelli, B. and Cysewski, G. (2006). Natural Astaxanthin: King of the Carotenoids. Cyanotech Corporation, California, USA.

Chew, B. P., Park, J. S., Wong, M. W., and Wong, T. S. (1999). A comparison of the anticancer activities of dietary beta-carotene, canthaxanthin and astaxanthin in mice *in vivo*. *Anticancer Res.* **19**, 1849–1853.

Chilliard, Y., Ferlay, A., and Doreau, M. (2001). Effect of different types of forages, animal fat or marine oils in cow's diet on milk fat secretion and composition, especially conjugated linoleic acid (CLA) and polyunsaturated fatty acids. *Livest. Prod. Sci.* **70**, 31–48.

Cho, M. L., Lee, B., and You, S. (2011). Relationship between oversulfation and conformation of low and high molecular weight fucoidans and evaluation of their *in vitro* anticancer activity. *Molecules* **16**, 291–297.

Choi, H. J., Ahn, J., Kim, N. C., and Kwak, H. S. (2006). The effect of microencapsulated chitooligosaccharides on physical and sensory properties of milk. *Asian Aus. J. Anim. Sci.* **19**, 1347–1353.

Davis, T. A., Volesky, B., and Mucci, A. (2003). A review of the biochemistry of heavy metal absorption by brown algae. *Water Res.* **37**, 4311–4330.

Dufosse, L., Galaup, P., Yaron, A., Arad, S. M., Blanc, P., Murthy, K. N. C., and Ravishankar, G. A. (2005). Microorganisms and microalgae as sources of pigments for food use: A scientific oddity or an industrial reality? *Trends Food Sci. Technol.* **16**, 389–406.

Franklin, S. T., Martin, K. R., Baer, R. B., Schingoethe, D. J., and Hippen, A. R. (1999). Dietary marine algae (*Schizochytrium sp.*) increases concentrations of conjugated linoleic, docosahexaenoic and transvaccenic acids in milk of dietary cows. *J. Nutr.* **129**, 2048–2054.

Furusaki, E., Ueno, Y., Sakairi, N., and Tokura, S. (1996). Facile preparation and inclusion ability of a chitosan derivative bearing carboxymethyl-β-cyclodextrin. *Carbohydr. Polym.* **9**, 29–34.

Geering, R., Engelke, C., and Chandler, T. (2006). Polyunsaturated fatty acids—Are they alter native anti-inflammatories? European Companion Animal Healthhttp://www.touchbriefings.com/pdf/2398/Bionovate_tech.pdf Accessed on 24.06.2011.

Granato, D., Branco, G. F., Cruz, A. G., Faria, J. A. F., and Shah, N. P. (2010). Probiotic dairy products as functional foods. *Compr. Rev. Food Sci. Food Saf.* **9**, 455–470.

Guerin, M., Huntley, M. L., and Olaizola, M. (2003). *Haematococcus* astaxanthin: Applications for human health and nutrition. *Trends Biotechnol.* **21**, 210–216.

Harris, K. A., Hill, A. M., and Kris-Etherton, P. M. (2010). Health benefits of marine derived omega-3 fatty acids. *ACSMS Health Fit. J.* **14**, 22–28.

Honypattarakere, T., Cherntong, N., Wickienchot, S., Kolida, S., and Rastall, R. A. (2010). In vitro prebiotic evaluation of exopolysaccharides produced by marine isolated lactic acid bacteria. *J. Biotechnol.* S1–S576.

Il'ina, A. V. and Varlamov, V. P. (2004). Hydrolysis of chitosan in lactic acid. *Appl. Biochem. Microbiol.* **40**, 354–358.

Itoi, S., Yuasa, K., Washio, S., Abe, T., Ikuno, E., and Sugita, H. (2009). Phenotypic variation in *Lactococcus lactis* subsp. Lactis isolates derived from intestinal tract of marine and freshwater fish. *J. Appl. Microbiol.* **107**, 867–874.

Jawanda, M. K. (2009). Antitumour activity of antioxidant—Overview. *Int. J. Clin. Dent.* **1**, 3–7.

Jiang, J., Björck, L., and Fondén, R. (1998). Production of conjugated linoleic acid by dairy starter cultures. *J. Appl. Microbiol.* **85**, 95–102.

Jiménez-Escrig, A. and Sánchez-Muniz, F. J. (2000). Dietary fiber from edible seaweeds: Chemical structure, physicochemical properties and effects on cholesterol metabolism. *Nutr. Res.* **4**, 585–598.

Jyonouchi, H., Sun, S., Iijima, K., and Gross, M. D. (2000). Antitumour activity of astaxanthin and its mode of action. *Nutr. Cancer* **36**, 59–65.

Kaats, G. R., Michalek, J. E., and Preuss, H. G. (2006). Evaluating efficacy of a chitosan product using a double-blinded, placebo-controlled protocol. *J. Am. Coll. Nutr.* **25**, 389–394.

Kailasapathy, K. (2006). Survival of free and encapsulated probiotic bacteria and their effect on sensory properties of yogurt. *LWT Food Sci. Technol.* **39**, 1221–1227.

Kim, M. M. and Kim, S. K. (2006). Chitooligosaccharides inhibit activation and expression of matrix mettaloproteinase-2 in human dermal fibroblast cells. *FEBS Lett.* **580**, 2661–2666.

Kim, M. M. and Kim, S. K. (2010). Antiinflammatory activity of chitin, chitosan, and their derivatives. *In* "Chitin, Chitosan, Oligosaccharides and Their Derivatives", (S. K. Kim, Ed.), pp. 215–221. Taylor & Francis, New York.

Kim, S. K. and Wijesekara, I. (2010). Development and biological activities of marine derived bioactive peptides: A review. *J. Funct. Foods* **2**, 1–9.

Kim, S. J., Kang, S. Y., Park, S. L., Shin, T. K., and Ko, Y. H. (1998). Effects of chitooligosaccharides on liver function in the mouse. *Korean J. Food Sci. Technol.* **30**, 693–696.

Kim, S. K., Ravichandran, Y. D., Khan, S. B., and Kim, Y. (2008). Prospective of the cosmeceuticals derived from marine organisms. *Biotechnol. Bioprocess Eng.* **13**, 511–523.

Kim, E. J., Park, S. Y., Lee, J. Y., and Park, J. H. I. (2010). Fucoidan present in brown algae induces apoptosis of human colon cancer cells. *BMC Gastroenterol.* **10**, 96–107.

Kong, C. S. and Kim, S. K. (2010). Antidiabetic activity and cholesterol-lowering effect of chitin, chitosan and their derivatives. *In* "Chitin, Chitosan, Oligosaccharides and Their Derivatives", (S. K. Kim, Ed.), pp. 285–292. Taylor & Francis, New York.

Kong, C. S., Kim, J. A., Yoon, N. Y., and Kim, S. K. (2009). Induction of apoptosis by phloroglucinol derivative from *Ecklonia cava* in MCF-7 human breast cancer cells. *Food Chem. Toxicol.* **47**, 1653–1658.

Lal, S. N. D., O'Connor, C. J., and Eyres, L. (2006). Application of emulsifiers/stabilizers in dairy products of high rheology. *Adv. Colloid Interface Sci.* **123–126**, 433–437.

Lee, H., Park, Y., Jung, J., and Shin, W. H. (2002). Chitosan oligosaccharides, DP2-8, have prebiotic effects on the *Bifidobacterium bifidum* and *Lactobacillus* species. *Anaerobe* **8**, 319–324.

Lee, J., Koo, N., and Min, D. B. (2004). Reactive oxygen species, aging, and antioxidative nutraceuticals. *Compr. Rev. Food Sci. Food Saf.* **3**, 21–33.

Lee, J. H., Seo, Y. B., Jeong, S. Y., Nam, S. W., and Kim, Y. T. (2007). Functional analysis of combinations in astaxanthin biosynthesis genes from *Paracoccus haeundaensis*. *Biotechnol. Bioprocess. Eng.* **12**, 312–317.

Let, M. B., Jacobsen, C., and Meyer, A. S. (2007a). Lipid oxidation in milk, yogurt, salad dressing enriched with neat fish-oil or pre-emulsified fish-oil. *J. Agric. Food Chem.* **55**, 7802–7809.

Let, M. B., Jacobsen, C., and Meyer, A. S. (2007b). Homogenization conditions affect the oxidative stability of fish oil enriched milk emulsions: Lipid oxidation. *J. Agric. Food Chem.* **55**, 1773–1780.

Levitan, E. B., Wolk, A., and Mittleman, M. A. (2010). Fatty fish, marine ω-3 fatty acids and incidence of heart failure. *Eur. J. Clin. Nutr.* **64**, 587–594.

Li, B., Lu, F., Wei, X., and Zhao, R. (2008). Fucoidan: Structure and bioactivity. *Molecules* **13**, 1671–1695.

Lin, S. B., Chen, S. H., and Pen, K. C. (2009). Preparation of antibacterial chitooligosaccharides by altering the degree of deacetylation of β-chitosan in a *Tricoderma harzianum* chitinase hydrolyzing process. *J. Sci. Food Agric.* **89**, 238–244.

Lyons, N. M. and O'Brien, N. M. (2002). Modulatory effects of an algal extract containing astaxanthin on UVA-irradiated cells in culture. *J. Dermatol. Sci.* **30**, 73–84.

Maeda, H., Hosokawa, M., Shashima, T., Funayama, K., and Miyashita, K. (2005). Fucoxanthin from edible seaweed, *Undaria pinnatifida*, shows antiobesity effect through UCP1 expression in white adipose tissues. *Biochem. Biophys. Res. Commun.* **332**, 392–397.

Maeda, H., Hosokawa, M., Sashima, T., Takahashi, N., Kawada, T., and Miyashita, T. (2006). Fucoxanthin and its metabolite, fucoxanthinol, suppress adipocyte differentiation in 3T3-L1 cells. *Int. J. Mol. Med.* **18**, 147–152.

Marinangeli, C. P. H. and Jones, P. J. H. (2010). Plant sterols, marine-derived omega-3 fatty acids and other functional ingredients: A new frontier for treating hyperlipidemia. *Nutr. Metab.* **7**, 76–83.

Mele, M., Contarini, G., Cercaci, L., Serra, A., Buccioni, A., Povolo, M., Conte, G., Funaro, A., Banni, S., Lercker, G., and Secchiari, P. (2011). Enrichment of pecorino cheese with conjugated linoleic acid by feeding dairy ewes with extruded linseed: Effect on fatty acid and triglycerides composition and on oxidative stability. *Int. Dairy J.* **21**, 365–372.

Miyamoto, Y. Y., Yamasaki, M., Tachibana, H., and Yamada, K. (2009). Fucoidan induces apoptosis through activation of caspase-8 on human breast cancer MCF-7 cells. *J. Agric. Food Chem.* **57**, 8677–8682.

Miyashita, K., Nishikawa, S., Beppu, F., Tsukui, T., Abe, M., and Hosokawa, M. (2011). The allenic carotenoid fucoxanthin, a novel marine nutraceutical from brown seaweeds. *J. Sci. Food Agric.* **91**, 1166–1174.

Moon, H. J., Lee, S. H., Ku, M. J., Yu, B. C., Jeon, M. J., Jeong, S. H., Stonick, V. A., Zvyagintseva, T. N., Ermakova, S. P., and Lee, Y. H. (2009). Fucoidan inhibits UVB-induced MMP-1 promoter expression and down regulation of type I procollagen synthesis in human skin fibroblasts. *Eur. J. Dermatol.* **19**, 129–134.

Morganti, P. and Morganti, G. M. (2008). Chitin nanofibrils for advanced cosmeceuticals. *Clin. Dermatol.* **26**, 334–340.

Murthy, K. N. C., Vanitha, A., Rajesh, M., Mahadewa, S. M., Sowmya, P. R., and Ravishankar, G. A. (2005). *In vitro* antioxidant activity of carotenoids from *Dunaliella salina*—A green microalga. *Life Sci.* **76**, 1381–1390.

Myers, S. P., O'Connor, J., Fitton, J. H., Brooks, L., Rofle, M., Connellan, P., Wohlmuth, H., Cheras, P. H., and Morris, C. (2010). A combined phase I and II open label study on the effects of a seaweed extract nutrient complex on osteoarthritis. *Biologics* **4**, 33–44.

Ngo, D. H., Wijesekara, I., Vo, T., Ta, Q. V., and Kim, S. K. (2011). Marine food derived functional ingredients as potential antioxidants in the food industry: An overview. *Food Res. Int.* **44**, 523–529.

Nomura, T., Kikuchi, M., Kubodera, A., and Kawakami, Y. (1997). Proton-donative antioxidant activity of fucoxanthin with 1,1-diphenyl-2-picrylhydrazyl (DPPH). *Biochem. Mol. Biol. Int.* **42**, 361–370.

O'Sullivan, L., Murphy, B., McLoughlin, P., Duggan, P., Lawlor, P. G., Hughes, H., and Gardiner, G. E. (2010). Prebiotics from marine macroalgae for human and animal health applications. *Mar. Drugs* **8**, 2038–2064.

Ozer, B. H. and Kirmaci, H. A. (2010). Functional milks and dairy beverages. *Int. J. Dairy Technol.* **63**, 1–15.

Pan, X., Chen, F., Wu, T., Tang, H. G., and Zhao, Z. U. (2009). Prebiotic oligosaccharides change the concentration of short chain fatty acids and the microbial population of mouse bowel. *J. Zhejiang Univ. Sci. B* **10**, 258–263.

Pangestuti, R. and Kim, S. K. (2004). Neuroprotective effect of marine algae. *Mar. Drugs* **9**, 803–818.

Park, H. Y., Han, M. H., Park, C., Jin, C. Y., Kim, G. Y., Choi, I. W., Kim, N. D., Nam, T. J., Kwon, T. K., and Choi, Y. H. (2011). Anti-inflammatory effects of fucoidan through inhibition of NF-κB, MAPK and Akt activation in lipopolysaccharide-induced BV2 microglia cells. *Food Chem. Toxicol.* **49**, 1745–1752.

Pavinatto, F. J., Santos, D. S., and Olivera, O. N. (2008). Interaction between cholesterol and chitosan in Langmuir monolayers. *Polimeros: Ciencia e Tecnologia* **15**, 91–94.

Peerapornpisal, Y., Amornlerdpison, D., Jamjai, U., Taesotikul, T., Pongpaibul, Y., Nualchareon, M., and Kanjanapothi, D. (2010). Antioxidant and anti-inflammatory activities of brown marine alga, Padina minor Yamada. *Chiang Mai J. Sci.* **37**, 507–516.

Pihalanto-Leppala, A. (2001). Bioactive peptides derived from bovine proteins: Opioid and ACE-inhibitory peptides. *Trends Food Sci. Technol.* **11**, 347–356.

Ra, S. R., Qiao, M., Immonen, T., Pujana, I., and Saris, P. E. J. (1996). Genes responsible for nisin synthesis, regulation and immunity form a regulation of two operons and are induced by nisin in *Lactobacillus lactis* N8. *Microbiology* **142**, 1281–1288.

Rajapakse, N., Mendis, E., Kim, M. M., and Kim, S. K. (2007). Inhibition of free radical-mediated oxidation of cellular biomolecules by carboxylated chitooligosaccharides. *Bioorg. Med. Chem.* **15**, 997–1003.

Rajapakse, N., Kim, M. M., Mendis, E., and Kim, S. K. (2008). Inhibition of inducible nitric oxide synthase and cyclooxygenase-2 in lipopolysaccharide-stimulated RAW264.7 cells by carboxybutyrylated glucosamine takes place via down-regulation of mitogen-activated protein kinase-mediated nuclear factor-jB signalling. *Immunology* **123**, 348–357.

Rajaram, G., Manivasagan, P., Thilagavathi, B., and Saravanakumar, A. (2010). Purification and characterization of a bacteriocin produced by *Lactobacillus lactis* isolated from marine environment. *Adv. J. Food Sci. Technol.* **2**, 138–144.

Rajasekaran, A., Sivagnanam, G., and Xavier, R. (2008). Nutraceuticals as therapeutic agents: A review. *Res. J. Pharm. Tech.* **1**, 328–340.

Sachindra, N. M., Sato, E., Maeda, H., Hosokawa, M., Niwano, Y., Kohno, M., and Miyashita, K. (2007). Radical scavenging and singlet oxygen quenching activity of marine carotenoid fucoxanthin and its metabolites. *J. Agric. Food Chem.* **55**, 8516–8522.

Samaraweera, A. M., Vidanarachchi, J. K., and Kurukulasuriya, M. S. (2011). Industrial applications of macroalgae. *In* "Handbook of Marine Macroalgae: Biotechnology and Applied Phycology", (S. K. Kim, Ed.), pp. 500–521. John Willey and Sons Ltd., West Sussex, England.

Shan, B. E., Yoshida, Y., Kuroda, E., and Yamashita, U. (1999). Brief communication immunomodulating activity of seaweed extract on human lymphocytes in vitro. *Int. J. Immunopharmacol.* **21**, 59–70.

Shibata, T., Ishimaru, K., Kawaguchi, S., Yoshikawa, H., and Hama, Y. (2008). Antioxidant activities of phlorotannins isolated from Japanese Laminariaceae. *J. Appl. Phycol.* **20**, 705–711.

Souza, M. C. R. D., Marques, C. T., Dore, C. M. G., Da Silva, F. R. F., and Rocha, H. A. O. (2007). Antioxidant activities of sulphated polysaccharides from brown and red seaweeds. *J. Appl. Phycol.* **19**, 153–160.

Sugano, M., Watanabe, S., Krishi, A., Izume, M., and Ohtakara, A. (1988). Hypocholesterolemic action of chitosans with different viscosity in rats. *Lipids* **23**, 187–193.

Synytsya, A., Kim, W. J., Kim, S. M., Pohi, R., Synytsya, A., Kvasnicka, F., Copikova, J., and Park, Y. (2010). Structure and antitumour activity of fucoidan isolated from sporophyll of Korean brown seaweed *Undaria pinnatifida*. *Carbohydr. Polym.* **81**, 41–48.

Tai, T. S., Sheu, W. H. H., Lee, W. J., Yao, H. T., and Chiang, M. T. (2000). Effect of chitosan on plasma lipoprotein concentrations in type-2 diabetic subjects with hypercholesterolemia. *Diabetes Care* **23**, 173–174.

Tsuchihashi, H., Kigoshi, M., Iwatsuki, M., and Niki, E. (1995). Action of β-carotene as an antioxidant against lipid peroxidation. *Arch. Biochem. Biophys.* **323**, 137–147.

Varga, L., Szigeti, J., and Ordog, V. (1999a). Effect of a spirulina platensis biomass enriched with trace elements on combinations of starter cultures strains employed in the dairy industry. *Milchwissenschaft* **54**, 247–248.

Varga, L., Szigeti, J., and Ordog, V. (1999b). Effect of a spirulina platensis biomass and that of its active components on single strains of dairy starter cultures. *Milchwissenschaft* **54**, 187–190.

Verum, K. M., Ottϕy, M. H., and Smidarϕd, O. (2001). Acid hydrolysis of chitosans. *Carbohydr. Polym.* **46**, 89–98.

Vidanarachchi, J. K. and Kurukulasuriya, M. S. (2011). Industrial applications of marine cosmeceuticals. *In* "Marine Cosmeceuticals: Trends and Prospects", (S. K. Kim, Ed.). Taylor & Francis, New York. Accepted for publication.

Vidanarachchi, J. K., Kurukulasuriya, M. S., and Kim, S. K. (2010). Chitin, chitosan, and their oligosaccharides in food industry. *In* "Chitin, Chitosan, Oligosaccharides and Their Derivatives", (S. K. Kim, Ed.), pp. 543–560. Taylor & Francis, New York.

Vijayaval, K. and Martinez, J. A. (2010). *In vitro* antioxidant and antimicrobial activities of two Hawaiian marine *Limu: Ulva fasciata* (Chlorophyta) and *Gracilaria salicornia* (Rhodophyta). *J. Med. Food* **13**, 1494–1499.

Wang, Y., Zhou, P., Yu, J., Pan, X., Wang, P., Lan, W., and Tao, S. (2007). Antimicrobial effects of chitooligosaccharides produced by chitosanases from *Pseudomonas* CUY 8. *Asia Pac. J. Clin. Nutr.* **16**, 174–177.

Xu, G., Huang, X., Qiu, L., Wu, J., and Hu, Y. (2007). Mechanism study of chitosan on lipid metabolism in hyperlipemic rats. *Asia Pac. J. Clin. Nutr.* **16**, 313–317.

Yan, X., Chuda, Y., Suzuki, M., and Nagata, T. (1999). Fucoxanthin as the major antioxidant in *Hijika fusiformis*, a common edible seaweed. *Biosci. Biotechnol. Biochem.* **63**, 605–607.

Ye, J., Li, Y., Teruya, K., Katakura, Y., Ichikawa, A., Eto, H., Hosoi, M., Hosoi, M., Nishimoti, S., and Shirahat, H. (2005). Enzyme-digested fucoidan extracts derived from *Mozuku* of *Cladosiphon novae-caledoniae kylin* inhibit invasion and angiogenesis of tumour cells. *Cytotechnology* **47**, 117–126.

Ye, A., Cui, J., Taneja, A., Zhu, X., and Singh, H. (2009). Evaluation of processed cheese fortified with fish oil. *Food Res. Int.* **42**, 1093–1098.

Yoon, H. J., Moon, M. E., Park, S. H., Im, S. Y., and Kim, Y. H. (2007). Chitosan oligosaccharide (COS) inhibits LPS-induced inflammatory effects in RAW 264.7 macrophage cells. *Biochem. Biophys. Res. Commun.* **358**, 954–959.

Yuan, H., Song, J., Li, X., Li, N., and Dai, J. (2006). Immunomodulation and antitumor activity of κ-carrageenan oligosaccharides. *Cancer Lett.* **243**, 228–234.

Zhang, C. and Kim, S. K. (2009). Matrix metalloproteinase inhibitors (MMPIs) from marine natural products: The current situation and future prospects. *Mar. Drugs* **7**, 71–84.

CHAPTER **31**

Upgrading of Sea By-products: Potential Nutraceutical Applications

B. Cudennec, T. Caradec, L. Catiau, and **R. Ravallec**[1]

Contents

Abstract

Since many years, numerous kinds of processes based on enzymatic hydrolysis at various pH, involving added plant or bacterial enzymes after inactivation by heating of endogenous enzymes present in the raw material or, alternatively, based on the action of endogenous enzymes, have contributed to the degradation of marine by-product proteins in order to produce fractions exerting biological activities. Peptides obtained by enzymatic hydrolysis of fish proteins exhibit not only nutritional but also biological properties of dietary uses, or even therapeutic potential. In this review, we have focused on the different enzymatic processes able to generate bioactive peptides from marine by-products and exerting high potential in nutraceutical applications to fight against important public health issues like obesity, stress, hypertension, and migraine. Beyond the nutraceutical and pharmaceutical aspects, this way of valorization

Laboratoire ProBioGEM, IUT A, EA 1026, Université Sciences et Technologies, Lille, Villeneuve d'Ascq, France
[1] Corresponding author: R. Ravallec, *E-mail address:* ravallec-ple@univ-lille1.fr

Advances in Food and Nutrition Research, Volume 65
ISSN 1043-4526, DOI: 10.1016/B978-0-12-416003-3.00031-7

is also included in the necessary development of by-product fishing industries for economic and ecological reasons in the worldwide context of marine resources depletion.

I. INTRODUCTION

Today, the effects of the food on human health are well established. The health food world market is in constant increase estimated between 135 and 150 billion dollars in 2005. In Europe, it represented 27% of the world market, against 29% in the United States which possessed the first place.

In France, the selected market of functional food and dietary products represented in 2006 about 2.15 billion euros, corresponding to 3.4% of the total agro-alimentary French market. This selected market knows today a consolidation step with the application of the European rules by l'EFSA (Autorité Européenne de Sécurité des Aliments). But a growing step is awaited in 2011 with the attended publications of a positive list by the EFSA.

In the diversity of the agro-alimentary sources for the production of functional food or ingredients, the sea by-products represent an important raw material. The world fisheries production including capture fisheries and aquaculture represented more than 140million tons in 2006 (FAO, 2009). This production generated 60% of by-products like heads, internal organs, fish skins, or cartilage already partially used for food or feed, cosmetics, and nutraceutical applications. After enzymatic hydrolysis by commercial or digestive enzymes, the cut proteins could generate peptides of different size and sequences able to exert interesting biological activities in *in vitro* assays (Ritchie and Mackie, 1982).

Due to new European rules or regulation concerning the control of fishing and, in particular, the amount of fishing, the fishermen will be subjected to the prohibition of the rejection of fishes (perhaps too small, of noncommercial species) or by-products in the sea. This regulation will have a considerable impact on the storage and the transport on boat, particularly due to the increase of the costs. The direct consequence will be an important need for by-products upgrading to counterbalance the overcost of this storage and of the transport through the port.

For many years, peptides derived from protein hydrolysates have been found of interest in feed and food industries or in pharmaceutical domains for hypoallergenic infant diets (Terracciano *et al.*, 2002) and nutritional therapy (Deal and Moskowitz, 1999). A wide range of biologically active peptides have already been isolated from milk proteins (Meisel, 2004). Recently, fish protein hydrolysates (FPH) and more broadly marine protein hydrolysates were identified to exert good functional and nutritional properties (Kristinsson and Rasco, 2000; Liaset *et al.*, 2003; Liceaga-Gesualdo and Li-Chan, 1999; Quaglia and Orban, 1987; Shahidi *et al.*, 1995). Moreover, FPH could provide a novel cheap source of bioactive molecules of real interest in therapeutic and nutraceutical domains. In this way, numerous works have identified many peptides with biological activities. Antihypertensive activity was highlighted by the characterization of peptides able to inhibit the angiotensin-I converting enzyme (ACE) (Bordenave *et al.*, 2002; Byun and Kim, 2001; Cinq-Mars and Li-Chan, 2007; Cinq-Mars *et al.*, 2008; Je *et al.*, 2005; Matsufuji and Matsui, 1994; Matsui and Matsufuji, 1993; Qian *et al.*, 2007). This *in vitro* activity was associated *in vivo* with a diminution of blood pressure in hypertensive rats (Fujita and Yoshikawa, 1999). Other works evidenced biological activies in marine byproducts (molluscs, crustacean, and fish) such as anticoagulant (Rajapakse *et al.*, 2005), antiproliferative on human cancer cell lines (Picot *et al.*, 2006), epithelial reparative (Fitzgerald *et al.*, 2005), antioxidant (Chuang *et al.*, 2000; Je *et al.*, 2008; Jun *et al.*, 2004; Kim *et al.*, 2001; Raghavan and Kristinsson, 2008; Suetsuna, 2002; Suetsuna and Nakano, 2000; Suetsuna *et al.*, 2004; Wu *et al.*, 2003), immunomodulative (Bogwald *et al.*, 1996; Gildberg *et al.*, 1996; Kotzamanis *et al.*, 2007; Tang *et al.*, 2008), secretaguogue (Cancre *et al.*, 1999; Guerard and Ravallec-Ple, 2001; Ravallec-Ple and Van Wormhoudt, 2003; Ravallec-Plé *et al.*, 2000, 2001), and cellular growth factor (Cancre *et al.*, 1999; Guerard and Ravallec-Ple, 2001) activities.

In this review, we first present an overview of the hydrolysis mechanisms for the production of protein hydrolysates. Second are presented the recent findings on biologically active peptides obtained from marine byproduct hydrolysates, which when incorporated in functional food may exert an important role to fight against and to prevent public health issues like obesity and diabetes (intestinal cholecystokinin (CCK) secretion stimulating peptides), hypertension (CGRP agonist), migraine (CGRP antagonist peptides), and stress (antioxidant peptides).

II. ENZYMATIC PROCESS

Enzymatic hydrolysis of protein generates peptides of various molecular weights that could harbor biological properties of important interest in several domains such as nutraceutics and pharmaceutics. Protein hydrolysates could be obtained by two main kinds of processes. The first one is the autolysis of the raw material by gastric endogenous protease like pepsin, trypsin, and chymotrypsin and by lysosomic enzymes like cathepsin (Raa and Gildberg, 1982). This process is used since antiquity in the South East Asia as fish sauce (Nuoc-mam in Vietnam, Nampla in Thailand, Mam-tom in China, and the Shiokara in Japan). Due to several factors like fish species, season as well as the selectivity and the amount of enzymes, the control of autolysis processes could be difficult (Benjakul and Morrissey, 1997). The second one is based on enzymatic hydrolysis in controlled conditions (pH, temperature, hydrolysis degree (DH), enzyme/substrate ratio), involving added commercial enzymes isolated from bacteria, fungi, plant, or animal sources, after inactivation by heating of endogenous enzymes present in the raw material, or alternatively, the action of endogenous enzymes contributes to the degradation of proteins (Diniz and Martin, 1997; Gildberg, 1994). In the case of hydrolysis, the process is reproducible, and the inactivation of the enzymes present in fishes permits to control reproducibility. In both cases, products harbor excellent flavor (Haard *et al.*, 1994) and possess a good nutritive value mainly based on a high level of free amino acids (Martin *et al.*, 1995).

Due to the length of the reaction (several hours) and the separation procedures, the different processes (hydrolysis, autolysis) currently used are not likely to generate molecules larger than small peptides and individual amino acids. In this context, one may wonder if shorter times and milder conditions of hydrolysis may release peptides of various lengths, some of them having conserved (or harboring new) biological properties. Among the different sources of enzyme, the bacterial enzymes are the most used because of their stability to high temperature, the important diversity of their catalytic capacity, and their price (Diniz and Martin, 1997). A nonexhaustive list of the commercial enzymes commonly used in hydrolysis of marine protein by-products for the obtaining of biologically active peptides are presented in Table 31.1.

III. CCK SECRETION STIMULATED PEPTIDES

Excess of food intake largely contributes to body weight increase and obesity that predisposes to numerous pathologies such as diabetes and hypertension (Arora and Anubhuti, 2006). The long-term regulation of maintenance of body weight is integrated by endocrine signals and coupled with short-term signals related to meal size, like CCK. This contributes to regulate the daily energy intake in synchronization with long-term body weight regulation (Arora and Anubhuti, 2006; Schwartz *et al.*, 2000). CCK was the first intestinal hormone evidenced in the regulation of food intake and is one of the most studied satiety peptide hormone (Gibbs *et al.*, 1973). CCK is mainly synthesized in two forms (CCK-8 and CCK-33) in response to a meal stimulation from I-cells localized in the duodenum and is also synthesized in the central nervous system (Beinfeld, 2000; Liddle, 1997). Intestinal CCK release leads to several gastrointestinal actions like intestinal motility, gall bladder contraction, inhibition of gastric emptying, and stimulation of pancreatic secretion (Strader and Woods, 2005). Among the individual alimentary compounds, proteins and their hydrolysis products (peptides of various sizes) appear to be

TABLE 31.1 Summary of the main enzymes used for the generation of bioactive peptides from different protein marine sources

Enzymes	Species	Activities	References
Alcalase	*Penaeus* spp.	Antihypertensive	Bordenave *et al.* (2002)
	Gadus morhua	Secretagogue	Ravallec-Plé *et al.* (2000)
	Gadus morhua, Pandealus borealis	CGRP-like	Fouchereau-Peron *et al.* (1999)
	Gadus morhua, Pandealus borealis	Secretaguogue; GF	Cancre *et al.* (1999)
	Gadus morhua, Penaeus aztecus	Secretaguogue	Ravallec-Ple and Van Wormhoudt (2003)
	Limanda aspera	Antioxidant	Jun *et al.* (2004)
	Limanda aspera	Anticoagulant	Rajapakse *et al.* (2005)
	Pollachius virens	CGRP-like	Martinez-Alvarez *et al.* (2008)
	Sardina pilchardus	Secretagogue, CGRP-like, and GF	Ravallec-Plé *et al.* (2001)
	Sardina pilchardus	CGRP-like	Rousseau *et al.* (2001)
	Sardinella aurita	Antihypertensive	Bougatef *et al.* (2008)
	Selaroides leptolepis	Antioxidant	Klompong *et al.* (2009)
	Thunnus obesus	Antihypertensive	Qian *et al.* (2007)
	Thunnus obesus	Antioxidant	Je *et al.* (2008)
Neutrase	*Thunnus obesus*	Antihypertensive	Qian *et al.* (2007)
	Limanda aspera	Antioxidant	Jun *et al.* (2004)
	Oreochromis niloticus	Antioxidant	Raghavan (2008)
	Oreochromis niloticus	Antioxidant	Raghavan and Kristinsson (2008)
	Thunnus obesus	Antioxidant	Je *et al.* (2008)
Protamex	*Merluccius productus*	Antihypertensive	Cinq-Mars *et al.* (2008)
	Merluccius products	Antihypertensive	Cinq-Mars and Li-Chan (2007)

(continued)

TABLE 31.1 *(continued)*

Enzymes	Species	Activities	References
Pronase	*Katsuwonus pelamis*	Antihypertensive	Astawan *et al.* (1995)
	Limanda aspera	Antioxidant	Jun *et al.*, 2004
	Pollachius virens	Antioxidant	Kim *et al.* (2001)
	Scomber austriasicus	Antioxidant	Wu *et al.* (2003)
Flavourzyme	*Oreochromis niloticus*	Antioxidant	Raghavan and Kristinsson (2008)
	Oreochromis niloticus	Antioxidant	Raghavan *et al.* (2008)
	Selaroides leptolepis	Antioxidant	Klompong *et al.* (2009)
Pepsin	*Limanda aspera*	Anticoagulant	Rajapakse *et al.* 2005
	Limanda aspera	Antioxidant	Jun *et al.* (2004)
	Merluccius productus	Antihypertensive	Cinq-Mars *et al.*, 2008
	Limanda aspera	Anticoagulant	Rajapakse *et al.* (2005)
Trypsin	*Limanda aspera*	Antioxidant	Jun *et al.* (2004)
	Magalaspis cordyla, Otolithes rubber	Antioxidant	
	Thunnus obesus	Antioxidant	Je *et al.* (2008)
Chymotrypsin	*Limanda aspera*	Anticoagulant	Rajapakse *et al.* (2005)
	Limanda aspera	Antioxidant	Jun *et al.* (2004)
	Thunnus obesus	Antioxidant	Je *et al.* (2008)
	Sardinella aurita	Antihypertensive	Bougatef *et al.* (2008)
	Thunnus obesus	Antihypertensive	Qian *et al.* (2007)
Papain	*Limanda aspera*	Anticoagulant	Rajapakse *et al.* (2005)
	Limanda aspera	Antioxidant	Jun *et al.* (2004)
	Thunnus obesus	Antioxidant	Je *et al.* (2008)
	Thunnus obesus	Antihypertensive	Qian *et al.* (2007)

GF: cell growth factor activity; CGRP-like: CGRP activity.

the main stimulant of the intestinal CCK secretion (Liddle *et al.*, 1985). In this context, it was shown that protein hydrolysates stimulate CCK secretion in STC-1 cells via the CCK gene transcription (Cordier-Bussat *et al.*, 1997). Several studies have shown the high potential of protein hydrolysates from meat and vegetable source to *in vitro* stimulate CCK secretion in STC-1 cells (Foltz *et al.*, 2008; Nemoz-Gaillard *et al.*, 1998; Sufian *et al.*, 2006). Mouse enteroendocrine STC-1 cell line shows high similarity with human CCK-secreting I-cells and was described as a good model for *in vitro* CCK-secretion study (Rindi *et al.*, 1990).

As shown in Fig. 31.1, *in vitro* study showed that fish and shellfish by-product hydrolysates from industrial origin were able to highly stimulate CCK secretion in STC-1 cells, and that this stimulation was mainly due to peptide molecules. The partial purification of these CCK-stimulating

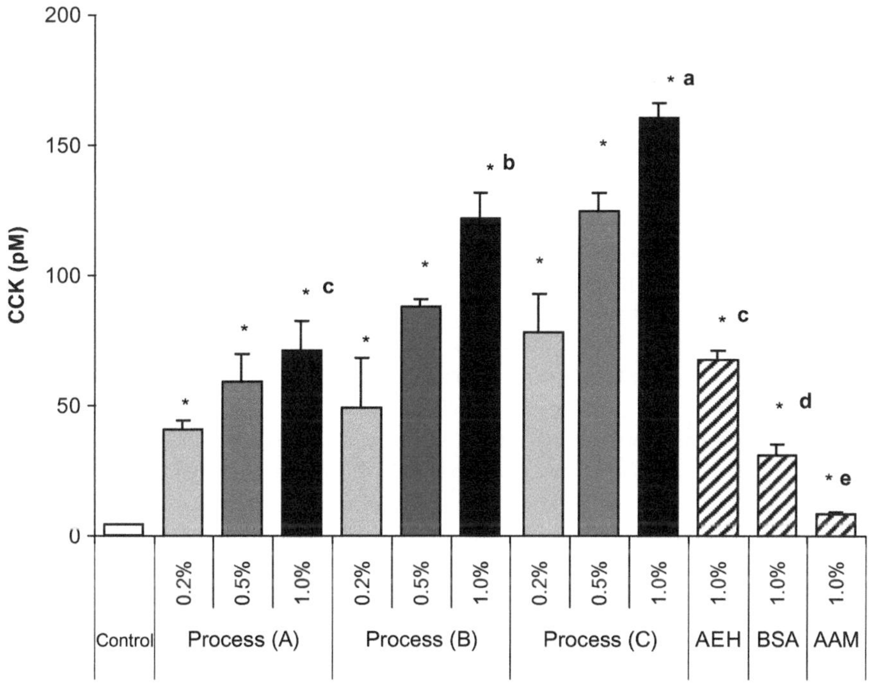

FIGURE 31.1 Effect of industrial product hydrolysates obtained from blue whiting (A and B) and brown shrimp (C) by-products on CCK-releasing activity of STC-1 cells. Hydrolysates were tested at different concentrations (0.2–1.0%, wt/vol) on cholecystokinin (CCK) release and compared to those obtained with albumin egg hydrolysate (AEH, a positive control), bovine serum albumin (BSA), and free amino acid mixture (AAM). Values are means±SEM of four repeated measurements. (*): $P<0.05$ versus control. CCK releases at 1.0% concentration (wt/vol) were compared: means without a common letter are different ($P<0.05$).

peptides showed that their apparent molecular weights were ranged between 1000 and 1500 Da for both fish and crustacean by-product hydrolysates (Cudennec *et al.*, 2008). Other works showed that CCK-stimulating action of peptones obtained from soybean and pork was able, *ex vivo*, to bind to the intestinal brush border membrane and *in vivo* to reduce food intake in rats (Nishi *et al.*, 2003a,b; Sufian *et al.*, 2006). Concerning hydrolysates from marine source, a recent study demonstrated that FPH (obtained from smooth hound, *Mustelus mustelus*) preload administration led, on the one hand, to a decrease of short-term food intake and, on the other hand, to a decrease in the body weight gain after 21 days of treatment (Bougatef *et al.*, 2010b). Despite promising therapeutically applications in health- and nutrition-related markets, few studies dealing with the use of functional food to prevent obesity and resulting pathologies have been published on sea by-products.

IV. CGRP-LIKE PEPTIDES

CGRP is a 37-amino acid neuropeptide, initially discovered as an mRNA splice variant of the calcitonin gene (Rosenfeld *et al.*, 1983). CGRP has a wide distribution throughout the central and peripheral nervous system and is usually colocalized in C-fibers (Brain and Cox, 2006). CGRP is involved in an increasing number of biological activities and was first described as a potent vasodilator. CGRP is active throughout the cardiovascular system, but with the most pronounced effects in microcirculation (Brain *et al.*, 1985). Additional biological functions mediated by CGRP include regulation of pituitary hormone secretion, release of pancreatic enzymes, control of gastric acid secretion, thermoregulation, decrease in food intake, antagonism of some actions of insulin growth factor-like functions, bone remodeling, and prevention of complication during pregnancy (Arulmani *et al.*, 2004; Yallampalli *et al.*, 2002).

One of the most important functions of CGRP is to regulate blood flow to various organs by its potent local vasodilatory actions (Tippins *et al.*, 1986; Watson *et al.*, 2005). Consequently, in the past year, several studies highlight the presence of CGRP-like molecules in marine processing wastes. Thus, CGRP-like molecules were identified in marine by-product protein hydrolysates obtained from cod (*Gadus morhua*), shrimp (*Penaeus* spp.), sardine (*Sardina pilchardus*), saithe (*Pollachius virens*), and siki (*Centroscymnius coelolepsis*; Fouchereau-Peron *et al.*, 1999; Martinez-Alvarez *et al.*, 2007, 2008; Ravallec-Plé *et al.*, 2001; Rousseau *et al.*, 2001). CGRP-like molecules obtained from siki hydrolysate were able to interact with specific receptors for CGRP in rat liver membranes and modulated the adenylate cyclase activity. This effect was reversed by the presence of $CGRP_{8-37}$, a specific antagonist of CGRP action, showing the antagonist nature of the identified peptides that presented after purification a molecular weight around 1500–2500 Da (Martinez-Alvarez *et al.*, 2007). Therefore, these molecules could be used in the formulation of functional foods to prevent pathologies like edema, hypertension, and increased cardiac workload hypertension. Otherwise, the CGRP is also involved in migraine neurovascular disorder, and several studies have shown the strong potential of CGRP receptor antagonist in the treatment of this pathology (Doods *et al.*, 2007; Edvinsson and Petersen, 2007). Consequently, the finding of CGRP antagonist in marine by-product hydrolysates could give their substantial ways of upgrading. In this approach, Rousseau *et al.* characterized CGRP-like molecules in sardine (*S. pilchardus*) by-product hydrolysates. These peptides, with an apparent molecular weight of 6000 Da, were characterized to be CGRP antagonists by their capacity to interact with specific receptors for CGRP in rat liver membrane and to specifically induce the inhibition of the CGRP-stimulated adenylate cyclase activity (Rousseau *et al.*, 2001). According to these recent published results, CGRP agonist and antagonist peptides will certainly play an important role in the pharmaceutical and food industries in the future.

V. ANTIOXIDANT PEPTIDES

An antioxidant is defined as a substance which, added in low doses to a naturally occurring air oxidation, is able to slow or inhibit the oxidation (Park and Hyun, 2002). This definition can be extended, and the term antioxidant will include all substances that protect biological system against the potential deleterious effects of processes or reactions that cause excessive oxidation. The lipid oxidation is of important interest in food industry and especially for better consumer acceptance because of the possibility of undesirable off-flavors, odors development, and of potentially toxic reaction products (Lin and Liang, 2002). Moreover, cancer, coronary heart disease, and Alzheimer's diseases are now known to be partially caused by oxidation or free radical reactions in human (Diaz *et al.*, 1997). In this context, the inhibition of lipid oxidation, occurring in food and in human physiology, becomes an important goal not only in the food prevention from undergoing such deterioration or discoloration but also in the protection against several diseases. Nowadays, synthetic antioxidants, such as butylated hydroxyanisole (BHA) and butylated hydroxytoluene (BHT), are used as additives to prevent food deterioration. Unfortunately, even if these synthetic antioxidants have demonstrated stronger antioxidant activities than those of natural antioxidants, such as ascorbic acid, a restriction in their use has begun because of their toxicity and their capacity to induce DNA damage (Barlow, 1990; Ito *et al.*, 1986; Yu *et al.*, 2000). Further researches are conducted to discover new naturally occurring antioxidants for use in food and medicinal materials in order to replace synthetic antioxidants. Several natural antioxidants such as vitamin C, α-tocopherol, and phenolic compounds, which are present naturally in vegetables, fruits, and seeds, possess the ability to reduce oxidative damage associated with diseases such as cancer, cardiovascular diseases, or atherosclerosis. Recent research showed the ability of phenolic substances, like flavonoids and phenolic acids, to act as antioxidants (Miraliakbari and Shahidi, 2008; Shahidi, 2009, Shahidi *et al.*, 2007; Zhong *et al.*, 2007). Further, many studies have described the antioxidant activity of protein hydrolysates such as casein milk (Suetsuna *et al.*, 2000), soybean (Chen *et al.*, 1998), egg proteins (Park *et al.*, 2001), egg white albumin (Uchida and Kawakishi, 1992), and pork protein (Saiga *et al.* 2003). Marine products and by-products become a source of natural antioxidant peptides. These peptides have been highlighted in marine proteins such as smooth hound (*M. mustelus*), Pollack (*P. virens*), or cod (*G. morhua*; Bougatef *et al.*, 2009; Jeon *et al.*, 1999; Kim *et al.*, 2001). Bougatef *et al.* investigated the antioxidative activity of sardinelle (*Sardinella aurita*) heads and viscera protein hydrolysates obtained after various enzymatic treatments. The authors present seven new antioxidative peptides isolated from these hydrolysates obtained by treatment with crude enzyme extract from sardine (*S. pilchardus*) viscera. In this chapter, the amino acid sequences of the active peptides were determined and presented in Table 31.2. The authors showed that the peptide Leu-His-Tyr displayed the highest radical-scavenging activity (Bougatef *et al.*, 2010a). All these results suggest that hydrolysates from marine by-products could be used as natural antioxidants not only in enhancing antioxidant properties of functional foods but also in preventing oxidation reactions in food processing.

TABLE 31.2 Peptides identified by MS/MS in sardine (*Sardinella aurita*) hydrolysates purified by HPLC (Bougatef *et al.*, 2010b)

MW (Da)	Sequence	DPPH scavenging activity (%)
538.2	Gly-Ala-Leu-Ala-Ala-His	54±1.38
528.2	Pro-His-Tyr-Leu	
471.3	Leu-Ala-Arg-Leu	51±1.31
431.2	Leu-His-Tyr	63±1.5
403.1	Gly-Ala-Trp-Ala	52±1.44
283.1	Gly-Ala-His	
263. 08	Gly-Gly-Glu	38±1.27

The DPPH radical-scavenging activity of the hydrolysates was determined as described by Bersuder *et al.* (1998).

VI. CONCLUSION

This review notices that FPH could be of important interest for their potential to reduce body weight, hypertension, stress, migraine, and for their effect as antioxidant. These hydrolysates from different sources and different processes are able to exert potential activities in relation with the peptide sequences present after the hydrolysis process. The relationship between the sequence and the activity is always studied and represent a fundamental research for the comprehension of the complex mechanism of stress or satiety. Another point to highlight is the necessity of a better understanding of the peptide's fate in the gut. Recently, many studies ask the question of the *in vivo* effect of the peptides having shown an *in vitro* activity, trying to reproduce a simulated digestion. In this context, the *in vivo* and clinical experiments are required to confirm and understand the mechanisms involved in the biological activity. These hydrolysates represent an important source of peptides, able to be generated after the hydrolysis of marine proteins by gastrointestinal enzymes. From an upgrading point of view, an industrial scale fraction exhibiting an activity able to fight against stress or weight problems would be of great interest in several domains such as feed or nutraceuticals.

REFERENCES

Arora, S. and Anubhuti, (2006). Role of neuropeptides in appetite regulation and obesity—A review. *Neuropeptides* **40**, 375–401.

Arulmani, U., MaassenVanDenBrink, A., Villalon, C. M., and Saxena, P. R. (2004). Calcitonin gene-related peptide and its role in migraine pathophysiology. *Eur. J. Pharmacol.* **500**, 315–330.

Astawan, M., Wahyuni, M., Yasuhara, T., Yamada, K., Tadokoro, T., and Maekawa, A. (1995). Effects of angiotensin I-converting enzyme inhibitory substances derived from Indonesian dried-salted fish on blood pressure of rats. *Biosci. Biotech. Biochem.* **59**, 425–429.

Barlow, S. M. (1990). Toxicological aspects of antioxidants used as food additives. *In* "Food Antioxidants", (B. J. F. Hudson, Ed.), pp. 253–307. Elsevier, Amsterdam.

Beinfeld, C. M. (2000). Cholecystokinin. *In* "Psychopharmacology—The Fourth Generation of Progress", (F. E. Bloom and D. Kupfer, Eds). Raven, New York.

Benjakul, S. and Morrissey, M. T. (1997). Protein hydrolysates from pacific whiting solid wastes. *J. Agric. Food Chem.* **45**, 3423–3430.

Bersuder, P., Hole, M., and Smith, G. (1998). Antioxidants from a heated histidine-glucose model system. I: Investigation of the antioxidant role of histidine and isolation of antioxidants by high-performance liquid chromatography. *J. Am. Oil Chem. Soc.* **75**, 181–187.

Bogwald, J., Dalmo, R., McQueen, L. R., Stenberg, E., and Gildberg, A. (1996). The stimulatory effect of a muscle protein hydrolysate from Atlantic cod, *Gadus morhua* L., on Atlantic salmon, *Salmo salar* L., head kidney leucocytes. *Fish Shellfish Immunol.* **6**, 3–16.

Bordenave, S., Fruitier, I., Ballandier, I., Sannier, F., Gildberg, A., Batista, I., and Piot, J. M. (2002). HPLC preparation of fish waste hydrolysate fractions. Effect on guinea pig ileum and ACE activity. *Prep. Biochem. Biotechnol.* **32**, 65–77.

Bougatef, A., Nedjar-Arroume, N., Ravallec-Plé, R., Leroy, Y., Guillochon, D., Barkia, A., *et al.* (2008). Angiotensin I-converting enzyme (ACE) inhibitory activities of sardinelle (Sardinellaaurita) by-products protein hydrolysates obtained by treatment with microbial and visceral fish serine-proteases. *Food Chem.* **111**, 350–356.

Bougatef, A., Hajji, M., Balti, R., Lassoued, I., Triki-Ellouz, Y., and Nasri, M. (2009). Antioxidant and free radical-scavenging activities of smooth hound (Mustelus mustelus) muscle protein hydrolysates obtained by gastrointestinal proteases. *Food Chem.* **114**, 1198–1205.

Bougatef, A., Ravallec, R., Nedjar-Arroume, N., Barkia, A., Guillochon, D., and Nasri, M. (2010a). Evidence of in vivo satietogen effect and control of food intake of smooth hound (Mustelus mustelus) muscle protein hydrolysate in rats. *J. Funct. Foods* **2**, 10–16.

Bougatef, A., Nedjar-Arroume, N., Manni, L., Ravallec, R., Barkia, A., Guillochon, D., and Nasri, M. (2010b). Purification and identification of novel antioxidant peptides from enzymatic hydrolysates of sardinelle (Sardinella aurita) by-products proteins. *Food Chem.* **118**, 559–565.

Brain, S. D. and Cox, H. M. (2006). Neuropeptides and their receptors: Innovative science providing novel therapeutic targets. *Br. J. Pharmacol.* **147**(Suppl. 1), S202–S211.

Brain, S. D., Williams, T. J., Tippings, J. R., Morris, H. R., and MacIntyre, I. (1985). Calcitonin gene related peptide is a potent vasodilator. *Nature* **313**, 54–56.

Byun, H. G. and Kim, S. K. (2001). Purification and characterization of angiotensin I converting enzyme (ACE) inhibitory peptides from Alaska pollack (*Theragra chalcogramma*) skin. *Process Biochem.* **36**, 1155–1162.

Cancre, I., Ravallec-Plé, R., Van Wormhoudt, A., and Le Gal, Y. (1999). Secretagogues and growth factors in fish and crustacean protein hydrolysates. *Mar. Biotechnol.* **1**, 489–494.

Chen, H.-M., Muramoto, K., Yamauchi, F., Fujimoto, K., and Nokihara, K. (1998). Antioxidative properties of histidine-containing peptides designed from peptide fragments found in the digests of a soybean protein. *J. Agric. Food Chem.* **46**, 49–53.

Chuang, W.-L., Pan, B. S., and Tsai, J.-S. (2000). Inhibition of lipoxygenase and blood thinning effects of mackerel protein hydrolysate. *J. Food Biochem.* **24**, 333–343.

Cinq-Mars, C. D. and Li-Chan, E. C. (2007). Optimizing angiotensin I-converting enzyme inhibitory activity of Pacific hake (Merluccius productus) fillet hydrolysate using response surface methodology and ultrafiltration. *J. Agric. Food Chem.* **55**, 9380–9388.

Cinq-Mars, C. D., Hu, C., Kitts, D. D., and Li-Chan, E. C. (2008). Investigations into inhibitor type and mode, simulated gastrointestinal digestion, and cell transport of the angiotensin I-converting enzyme-inhibitory peptides in Pacific hake (Merluccius productus) fillet hydrolysate. *J. Agric. Food Chem.* **56**, 410–419.

Cordier-Bussat, M., Bernard, C., Haouche, S., Roche, C., Abello, J., Chayvialle, J. A., and Cuber, J. C. (1997). Peptones stimulate cholecystokinin secretion and gene transcription in the intestinal cell line STC-1. *Endocrinology* **138**, 1137–1144.

Cudennec, B., Ravallec-Plé, R., Courois, E., and Fouchereau-Peron, M. (2008). Peptides from fish and crustacean by-products hydrolysates stimulate cholecystokinin release in STC-1 cells. *Food Chem.* **111**, 970–975.

Deal, C. L. and Moskowitz, R. W. (1999). Nutraceuticals as therapeutic agents in osteoarthritis: The role of glucosamine, chondroitin sulfate, and collagen hydrolysate. *Rheum. Dis. Clin. North Am.* **25**, 379–395.

Diaz, M. N., Frei, B., Vita, J. A., and Keaney, J. F., Jr. (1997). Antioxidants and atherosclerotic heart disease. *N. Engl. J. Med.* **337**, 408–416.

Diniz, F. M. and Martin, A. M. (1997). Fish protein hydrolysates by enzymatic processing. *Agro Food Ind. Hi Tech* **8**, 9–13.

Doods, H., Arndt, K., Rudolf, K., and Just, S. (2007). CGRP antagonists: Unravelling the role of CGRP in migraine. *Trends Pharmacol. Sci.* **28**, 580–587.

Edvinsson, L. and Petersen, K. A. (2007). CGRP-receptor antagonism in migraine treatment. *CNS Neurol. Disord. Drug Targets* **6**, 240–246.

FAO, f. a. A. D. (2009). The State of World Fisheries and Aquaculture 2008. *In* FAO (Ed.), Rome.

Fitzgerald, A. J., Rai, P. S., Marchbank, T., Taylor, G. W., Ghosh, S., Ritz, B. W., and Playford, R. J. (2005). Reparative properties of a commercial fish protein hydrolysate preparation. *Gut* **54**, 775–781.

Foltz, M., Ansems, P., Schwarz, J., Tasker, M. C., Lourbakos, A., and Gerhardt, C. C. (2008). Protein hydrolysates induce CCK release from enteroendocrine cells and act as partial agonists of the CCK1 receptor. *J. Agric. Food Chem.* **56**, 837–843.

Fouchereau-Peron, M., Duvail, L., Michel, C., Gildberg, A., Batista, I., and le Gal, Y. (1999). Isolation of an acid fraction from a fish protein hydrolysate with a calcitonin-gene-related-peptide-like biological activity. *Biotechnol. Appl. Biochem.* **29**, 87–92.

Fujita, H. and Yoshikawa, M. (1999). LKPNM: A prodrug-type ACE-inhibitory peptide derived from fish protein. *Immunopharmacology* **44**, 123–127.

Gibbs, J., Young, R. C., and Smith, G. P. (1973). Cholecystokinin elicits satiety in rats with open gastric fistulas. *Nature* **245**, 323–325.

Gildberg, A. (1994). Enzymic processing of marine raw materials. *Process Biochem.* **28**, 1–15.

Gildberg, A., Bøgwald, J., Johansen, A., and Stenberg, E. (1996). Isolation of acid peptide fractions from a fish protein hydrolysate with strong stimulatory effect on atlantic salmon (*Salmo salar*) head kidney leucocytes. *Comp. Biochem. Physiol.* **114**, 97–101.

Guerard, F. and Ravallec-Ple, R. (2001). Enzymic solubilisation of proteins from tropical tuna using alcalase and some biological properties of the hydrolysates. *In* "Focus in Biotechnology", (P. Thonart and M. Hofman, Eds), pp. 39–50. Kluwer Academic Publishers, Dordrecht.

Haard, N. F., Simpson, B. K., and Sikorski, Z. E. (1994). Biotechnological applications of seafood proteins and other nitrogenous compounds. *In* "Seafood Proteins", (Z. E. Sikorski, *et al.*, Eds), pp. 194–216. Chapman & Hall, New York.

Ito, N., Hirose, M., Fukushima, S., Tsuda, H., Shirai, T., and Tatematsu, M. (1986). Studies on antioxidants: Their carcinogenic and modifying effects on chemical carcinogenic. *Food Chem. Toxicol.* **24**, 1099–1102.

Je, J. Y., Park, P. J., Byun, H. G., Jung, W. K., and Kim, S. K. (2005). Angiotensin I converting enzyme (ACE) inhibitory peptide derived from the sauce of fermented blue mussel, *Mytilus edulis*. *Bioresour. Technol.* **96**, 1624–1629.

Je, J. Y., Qian, Z. J., Lee, S. H., Byun, H. G., and Kim, S. K. (2008). Purification and antioxidant properties of bigeye tuna (*Thunnus obesus*) dark muscle peptide on free radical-mediated oxidative systems. *J. Med. Food* **11**, 629–637.

Jeon, Y. J., Byun, H. G., and Kim, S. K. (1999). Improvement of functional properties of cod frame protein hydrolysates using ultrafiltration membranes. *Process Biochem.* **35**, 471–478.

Jun, S.-Y., Park, P.-J., Jung, W.-K., and Kim, S.-K. (2004). Purification and characterization of an antioxidative peptide from enzymatic hydrolysate of yellowfin sole (*Limanda aspera*) frame protein. *Eur. Food Res. Technol.* **219**, 20–26.

Kim, S. K., Kim, Y. T., Byun, H. G., Nam, K. S., Joo, D. S., and Shahidi, F. (2001). Isolation and characterization of antioxidative peptides from gelatin hydrolysate of Alaska pollack skin. *J. Agric. Food Chem.* **49**, 1984–1989.

Klompong, V., Benjakul, S., Yachai, M., Visessanguan, W., Shahidi, F., and Hayes, K. D. (2009). Amino acid composition and antioxidative peptides from protein hydrolysates of yellow stripe Trevally (Selaroides leptolepis). *J. Food Sci.* **74**, C126–133.

Kotzamanis, Y. P., Gisbert, E., Gatesoupe, F. J., Zambonino Infante, J., and Cahu, C. (2007). Effects of different dietary levels of fish protein hydrolysates on growth, digestive enzymes, gut microbiota, and resistance to *Vibrio anguillarum* in European sea bass (*Dicentrarchus labrax*) larvae. *Comp. Biochem. Physiol.* **147**, 205–214.

Kristinsson, H. G. and Rasco, B. A. (2000). Fish protein hydrolysates: Production, biochemical, and functional properties. *Crit. Rev. Food Sci. Nutr.* **40**, 43–81.

Liaset, B., Julshamn, K., and Espe, M. (2003). Chemical composition and theoretical nutritional evaluation of the produced fractions from enzymic hydrolysis of salmon frames with protamex. *Process Biochem.* **38**, 1747–1759.

Liceaga-Gesualdo, A. M. and Li-Chan, E. C. Y. (1999). Functional properties of fish protein hydrolysate from Herring (*Clupea harengus*). *J. Food Sci.* **64**, 1000–1004.

Liddle, R. A. (1997). Cholecystokinin cells. *Annu. Rev. Physiol.* **59**, 221–242.

Liddle, R. A., Goldfine, I. D., Rosen, M. S., Taplitz, R. A., and Williams, J. A. (1985). Cholecystokinin bioactivity in human plasma. Molecular forms, responses to feeding, and relationship to gallbladder contraction. *J. Clin. Invest.* **75**, 1144–1152.

Lin, C. C. and Liang, J. H. (2002). Effect of antioxidants on the oxidative stability of chicken breast meat in a dispersion system. *J. Food Sci.* **67**, 530–533.

Martin, A. M., Porter, D., and George, C. (1995). Studies on the hydrolysis of fish protein by enzymatic treatment. Developments in Food Science. Vol. 37, pp. 1395–1404. Elsevier, Amsterdam.

Martinez-Alvarez, O., Guimas, L., Delannoy, C., and Fouchereau-Peron, M. (2007). Occurrence of a CGRP-like molecule in siki (*Centroscymnus coelolepsis*) hydrolysate of industrial origin. *J. Agric. Food Chem.* **55**, 5469–5475.

Martinez-Alvarez, O., Guimas, L., Delannoy, C., and Fouchereau-Peron, M. (2008). Use of a commercial protease and yeasts to obtain CGRP-like molecules from saithe protein. *J. Agric. Food Chem.* **56**, 7853–7859.

Matsufuji, H. and Matsui, T. (1994). Angiotensin I-converting enzyme inhibitory peptides in an alkaline protease hydrolysate derived from sardine muscle. *Biosci. Biotechnol. Biochem.* **58**, 2244–2245.

Matsui, T. and Matsufuji, H. (1993). Inhibition of angiotensin I-converting enzyme by *Bacillus licheniformis* alkaline protease hydrolysates derived from sardine muscle. *Biosci. Biotechnol. Biochem.* **56**, 922–925.

Meisel, H. (2004). Multifunctional peptides encrypted in milk proteins. *Biofactors* **21**, 55–61.

Miraliakbari, H. and Shahidi, F. (2008). Antioxidant activity of minor components of tree nut oils. *Food Chem.* **111**, 421–427.

Nemoz-Gaillard, E., Bernard, C., Abello, J., Cordier-Bussat, M., Chayvialle, J. A., and Cuber, J. C. (1998). Regulation of cholecystokinin secretion by peptones and peptidomimetic antibiotics in STC-1 cells. *Endocrinology* **139**, 932–938.

Nishi, T., Hara, H., Asano, K., and Tomita, F. (2003a). The soybean β-conglycinin β 51–63 fragment suppresses appetite by stimulating cholecystokinin release in rats. *J. Nutr.* **133**, 2537–2542.

Nishi, T., Hara, H., and Tomita, F. (2003b). Soybean β-conglycinin peptone suppresses food intake and gastric emptying by increasing plasma cholecystokinin levels in rats. *J. Nutr.* **133**, 352–357.

Park, K.-J. and Hyun, C.-K. (2002). Antigenotoxic effects of the peptides derived from bovine blood plasma proteins. *Enzyme Microb. Technol.* **30**, 633–638.

Park, S. Y., Lee, J.-S., Baek, H.-H., and Lee, H. G. (2001). Purification and characterization of antioxidant peptides from soy protein hydrolysate. *J. Food Biochem.* **34**, 120–132.

Picot, L., Bordenave, S., Didelot, S., Fruitier-Arnaudin, I., Sannier, F., Thorkelsson, G., Bergé, J. P., Guérard, F., Chabeaud, A., and Piot, J. M. (2006). Antiproliferative activity of fish protein hydrolysates on human breast cancer cell lines. *Process Biochem.* **41**, 1217–1222.

Qian, Z. J., Je, J. Y., and Kim, S. K. (2007). Antihypertensive effect of angiotensin I converting enzyme-inhibitory peptide from hydrolysates of Bigeye tuna dark muscle, *Thunnus obesus*. *J. Agric. Food Chem.* **55**, 8398–8403.

Quaglia, G. B. and Orban, E. (1987). Influence of the degree of hydrolysis on the solubility of the protein hydrolysates from sardine (*Sardina pilchardus*). *J. Sci. Food Agric.* **38**, 271–276.

Raa, J. and Gildberg, A. (1982). Fish silage: A review. *Crit. Rev. Food Sci. Nutr.* **16**, 383–419.

Raghavan, S. and Kristinsson, H. G. (2008). Antioxidative efficacy of alkali-treated tilapia protein hydrolysates: a comparative study of five enzymes. *Journal of Agricultural and Food Chemistry* **56**, 1434–1441.

Raghavan, S., Kristinsson, H. G., and Leeuwenburgh, C. (2008). Radical scavenging and reducing ability of tilapia (Oreochromis niloticus) protein hydrolysates. *J. Agric. Food Chem.* **56**, 10359–10367.

Rajapakse, N., Jung, W. K., Mendis, E., Moon, S. H., and Kim, S. K. (2005). A novel anticoagulant purified from fish protein hydrolysate inhibits factor XIIa and platelet aggregation. *Life Sci.* **76**, 2607–2619.

Ravallec-Ple, R. and Van Wormhoudt, A. (2003). Secretagogue activities in cod (*Gadus morhua*) and shrimp (*Penaeus aztecus*) extracts and alcalase hydrolysates determined in AR4-2J pancreatic tumour cells. *Comp. Biochem. Physiol.* **134**, 669–679.

Ravallec-Plé, R., Gilmartin, L., Van Wormhoudt, A., and Le Gal, Y. (2000). Influence of the hydrolysis process on the biological activities of protein hydrolysates from cod (*Gadus morhua*) muscle. *J. Sci. Food Agric.* **80**, 2176–2180.

Ravallec-Plé, R., Charlot, C., Batista, I., Van Wormhoudt, A., Le Gal, Y., and Fouchereau Peron, M. (2001). The presence of bioactive peptides in hydrolysates prepared from processing waste of sardine (*Sardina pilchardus*). *J. Sci. Food Agric.* **81**, 1120–1125.

Rindi, G., Grant, S. G., Yiangou, Y., Ghatei, M. A., Bloom, S. R., Bautch, V. L., Solcia, E., and Polak, J. M. (1990). Development of neuroendocrine tumors in the gastrointestinal tract of transgenic mice. Heterogeneity of hormone expression. *Am. J. Pathol.* **136**, 1349–1363.

Ritchie, A. H. and Mackie, I. M. (1982). Preparation of fish protein hydrolysates. *Anim. Feed Sci. Technol.* **7**, 125–133.

Rosenfeld, M. G., Mermod, J. J., Amara, S. G., Swanson, L. W., Rivier, J., Vale, W. W., and Evans, R. M. (1983). Production of a novel neuropeptide encoding by the calcitonin gene via tissue specific RNA processing. *Nature* **304**, 129–135.

Rousseau, M., Batista, I., Le Gal, Y., and Fouchereau-Peron, M. (2001). Purification of a functional antagonist for calcitonin gene related peptide action from sardine hydrolysates. *Electron. J. Biotechnol.* **4**, 25–32.

Saiga, A., Tanabe, S., and Nishimura, T. (2003). Antioxidant activity of peptides obtained from porcine myofibrillar proteins by protease treatment. *J. Agric. Food Chem.* **51**, 3661–3667.

Schwartz, M. W., Woods, S. C., Porte, D., Jr., Seeley, R. J., and Baskin, D. G. (2000). Central nervous system control of food intake. *Nature* **404**, 661–671.

Shahidi, F. (2009). Nutraceuticals and functional foods: Whole versus processed foods. *Trends Food Sci. Technol.* **20**, 376–387.

Shahidi, F., Han, X.-Q., and Synowiecki, J. (1995). Production and characteristics of protein hydrolysates from capelin (*Mallotus villosus*). *Food Chem.* **53**, 285–293.

Shahidi, F., Alasalvar, C., and Liyana-Pathirana, C. M. (2007). Antioxidant phytochemicals in hazelnut kernel (Corylus avellana L.) and hazelnut byproducts. *J. Agric. Food Chem.* **55**, 1212–1220.

Strader, A. D. and Woods, S. C. (2005). Gastrointestinal hormones and food intake. *Gastroenterology* **128**, 175–191.

Suetsuna, K. (2002). Identification of antihypertensive peptides from peptic digest of the short-necked clam *Tapes philippinarum* and the pearl oyster *Pinctada fucata martensii*. *Fish. Sci.* **68**, 233–235.

Suetsuna, K. and Nakano, T. (2000). Identification of an antihypertensive peptide from peptic digest of wakame (*Undaria pinnatifida*). *J. Nutr. Biochem.* **11**, 450–454.

Suetsuna, K., Ukeda, H., and Ochi, H. (2000). Isolation and characterization of free radical scavenging activities peptides derived from casein. *J. Nutr. Biochem.* **11**, 128–131.

Suetsuna, K., Maekawa, K., and Chen, J. R. (2004). Antihypertensive effects of *Undaria pinnatifida* (wakame) peptide on blood pressure in spontaneously hypertensive rats. *J. Nutr. Biochem.* **15**, 267–272.

Sufian, M. K., Hira, T., Miyashita, K., Nishi, T., Asano, K., and Hara, H. (2006). Pork peptone stimulates cholecystokinin secretion from enteroendocrine cells and suppresses appetite in rats. *Biosci. Biotechnol. Biochem.* **70**, 1869–1874.

Tang, H. G., Wu, T. X., Zhao, Z. Y., and Pan, X. D. (2008). Effects of fish protein hydrolysate on growth performance and humoral immune response in large yellow croaker (*Pseudosciaena crocea* R.). *J. Zhejiang Univ. Sci. B* **9**, 684–690.

Terracciano, L., Isoardi, P., Arrigoni, S., Zoja, A., and Martelli, A. (2002). Use of hydrolysates in the treatment of cow's milk allergy. *Ann. Allergy Asthma Immunol.* **89**, 86–90.

Tippins, J. R., Di Marzo, V., Panico, M., Morris, H. R., and MacIntyre, I. (1986). Investigation of the structure/activity relationship of human calcitonin gene-related peptide (CGRP). *Biochem. Biophys. Res. Commun.* **134**, 1306–1311.

Uchida, K. and Kawakishi, S. (1992). Sequence-dependent reactivity of histidine-containing peptides with copper (II)/ascorbate. *J. Agric. Food Chem.* **40**, 13–16.

Watson, R. E., DiPette, D. J., Supowit, S. C., Katki, K. A., and Zhao, H. (2005). Vasodilator peptides: CGRP, substance P, and adrenomedullin. *In* "Hypertension", (O. Suzanne and A. W. Michael, Eds), 2nd edn., pp. 193–202. W.B. Saunders, Philadelphia.

Wu, H.-C., Chen, H.-M., and Shiau, C.-Y. (2003). Free amino acids and peptides as related to antioxidant properties in protein hydrolysates of mackerel (*Scomber austriasicus*). *Food Res. Int.* **36**, 949–957.

Yallampalli, C., Chauhan, M., Thota, C. S., Kondapaka, S., and Wimalawansa, S. J. (2002). Calcitonin gene-related peptide in pregnancy and its emerging receptor heterogeneity. *Trends Endocrinol. Metab.* **13**, 263–269.

Yu, R., Mandlekar, S., and Kong, A. N. (2000). Molecular mechanisms of butylated hydroxylanisole-induced toxicity: Induction of apoptosis through direct release of cytochrome C. *Mol. Pharmacol.* **58**, 431–437.

Zhong, Y., Khan, M. A., and Shahidi, F. (2007). Compositional characteristics and antioxidant properties of fresh and processed sea cucumber (Cucumaria frondosa). *J. Agric. Food Chem.* **55**, 1188–1192.

CHAPTER 32

Utilization of Seafood Processing By-products: Medicinal Applications

Mahinda Senevirathne* and **Se-Kwon Kim***,†,1

Contents

* Marine Bioprocess Research Center, Pukyong National University, Busan, Republic of Korea
† Department of Chemistry, Marine Biochemistry Laboratory, Pukyong National University, Daeyoung 3-dong, Nam-Gu, Busan, Republic of Korea
[1] Corresponding author: Se-Kwon Kim, *E-mail address:* sknkim@pknu.ac.kr

Advances in Food and Nutrition Research, Volume 65
ISSN 1043-4526, DOI: 10.1016/B978-0-12-416003-3.00032-9

Abstract Large amount of underutilized by-products are generated from the seafood processing plants annually. Consequently, researches have been initiated to investigate those discarded materials and have identified a number of bioactive compounds including bioactive peptides, collagen and gelatin, oligosaccharides, fatty acids, enzymes, calcium, water-soluble minerals, and biopolymers. Bioactive peptides derived from fish by-products have shown various biological activities including antihypertensive and antioxidant activities and hence may be a potential material for biomedical and food industries. Collagen and gelatin are currently used in diverse fields including food, cosmetic, and biomedical industries. Other than that, they are promising drug carriers for the treatment of cancer. Many studies have reported that chitin, chitosan, and their derivatives possess biologically active polysaccharides and hence they are potential agents for many applications. Further, those compounds have also showed potential activities such as antioxidant, antibacterial, antiviral, antihypertensive, anticancer, etc. Hence, seafood by-products are valuable natural resources that show range of functionalities and hence potential materials for biomedical and nutraceutical industries.

I. INTRODUCTION

Fish catch is mainly used for human consumptions and other minor uses such as meal production and bait. Fish used as human food accounts for 78% of the total fish catch in developing and developed countries leaving some 21% as waste materials from the food processing industries (Vannuccini, 2004). Processing leads to the generation of large amount of biomass of fish waste (e.g., skin, bones, and fins) which is annually discarded (~7.3 million ton/year; Kelleher, 2005). Consequently, research has been initiated to investigate an increased utilization of those discarded materials and found that most of them can be reused as potential bioactive materials such as gelatin, collagen, and fish oil in various industries (Wassawa *et al.*, 2007).

Most of these recycled products possess low economic value. Recent studies have identified a number of bioactive compounds from remaining fish muscle proteins, collagen and gelatin, fish oil, fish bone, internal organs, and shellfish and crustacean shells (Je *et al.*, 2005; Jeon and Kim, 2002; Kim *et al.*, 2001). Generally, a far better profitability is obtained by producing human consumables and the highest profitability is currently expected from bioactive compounds. These bioactive compounds can be extracted and purified with technologies varying from simple to complex, and such compounds may include preparation and isolation of bioactive peptides, oligosaccharides, fatty acids, enzymes, water-soluble minerals, and biopolymers for biotechnological and pharmaceutical applications.

Further, some of these bioactive compounds have been identified to possess nutraceutical potentials that are beneficial in human health promotion. Therefore, development of new technologies in search of novel bioactive compounds from marine processing by-products will bring more value out of what is today considered a waste and represents unique challenges and opportunities for the seafood industry.

II. PROTEIN AND PEPTIDES FROM VARIOUS FISH WASTES AND THEIR BIOACTIVITIES

A. Protein from fish muscle

Discarded fish bones and cutoffs may contain considerable amounts of muscle proteins. These muscle proteins are nutritionally valuable and easily digestible with well-balanced amino acid composition (Venugopal *et al.*, 1996). Therefore, fish proteins derived from seafood processing by-products can be hydrolyzed enzymatically to recover protein. Protein hydrolysates from seafood by-products of several marine species have been analyzed for their nutritional and functional properties, and researches have mainly explored the possibility of obtaining biologically active peptides (Benkajul and Morrissey, 1997). Moreover, several hydrolytic enzymes from microbes, plants, and animals are employed for the hydrolysis of seafood processing by-products (Simpson *et al.*, 1998). Further, peptides present in enzymatically digested hydrolysates have exhibited various physicochemical properties and biological activities. It is a well-known fact that molecular weights of peptide fragments are crucial for their biological activities. Therefore, it is a key step to develop methods to separate peptides with different molecular weights. An ultrafiltration membrane system equipped with the appropriate molecular weight cutoff has been effectively used in separating peptides having desired molecular weights (Jeon *et al.*, 2000). In order to obtain functionally active peptides, it is a common method to use a sort of enzymes letting sequential enzymatic digestions.

B. Peptides from fish skin and back bone

ACE inhibitory peptides have been separated from the skin of skate by Lee *et al.* (2011). The purified peptides showed IC_{50} values of 95 and 148 μM, respectively, for both peptides isolated from the skin of skate. Further, Lineweaver–Burk plots indicated that the peptides act as noncompetitive inhibitor against ACE. Zhao *et al.* (2009) have used body wall protein of sea cucumber to make a hydrolysate and to obtain peptides. In their study, they have used membrane bioreactor for the fractionation of hydrolysate. Isolated peptides have been reported to have strong ACE inhibitors with the IC_{50} of 15.9–4.5 μM.

Recently, many inhibitory peptides against ACE are reported as natural alternative biofunctional peptides that are safer than that of the existing artificial ACE inhibitory compounds in the market that show some side effects. Various peptides have been isolated from seafood by-products from the fisheries industry such as backbones from tuna (Lee *et al.*, 2010). Tuna backbone has hydrolyzed using various proteases such as alcalase, α-chymotrypsin, neutrase, papain, pepsin, and trypsin to obtain an antioxidant peptide (Je *et al.*, 2007). In another study, an antioxidant peptide has been isolated from horki frame protein by gastrointestinal digestion (Kim *et al.*, 2007). They also have used number of enzymes for the hydrolysis process, and the isolated peptides have showed strong antioxidant activities in various model systems. Batista *et al.* (2010) have prepared a protein hydrolysate from black scabbard fish by-products as an alternative raw material that can be used in various industries. The hydrolysate has shown potential antioxidant activity against various models.

C. Biological activities of peptides and fish protein hydrolysates

Peptides isolated from various fish protein hydrolysates have shown a different biological activities such as antihypertensive, antioxidative, antithrombotic, and immunomodulatory activities. Enzymatically prepared fish muscle peptides also have shown anticoagulant and antiplatelet properties tested *in vitro*, and the results have suggested the capability of fish peptides to inhibit coagulation factors in the intrinsic pathway of coagulation (Rajapakse *et al.*, 2005).

Peptides derived from fish proteins have shown the ability of exerting potent antioxidative activities in different oxidative systems (Rajapakse *et al.*, 2005). An increasing interest exists currently to explore natural antioxidative substances without side effects, and these identified antioxidative activities have a potential to develop safe and nonhazardous natural antioxidants for the complications arose from oxidation of biomolecules. Byun *et al.* (2009) purified an antioxidative peptide from marine rotifer which is very important regarding the functional foods. It was reported that the derived peptides were good radical scavengers and antioxidant. Development of methods to increase the solubility of calcium compounds can significantly improve the calcium availability in biological systems. Kim *et al.* (1999) and Jung *et al.* (2005) reported that fish peptides are capable of accelerating calcium absorption. Further, researchers have identified that fish protein hydrolysates possess hormone-like peptides and growth factors to accelerate calcium absorption (Fouchereau-Peron *et al.*, 1999). These peptides are capable of binding to cell surface receptors on osteoclasts and involved mainly in calcium metabolism by decreasing the number of osteoclasts. Therefore, these peptides could be used in the treatment of osteoporosis and Paget's disease. Further, acidic peptide fractions from Atlantic cod hydrolysate have shown strong immunostimulatory effects, and treatment of these peptides has stimulated the oxidative burst of Atlantic salmon leucocytes (Gildberg *et al.*, 1996). Basically, immunomodulators enhance the production of oxygen metabolites in macrophages that are responsible for these oxygen metabolites determining the oxidative burst. Oxidative burst reactions are of major importance for the bactericidal power of phagocytes.

Many studies have been conducted with protein hydrolysates from fish muscle protein, and those hydrolysates have been proved to be bioactive agents in many models. Picot *et al.* (2006) have analyzed 16 protein hydrolysates against two kinds of human breast cancer cell lines. Three blue whitings, three cods, three plaices, and one salmon hydrolysate were identified as significant growth inhibitors against two cancer cell lines. Slizyte *et al.* (2009) also evaluated the hydrolysates from cod backbones on antioxidant activities.

III. FISH SKIN COLLAGEN AND GELATIN

Gelatin is derived from the fibrous protein collagen, which is the principal constituent of animal skin, bone, and connective tissue. Gelatin is produced via the partial hydrolysis of native collagen. During the manufacturing of gelatin, raw animal material is treated with dilute acid or alkali, resulting in partial cleavage of the cross-links: the structure is broken down to such an extent that "warm-water-soluble collagen," that is, gelatin, is formed (Schrieber and Gareis, 2007).

Fish skin waste could be used as a potential source to isolate collagen and gelatin. They are currently used in various fields such as food, cosmetic, and biomedical industries. Collagen is structurally formed as a triple helix by three extended protein chains that wrap around one another. Collagen and gelatin are different forms of the same macromolecule, and gelatin is the partially hydrolyzed form of collagen. Heat denaturation easily converts collagen into gelatin. In addition to fish skin, collagen and gelatin could be isolated from bone and fins of fish processing by-products. Collagen and gelatin are unique proteins compared to fish muscle proteins, and this uniqueness of fish lies in the amino acid content and they are rich in nonpolar amino acids (above 80%) such as Gly, Ala, Val, and Pro. Even though main industrial sources of collagen and gelatin are bovine and porcine skin, many studies have been conducted to extract collagen and gelatin from fish skin and have used to screen their potential industrial applications (Supavititpatana *et al.*, 2008; Wassawa *et al.*, 2007). The utilization of fish skin collagen and gelatin is expected to attract the interest of the industry as an alternative source.

Collagen is generally extracted with acid treatment and solubilized without altering its triple helix. However, thermal treatment cleaves hydrogen and covalent bonds that stabilizes the triple helix configuration of collagen and converts its helical conformation into coiled conformation resulting gelatin state (Djabourov *et al.*, 1993). The properties of extracted gelatin may vary slightly with the extraction conditions such as temperature and pH. Optimum conditions for extraction are dependent on fish species and type of collagen (Wassawa *et al.*, 2007).

Gelatin from marine-discarded sources such as fish skins, bones, and fins is a significant alternative to bovine gelatin (Kim and Mendis, 2006; Rustad, 2003; Wassawa *et al.*, 2007). One of the major advantages of marine gelatin sources is that they are not associated with the risk of outbreaks of bovine spongiform encephalopathy. In addition, fish skin, which is a major by-product of the fish-processing industry, causing waste and pollution, could provide a valuable source of gelatin (Badii and Howell, 2006). Fish skin contains a large amount of collagen: Nagai and Suzuki (2000) reported that the collagen contents in the fish skin waste of Japanese sea-bass, chub mackerel, and bullhead shark were 51.4%, 49.8%, and 50.1% (dry basis), respectively.

A. Bioactivities of collagen and gelatin peptides

Enzymatically hydrolyzed fish skin gelatin has shown better biological activities compared to that of the peptides derived from fish muscle protein to act as potential antioxidants and antihypertensive agents (Byun and Kim, 2001; Kim *et al.*, 2001; Mendis *et al.*, 2005). Gelatin peptides have repeated unique Gly–Pro–Ala sequence in their structure, and it is presumed that the observed antioxidative and antihypertensive properties of gelatin peptides can be associated with their unique amino acid compositions. In addition, these peptides have shown to accelerate absorption of dietary calcium in animal models increasing calcium bioavailability (Kim *et al.*, 1998). Previous studies with skate skin included the production of collagen and the purification of an antimicrobial peptide (Cho *et al.*, 2004). Hence, it is proved that peptides from various fish by-products not only show ACE or antioxidant activities but also show antimicrobial activities.

B. Biomedical and nutraceutical applications

Collagen is commonly used in medical and pharmaceutical industries as carrier molecules for drugs, proteins, and genes (Lee *et al.*, 2001). Especially, microfibrous collagen sheets are used as promising drug carriers for the treatment of cancer (Sato *et al.*, 1996). Long-term maintenance of drug concentration and controlled release at target sites promote the utilization of collagen as efficient drug carriers. In addition, recently, collagen film/matrix is used as gene delivery agents promoting bone and cartilage formation (Nakagawa and Tagawa, 2000). Clinical investigations suggest that ingestion of collagen/gelatin hydrolysates reduces pain in patients suffering from osteoarthritis, and hydrolyzed collagen have been involved in cartilage matrix synthesis (Moskowitz, 2000). Moreover, collagen/gelatin is currently marketed as supplement for the maintenance of normal bone integrity, treatment of brittle nails, and nourishment of scalp hair.

Gelatin possesses a characteristic melt-in-the-mouth property that makes them used in the wide range of applications in the food and pharmaceutical industries. Fish collagen releases a kind of aroma and shows higher digestibility than that of animal gelatin (Gomez-Guillen *et al.*, 2002). Fish collagen is widely used in cosmeceutical industry in cosmetic products formulation (Bae *et al.*, 2008).

Some studies on the food properties of fish gelatin have been conducted (Choi and Regenstein, 2000; Norland, 1990). Moreover, physicochemical and functional properties of fish gelatin have been evaluated, especially with respect to its rheological properties (Badii and Howell, 2006) as well as its emulsifying, foaming (Surh *et al.*, 2006), film-forming (Avena-Bustillos *et al.*, 2006), and sensory properties (Choi and Regenstein, 2000).

IV. CHITIN, CHITOSAN, AND THEIR DERIVATIVES FROM CRUSTACEAN SHELLS AND SHELLFISH

Crustacean shells and shellfish wastes are another important category of by-products released from processing plants. Efficient utilization of these by-products helps to keep the environment clean due to increased quantity of accumulation from processing plants as well as slow natural degradation of these materials. Chitin is one of the major structural components of these shells and many studies have reported that they possess biologically active polysaccharides and hence they are potential agents for many applications. These shell wastes are potential sources for the isolation of chitin and are currently utilized for the commercial-scale chitin production, as well as production of chitosan and their oligomers.

Chitin is a linear polymer that can be converted into other derivatives such as chitosan. Chitosan contains less acetyl groups compared to that of chitin, and they can be prepared either by chemical or by microbiological treatments. Chitin, chitosan, and their derivatives have gained much attention due to their nontoxicity, biocompatibility, biodegradability, and variety of bioactivities.

A. Biological activities of chitin, chitosan, and their derivatives

Chitosan and its derivatives are reported as potential antioxidants by scavenging various radicals including oxygen radicals, and it is also reported that their activities depend on the molecular weight and the degree of deacetylation. Recently, the antioxidant activity of chitosan and its derivatives attracted a greater attention. The antioxidant activity of chitosan and its various derivatives have been reported by many researchers (Aytekin *et al.*, 2010; Guo *et al.*, 2008; Zhong *et al.*, 2007, 2010). In some studies, the antioxidant activities of the chitosan and its derivatives have been increased by conjugation of those compounds with various phenolic compounds (Aytekin *et al.*, 2010; Cho *et al.*, 2011; Pasanphan and Chirachanchai, 2008). DPPH, hydroxyl, superoxide, and alkyl radical scavenging, metal chelating, reducing power activity, and lipid peroxidation have been studied with chitosan and its derivatives and showed significant activities.

Chitosan and its derivatives have been reported as good antimicrobial agents. Various groups have been conjugated with chitosan main body to increase its water solubility and activity, and also they have proved to be good antimicrobial agents due to this conjugation. It is believed that the antibacterial action is due to adsorption of the positive charge of quaternary ammonium groups onto the negative charge of the bacterial cell surfaces and membrane with subsequent disruption of membrane integrity (Rabea *et al.*, 2003). Many studies have proved that chitosan and its derivatives showed antibacterial, antifungal, and antiviral activities (Artan *et al.*, 2010; Jayakumar *et al.*, 2007; Li *et al.*, 2010; Runarsson *et al.*, 2010; Sajomsang *et al.*, 2009, 2010; Xu *et al.*, 2010; Yang *et al.*, 2010; Zhong *et al.*, 2009).

Chitin, chitosan, and their derivatives also act as the inhibitors of angiotensin-converting enzyme, an enzyme which is associated with hypertension. As chloride ion can activate the angiotensin-converting enzyme, high-salt food can raise the blood pressure but chitosan and its derivatives can bind to the chloride ion and can alleviate the problem. Many researchers have investigated the ACE inhibitory activity of chitosan and its derivatives and have reported that the activity depends on the degree of deacetylation (Park *et al.*, 2003). Further, they have reported that chitooligosaccharide (COS) with low degree of deacetylation showed higher ACE inhibitory activity. Substitution of hydrogen atom at C-6 position of pyranose residue with aminoethyl group has promoted the ACE inhibitory activity significantly (Ngo *et al.*, 2008). In addition, potent ACE inhibitory activity of enzymatically produced hydrolysate of shrimp was reported by Cheung and Li-Chan (2010). They have reported that protease type that is used for hydrolysis of shells is responsible for the ACE activity, and the hydrolysate obtained by alcalase and protomex showed the highest ACE inhibitory activity.

The immunostimulatory activity of chitosan and its derivatives has been reported for decades, and many studies have done so far with chitosan and its derivatives to evaluate the anticancer and antitumor activity. Chitosan also showed an immunity-enhancing effect by enhancing the response of antibody. The effect of chitosan as a novel adjuvant to an inactivated influenza vaccine was studied by Chang *et al.* (2004). Even if several studies have reported the importance of chitosan derivatives for their anticancer activity, no clear information has been reported about the relationship between their charge properties and the observed activities. Huang *et al.* (2006) studied the anticancer activities of COS derivatives using various cancer cell lines such as HeLa, Hep3B, and SW480. The conjugates of mitomycin C with *N*-succinyl chitosan showed significant antitumor activities due to their predominant distribution into the tumor tissue and sustained-release characteristics with both water-insoluble and -soluble formulations (Kato *et al.*, 2005).

Several studies have been reported that chitosan can lower plasma and live triacylglycerol and total cholesterol. Further, it has been shown that chitosan has the ability to reduce the risk of cardiovascular diseases and also has potent fat binding capacity (Zhou *et al.*, 2006). Several studies have been conducted to evaluate the effect of degree of deacetylation, viscosity, and the molecular weight of chitosan on fat binding, cholesterol, and bile salt-binding capacities (Liu *et al.*, 2008a,b; Zhou *et al.*, 2006). Chitosans with higher molecular weights limited the body weight gain of adult rats significantly, reduced the food-efficiency ratio, and lowered plasma lipids (Liu *et al.*, 2008a,b). In addition, the powdered form of chitosan exhibited a greater rate of absorption that of flake chitosan (Ahmad *et al.*, 2005).

B. Application of chitin, chitosan, and their derivatives

Chitin is widely used to immobilize enzymes and whole cells. Enzyme immobilization can be used in the food industry to clarify the fruit juices and process the milk when α- and β-amylases or invertase is grafted on chitin (Krajewska, 2004). On account of its biodegradability, nontoxicity, physiological inertness, antibacterial properties, hydrophilicity, gel-forming properties, and affinity for proteins, chitin has found applications in many areas other than food such as in biosensors. The main development of chitin film and fiber is in medical and pharmaceutical applications as wound-dressing material (Yusof *et al.*, 2003) and controlled drug release (Kato *et al.*, 2003). It is also used as an excipient and drug carrier in film, gel, or powder form for applications involving mucoadhesivity. Hydroxyapatite–chitin–chitosan composite is reported to be used as a bone-filling material, which forms a self-hardening paste for guided tissue regeneration in treatment of periodontal bony defects (Ito *et al.*, 1998).

Medical application of chitosan and its derivatives has gained much interest among many researchers. Chitosan is a natural pseudocationic polymer, and due to this character, it is suitable in cosmetic applications, especially for hair care in relation to its electrostatic interactions. Moreover, chitosan and its derivatives are potential agents for pharmaceutical and medical applications due to their characteristic features. The transmucosal administration of drugs has been discussed recently (Illum and Davis, 2005). Mucoadhesivity of chitosan and cationic derivatives is recognized and has been proved to enhance the adsorption of drugs especially at neutral pH; *N*-trimethyl chitosan chloride interacts with negatively charged cell membranes (Venter *et al.*, 2006). Chitosan films incorporated with prednisolone have been used as controlled drug release. Chitosan gels and layer-by-layer polyelectrolyte capsules are frequently used in controlled releasing drugs.

V. SEPARATION AND APPLICATION OF FISH OIL AS HEALTH-PROMOTING AGENT

Scientific data show that the consumption of fish or fish oil containing omega-3 polyunsaturated fatty acids (PUFAs) reduces the risk of coronary heart disease, decreases mild hypertension, prevents certain cardiac arrhythmia and sudden death, lowers the incidence of diabetes, and appears to alleviate symptoms of rheumatoid arthritis. Therefore, utilization of marine fish processing by-products by converting them into fish oil would be great benefit. In general, the fat content of fish is varied from 2% to 30%, and it basically depends on the type of species, dietary, geographic, environmental, reproductive, and seasonal variations. However, with the depleting marine fisheries resources, fishing for oil is not encouraged. Therefore, a large amount of offal generated from processing would be a potential source to produce good quality fish oil for human consumption, especially from fatty fish processing by-products. Fish oil is mainly composed of two types of fatty acids, eicosapentaenoic acid (EPA) and docosahexaenoic acid (DHA). These are PUFAs classified as omega-3 fatty acids and predominantly found in many marine animals including cold-water fish species with a higher unsaturated fat content. Compared to saturated fats, PUFAs in fish oil are readily digested for energy production and have been reported to have various bioactivities. Methods and conditions of fish oil extraction are vital for quality of fish oil. The traditional extraction technique involves heating or steam stripping of the raw material in order to release the lipids. Presently, several methods such as high-speed centrifugation, low-temperature solvent extraction, and supercritical fluid extraction are utilized to extract fish oil (Hirata *et al.*, 1993). In addition, wet and steam rendering methods are also utilized to extract fish oil from fish processing by-products. Separation of omega-3 fatty acids is one of the main concerns in the current research. However, due to the presence of complex mixture of triacylglycerols, separation of omega-3 fatty acids becomes complicated. Many separation methods have been developed to isolate them at higher purity including crystallization, distillation, supercritical fluid extraction, and chromatography. Especially, column liquid chromatography is frequently used to analyze and separate fatty acids. Sahena *et al.* (2010) have used supercritical fluid extraction to separate oil from the skin of Indian mackerel. The largest recoveries of PUFA, especially the omega-3 family, were achieved from the soaking and pressure swing techniques in the same study. In addition, combination of chromatographic method with urea crystallization enables successful separation of fatty acids (Hidajat *et al.*, 1995). Proper separation techniques and advanced processing methods could promote the production of higher-quality fish oil, and fish oil might play an even more important role in the pharmaceutical and food industry in the near future. Liu *et al.* (2008a,b) used response surface method to determine the optimum operating conditions for extracting the gelatin from channel catfish skin. The optimal conditions for maximum gel strength are 68.8h for the time of treatment with calcium hydroxide solution, 43.2°C for the extraction temperature, and 5.73h for the extraction time with hot water. The gelatin from channel catfish skin showed high gel strength, 276±5g. Intake of fish oil, which is an excellent source of omega-3 fatty acids, has been linked to promotion of human health to fight against numerous diseases. Moreover, researchers are convinced that EPA and DHA are the main protective components of fish oil that act against certain types of diseases. Omega-3 PUFAs play a vital role in the development and function of the nervous system (brain), photoreception (vision), and the reproductive system (Kirpal, 2003; Simopoulos, 1999).

Moreover, fish oil is used for various industries and it is mainly used for the production of margarine. EPA is produced, for instance, from sardine oil which contains EPA as well as DHA in some degree. Sardine oil is concentrated to rich oil containing EPA 25% and DHA 15% through alkali treatment, decolorization with carbon, with active white soil, and with wintering. This, so called, EPA-rich oil (which also contains DHA) is capsulated for oral administration or directly assorted to canned foods, powdered milk, and others. EPA-rich oil is effective against arteosclerosis and for improving the fatty acid quality in blood. DHA is produced from the eye fat of tuna fish which contains DHA 27% and EPA 7%. About 10 ton/year of DHA-rich oil is used for assorting to powdered milk, canned food, etc., and the rest in the capsule for the oral administration. It is believed to be effective for the health of brain and nerve, and for remembrance.

VI. APPLICATION OF FISH BONE AS POTENTIAL CALCIUM SOURCE

Fish bone has been identified as another important source from seafood by-products that can be used in the production of calcium. Since annually huge amount of bone by-products are removed from seafood processing industries, they can be used as an important source to produce inorganic calcium than can be used in various applications such as biomedical and food industries. The organic component of fish bone represents 30% of collagen while inorganic component mainly consists of calcium phosphate and hydroxyapatite and it was around 60–70% (Aronove *et al.*, 2007; Barakat *et al.*, 2008). Fish bone materials derived from processing of large fish are a source of calcium when the quantity is considered. Inorganic calcium can be used to fortify the milk powder or other food items. In a study, Heu *et al.* (2003) have reported that analysis of shrimp waste for minerals has revealed that it has 3000mg/100g of calcium. Not only calcium but also phosphorous which is a very important mineral has been analyzed by the same authors. However, many studies have not done to recover this important mineral. Jung *et al.* (2006) reported that fish bone peptides (FBP) II could inhibit the formation of insoluble Ca salts in neutral pH. During the experimental period, Ca retention was increased and loss of bone mineral was decreased by FBP II supplementation in ovariectomized rats. The levels of femoral total Ca, bone mineral density, and strength were also significantly increased by the FBP II diet to levels similar to those of the casein phosphopeptide diet group. Fish bones discarded from industrial processing of *Johnius belengerii* were digested by a heterogeneous enzyme extracted from the intestine of a carnivorous fish, bluefin tuna in order to utilize the bone to produce nutraceuticals with a high Ca bioavailability (Jung *et al.*, 2005). As reported by Jiang and Mine (2000), Ca-binding phosphoproteins such as osteocalcin, phosvitin, and casein phosphoprotein mainly consist of Ser, Thr, Ala, and Tyr residues phosphorylated or bound to Ca (Houben *et al.*, 1999). In addition, calcium-binding phosphoproteins derived from noncollagenous materials in the bone have a high affinity to Ca^{2+} on the surface of hydroxyapatite (Hoang *et al.*, 2003). Further, microcrystalline hydroxyapatite has been successfully prepared from fish bone by sinter treatment (Jung *et al.*, 2005; Ozawa and Suzuki, 2002). Hydroxyapatite is the principle calcium compound existing in bone and is high in calcium. Bioavailability of this compound has been proposed by several researchers (Kim *et al.*, 1997; Larsen *et al.*, 2000). Also, Tongchan *et al.* (2009) have investigated the effect of calcium compounds obtained from fish by-products on calcium metabolism in rats. Supplementation of calcium increased mean body weight of rats whereas a 9% loss of mean body weight of the control group was observed by week seven. Absorption and retention rate of the supplemented calcium increased significantly with increase of calcium intake. Femur weight of rats was increased by increasing the amount of calcium supplement with respect to that of the control rats. Results revealed that high bioavailability of calcium compound could be prepared by sintering process from fish bone.

VII. CONCLUSIONS

Marine seafood by-products have recently attracted much attention due to their potential biological activities and natural abundance. Also, they are underutilized natural resources that have potential to be used in various industries. Annually, large amount of wastes is discarded from seafood processing plants, and they have been recognized as potential materials in various fields. Various bioactive materials such as peptides, chitin chitosan, calcium, gelatin, collagen, and fish oil have been isolated from the fish wastes that have reported to have range of biological activities such as antioxidant, antihypertensive, antitumor, anticancer, antibacterial, anti-HIV, etc. Hence, seafood by-products are potential natural resources for range of industries.

REFERENCES

Ahmad, A. L., Sumathi, S., and Hameed, B. H. (2005). Adsorption of residue oil from palm oil mill effluent using powder and flake chitosan: Equilibrium and kinetic studies. *Water Res.* **39**, 2483–2494.

Aronove, D., Karlov, A., and Rosenman, G. (2007). Hydroxyapatite nanoceramics: Basic physical properties and biointerface modification. *J. Eur. Ceramic Soc.* **27**, 4181–4186.

Artan, M., Karadeniz, F., Karagozlu, M. Z., Kim, M. M., and Kim, S. W. (2010). Anti-HIV-1 activity of low molecular weight sulfated chitooligosaccharides. *Carbohydr. Res.* **345**, 656–662.

Avena-Bustillos, R. J., Olsen, C. W., Chiou, B., Yee, E., Bechtel, P. J., and McHugh, T. H. (2006). Water vapor permeability of mammalian and fish gelatin films. *J. Food Sci.* **71**, E202–E207.

Aytekin, A. O., Morimura, S., and Kida, K. (2010). Synthesis of chitosan–caffeic acid derivatives and evaluation of their antioxidant activities. *J. Biosci. Bioeng.* **111**, 212–216.

Badii, F. and Howell, N. K. (2006). Fish gelatin: Structure, gelling properties and interaction with egg albumen proteins. *Food Hydrocolloids* **20**, 630–640.

Bae, I., Osatomi, K., Yoshida, A., Osako, K., Yamaguchi, A., and Hara, K. (2008). Biochemical properties of acid-soluble collagens extracted from the skins of underutilized fishes. *Food Chem.* **108**, 49–54.

Barakat, A. M. N., Khalil, K. A., Sheikh, F. A., Omran, A. M., Gaihre, B., Khil, S. M., and Kim, H. Y. (2008). Physicochemical characterization of hydroxyapatite extracted from bones by three different methods: Extraction of biological desirable HAp. *Mater. Sci. Eng. C* **28**, 1381–1387.

Batista, I., Ramos, C., Coutinho, J., Bandarra, N. M., and Nunes, M. L. (2010). Characterization of protein hydrolysates and lipids obtained from black scabbardfish (Aphanopus carbo) by-products and antioxidative activity of the hydrolysates produced. *Process Biochem.* **45**, 18–24.

Benkajul, S. and Morrissey, M. T. (1997). Protein hydrolysates from Pacific whiting solid wastes. *J. Agric. Food Chem.* **45**, 3423–3430.

Byun, H. G. and Kim, S. K. (2001). Purification and characterization of angiotensin I converting enzyme (ACE) inhibitory peptides from Allaska Pollack (*Theragra chalcogramma*) skin. *Process Biochem.* **36**, 1155–1162.

Byun, H. G., Lee, J. K., Park, H. G., Jeon, J. K., and Kim, S. K. (2009). Antioxidant peptides isolated from the marine rotifer, Brachionus rotundiformis. *Process Biochem.* **44**, 842–846.

Chang, H. Y., Chen, J. J., Fang, F., and Chen, Z. (2004). Enhancement of antibody response by chitosan, a novel adjuvant of inactivated influenza vaccine. *Chin. J. Biol.* **17**, 21–24.

Cheung, I. W. Y. and Li-Chan, E. C. Y. (2010). Angiotensin-I-converting enzyme inhibitory activity and bitterness of enzymatically-produced hydrolysates of shrimp (*Pandalopsis dispar*) processing byproducts investigated by Taguchi design. *Food Chem.* **122**, 1003–1012.

Cho, S. H., Michael, L. J., and Eun, J. B. (2004). Nutritional composition and microflora of the fresh and fermented skate (*Raja Kenojei*) skins. *Int. J. Food Sci. Nutr.* **55**, 45–51.

Cho, Y. S., Kim, S. K., Ahn, C. B., and Je, J. Y. (2011). Preparation, characterization, and antioxidant properties of gallic acid-grafted-chitosans. *Carbohydr. Polym.* **83**, 1617–1622.

Choi, S. S. and Regenstein, J. M. (2000). Physicochemical and sensory characteristics of fish gelatin. *J. Food Sci.* **65**, 194–199.

Djabourov, M., Lechaire, J. P., and Gaill, F. (1993). Structure and rheology of gelatin and collagen gels. *Biorheology* **30**, 191–205.

Fouchereau-Peron, M., Duvail, L., Michel, C., Gildberg, A., Batista, I., and Gal, Y. I. (1999). Isolation of an acid fraction from a fish protein hydrolysate with a calcitonin-gene-related-peptide-like biological activity. *Biotechnol. Appl. Biochem.* **29**, 87–92.

Gildberg, A., Bogwald, J., Johansen, A., and Stenberg, E. (1996). Isolation of acid peptide fractions from a fish protein hydrolysate with strong stimulatory effect on Atlantic salmon (*Salmo salar*) head kidney leucocytes. *Comp. Biochem. Physiol.* **11**, 97–101.

Gomez-Guillen, M. C., Turnay, J., Fernandez-Diaz, M. D., Ulmo, N., Lizarbe, M. A., and Montero, P. (2002). Structural and physical properties of gelatin extracted from different marine species: A comparative study. *Food Hydrocolloids* **16**, 25–34.

Guo, Z., Xing, R., Liu, S., Zhong, Z., and Li, P. (2008). Synthesis and hydroxyl radicals scavenging activity of quaternized carboxymethyl chitosan. *Carbohydr. Polym.* **73**, 173–177.

Heu, M. S., Kim, J. S., and Shahidi, F. (2003). Components and nutritional quality of shrimp processing by-products. *Food Chem.* **82**, 235–242.

Hidajat, K., Ching, C. B., and Rao, M. S. (1995). Preparative-scale liquid chromatographic separation of x-3 fatty acids from fish oil sources. *J. Chromatogr. A* **702**, 215–221.

Hirata, H., Saeki, H., Nonaka, M., Kawasaki, K., Ooizumi, T., and Motoe, K. (1993). Recovery of fish oil from the manufacturing process of highly nutritional fish meat for foodstuffs from sardine. *Nippon Suisan Gakkaishi* **59**, 111–116.

Hoang, Q. Q., Sicheri, F., Howard, A. J., and Yang, D. S. C. (2003). Bone recognition mechanism of porcine osteocalcin from crystal structure. *Nature* **425**, 977–980.
Houben, R., Jin, D., Stafford, D., Proost, P., Ebberink, R., and Vermeer, C. (1999). Osteocalcin binds tightly to the γ-glutamylcarboxylase at a site distinct from that of the other known vitamin K-dependent proteins. *Biochem. J.* **344**, 265–269.
Huang, R. H., Mendis, E., Rajapakse, N., and Kim, S. K. (2006). Strong electronic charge as an important factor for anticancer activity of chitooligosaccharides (COS). *Life Sci.* **78**, 2399–2408.
Illum, L. and Davis, S. (2005). Chitosan as a delivery system for the transmucosal administration of drugs. *In* "Polysaccharides. Structural Diversity and Functional Versatility", (S. Dumitriu, Ed.), 2nd edn., pp. 643–660. Marcel Dekker Publishing, New York.
Ito, M., Matahira, Y., and Sakai, K. (1998). The application of chitin chitosan to bone filling materials. Vol. 4. (pp. 142–143). Nippon Kichin Publishing Co., Kichin.
Jayakumar, R., Nwe, N., Tokura, S., and Tamura, H. (2007). Sulfated chitin and chitosan as biomaterials. *Int. J. Biol. Macromol.* **40**, 175–181.
Je, J. Y., Park, P. J., Jung, W. K., and Kim, S. K. (2005). Isolation of angiotensin I converting enzyme (ACE) inhibitor from fermented oyster sauce, *Crassostrea gigas*. *Food Chem.* **90**, 809–814.
Je, J. Y., Qian, Z. J., Byun, H. G., and Kim, S. K. (2007). Purification and characterization of an antioxidant peptide obtained from tuna backbone protein by enzymatic hydrolysis. *Process Biochem.* **42**, 840–846.
Jeon, Y. J. and Kim, S. K. (2002). Antitumor activity of chitosan oligosaccharides produced in an ultra filtration membrane reactor system. *J. Microbiol. Biotechnol.* **12**, 503–507.
Jeon, Y. J., Byun, H. G., and Kim, S. K. (2000). Improvement of functional properties of cod frame protein hydrolysates using ultrafiltration membranes. *Process Biochem.* **35**, 471–478.
Jiang, B. and Mine, Y. (2000). Preparation of novel functional oligophosphopeptides from hen egg yolk phosvitin. *J. Agric. Food Chem.* **48**, 990–994.
Jung, W. K., Park, P. J., Byun, H. G., Moon, S. H., and Kim, S. K. (2005). Preparation of hoki (*Johnius belengerii*) bone oligophosphopeptide with a high affinity to calcium by carnivorous intestine crude proteinase. *Food Chem.* **91**, 333–340.
Jung, W. K., Lee, B. J., and Kim, S. K. (2006). Fish-bone peptide increases calcium solubility and bioavailability in ovariectomised rats. *Br. J. Nutr.* **95**, 124–132.
Kato, Y., Onishi, H., and Machida, Y. (2003). Application of chitin and chitosan derivatives in the pharmaceutical field. *Curr. Pharm. Biotechnol.* **4**, 303–309.
Kato, Y., Onishi, H., and Machida, Y. (2005). Contribution of chitosan and its derivatives to cancer chemotherapy. *In Vivo* **19**, 301–310.
Kelleher, K. (2005). Discards in the world's marine fisheries. An update. FAO Fisheries Technical Paper N 470, Rome. p. 131.
Kim, S. K. and Mendis, E. (2006). Bioactive compounds from marine processing byproductsA review. *Food Res. Int.* **39**, 383–393.
Kim, S. K., Jeon, Y. J., Byun, H. G., Kim, Y. T., and Lee, C. K. (1997). Enzymatic recovery of cod frame proteins with crude proteinase from tuna pyloric caeca. *Fish. Sci.* **63**, 421–427.
Kim, G. H., Jeon, Y. J., Byun, H. G., Lee, Y. S., and Kim, S. K. (1998). Effect of calcium compounds from oyster shell bound fish skin gelatin peptide in calcium deficient rats. *J. Korean Fish. Soc.* **31**, 149–159.
Kim, S. K., Jeon, Y. J., Byun, H. G., Park, P. J., Kim, G. H., Choi, Y. R., and Lee, Y. S. (1999). Calcium absorption acceleration effect on phosphorylated and non-phosphorylated peptides from hoki (*Johnius belengerii*) frame. *J. Korean Fish. Soc.* **32**, 713–717.
Kim, S. K., Kim, Y. T., Byun, H. G., Nam, K. S., Joo, D. S., and Shahidi, F. (2001). Isolation and characterization of antioxidative peptides from gelatin hydrolysate of Allaska Pollack skin. *J. Agric. Food Chem.* **49**, 1984–1989.

Kim, S. Y., Je, J. Y., and Kim, S. K. (2007). Purification and characterization of antioxidant peptide from hoki (*Johnius belengerii*) frame protein by gastrointestinal digestion. *J. Nutr. Biochem.* **18**, 31–38.

Kirpal, S. S. (2003). Health benefits and potential risks related to consumption of fish or fish oil. *Regul. Toxicol. Pharmacol.* **38**, 336–344.

Krajewska, B. (2004). Application of chitin- and chitosan-based materials for enzyme immobilizations: A review. *Enzyme Microbiol. Technol.* **35**, 126–139.

Larsen, T., Thilsted, S. H., Kongsbak, K., and Hansen, M. (2000). Whole small fish as a rich calcium source. *Br. J. Nutr.* **83**, 191–196.

Lee, H. C., Singla, A., and Lee, Y. (2001). Biomedical applications of collagen. *Int. J. Pharm.* **221**, 1–22.

Lee, S. H., Qian, Z. J., and Kim, S. K. (2010). A novel angiotensin I converting enzyme inhibitory peptide from tuna frame protein hydrolysate and its antihypertensive effect in spontaneously hypertensive rats. *J. Agric. Food Chem.* **118**, 96–102.

Lee, J. K., Jeon, J. K., and Byun, H. G. (2011). Effect of angiotensin I converting enzyme inhibitory peptide purified from skate skin hydrolysate. *Food Chem.* **125**, 495–499.

Li, R., Guo, Z., and Jiang, P. (2010). Synthesis, characterization, and antifungal activity of novel quaternary chitosan derivatives. *Carbohydr. Res.* **345**, 1896–1900.

Liu, H. Y., Li, D., and Guo, S. D. (2008a). Extraction and properties of gelatin from channel catfish (*Ietalurus punetaus*) skin. *Food Sci. Technol.* **41**, 414–419.

Liu, J. N., Xia, W. S., and Zhang, J. L. (2008b). Effects of chitosans physico-chemical properties on binding capacities of lipid and bile salts in vitro. *Chin. Food Sci.* **29**, 45–49.

Mendis, E., Rajapakse, N., Byun, H. G., and Kim, S. K. (2005). Investigation of jumbo squid (Dosidicus gigas) skin gelatin peptides for their in vitro antioxidant effects. *Life Sci.* **77**, 2166–2178.

Moskowitz, R. W. (2000). Role of collagen hydrolysate in bone and joint disease. *Semin. Arthritis Rheum.* **30**, 87–99.

Nagai, T. and Suzuki, N. (2000). Isolation of collagen from fish waste material-skin, bone and fins. *Food Chem.* **68**, 277–281.

Nakagawa, T. and Tagawa, T. (2000). Ultrastructural study of direct bone formation induced by BMPs-collagen complex implanted into an ectopic site. *Oral Dis.* **6**, 172–179.

Ngo, D. N., Kim, M. M., and Kim, S. K. (2008). Chitin oligosaccharides inhibit oxidative stress in live cells. *Carbohydr. Polym.* **74**, 228–234.

Norland, R. E. (1990). Fish gelatin. *In* "Advances in Fisheries Technology and Biotechnology for Increased Profitability", (M. N. Voight and J. K. Botta, Eds), pp. 325–333. Technomic Publishing Co., Lancaster.

Ozawa, M. and Suzuki, S. (2002). Microstructural development of natural hydroxyapatite originated from fish-bone waste through heat treatment. *J. Am. Ceramic Soc.* **85**, 1315–1317.

Park, P. J., Je, J. Y., and Kim, S. K. (2003). Angiotensin I converting enzyme (ACE) inhibitory activity of hetero-chitooligosaccharides prepared from partially different deacetylated chitosans. *J. Agric. Food Chem.* **51**, 4930–4934.

Pasanphan, W. and Chirachanchai, S. (2008). Conjugation of gallic acid onto chitosan: An approach for green and water-based antioxidant. *Carbohydr. Polym.* **72**, 169–177.

Picot, L., Bordenave, S., Didelot, S., Fruitier-Arnaudin, I., Sannier, F., Thorkelsson, G., Berge, J. P., Guerard, F., Chabeaud, A., and Piot, J. M. (2006). Antiproliferative activity of fish protein hydrolysates on human breast cancer cell lines. *Process Biochem.* **41**, 1217–1222.

Rabea, E. I., Badawy, M. E. T., Stevens, C. V., Smagghe, G., and Steurnaut, W. (2003). Chitosan as antimicrobial agent: Applications and mode of action. *Biomacromolecules* **4**, 1457–1465.

Rajapakse, N., Jung, W. K., Mendis, E., Moon, S. H., and Kim, S. K. (2005). A novel anticoagulant purified from fish protein hydrolysate inhibits factor XIIa and platelet aggregation. *Life Sci.* **76**, 2607–2619.

Rajapakse, N., Mendis, E., Byun, H. G., and Kim, S. K. (2005). Purification and in vitro antioxidative effect of giant squid muscle peptides in free radical -mediated oxidative systems. *J. Nutr. Biochem.* **16**, 562–569.

Runarsson, O. V., Holappa, J., Malainer, C., Steinsson, H., Hjalmarsdottir, M., Nevalainen, T., and Masson, M. (2010). Antibacterial activity of N-quaternary chitosan derivatives: Synthesis, characterization and structure activity relationship (SAR) investigations. *Eur. Polym. J.* **46**, 1251–1267.

Rustad, T. (2003). Utilization of marine by-products. *Electronic J. Environ. Agric. Food Chem.* **2**, 458–463.

Sahena, F., Zaidul, I. S. M., Jinap, S., Jahurul, M. H. A., Khatib, A., and Norulaini, N. A. N. (2010). Extraction of fish oil from the skin of Indian mackerel using supercritical fluids. *J. Food Eng.* **99**, 63–69.

Sajomsang, W., Gonil, P., and Tantayanon, S. (2009). Antibacterial activity of quaternary ammonium chitosan containing mono or disaccharide moieties: Preparation and characterization. *Int. J. Biol. Macromol.* **44**, 419–427.

Sajomsang, W., Ruktanonchai, U. R., Gonil, P., and Warin, C. (2010). Quaternization of N-(3-pyridylmethyl) chitosan derivatives: Effects of the degree of quaternization, molecular weight and ratio of N-methylpyridinium and N,N,N-trimethyl ammonium moieties on bactericidal activity. *Carbohydr. Polym.* **82**, 1143–1152.

Sato, H., Kitazawa, H., Adachi, I., and Horikoshi, I. (1996). Microdialysis assessment of microfibrous collagen containing a p-glycoproteinmediated transport inhibitor, cyclosporine A, for local delivery of etopocide. *Pharmacol. Res.* **13**, 1565–1569.

Schrieber, R. and Gareis, H. (2007). Gelatine Handbook. Wiley-VCH GmbH & Co., Weinhem.

Simopoulos, A. P. (1999). Essential fatty acids in health and chronic disease. *Am. J. Clin. Nutr.* **70**, 560S–569S.

Simpson, B. K., Nayeri, G., Yaylayan, V., and Ashie, I. N. A. (1998). Enzymatic hydrolysis of shrimp meat. *Food Chem.* **61**, 131–138.

Slizyte, R., Mozuraityte, R., Martınez-Alvarez, O., Falch, E., Fouchereau-Peron, M., and Rustad, T. (2009). Functional, bioactive and antioxidative properties of hydrolysates obtained from cod (*Gadus morhua*) backbones. *Process Biochem.* **44**, 668–677.

Supavititpatana, P., Wirjantoro, T. I., Apichartsrangkoon, A., and Raviyan, P. (2008). Addition of gelatin enhanced gelation of corn–milk yogurt. *Food Chem.* **106**, 211–216.

Surh, J., Decker, E. A., and McClements, D. J. (2006). Properties and stability of oilin-water emulsions stabilized by fish gelatin. *Food Hydrocolloids* **20**, 596–606.

Tongchan, P., Prutipanlai, S., Niyomwas, S., and Thongraung, C. (2009). Effect of calcium compound obtained from fish by-product on calcium metabolism in rats. *Asian J. Food Ag-Ind.* **2**, 669–676.

Vannuccini, S. (2004). Overview of fish production, utilization, consumption and trade. FAO fishery information, data and statistic unit report, ftp://ftp.fao.org/fi/stat/overview/2001/commodit/2001fisheryoverview.pdf.

Venter, J. P., Kotze, A. F., Auzély-Velty, R., and Rinaudo, M. (2006). Synthesis and evaluation of the mucoadhesivity of a CD-chitosan derivative. *Int. J. Pharm.* **313**, 36–42.

Venugopal, V., Chawla, S. P., and Nair, P. M. (1996). Spray dried protein powder from threadfin beam: Preparation, properties and comparison with FPC type B. *J. Muscle Foods* **7**, 55–58.

Wassawa, J., Tang, J., and Gu, X. (2007). Utilization of fish processing by-products in the gelatin industry. *Food Rev. Int.* **23**, 159–174.

Xu, T., Xin, M., Li, M., Huang, H., and Zhou, S. (2010). Synthesis, characteristic and antibacterial activity of N,N,N-trimethyl chitosan and its carboxymethyl derivatives. *Carbohydr. Polym.* **81**, 931–936.

Yang, L., Chen, L., Zeng, R., Li, C., Qiao, R., Hu, L., and Li, Z. (2010). Synthesis, nanosizing and in vitro drug release of a novel anti-HIV polymeric prodrug: Chitosan-O-isopropyl-50-O-d4T monophosphate conjugate. *Bioorg. Med. Chem.* **18**, 117–123.

Yusof, N. L., Wee, A., Lim, L. Y., and Khor, E. (2003). Flexible chitin films as potential wound-dressing materials: Wound model studies. *J. Biomed. Mater. Res. A* **66**, 224–232.

Zhao, Y., Li, B., Dong, S., Liu, Z., Zhao, X., Wang, J., and Zeng, M. (2009). A novel ACE inhibitory peptide isolated from *Acaudina molpadioidea* hydrolysate. *Peptides* **30**, 1028–1033.

Zhong, Z., Ji, X., Xing, R., Liu, S., Guo, Z., Chen, X., and Li, P. (2007). The preparation and antioxidant activity of the sulfanilamide derivatives of chitosan and chitosan sulfates. *Bioorg. Med. Chem.* **15**, 3775–3782.

Zhong, Z., Li, P., Xing, R., Chen, X., and Liu, S. (2009). Preparation, characterization and antifungal activities of 2-(α-arylamino phosphonate)-chitosan. *Int. J. Biol. Macromol.* **45**, 255–259.

Zhong, Z., Zhong, Z., Xing, R., Li, P., and Mo, G. L. (2010). The preparation and antioxidant activity of 2-[phenylhydrazine (or hydrazine)-thiosemicarbazone]-chitosan. *Int. J. Biol. Macromol.* **47**, 93–97.

Zhou, K., Xia, W., Zhang, C., and Yu, L. (2006). In vitro binding of bile acids and triglycerides by selected chitosan preparations and their physicochemical properties. *LWT- Food Sci. Technol.* **39**, 1087–1092.

INDEX

Note: Page numbers followed by "*f*" indicate figures, and "*t*" indicate tables.

U

V

W

Kolibri 360 GmbH
Ferdinand-Jühlke-Straße 7
99095 Erfurt, Deutschland
produktsicherheit@kolibri360.de